SUSTAINABLE MANAGEMENT AND CONSERVATION OF ENVIRONMENTAL RESOURCES IN INDIA

SUSTAINABLE MANAGEMENT AND CONSERVATION OF ENVIRONMENTAL RESOURCES IN INDIA

Edited by

Sumit Chakravarty, PhD

Vineeta, PhD

Jahangeer A. Bhat, PhD

Munesh Kumar, PhD

Gopal Shukla, PhD

AAP | APPLE ACADEMIC PRESS

First edition published 2025

Apple Academic Press Inc.
1265 Goldenrod Circle, NE,
Palm Bay, FL 32905 USA

760 Laurentian Drive, Unit 19,
Burlington, ON L7N 0A4, CANADA

CRC Press
2385 NW Executive Center Drive,
Suite 320, Boca Raton FL 33431

4 Park Square, Milton Park,
Abingdon, Oxon, OX14 4RN UK

Library and Archives Canada Cataloguing in Publication

CIP data on file with Canada Library and Archives

Library of Congress Cataloging-in-Publication Data

CIP data on file with US Library of Congress

ISBN: 978-1-77491-592-9 (hbk)
ISBN: 978-1-77491-593-6 (pbk)
ISBN: 978-1-00346-927-8 (ebk)

About the Editors

Sumit Chakravarty, PhD
Professor, Department of Forestry, Uttar Banga Krishi Viswavidyalaya, Pundibari, Cooch Behar, West Bengal, India

Sumit Chakravarty, PhD, has a wide experience of more than 20 years in forestry training, research and development. He is currently working as a Professor and Head of the Department of Forestry, Uttar Banga Krishi Viswavidyalaya, Pundibari, Cooch Behar, West Bengal, India. He holds a master's degree in forestry and a doctoral degree in Agronomy from Punjab Agricultural University, Ludhiana. He has conducted research on several aspects of forestry, agroforestry, medicinal plants and climate change with national and international reputed publications. This includes 120 research papers, 48 book chapters, and eight books. He has successfully completed and continues work on 15 research projects on forest ecology, medicinal plants, and agroforestry. He has trained more than 50 master's and doctoral students in the field of forestry. Dr. Chakravarty has attended more than 50 national and international conferences as a capacity of presenter and invited speaker.

Vineeta, PhD
Assistant Professor, Forestry Department, Uttar Banga Krishi Viswavidyalaya, Pundibari, Cooch Behar, West Bengal, India

Vineeta, PhD, is an emerging researcher, currently holding the position of Assistant Professor in Forestry at Uttar Banga Krishi Viswavidyalaya, Pundibari, Cooch Behar, West Bengal, India. She has more than eight years of teaching and research experience in the field of forestry. Her major contribution as a researcher is in the field of ethnobotany, non-timber forest products (NTFPs), agroforestry, and medicinal plants. She currently has 33 research papers, eight book chapters, and two books. She is holding a degree of BSc (Forestry) and MSc (Forestry) from HNB Garhwal Central University, Srinagar, Garhwal, Uttarakhand, India. She also has a PhD.

Jahangeer A. Bhat, PhD
Piramal Foundation for Education Leadership, Piramal University, Mumbai, India

Jahangeer A. Bhat, PhD, is currently associated with the Piramal Foundation for Education Leadership at Piramal University, Mumbai, India. Before taking on his current position, he was associated with the College of Horticulture and Forestry at Rani Lakshmi Bai Central Agricultural University, India. Earlier, he served as Head of the Department of Forestry at the College of Agriculture, Fisheries and Forestry at Fiji National University, Republic of Fiji Islands. His research work in India and Fiji Islands spans 12 years and encompasses emerging issues in forestry, including conservation of biodiversity, traditional knowledge of plants, and sustainable management of forest resources, particularly in the light of climate change scenarios. He has worked as counselor, mentor, and coordinator for forestry academic programs and has been instrumental in developing higher education and technical and vocational education and training streams of forestry and allied programs. He worked closely for accreditation with the Fiji Higher Education Commission and forestry stakeholders. He is the first researcher to report the first time growing of Acacia dealbata Link. (silver wattle), an invasive species in the high altitudes of the Himalayas. His focus of research is on vegetation ecology, ethnobotany, and evaluation of ecosystem services, forest plant biodiversity, climate change, and sociocultural issues in forestry.

Munesh Kumar, PhD
Professor, Department of Forestry and Natural Resources, H.N.B. Garhwal University, Srinagar Garhwal, Uttarakhand, India

Munesh Kumar, PhD, is Professor in the Department of Forestry and Natural Resources, H.N.B. Garhwal University, Srinagar Garhwal, Uttarakhand, India. He also served at Mizoram University, Aizawl, India, as Assistant Professor in the Department of Forestry. His areas of research interest are forest ecology and biodiversity, agroforestry, and ethnobotany. He has more than 17 years of teaching and research experience and has published more than 135 research papers in international and national journals. Recently, he published an edited book as lead editor from CRC Press titled *Diversity and Dynamics in Forest Ecosystem*. Dr. Kumar is presently working as a guest editor for two different issues of the Journal Land of MDPI and guest editor for *Frontier of Environmental Sciences* and *Frontiers in Forests and Global Change*.

Gopal Shukla, PhD

Professor and Head, Department of Forestry, North Eastern Hill University, Tura Campus, Tura, Meghalaya, India

Gopal Shukla, PhD, is currently working as an Professor of Forestry at North Eastern Hill University, Tura Campus, Tura, Meghalaya, India. Before joining as Professor, he was working as Assistant Professor (Senior Scale) of Forestry at Uttar Banga Krishi Viswavidyalaya, Pundibari, Cooch Behar, West Bengal, India. Dr. Shukla has teaching and research experience of more than 12 years. He has worked on different projects, including the National Agricultural Innovation Project (ICAR), and National Innovations in Climate Resilient Agriculture (ICAR) and also as a young scientist on the project on carbon stock of different land uses (Science and Engineering Research Board, Department of Science and Technology, Government of India). The focus of his research and development work is on agroforestry, medicinal plants, forest ecology, and conservation. He has national and international publications that include 90 research papers, 30 book chapters, and eight books. He has successfully completed several research projects on forest ecology, medicinal plants, and agroforestry. Dr. Shukla also guided four doctoral and six master's students.

Contents

Contributors

S. K. Aditya
ESSO–National Center for Earth Science Studies (NCESS), Ministry of Earth Sciences, Government of India, Akkulam, Thiruvananthapuram, Kerala, India

K. Anoopkrishnan
ESSO–National Center for Earth Science Studies (NCESS), Ministry of Earth Sciences, Government of India, Akkulam, Thiruvananthapuram, Kerala, India

Basabduttaa Bhabai
Department of Soil Science and Agricultural Chemistry, Uttar Banga Krishi Viswavidyalaya, Pundibari, Cooch Behar, West Bengal, India

Arpan Bhowmik
Indian Agricultural Research Institute, Gogamukh, Assam, India

K. G. Bhutia
Rain Forest Research Institute, Jorhat, Assam, India

P. L. Bhutia
ICAR–Indian Institute of Soil and Water Conservation, Madhya Marg, Chandigarh, India;
ICAR–Research Complex for NEH Region, Nagaland Center, Medziphema, Nagaland, India

Bisworanjita Biswal
Department of Agronomy, National Dairy Research Institute, Karnal, Haryana, India

Nirmali Bordoloi
Department of Environmental Sciences, Central University of Jharkhand, Brambe, Ranchi, Jharkhand, India

Debasish Chakraborty
ICAR–Research Complex for NEH Region, Umiam, Meghalaya, India

R. S. Chaudhary
ICAR–Indian Institute of Soil Science, Nabibagh, Bhopal, Madhya Pradesh, India

S. B. Chavan
ICAR–National Institute of Abiotic Stress Management, Malegaon, Baramati, Pune, Maharashtra, India

Santosh K. Chettri
G. B. Pant National Institute of Himalayan Environment (NIHE), Sikkim Regional Center, Pangthang, Gangtok, Sikkim, India

Geetamani Chhetri
G. B. Pant National Institute of Himalayan Environment, Sikkim Regional Center, Pangthnag, Gangtok, Sikkim, India

A. R. Chichaghare
Department of Silviculture and Agroforestry, Kerala Agricultural University, Thrissur, Kerala, India

Manoj Choudhary
ICAR–Indian Agricultural Research Institute, Hazaribag, Jharkhand, India

Mukesh Choudhary
ICAR–Indian Grassland and Fodder Research Institute, Jhansi, India

Ashesh Kumar Das
Department of Ecology and Environmental Science, Assam University, Silchar, Assam, India

Rajeswari Das
Department of Soil Science, Dr. Rajendra Prasad Central Agricultural University, Pusa, Samastipur, Bihar, India

Dibyendu Deb
ICAR–Indian Agricultural Research Institute, Gogamukh, Assam, India

Rajendra Gadhwal
Department of Soil Science and Agricultural Chemistry, Institute of Agricultural Science, Banaras Hindu University, Varanasi, Uttar Pradesh, India

Kailash S. Gaira
G. B. Pant National Institute of Himalayan Environment (NIHE), Sikkim Regional Center, Pangthang, Gangtok, Sikkim, India

Debnirmalya Gangopadhyay
Department of Sericulture, Raiganj University, Raiganj, North Dinajpur, West Bengal, India

A. K. Ghosh
Department of Soil Science and Agricultural Chemistry, Institute of Agricultural Science, Banaras Hindu University, Varanasi, Uttar Pradesh, India

Prabhu Govindasamy
ICAR–Indian Grassland and Fodder Research Institute, Jhansi, India; ICAR–National Research Centre for Banana, Tiruchirappalli, Tamil Nadu, India

Khushboo Gupta
Department of Environmental Sciences, Central University of Jharkhand, Brambe, Ranchi, Jharkhand, India

Bholuram Gurjar
ICAR–Indian Grassland and Fodder Research Institute, Jhansi, India

Hanamant M. Halli
ICAR–Indian Grassland and Fodder Research Institute, Jhansi, India; ICAR–National Institute of Abiotic Stress Management, Baramati, Maharashtra, India

Rajesh Joshi
G. B. Pant National Institute of Himalayan Environment (NIHE), Sikkim Regional Center, Pangthang, Gangtok, Sikkim, India

Vijaysinh D. Kakade
ICAR–National Institute of Abiotic Stress Management, Malegaon, Baramati, Pune, Maharashtra, India

H. Kalita
ICAR–RC for NEH Region Arunachal Pradesh Center, Basar, Arunachal Pradesh, India

M. C. Keerthi
ICAR–Indian Grassland and Fodder Research Institute, Jhansi, India; ICAR–Indian Institute of Horticulture Research, Bengaluru, Karnataka, India

O. P. S. Khola
ICAR–Indian Institute of Soil and Water Conservation, Madhya Marg, Chandigarh, India

A. Krishnakumar
ESSO–National Center for Earth Science Studies (NCESS), Ministry of Earth Sciences, Government of India, Akkulam, Thiruvananthapuram, Kerala, India

Amit Kumar
Department of Geoinformatics, Central University of Jharkhand, Ranchi, India;
Department of Geography, Sikkim University, Tadong, India

Dhiraj Kumar
ICAR–Indian Institute of Soil Science, Bhopal, Madhya Pradesh, India

Jitendra Kumar
ICAR–Indian Institute of Soil Science, Nabibagh, Bhopal, Madhya Pradesh, India

Raushan Kumar
Department of Environmental Sciences, Central University of Jharkhand, Brambe, Ranchi, Jharkhand, India

Rima Kumari
Department of Environmental Science and Engineering, Indian Institute of Technology (Indian School of Mines), Dhanbad, India

Vandana Kumari
Department of Soil Science, Dr. Rajendra Prasad Central Agricultural University, Pusa, Samastipur, Bihar, India

Preet Lal
Department of Civil and Environmental Engineering, Michigan State University, East Lansing, MI, United States

A. Madhumathi
Vellore Institute of Technology, Vellore, Tamil Nadu, India

Sonu Kumar Mahawer
ICAR–Indian Grassland and Fodder Research Institute, Jhansi, India

Manish Mathur
ICAR–Central Arid Zone Research Institute, Jodhpur, Rajasthan, India

Preet Mathur
Jodhpur Institute of Engineering and Technology, Jodhpur, Rajasthan, India

Ajay Kumar Mishra
Terrestrial Ecosystem Management Laboratory, Graduate School of Global Environmental Studies, Kyoto University, Kyoto, Japan

Deepak Mishra
Division of Forest Ecology, Arid Forest Research Institute, Jodhpur, Rajasthan, India

Sarita Mishra
Department of Chemistry, Udai Pratap College, Varanasi, Uttar Pradesh, India

Biswajita Mohanty
Vellore Institute of Technology, Vellore, Tamil Nadu, India

Nilimesh Mridha
National Institute of Natural Fiber Engineering and Technology, Kolkata, West Bengal, India

Dibyendu Mukhopadhyay
Department of Soil Science and Agricultural Chemistry, Uttar Banga Krishi Viswavidyalaya, Pundibari, Cooch Behar, West Bengal, India

Amitabha Nath
Department of Information Technology, North-Eastern Hill University, Shillong, Meghalaya, India

Arun Jyoti Nath
Department of Ecology and Environmental Science, Assam University, Silchar, Assam, India

Debabrata Nath
Department of Soil Science, Dr. Rajendra Prasad Central Agricultural University, Pusa, Samastipur, Bihar, India

Nirmal
ICAR–Indian Institute of Farming System Research, Modipuram, Meerut, Uttar Pradesh, India

S. Pal
ICAR–Indian Institute of Soil and Water Conservation, Madhya Marg, Chandigarh, India

Aseesh Pandey
G. B. Pant National Institute of Himalayan Environment (NIHE), Sikkim Regional Center, Pangthang, Gangtok, Sikkim, India

Avinash Pandey
ICAR–Indian Institute of Agricultural Biotechnology, Ranchi, Jharkhand, India

P. Panwar
ICAR–Indian Institute of Soil and Water Conservation, Madhya Marg, Chandigarh, India

A. K. Patra
ICAR–Indian Institute of Soil Science, Nabibagh, Bhopal, Madhya Pradesh, India

B. Prabadevi
Vellore Institute of Technology, Vellore, Tamil Nadu, India

Chandini Pradhan
Department of Soil Science and Agricultural Chemistry, Institute of Agricultural Science, Banaras Hindu University, Varanasi, Uttar Pradesh, India

Alisha Prasad
Department of Geoinformatics, Central University of Jharkhand, Ranchi, India

Mahendra Prasad
ICAR–Indian Grassland and Fodder Research Institute, Jhansi, India

K. G. Prijilal
Ministry of Environment, Forests, and Climate Change, Government of India, Regional Office, Chennai, Tamil Nadu, India

Kitdor Pyngrope
ICAR–Research Complex for NEH Region, Umiam, Meghalaya, India

Sivakumar Rajagopal
Vellore Institute of Technology, Vellore, Tamil Nadu, India

V. Rajagopal
ICAR–National Institute of Abiotic Stress Management, Baramati, Maharashtra, India

K. Rajarajan
ICAR–Central Agroforestry Research Institute, Jhansi, India

S. Suresh Ramanan
ICAR–National Research Centre for Banana, Tiruchirappalli, Tamil Nadu, India

Demsai Reang
Department of Ecology and Environmental Science, Assam University, Silchar, Assam, India

Purabi Saikia
Department of Environmental Sciences, Central University of Jharkhand, Ranchi, India

G. Singh
Division of Forest Ecology, Arid Forest Research Institute, Jodhpur, Rajasthan, India

Kaushal Singh
Department of Silviculture and Agroforestry, College of Forestry, Banda University of Agriculture and Technology, Banda, Uttar Pradesh, India

N. R. Singh
ICAR–Research Complex for NEH Region, Umiam, Meghalaya, India

Preeti Singh
Department of Soil Science and Agricultural Chemistry, Institute of Agricultural Science, Banaras Hindu University, Varanasi, Uttar Pradesh, India

Nishant Sinha
ICAR–Indian Institute of Soil Science, Nabibagh, Bhopal, Madhya Pradesh, India

Chhavi Sirohi
Department of Forestry, CCS Haryana Agricultural University, Hisar, Haryana, India

R. Srinivasan
ICAR–Indian Grassland and Fodder Research Institute, Jhansi, India

Princy Thakur
Department of Soil Science and Agricultural Chemistry, Uttar Banga Krishi Viswavidyalaya, Pundibari, Cooch Behar, West Bengal, India

Rakhi Tyagi
Department of Environmental Sciences, Sri Guru Govind Singh Tricentenary University, Gurgaon, Haryana, India

V. C. Tyagi
ICAR–Indian Grassland and Fodder Research Institute, Jhansi, India

A. R. Uthappa
ICAR–National Institute of Abiotic Stress Management, Malegaon, Baramati, Pune, Maharashtra, India

Abbreviations

$(CH_3)_2S$	dimethyl sulfide
AF	agroforestry
AFS	agroforestry systems
AGB	aboveground biomass
AGBC	aboveground biomass carbon
AICRP	All India Coordinated Research Project
AMF	arbuscular mycorrhizal fungi
AVHRR	advanced very high-resolution radiometer
BA	burnt area
BGA	blue green algae
BNF	biological nitrogen fixation
Ca	calcium
CAPE	crop acreage and production estimation
CBD	convention of biological diversity
CC	climate change
CCD	charged coupled device
CDM	clean development mechanism
CEC	cation exchange capacity
CEER	certified emission reductions
Cf	contamination factor
CFCs	chlorofluorocarbons
CFFDRS	Canadian Forest Fire Danger Rating System
CG	cloud-to-ground
CH_4	methane
CHR	cardamom hill reserves
CI-TARS	crop identification technology assessment for remote sensing
Cl	chlorine
CO_2	carbon dioxide
COP	conferences of the parties
COS	carbonyl sulfide
CS_2	carbon disulfide

CSA	climate-smart agriculture
CSSRI	Central Soil Salinity Research Institute
DA	discriminant analysis
DMSP	defense meteorological satellite program
DNA	Designated National Authority
DOC	dissolve organic carbon
DRS	diffuse reflectance spectroscopy
DT	decision tree
EAV	equivalent annual value
EC	electrical conductivity
EF	enrichment factor
EHR	Eastern Himalayan Region
EME	electromagnetic energy
EMR	electromagnetic
EMS	electromagnetic spectrum
ENSO	El Nino southern oscillation
ES	ecosystem services
ESA	European Space Agency
ESP	exchangeable sodium percentage
ESV	ecological service value
ET	electronic transition
ET	evapotranspiration
FAST 3.0	forest fire alert system 3.0
FDRS	fire danger rating systems
FSI	forest survey of India
GBPNIHE	G.B. Pant Institute of Himalayan Environment
GC	global carbon
GHG	greenhouse gas
GI	geographical indicators
GIS	geographic information system
GL	grasslands
GOES	geostationary operational environmental satellite
GPS	geographic positioning system
GPS	global positioning system
H_2S	hydrogen sulfide
HER	Eastern Himalayan Region
HI	hedgerow intercropping

HRML	high range mountain landscape
HRS	hyperspectral remote sensing
HYVs	high yielding varieties
ICIMOD	International Center for Integrated Mountain Development
ICRAF	International Center for Research in Agroforestry
ICRI	Indian Cardamom Research Institute
IGP	Indo-Gangetic Plains
IK	indigenous knowledge
INDC	intended nationally determined contribution
INFFRAS	Indian forest fire response and assessment system
IoT	internet of things
IPCC	Intergovernmental Panel on Climate Change
IR	infrared
IS	imaging spectrometer
ISFR	India State of Forest Report
ISRO	Indian Space Research Organization
IUCN	International Union for the Conservation of Nature and Natural Resources
IVI	importance value index
KLCDI	Khangchendzonga Landscape Conservation and Development Initiative
KML	keyhole mark-up language
LACA	large area crop acreage
LACIE	large area crop inventory experiment
LER	land equivalent ratio
LGP	lower Gangetic plain
LISS-1	linear imaging self-scanning system
LST	land surface temperature
LU	land use
LULUCF	land use, land use change, and forestry
LUS	land-use systems
MARS	multivariate adaptive regression spline
MEA	millennium environment assessment
Mg	magnesium
MGP	middle Gangetic plain
MIR	mid-infrared
ML	machine learning

MLR	multiple linear regression
MMU	minimum mappable unit
MNCFC	Mahalanobis National Crop Forecast Centre
MODIS	moderate resolution imaging spectroradiometer
MPTs	multipurpose trees
MRV	measured, reported, and verified
MSS	multi-spectral scanner
N_2O	nitrous oxide
Na	sodium
NASA	National Aeronautics and Space Administration
NASS	National Agricultural Statistics Service
NB	Naïve Bayes
NBSS & LUP	National Bureau of Soil Survey and Land Use Planning
NDCs	nationally determined contributions
NDVI	normalized difference vegetation index
NEH	northeastern hill
NEI	North East India
NER	North Eastern Region
NERAMAC	North Eastern Agricultural Marketing Corporation
NF	natural forest
NFDRS	National fire danger rating system
NIHE	National Institute of Himalayan Environment
NO_2	nitrogen dioxide
NOAA	National Oceanic and Atmospheric Administration
NPP	net primary productivity
NPS	non-point source
NPV	net present value
NRSC	National Remote Sensing Center
NTFP	non-timber forest produce
NUE	nitrogen use efficiency
O_3	ozone
OLS	optical line-scan system
PA	precision agriculture
PA	protected areas
PCA	principal component analysis
PCR	principal component regression
PES	payment of ecosystem services

PF	plantation forest
PLI	pollution load index
PLSR	partial-least-squares regression
PM	poultry manure
PRB	Periyar River Basin
PSB	phosphate solubilizing bacteria
RF	random forest
RS	remote sensing
RSI	reciprocal of Simpson's diversity index
SAC	space application center
SAR	sodium adsorption ratio
SCR	selective catalytic reduction
SEV	soil expectation value
SF_6	sulfur hexafluoride
SIC	soil inorganic carbon
SMBC	soil microbial biomass carbon
SMC	soil moisture content
SMLR	stepwise multiple linear regression
SNCR	selective non-catalytic reduction
SO_2	sulfur dioxide
SO_4^{2-}	sulfate
SOC	soil organic carbon
SOM	soil organic matter
SPM	suspended particulate matter
SRF	slow-release fertilizer
SVM	support vector machine
SWC	soil-water characteristic
SWM	South-West Monsoon
TBI	tree-based intercropping
TBOs	tree borne oilseeds
TCP	tricalcium phosphate
TDS	total dissolved solids
TGA	total geographical area
TGP	trans Gangetic plain
TOC	total organic carbon
UAV	unmanned aerial vehicles
UGP	upper Gangetic plain

UNDP	United Nations Development Program
UNFCCC	United Nations Framework Convention on Climate Change
UPA	urban and peri-urban agriculture
USD	United States Department of Agriculture
VHR	very high resolution
VOCs	volatile organic compounds
WB	West Bengal
WHC	water holding capacity
WIMS	weather information management system
WMO	World Meteorological Organization
WTP	willing to pay
WUE	water use efficiency

Foreword

It gives me immense pleasure to know that five distinguished professionals associated with forestry research and education, including three from my university, have joined hands to compile this valuable and very explanatory book titled *Sustainability, Development, and Conservation of Natural Resources*. This book edition contains 18 chapters covering various aspects of sustainable land management through tree-based land use systems, both in natural and agroecosystems.

Resource management through forestry and agroforestry is a sustainable option to increase carbon sink for mitigating adverse effects of climate change, as well as to increase land productivity through enhanced nutrient inputs, reduced nutrient losses, and improved physicochemical properties of the soil. Human population pressure, coupled with advancements in lifestyle and other endless needs arising out of modernization, has increased the pressure on forest resources multifold. Therefore, there is an urgent need to understand the forest ecosystem and its dynamics in terms of conservation, utilization, and sustainable management of resources, while also encouraging tree-based land use management in human-dominated and agricultural landscapes to supplement and conserve the forest resources. This edition has compiled all the important and necessary contents for sustainable development and conservation of forest and agricultural land resources.

This book highlights the importance of agricultural landscape management through agroforestry for biodiversity conservation and nutritional security, as well as features of climate-resilient farming methods for improved livelihoods. It was indeed my pleasure to go through the in-depth information provided in this book, and I would like to congratulate the editors as well as the contributing authors for bringing out this timely compilation. I also congratulate the publisher, Press, for accepting this edition for publication.

I am sure that this edition with updated and latest information will be of great assistance to the targeted readers, particularly students, scholars, academicians, and policymakers.

—S. K. Chakrabarti, PhD
Vice Chancellor
Uttar Banga Krishi Viswa Vidyalaya, Cooch Behar, West Bengal, India

Preface

We are thrilled to announce the publication of *Sustainable Management and Conservation of Environmental Resources in India*, an edited book. The concept of natural resource sustainability has taken on new connotations in today's society as required for conservation of natural resources. Recognizing community participation in conservation for sustainable forest resource management is essential for generating innovative climate change (CC) and mitigation measures. As a result, it is critical to study the many aspects of forestry in terms of sustainable natural resource conservation at temporal and spatial ecosystem levels in order to design policies for sustainable development.

The majority of the research topics important to natural resource conservation, from traditional knowledge to environmental resources covering various forest settings, have been collected together in this edition's 18 chapters. This book also discusses the possibilities of agroforestry (AF) for biodiversity conservation and nutritional security, as well as features of climate-resilient farming methods for improved livelihoods. The importance of sericulture-based agroforestry systems in livelihood diversification for local communities and ethnic foods, as well as livelihood generating and forest resource extraction by indigenous populations, is discussed. This book discusses topics such as carbon sequestration in agroforestry systems, an integrated approach to repairing salty and sodic soil through the use of halophytes, and the lowland tropical forests of the Eastern Himalaya in connection to the environment and tree features.

The editors anticipate that this compilation will eventually help to spread the word about the necessity of sustainable development and natural resource conservation. This publication will help scholars, academicians, and intellectuals interested in this subject.

—Editors

Introduction

Agro-ecosystem is essentially a man-made ecosystem and reflects the evolution of human culture. Resource management through forestry and agroforestry (AF) is an important concern for the present-day scenario, and it enhances productivity through increased nutrient inputs, reduced nutrient losses, and improved physicochemical properties of the soil in many geographical regions around the world. It is now globally viewed as part of a multifunctional working landscape that can be a viable land-use option, offering a number of ecosystem services and environmental benefits (improved soil fertility and physical properties, improved water quality, reduced erosion, increased biodiversity and aesthetics, and carbon sequestration), including alleviation of poverty over a range of spatial and temporal scales.

The present edited book focuses on livelihood upliftment and its sustainability, forest biodiversity conservation strategies for climate change mitigation, ecosystem services, and the importance and role of traditional knowledge in conservation strategies.

Chapter 1 discusses the importance and conservation of natural resources in the environment and the livelihood security for sustainable agricultural production and development, and also about how the imbalance of natural resources may lead to the deposition of some toxic elements or sometimes may be deficient in some essential elements in the system.

Chapter 2 discusses a study that was conducted to analyze the change in vegetation vigor on the basis of the normalized difference vegetation index (NDVI) from 2001 to 2017 in the Himalayan regions. The insights in the study will aid in addressing the challenges being faced while dealing with the issues associated with the change in vegetation dynamics in the Himalayan region and also in formulating adaptation strategies towards climate change.

Chapter 3 highlights the importance of the tree-based ecosystems in carbon storage potential and as large carbon pools. The higher levels of biodiversity under these systems also support greater levels of ecosystem service production and ecosystem properties.

Chapter 4 discusses a study that was conducted to assess the effect of five (forest, agriculture, sacred groves, pasture, and roadside) land use types on plant diversity in Jodhpur district of Western Rajasthan and observed the sacredness and involvement of the community in the protection and conservation of sacred groves; species diversity, richness, and evenness were generally high as compared to other land use types.

Chapter 5 discusses how forest fires incited by human actions, such as land clearing, extreme drought, or in rare cases by lightning, also affect the entire flora and fauna in the forests, thereby also affecting the biodiversity, ecology, and environment of the forest region.

Chapter 6 discusses the incorporation of agroforestry in a land-use can prevent degradation of the environment, enhance agricultural productivity, sequester carbon, enhance the quality of water, alleviate soil health, and support healthy ecosystems while generating stable earnings and other benefits to human welfare.

Chapter 7 highlights the role of agroforestry systems in enhancing existing biodiversity and its conservation and explains how the adoption of agroforestry practices in areas close to the large natural habitat or between protected areas has shown to improve flora and faunal diversity by linking landscapes. All this information will serve as useful guidelines for achieving the United Nations Sustainable Development Goal 15 of the 2030 agenda.

Chapter 8 encapsulates brief definitions, extent, and types of agroforestry systems present across the world, mechanisms of C-sequestration by agroforestry systems, and ancillary benefits associated with them. In addition, some light has been shed on international policies for promoting agroforestry all over the world, their execution, loophòles, and recommendations.

Chapter 9 reviews various traditional agroforestry systems practiced in the northeastern region of India and actions to be taken with regard to scientific research and administrative reforms in order to make these traditional agroforestry systems more sustainable and productive.

Chapter 10 highlights ecosystem services, aesthetic value, and climate change mitigation of agroforestry systems adapted and suited to the Indo-Gangetic Plains of India. Furthermore, agro-ecology specific rotation years, temporal variations, and synergy of the crops are crucial for securing higher returns in agroforestry models.

Chapter 11 would be helpful in recognizing the potential of sericulture-based agroforestry systems to formulate a strategy for climate change responses that are consistent with safeguarding food security and livelihood in rural India.

Chapter 12 provides information about increasing carbon sequestration potential and reducing greenhouse gas emissions through the application of organic amendments (biochar, manure, crop residue, and compost) in land ecosystems, particularly in agroforestry land use management.

Chapter 13 belongs to the global problem of saline and sodic soils that affect the various ecosystem services provided by the land and soil.

Chapter 14 discusses how agroforestry can help in improving the quality, fertility status, and health of soils through closed nutrient cycling while simultaneously reducing the ecological footprint of agriculture.

Chapter 15 discusses how *Jhum* (shifting) cultivation of Arunachal Pradesh, which was considered a sustainable system, has now declined due to increased population pressure, unplanned, and faulty agricultural practices. Along with these, the steep slope and high rainfall during a short rotation period or *Jhum* cycle make the soil more vulnerable to erosion through runoff. This has made researchers rethink these problems and their solutions.

Chapter 16 discusses the present status, production trends of the declining phase, problems, and issues highlighting the changing socio-cultural scenario and deadly viral (*chirkey/foorkey*) diseases, and future opportunities under large cardamom farming of the Sikkim Himalayan region.

Chapter 17 discusses the conservation and promotion of nature-based indigenous traditions of the Lepcha ethnic community, which can be achieved by encouraging younger generations to take up traditional practices as a means of their livelihood.

Chapter 18 describes the solutions by identifying the root cause and assessing the nutrient inflows and outflows from the system, i.e., at the field level, farm level, and regional level. High-resolution data like multispectral data, hyperspectral data, and geographic information system (GIS) approach for spatial variability-based methods are proposed in this chapter.

CHAPTER 1

Nurturing Natural Resources for Reducing Environmental Pollution

DIBYENDU MUKHOPADHYAY, PRINCY THAKUR, and BASABDUTTAA BHABAI

Department of Soil Science and Agricultural Chemistry, Uttar Banga Krishi Viswavidyalaya, Pundibari, Cooch Behar, West Bengal, India

ABSTRACT

The most important natural resources like soil, water, and air are available in the environment for providing support to the life processes both at the biotic and abiotic systems to maintain the relevant factors for the livelihood security. The natural resources on earth, plant, and animal kingdom could affect the ecological balance in the environment until the resources are being depleted at a faster rate due to indiscriminate use. Both the natural resources (renewable and non-renewable) are directly or indirectly related to water use, agriculture, fossil fuel consumption, fishing, and mining. Hence, natural resources are to be conserved for sustainable agriculture. The imbalance of natural resources may lead to the deposition of some toxic elements or sometimes may be deficient in some essential elements in the system. The deposition of pollutants in the soil-water-plant continuum has a negative impact when it is being translocated in the food chain. The air pollution, water pollution, and soil pollution are the major determining factors to achieve a pollution-free ecosystem. The conservation of natural resources can ensure the maintenance of ecological balance and thus reduce the level of the toxicants in the environment as well.

Sustainable Management and Conservation of Environmental Resources in India.
Sumit Chakravarty, Vineeta, Jahangeer A. Bhat, Munesh Kumar, Gopal Shukla (Eds.)

1.1 INTRODUCTION

It is evident that a favorable environmental condition is required to sustain an organism. It is the environment which can create a situation for nurturing the growth and development of its living forms. The environment surrounding the living bodies can influence the life cycle as a whole. Hence, our environment refers to those agents or factors which, although distinct from us, could regulate our life or activities. These factors may be natural, artificial, social, biological, and psychological, affecting the ecological balance in the environment. There are different kinds of gases (sulfur, nitrogen, carbon dioxide, etc.) in the air which may act as pollutants based on the pattern of utilization in the ecosystem. Sulfur, being one of the pollutants in the air, may deposit an excess amount of acid in surface water bodies and streams and thus degrade the quality of soil. Similarly, nitrogen in the atmosphere when deposited on surface waters can affect the aquatic life. Our natural resources are facing challenges from the potential threat due to overpopulation, alterations in the global climatic pattern leading to environmental pollution. Hence, proper policies should be framed to maintain the balance of the natural resources in the environment. The changing environmental scenario has an impact on natural resources as input to production and also as reciprocal discharges back to the environment.

In order to sustain the livelihood security, the output-processing parts in the agricultural sector need to be attached to the economy-environment linkages. It requires understanding the long-term feedback on natural resources as well as the environment. Soil, being a complex heterogeneous entity, could affect the physical and biological conditions of the ecosystem and thus becomes an important quasi-renewable input to agriculture. There should be mechanisms to identify "soil rejuvenation inputs" and "soil management options" to regulate various unwanted and environmentally deleterious wastes related to food, fiber, and animal products. The water-borne effluents derived from runoff and reverse irrigation flows are also included. The management strategies should be in the active monitoring phase to control the damages caused by the environment, where water may be the most limiting natural resource in the long run if not used judiciously. Groundwater recharge has become an important issue for increasing the supplying capacity of the aquifer. More attention should be given to water use efficiency (WUE) for crops and developing water harvesting structures where there is an increasing scarcity of water. The conservation of biodiversity under the natural ecosystem should be prioritized for sustaining the ecological balance in the long run. Thus, a system of controlling the use of natural resources for effective utilization in the environment is the management of natural resources. The management should be prioritized because of the limited resources of the earth, over-exploitation of the natural resources to get short-term benefit, ensuring the equitable distribution of the natural resources to minimize the damage caused for societal development. Preventing environmental pollution by conserving natural resources and also by utilizing judiciously the non-renewable energy sources could meet our challenges to restrict environmental pollution.

1.2 ATMOSPHERE AND ITS COMPOSITION

Atmosphere is the blanket of air covering the earth. The biotic and abiotic components of the biosphere provide resources to meet the basic needs for sustenance. In this regard, we need to understand the types of natural resources and utilities of each of these natural resources to maintain our livelihood security.

1.2.1 TYPES OF NATURAL RESOURCES

1. **Inexhaustible:** They are not likely to get exhausted despite continuous use (air, water, solar energy).
2. **Exhaustible:** It can be depleted by continuous and indiscriminate human consumption. They are either renewable (forest, wildlife, soil, ground water) or non-renewable (minerals, fossil fuels).

1.2.2 EARTH AND ITS SURROUNDINGS

Earth gives us the most favorable environmental support for sustaining the life. The gaseous mixture could make an eco-friendly atmosphere by protecting us from heat and radiation emitting from the sun. The thickness of the Earth's atmosphere is about 300 miles, but most of it is within 10 miles (16 km) from the surface. The air pressure may vary from 14.7 to 10 pounds per square inch depending upon the level of elevation. Due to the reducing air pressure at higher altitude could cause breathing problem for limited supply of oxygen.

1.2.3 COMPOSITION OF AIR IN THE ATMOSPHERE

The air in the atmosphere is composed of nitrogen (78%), oxygen (21%), argon (0.93%), carbon dioxide (0.04%) and traces of hydrogen, neon, helium, methane, krypton, and water vapor (Sharp, 2017).

1.2.4 LAYERS OF ATMOSPHERE

The atmosphere is divided into five layers namely:

- Exosphere;
- Thermosphere;
- Mesosphere;
- Stratosphere; and
- Troposphere.

A change in density of air in each higher layer till the gases dissipate in space is observed. An imaginary line (Karman line) of about 62 miles from the surface is there where the interphase between atmosphere and outer space is visible.

The troposphere is close to Earth's surface ranging 4–12 miles constituting half of the earth's atmosphere. Air is warmer at the vicinity of ground surface and temperature is reduced with increasing elevation. The water vapor and dust particles present in this layer result cloud formation. The **stratosphere** is above the troposphere and extends up to about 31 miles from the ground level. The presence of ozone layer absorbs harmful radiation from the sun and the air is dry and thinner than what is available at sea level. This layer is favorable for flying the jet and weather-balloons. The **mesosphere** starts at 31 miles and extends up to 53 miles in the atmosphere. The layer Mesopause, over the top of the mesosphere is the coldest part of earth's atmosphere, with temperatures averaging at – 130°F. The t**hermosphere** extends from about 56 miles to 620 miles where the temperatures could increase up to 2,700°F at this altitude. The density of air is very low, and the auroras occur in this layer. The charged particles from space when collide with atoms and molecules in the thermosphere, could move to their higher states of energy on excitement. The atoms when shed the excess energy by emitting photons, is observed as the colorful Aurora of Borealis and Aurora of Australis. The **exosphere** which is basically composed of dispersed particles of hydrogen and helium is a thin layer where the atmosphere apparently merges into outer space.

1.3 CLIMATE AND WEATHER

The ecosystem enables to accommodate wide varieties of living being in its diverse regional environment ranging from extreme cold at the poles to tropical heat at the equator. The climate of a region is basically described as the average weather report in a place over 30 years or so. A region's climate is often described as the weather condition in a certain place (i.e., sunny, windy, humid or dry). The weather can change within a few hours while, the climate-change is considered over a longer period of time.

It is reported that the global climate is an average of regional climates where the earth may be cooled and warmed within the span of a year. There are abrupt changes in global climate in recent times, because of the warming situation in the atmosphere. The higher emission of the greenhouse gases (NO_2, CO_2, CH_4, etc.) has caused depletion of ozone layers and thus, there is creation of ozone hole as well as global warming in the ecosystem.

1.4 CONTAMINANTS IN ENVIRONMENT

Our environment is endowed with natural resources like air, water, and soil. But the ecological balance is getting disturbed by the pollutants dispersed in air, dissolved in water or adsorbed in soil.

1.4.1 AIR POLLUTION

The problem of air pollution is enhanced due to the presence of ample amount of air pollutants which are usually the suspended particulate matter (SPM) (dusts, fumes, mists, smokes, gaseous pollutants).

1.4.1.1 SUSPENDED PARTICULATE MATTER (SPM)

Based on the total suspended particles, the SPM is categorized as:

i. PM10 – it is the most hazardous finer fractions which can reach the alveoli of the lungs.
ii. PM2.5 – the particles with median aerodynamic diameters of less than 10.0 microns and 2.5 microns, respectively come under this category. The PM2.5 consists of secondary pollutants developed

through the condensation of gaseous pollutants like sulfur dioxide (SO_2) and nitrogen dioxide (NO_2).

iii. The suspended particulate matter under this category includes particles coming out from diesel, fly ash, wood smoke, mineral dusts (coal, asbestos, limestone, cement, metal dusts, fumes, acid mists) and pesticidal mists.

1.4.1.2 SOURCES OF AIR POLLUTION

In order to reduce the air pollution, it is required to find out the possible sources through which air at different pockets of the ecosystem is getting polluted. Both the stationary and mobile sources (Figure 1.1) are responsible for the air pollution in the environment.

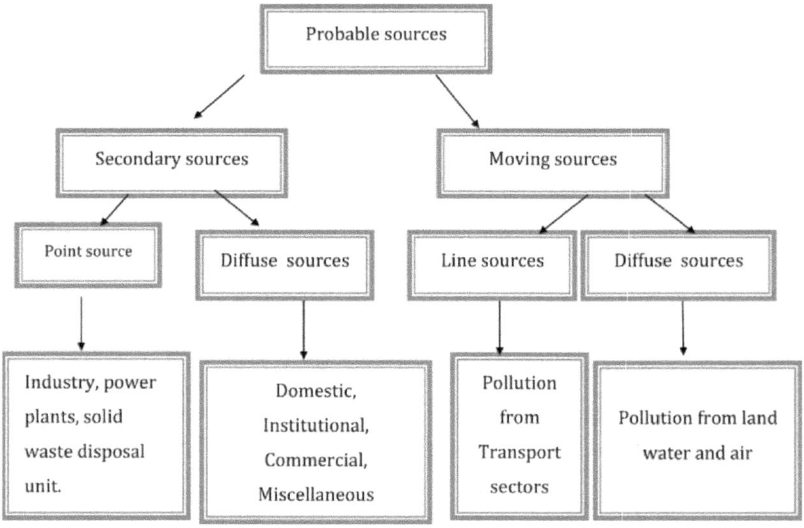

FIGURE 1.1 Sources of air pollution.

Source: Adapted from: Babu (2015).

The air and water are getting polluted by the industrial by-products comprising of harmful pollutant elements (Pb, Cd, Cr, As, Ni, F, etc.) and their oxides and hydroxides to act as toxicants for the human continuum as a result of which the workplaces are getting susceptible zones of pollution (Table 1.1).

TABLE 1.1 Important Pollutants in Air and Water at Workplace

Industrial Sector	Air	Water	Workplace
Cement industry	Particulate materials	Sludge	Silica
Coal mining	Coal dust	Sludge coal dust	Silica
Copper smelting	Arsenic	Arsenic	Arsenic, cadmium
Oil and natural gases	SO_2, carcinogens	Oil	Hydrocarbons
Pesticide manufacturing sectors	Pesticides and toxic intermediates	Pesticides and toxic intermediates	Pesticides and toxic intermediates
Tannery	Odor	Chromium, acids	Chromium, acids

Source: Adapted from: World Bank (1999).

1.4.1.3 CLASSIFICATION OF AIR POLLUTANTS

The air pollutants can be divided into primary and secondary air pollutants based on the origin of source. The primary air pollutants are directly emitted to the atmosphere and are found in the form of its emission. Some of the important air pollutants are:

i. Carbon compounds – carbon dioxide, carbon monoxide.
ii. Sulfur compounds – carbonyl sulfide (COS), carbon disulfide (CS_2) dimethyl sulfide [$(CH_3)_2S$], hydrogen sulfide (H_2S), sulfur dioxide (SO_2), and sulfate (SO_4^{2-}).
iii. Hydrocarbons – benzene, methane (marsh gas).
iv. Oxides of nitrogen – (NO_x) NO, NO_2, N_2O.
v. Metals – zinc, cadmium, lead, mercury.
vi. Toxic substances – arsenic, carbon tetra chloride, beryllium, chromium, copper, nickel, polycyclic aromatic hydrocarbons.
vii. Organic compounds – aldehydes, ketones, carboxylic acids, organic sulfur compounds comprising of finer particles (less than 100 μ in diameter) and coarse particles (greater than 100 μ in diameter).
viii. Radioactive compounds of radium–222, uranium–232, strontium–90, and plutonium–239.

The sources of primary pollutants may be enhanced by the human activities (i.e., combustion, chemical, atomic, heating, refining, mining or farming activities) or by the natural phenomenon (volcanoes, seas, pollen, terpenes, fire, dust, bacteria, viruses, etc.).

On interaction among two or more primary pollutants or by reaction with the existing atmospheric constituents (chemical or photochemical reactions) the secondary air pollutant [ozone formaldehyde, peroxyacetyl nitrate, photochemical smog and acid mist (H_2SO_4)] are produced.

1.4.1.4 GREEN HOUSE GASES

The greenhouse gases (NO_2, CO_2, CH_4, water vapor) which are formed on reaction with the primary gaseous element have the ability to trap energy or sometimes by absorbing infrared (IR) light by the water molecules for subsequent transfer of the energy to the surrounding air molecule. Sometimes, it becomes essential to know the atmosphere of the terrestrial world (Table 1.2) other than the earth, so as to compare the composition of air in the atmosphere along with the surface pressure and weather patterns.

TABLE 1.2 Atmospheres of the Terrestrial World

Terrestrial System	Composition	Surface Pressure*	Clouds/Haze
Mercury	He, Na, O_2	10^{-14} bar	None
Venus	96% CO_2, 3.5% N_2	90 bars	Clouds of sulfuric acid
Earth	77% N_2, 21% O_2, 1% Ar, H_2O (variable)	1 bar	Clouds of water vapor, pollutants
Moon	He, Na, Ar	10^{-14} bar	None
Mars	95% CO_2, 2.7% N_2, 1.6% Ar	0.007 bar	H_2O and CO_2 clouds, dust particles

**bar ≈ the pressure at sea level on earth.*

Source: Adapted from: Hulburt (2015).

The CO_2 is one of the important greenhouse gases (GHGs) bearing the following properties:

i. The CO_2 is colorless, which condenses into dry ice at – 78°C in atmospheric pressure and becomes liquid at – 57°C at pressure above atmospheric pressure 5.1.
ii. The CO_2 is derived from volcanic eruption, organic matter decomposition and respiration of microbes.
iii. The dissolved CO_2 in water can react with silicate minerals to produce carbonated minerals.

The incoming solar heating arising out for the greenhouse effect could cause Earth's atmosphere slightly warmer than what it should be, and the ground surface is heated by visible and some infrared (IR) light coming from the Sun. The heated surface emits IR light to some extend and the GHGs (H_2O, CO_2) in small amount can trap the IR radiation and increase the temperature of the atmosphere. The entrapment of energy by the CO_2 in its vibrational (four modes are there in CO_2) mode is one of the important factors for increasing the global warming.

1.4.1.5 GLOBAL WARMING

The average temperature of the Earth's atmosphere is increasing during the last couple of years to the tune of about 1°C since 1900. There may be human/non-human factors causing the global warming that has affected both the animal and plant kingdom. The atmospheric concentration of CO_2 has increased to the tune of 30% during last few years. The contribution of some natural GHGs like water vapor (36–70%); carbon dioxide (9–26%), methane (4–9%) and ozone (3–7%) lead to the warming in the earth's atmosphere. The melting of ice in polar region is due to the gradual heating in the environment.

1.4.1.6 ACID RAIN

The gases like SO_2 and NO_2 emitted from the industrial sectors or from the natural sources when react with water can form H_2SO_4 and HNO_3 that comes down as acid rain on earth. It has been observed that, sulfur and nitrogen oxides are commonly emitted as greenhouse gas in one location while, the acid deposition occurs in a distant area. There may be dry deposition of gases, particles, and aerosols or may be deposition of dissolved acids as wet deposition (Figure 1.2).

There are two processes involved in acid deposition. The hydrochloric acid can sometimes be expelled directly into the atmosphere. The secondary pollutants formed on oxidation of nitrogen oxides or SO_2 are also released into the atmosphere. The nitric acid or sulfuric acid formed on reaction can be transported to a distance apart from their original source. Again, the oxides of nitrogen and SO_2 may settle on the landscape and interact with dew or frost to develop acid precipitation. The emissions of SO_2 could cause 60–70% of the acid deposition globally where about 90% of the sulfur in the atmosphere is anthropogenic. The sulfur is obtained by burning of coal, smelting of metal sulfide ores, volcanic eruptions, organic matter decomposition, ocean spray, etc. The SO_2 can either be deposited on the Earth's surface in the form of dry deposition or it can undergo reactions to produce acids as components of wet deposition.

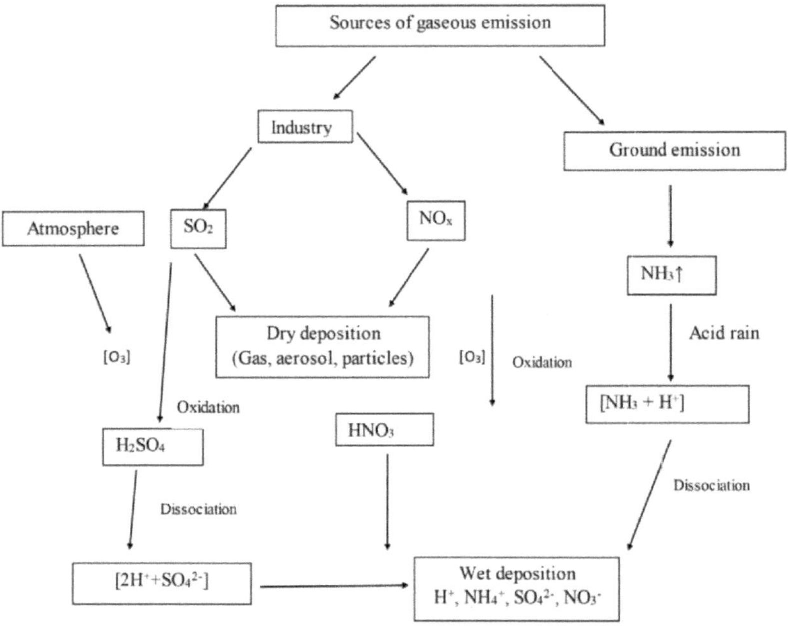

FIGURE 1.2 Mechanism of acid deposition caused by the oxides of nitrogen and sulfur.

Source: Adapted from: Ford (2020).

$$SO_2 + H_2O \rightarrow H_2SO_3$$

$$H_2SO_3 + 1/2O_2 \rightarrow H_2SO_4$$

About 95% of the increment of nitrogen in the atmosphere are anthropogenic while, the remaining 5% comes from different natural processes. The major sources of nitrogen oxides are lightening, combustion of oil, coal, and gas, reaction in soil, volcanic eruptions, etc. The following atmospheric reactions occur for the formation of HNO_3.

$$NO + 1/2O_2 \rightarrow NO_2$$

$$2NO_2 + H_2O \rightarrow HNO_2 + HNO_3$$

$$NO_2 + OH \rightarrow HNO_3$$

The relatively lower concentration of both nitrogen and SO_2 than the existing atmospheric carbon dioxide (CO_2) could make natural rainwater slightly acidic. The higher solubility of NO_2 and SO_2 than CO_2 can reduce more the pH of the precipitation.

The available calcium and/or magnesium (Mg) in ponds or lakes can naturally buffer the effects of acid deposition. The solubility of heavy metals like Hg, Al, and Cd which are locked in clay particles, minerals, and rocks is increased on acidification and are leached with infiltrating water for gradual deposition into the aquatic systems. The application of liming materials ($CaCO_3$, CaO, etc.) are more effective in correcting soil acidity than those which are made up of siliceous sand or weathered acidic bedrock. Vegetation is also affected due to the acid deposition. Due to the acid deposition in soil, there is reduction of soil pH which can regulate the availability of plant nutrients like calcium, potassium, and magnesium affecting the rate of plant growth. The availability of magnesium and potassium will be less available to the plants due to higher mobility of aluminum under lower pH of the soil. The decomposition rate will be affected due to the lowering of microbial population under extremely acidic situation in soil. The possibility of N-saturation in soil may occur due to high concentration of HNO_3. Besides, the dry deposition of SO_2 will reduce the ability of leaf to retain water under stress situation. The structure of the plant tissues also becomes weaker due to the acidic deposition.

1.4.1.7 OZONE LAYER DEPLETION

Ozone (O_3) having strong odor and blue in color exists within the tropospheric and stratospheric zones of the Earth's atmosphere. The ozone layer protects life on earth by absorbing harmful UV radiation before it reaches the earth in the stratosphere. The chlorofluorocarbons (CFCs) once released into the atmosphere are destroyed by ultraviolet rays and chlorine is released to combat the ozone molecules. Molecular oxygen is broken down to atomic oxygen in the stratosphere by solar radiation and combines with molecular oxygen to produce ozone. The ozone is then again destroyed by chlorine atoms. The UV radiation from the sun releases the radicals Cl and ClO. The unstable Ozone molecule readily donates its extra oxygen molecule to free radical species such as hydrogen, bromine, and chlorine to make compounds that can act as catalysts in the breakdown of ozone molecules (Figure 1.3).

$$Cl_2 \rightarrow Cl + Cl$$

$$O_2 + Cl \rightarrow OCl + O$$

$$2OCl \rightarrow O_2 + Cl_2 \rightarrow Cl + Cl$$

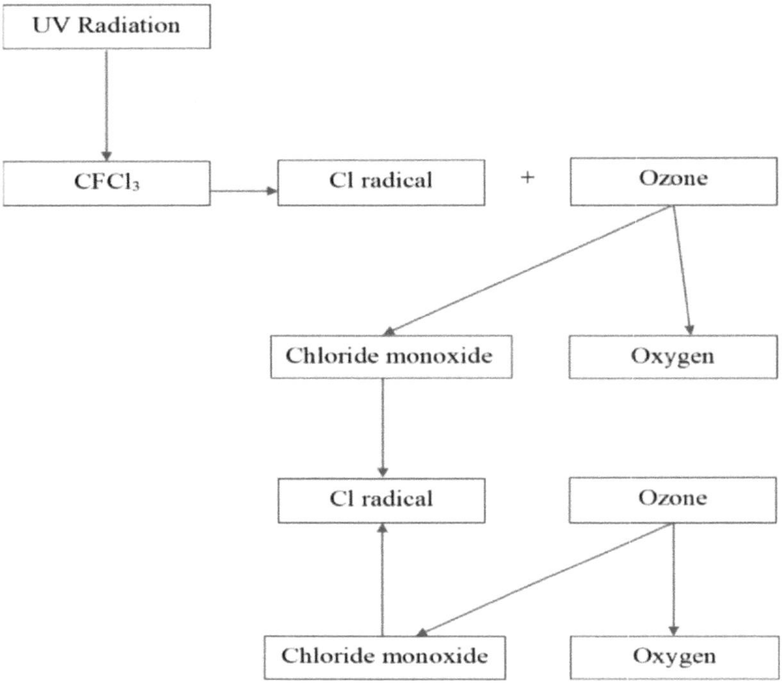

FIGURE 1.3 Flow chart of regeneration of chlorine radical.

Depletion of ozone layer will create ozone hole that permits the entry of UV rays to earth surface causing drastic effects on aquatic ecosystems. UV radiation damages DNA of the plant cells causing susceptibility to pest and disease infestation. It can cause mutation in mammalian cells and can damage the cell membrane as well. It is carcinogenic to humans, cause premature aging of the skin, cataracts, and eye disorders.

1.5 DYNAMICS OF ATMOSPHERIC CARBON, NITROGEN, AND SULFUR

1.5.1 CARBON CYCLE

The CO_2 emitted from the volcanic eruption may be dissolved in ocean or there may be formation of carbonated rocks when react with the eroded minerals (e.g., limestone). The atmospheric CO_2 concentration when reduced will lower down the effect of greenhouse. The carbonate minerals although formed during the cooling process of earth in the oceans yet, the increased CO_2 concentration warms the planet again through the greenhouse effect. It has been observed that the lighter gases have higher probability of getting the velocity more than escape velocity and can go away from the earth. It is pertinent to mention that there should be increase in the carbon stock on the planet under different pools which are present:

- as organic molecules of organisms;
- as carbon dioxide in the atmosphere;
- as component of soil organic matter (SOM);
- as fossil fuels and sedimentary rock in the lithosphere;
- as dissolved atmospheric carbon dioxide and calcium carbonate in marine organisms.

The microorganisms are also involved in the carbon cycle (Figure 1.4) through different biochemical reactions. The carbon-fixing organisms like photosynthetic algae which consume CO_2 and organic carbon, thus fixed carbon by microorganisms is transformed through the biogeochemical processes to fossil fuels. The organically produced substances are degraded by the microbes and available CO_2 in presence of sunlight and water and chlorophyll gets fixed by photosynthesis. Biodegradation of organic substances prevents accumulation of excess waste residue and converts complex organic substances into simple organic forms for further utilization by plants. The humus being one of the residues on biodegradation of organic substances can improve the physical condition of soils.

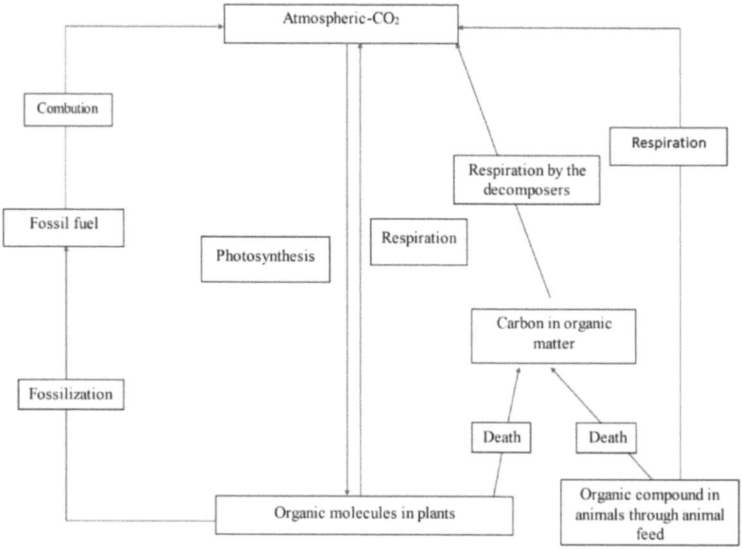

FIGURE 1.4 Carbon cycle.

Source: Adapted from: Balasubramanian (2017).

1.5.2 NITROGEN CYCLE

In the nitrogen cycle (Figure 1.5), nitrogen is converted into different forms through the biogeochemical processes by which nitrogen will be recycled from atmosphere to the soil involving five important steps:

- Nitrogen fixation (conversion of N_2 to NH_3/NH_4 + or NO_3^-);
- Nitrification (conversion of NH_3 to NO_3^-);
- Assimilation (incorporation of NH_3 and NO_3^- into biological systems);
- Ammonification (conversion of organic nitrogen compounds to NH_3); and
- Denitrification (conversion of NO_3^- to N_2).

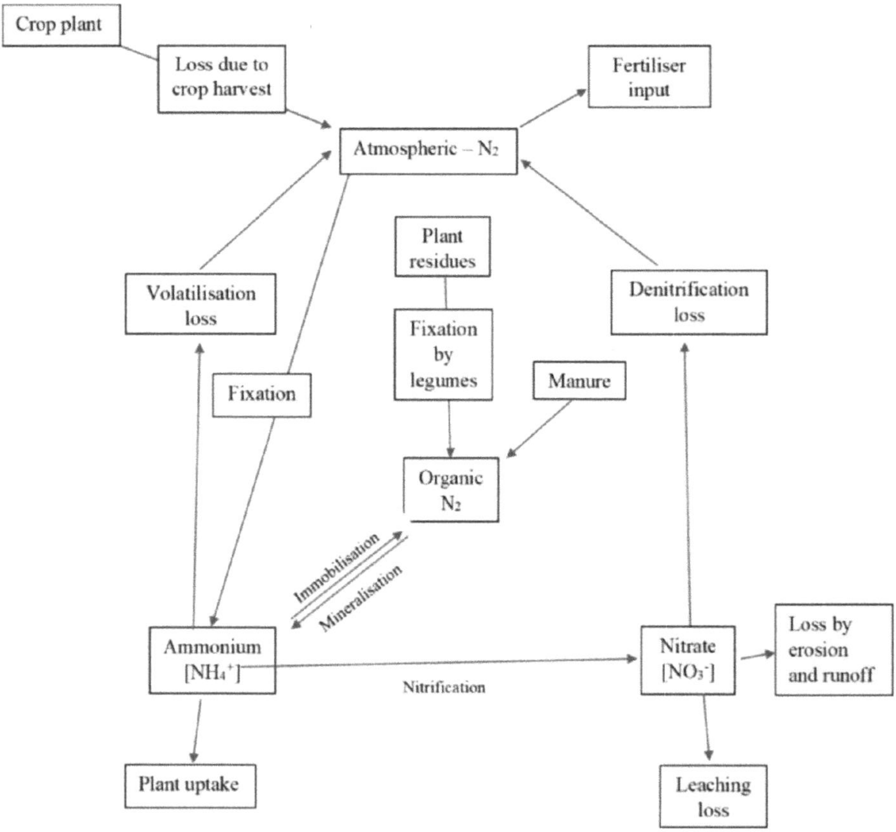

FIGURE 1.5 Nitrogen cycle.

Source: Adapted from: Reitsma et al. (2011).

1.5.3 SULFUR CYCLE

The movement of sulfur among the rocks, soils, and living systems encompasses the sulfur cycle (Figure 1.6). Sulfur is the important functional element of proteins and enzymes. The water-soluble sulfur is absorbed by the plants which are consumed by the animals to maintain their health. Besides, there are sources like volcanic eruptions, soil dust, biogenic gases [H_2S, dimethyl-sulfide $(CH_3)_2S$, carbonyl sulfide (COS)] and anthropogenic emissions. Sulfur is deposited in rivers through the oxidation of pyrite and also by the application of gypsum and is derived also from air pollution, mining, erosion, industrial sectors and other human activities.

$$4FeS_2 + 15O_2 + 8H_2O \rightarrow 2Fe_2O_3 + 8H_2SO_4$$

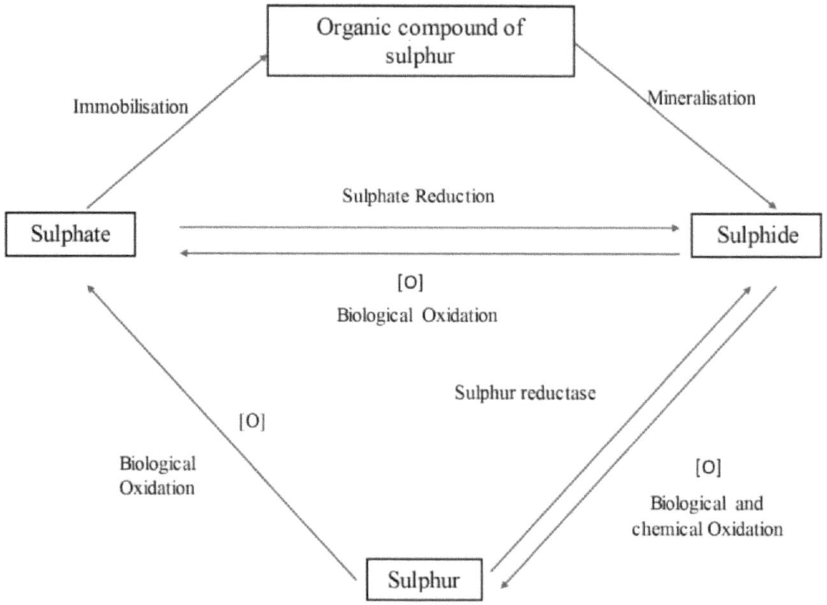

FIGURE 1.6 Sulfur cycle.

Source: Adapted from: Janssen et al. (1999).

1.6 ORGANIC POLLUTANTS

It has been observed that humans are exposed daily to numerous organic chemicals which are hazardous to health. There are harmful organic compounds which are quite stable in the environment for long periods are called persistent organic pollutants (insecticides, fungicides, herbicides).

These are also harmful to human health and also cause ecological damage. A biocide is a substance toxic to varying degrees of life forms in the form of pesticides to kill the harmful organisms in the environment. The biocides have the degradation ability, the products so generate may again have some toxic effect. The water-insoluble polychlorinated biphenyls (soluble in fat) readily degrade to less toxic forms and persist in the environment for a longer period of time and subsequently can enter in the food chains.

1.7 SOIL EROSION

The soil particles are removed from the top layer of soil due to soil erosion *via* water and wind by the process of detachment, movement, and deposition. Soil erosion and degradation is mainly caused by deforestation, intensive agriculture, overgrazing, excessive use of agrochemicals and construction activities. Soil erosion may cause land degradation, desertification, eutrophication in water bodies and sedimentation of water ways. Various soil conservation measures such as contour bunds, graded bunds, bench terraces, grassed waterways, water harvesting and surface cover by vegetation can be effective to prevent soil erosion.

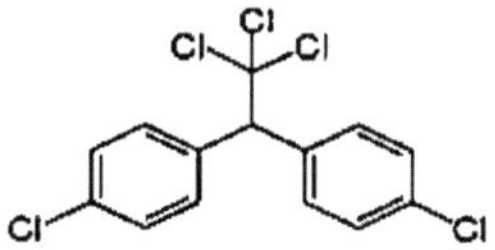

DDT
(Dichlorodiphenyltrichloroethane)

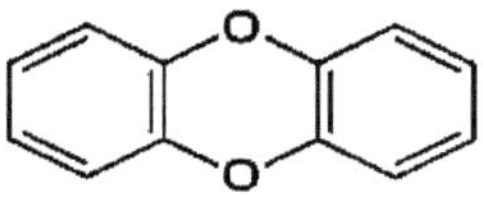

Dioxins
(Dibenzo-1,4-dioxin)

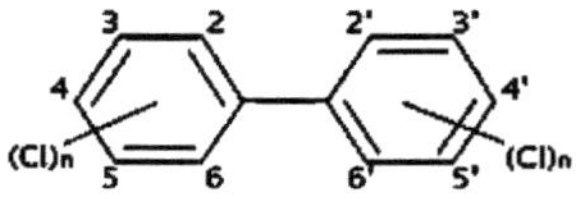

PCBs
(Polychlorinated biphenyl)

1.8 POLLUTION

1.8.1 AIR POLLUTION

The oxides of nitrogen which are commonly found in environment are NO, NO_2, N_2O, N_2O_3, and N_2O_5. Among them, NO (nitric oxide) and NO_2 are the two most important air pollutants with maximum emission in the environment. Oxides of nitrogen are released from different sources such as automobiles, industries, and coal-fired power plants. Reducing NOx emissions effectively will need control of both stationary sources and mobile transport sources. Use of low-nitrogen fuel decreases these emissions. Besides, modifications to the combustion process or installing air pollution control tools [such as selective non-catalytic reduction (SNCR) or selective catalytic reduction (SCR)] can generate low nitrogen oxides. High levels of NO_2 can have a negative impact to lung which causes chronic respiratory disease when exposed for longer duration. Elevated levels of NO_2 can damage foliage, reduce growth and crop yields. Any material containing sulfur is burned releasing SO_2 into the atmosphere. SO_2 is associated with respiratory illness and irritation in throat, eye, and nose. SO_2 also damages chlorophyll in green plants.

Emission of volatile organic compounds (VOCs) from motor vehicles and vessels also cause air pollution and smog. In the air, VOCs play an important role for formation of ozone gas and fine particulates. Nitrogen oxides which are emitted from vehicles, power plants, industrial activities react with VOC_S under sunlight and form ozone. Smog which is actually accumulation of ozone, fine particulates and other gaseous pollutants reduces visibility. It can irritate nose, eyes, and throat; even worsen the existing heart and respiratory problems. Emission of VOC and organic-hazardous air pollutant are being controlled and their treatment is done by the process of adsorption or absorption. Care should be taken to control emissions from motor vehicles, industrial, and commercial processes for reducing VOC-containing products.

1.8.2 SOIL POLLUTION

Soil may act as a major sink for the plant – available nutrients as well as for the pollutants. Still soil can play important roles in stabilizing natural imbalances on earth. Soil is a natural medium for growth of plant as it provides anchorage, oxygen, water, and nutrients. The soil provides a physical support for the crop. The plant draws the necessary water through its roots from the soil pore that provide strong buffering of the water supply that comes from rain on crops or is applied by humans as irrigation. Plant takes all necessary nutrients from soil for their growth and development. Soil is the habitat of nitrogen fixing organisms and organic matter. Nematodes, micro-arthropods (mites and collembola), enchytraeids and earthworms are inhabitants of rhizosphere. The soil surface and litter layer also contain several species of macro-fauna (mainly arthropods, beetles, spiders, diplopods, chilopods, and snails).

The filtering capacity of soil has an important role to play for the protection of groundwater and surface water. It depends on the hydrological transport processes and the behavior of pollutants in the soil. Soil particles act like a filter and their arrangement regulates overall filtration. The negatively charged clay particles can attract cations, such as calcium, magnesium, and potassium and thus remove ions from the contaminated water by selective adsorption of the ions. Plant absorbs these cations as nutrients and also some pollutants are removed by this process. Rainwater takes active role in groundwater recharge. Soil recycling reduces the extracting, refining, and processing of raw materials (solid waste, minerals, and water), helps for increment of economic security through tapping of domestic sources and also saves energy and reduces the GHG emissions.

To build up a modern infrastructure, accessibility of good soil is very important. The basic material for building foundation is brick and cement, which are made up of soil. In the building process, selection of sites with the best soil is an important engineering decision where soil maps are also used for site selection for the infrastructure of building. Engineer uses soil map for best site selection considering the slope of the land, physical, chemical, and biological properties of soil which are also the potential for water runoff, drainage, or storage. The soil-water characteristic (SWC) curve which describes the amount of water retained in a soil at equilibrium at a given matric potential is an important hydraulic property related with the pore space, soil texture, structure, and organic matter. The SWC is required for water management and in prediction of transport of contaminants in the environment. Soil can also influence climate. Plant is decomposed by soil microorganisms and releases CO_2 to the air. Release of CO_2 to the air can be reduced through application of soil conditioners with slower decomposition. Emission of GHGs such as CO_2 and methane from soil can be reduced by changing in soil management practices.

1.8.2.1 CAUSES OF SOIL POLLUTION

The substance which on application to the soil impacts adversely on plant growth and animal health is called soil contaminant or pollutant and when such loading exhibits the adverse effect on the buffering capacity of the soil then the soil is said to be polluted and the process as soil pollution. Soil pollution occurs due to the following reasons:

- Indiscriminate use of fertilizers and pesticides.
- Toxic substances coming out from the air, water, and from the industrial effluents and hydrocarbons.
- Dumping of waste materials, sewage/sludge materials due to rapid urbanization.

1.8.2.2 SOIL MANAGEMENT

Soil may act both as a source of nutrients and sink at the same time. When soil is saturated on adsorption of so many contaminants, then it rejects the pollutant elements and becomes sources of the toxicants for the environment.

1.8.2.2.1 *Use of Fertilizers*

Excess use of fertilizer in soil may cause toxicity to the plants and animal health. Application of excess amount of nitrogenous fertilizers to the crops may cause lodging, pest infestation or toxic effect on the economic produce.

1.8.2.2.2 *Heavy Metal Accumulation*

Heavy metal (specific gravity > 5.0 or atomic number > 20) sometimes get deposited in soil as impurities in fertilizers, manure, fungicide, and waste materials which then gets gradually deposited in soil and may act as a source of pollutant element for the crops. The heavy metals (Pb, Cd, Cr, Ni, As, Se, Cu, and Zn) behave in the soil systems on the basis of the nature of soil and on soil physicochemical properties. The shift in redox potential can influence the mobility of such heavy metals from the adsorbed phase to the soil solution phase.

1.8.2.2.3 *Use of Pesticides*

In modern system of agriculture, the use of pesticides has become inevitable to get rid of unwanted plants and animal population living in or on the crop. The most available pesticides which are used in the environment are insecticides, herbicides, fungicides, rodenticides, and nematicides. The synthetically prepared pesticides when applied into the soil may get entry into the plant and also to the animal kingdom depending upon the period of persistence in the environment. Some of the common pesticides which are used:

- DDT, BHC, malathion, carbamates (insecticides);
- Organo-mercurial fungicide, copper fungicides, sulfur fungicide (fungicide);
- 2,4-D, glyphosate (herbicide).

Pesticides when applied in soil may undergo adsorption, leaching, volatility, chemical degradation, plant absorption and microbial transformation. The applied pesticides may enter into the water bodies when the treated crops are found close to the river and or streams. Improper storage of pesticides and or spillage may thus enter and contaminate groundwater. Release of soluble salts to the soil due to high evapotranspiration (ET) process or indiscriminate use of fertilizers and effluents of sewage and sludge may cause pollution to a large extend.

1.8.2.2.4 Soil Salinity

Water containing high concentration of dissolved salts (sulfates, carbonate, bicarbonates, chlorides, sodium, etc.) used for irrigation purpose makes the soil saline. There is deposition of salt at the surface soil from the subsoil when the rate of evaporative demand exceeds precipitation.

1.8.2.2.5 Industrial Activity

The contaminated by products coming out from the industrial belts when are not properly disposed of in the land or in water bodies cause soil and water pollution. The scarce in having good quality water is sometimes replaced by the industrial wastewater for the production of crops. This effluent-water not only contain traces of crop nutrients but also organic and inorganic materials along with heavy metals which may be deposited on soil to enter into the food chain due to the prolonged use of this contaminated wastewater in the field and the soils become polluted.

1.8.2.2.6 Domestic Activities and Urbanization

The domestic refuse, municipal waste materials, city wastes, and sludge are generated by the domestic and urban activities. The solid wastes include the biodegradable (70%) and non-biodegradable (30%) materials. These are deposited into the soil and pollutant elements and are diffused into the food web.

1.8.3 WATER POLLUTION

Water pollution is caused due to the ground and surface water contamination through the chemical and bio-physical changes in water quality which has a harmful effect on living organisms or makes water unusable. The water may be contaminated by infectious agents (bacteria, viruses, and fungus); inorganic chemicals, organic chemicals (oil, detergents), plant nutrients (water soluble nitrates, phosphates) and radioactive substances (power plants).

1.8.3.1 AGRICULTURAL ACTIVITIES

The chemical fertilizers are applied to soil for maximizing the yield potential of the crops. The indiscriminate use of nitrogenous fertilizers has increased the nitrate level in the groundwater by leaching which when taken through drinking water (concentration of nitrate > 10 mg L^{-1}, WHO) may cause methemoglobinemia (blue baby syndrome) to the infants in particular. Besides, formation of nitrosamine in stomach on reaction with ingested amine and nitrate can lead to the cancer due to the nitrate contamination. The heavy withdrawal of groundwater for summer paddy cultivation during dry season (January to May) when the groundwater recharge is at its minimum may cause pollution of irrigation water with arsenic, fluoride, selenium, etc. When polluted groundwater is used for irrigation to the crops, the heavy metals may enter into the economic produce of the crops and subsequently enter into the food chain.

1.8.3.2 EUTROPHICATION

It has been observed that there is algal bloom and other aquatic plant growth due to the nutrient enhancement in waterbodies with phosphorus and nitrate coming through runoff, from dumping ground and sewage-sludge which is known as eutrophication. The massive proliferation of these aquatic plants deteriorates the quality of water in lakes, ponds, rivers, and other water bodies making it unusable for domestic and other uses. The sources of the indiscriminate pollution in such cases (> 80%) due to the use of fertilizers are observed mainly at the agricultural fields. There are different types of water pollution based on the sources:

1. Pollution Caused by Point Source: Contamination of water bodies is caused by substances released from a single, identifiable source through pipelines. Many harmful substances are released into water bodies from the industrial belts also.
2. Pollution by Non-Point Sources (NPS): Contamination of water bodies by substances that do not come from a discrete and specified source (e.g., pollutants coming from farmlands that mixes into the streams).

The origin of NPS of pollution (atmospheric deposition and hydro-modification) are from different diffuse sources either may be caused by rainfall or snowmelt moving over through the ground surface or due to the runoff which picks up and pollutants elements and deposits into the surface waterbodies as well as groundwater. These pollutants are:

- Residue of fertilizers and pesticides from agricultural sectors;
- Chemicals from urban runoff and energy production;
- Sediments from construction sites;
- Saline irrigation water;
- Residue from abandoned mines;

The NPS of pollution comes in large amounts from a single source, such as from the industrial operation or from the wastewater recycle plant.

1.8.3.3 GROUND WATER POLLUTION

Groundwater pollution is an environmental concern. The water may be polluted by indiscriminate use of fertilizers and pesticides. The waste disposal from the industrial belts as well as from the domestic usages and also for the leaking out of the petroleum oil from the sources may cause pollution to the water bodies. The gradual accumulation of nitrate and phosphate coming out from the fertilizers and arsenic from the arseno-pyrite minerals can add as contaminants into the groundwater. The soil sometimes becomes polluted when exposed with contaminated irrigation water for long period of time and thus may act as an effective source of pollutants for the groundwater. This may happen due to the leaching of heavy metals (As, F, Cd, Cr, etc.) from the contaminated soil to the groundwater to make the water unsuitable for drinking and other usages.

1.8.3.4 ACID MINE DRAINS

Sometimes, ferric ions (Fe^{3+}) liberated from the iron pyrite (FeS_2) reacts with water, releases more acid to the water and thus insoluble ferric hydroxide is formed which is further deposited as a yellow-orange coating on the stream-beds, known as yellow boy.

$$Fe^{3+} + 3H_2O \rightleftharpoons Fe(OH)_3 \downarrow + 3H^+$$

1.8.3.5 PESTICIDAL POLLUTION

Pesticides are applied for preventing, destroying, and repelling harmful pests. The pesticides are used in the agricultural farms as well as for other purposes which are synthesized in the laboratory. Excessive applied pesticides are absorbed by soil and runoff into nearby waterbodies contaminate both land and water. The groundwater contamination by pesticides application is governed mainly on the solubility in water, rainfall intensity and mobility of the applied pesticides.

1.8.3.6 CONTROL OF WATER POLLUTION

There are different methods of purifying polluted water which may further be used for drinking and irrigation purposes. Initial water purification or primary water treatment process can be of physical or chemical that includes methods related to screening and microfiltration, sedimentation, chemical precipitation, coagulation, gravity, and flocculation although, the efficiency is much less than that of the tertiary treatment process.

1.8.3.6.1 Microfiltration

The membrane filtration technology which contains pore size from 100–1,000 nm can be made on both ceramic and polymer membranes for separation of TDS (total dissolved solids) and microorganisms larger than viruses and micro-pollutants. Pre-treated water can also be purified through adsorption, ion exchange and membrane filtration processes.

1.8.3.6.2 Chemical Precipitation

The most extensively and effectively used purification process in industry is chemical precipitation due to its simple and low-cost operative systems. In this process, chemical reagents (lime, hydroxides, and sulfides) react with heavy metals with the changing pH to form insoluble particles for further removal by sedimentation.

1.8.3.6.3 Coagulation and Flocculation

This method basically destabilizes the suspended particles in the contaminated water by adding coagulants or flocculants that lead to the sedimentation of the particles. The coagulant and flocculating agents (ferric and aluminum salts) can improve settling of sludge along with removal of biological pollutants.

1.8.3.6.4 Secondary Water Treatment

Secondary water treatment is regulated mainly by the microorganisms (anaerobic and aerobic) to convert pollutants into other acceptable substances in the environment. Biogas produced through anaerobic processes can also be used in renewable energy sources. Removal of organic pollutant is mainly done by secondary water treatment.

1.8.3.6.5 Tertiary Water Treatment

An oxidizing agent for transfer of electrons from the oxidant to the pollutants is introduced for structural modification in the waste materials. The process of oxidation can remove organic and some inorganic compounds from polluted water in water treatment plants. Steam-stripping or activated carbon adsorption process can remove toxic by-products through the process of chemical oxidation.

1.9 BIOLOGICAL EFFECTS OF POLLUTANTS

1.9.1 BIOACCUMULATION

Sometimes the pollutants when being accumulated by the organisms during its life span may be a process of detoxification in the environment. The toxicants which are coming into the food chain due to continuous exposure to the chemicals are detoxified by the microbial assimilation prior to their entry into the biological systems.

1.9.2 BIOMAGNIFICATION

The term biomagnification is classified as a definition where the contaminant concentration of an organism exceeds the contaminant concentration of its food system in its diet when the main chemical exposure route to the organism is from food. By extension, the term food web biomagnification has been defined as the increase in contaminant concentration with increasing trophic status of organisms sampled from the same food web. Biomagnification and food web biomagnification was originally created from the observation of the bioaccumulation of chlorinated pesticides in the aquatic food webs. However, other contaminants like mercury, heavy metals and certain compounds of biogenic origin are also applied in biomagnification.

1.10 REMEDIATION OF SOIL AND WATER POLLUTION

In order to control the over-use of fertilizers and pesticides in soil, we have to concentrate on three 'R's, i.e., reduce, reuse, and recycle. However, the following measures may be adopted to control the soil and water pollution in the environment:

- The native status of nutrients may be analyzed to recommend the soil-test based fertilizer application according to the demand of the crop. The split application of nitrogenous fertilizer will not only reduce the loss of nitrogen from the soil but also will reduce the pollution of nitrate.
- In order to reduce the massive use of pesticides, farmers are adopting organic farming in some areas and also using biodegradable substances to control pests. Crop diversification, selection of pest-resistant crop varieties, biological method of pest control are the alternative methods to reduce the use of chemical pesticides and thus lower down the soil pollution.
- The non-biodegradable materials are to be separated from the complex solid waste and to be processed for further reuse at domestic levels.
- The non-biodegradable materials can also be recycled to save the natural resources which will conserve the natural resources.
- Solid waste treatment through physical, chemical, and biological processes will convert the toxic materials into the less toxic products for further agricultural use. The unutilized and treated solid waste can be used for land filling in the low-land areas for developing facilities in the urban areas.
- Firewood, palm-waste, paddy-straw, saw-dust, or any biological waste can be burnt under controlled atmosphere (pyrolysis) to yield combustible by-products (tar, fuel gas) for further use.

However, for the remediation of a contaminated soil, the information on kind and amount of materials is to be added. The C:N ratio of the materials, type of organisms present in the soil and impact on the environment for addition of the materials is also to be considered.

1.10.1 OTHER METHODS OF REMEDIATION OF SOIL CONTAMINATION

There are different methods of remediation for the contaminated soils. These methods are specific and if applied effectively can reduce the extent of contamination of the toxicants in soil-water-plant-animal continuum.

1.10.1.1 BIOREMEDIATION

The biological method to incinerate the contaminated soils is referred to as bioremediation where there is involvement of the microbial activities to degrade organic pollutants into the less toxic metabolic products. The biological transformation involves:

- Transforming less toxic or volatile compound by changing valency, for example, As^{+3} (more mobile, more soluble and more toxic) than As^{+5} (less mobile, less soluble and less toxic) and conversion of Hg ion to metallic mercury.
- Complexation reaction to form organo-metallic compound by methylation.

1.10.1.2 BIOSTIMULATION

The technology adopted to assist the naturally occurring microbial population in breaking down the contaminants in a faster rate is called biostimulation such as emulsion stabilizer, some special fertilizers and lauryl phosphate.

1.10.1.3 BIOAUGMENTATION

The selected microorganisms when applied into a contaminated soil can augment the natural microbial population and activities while, degrading the contaminants in soil. The technique is called bioaugmentation. Microorganisms commonly used in bioaugmentation are *Bacillus licheniformis*, *Bacillus thuringiensis*, *Penicillium* spp., *Aspergillus* spp., *Flavobacterium*, *Arthrobacter*, and *Pseudomonas.*

1.10.1.4 PHYTOREMEDIATION

Sometimes, higher plants are used for bioremediation, the process is called phytoremediation. The organic and inorganic pollutants can accumulate at different plant parts or convert them into less toxic by-products. The process of accumulation of high concentration of contaminants in plant biomass is called hyper-accumulation. For example, arsenic is absorbed by brake ferns (*Pterris vitata*) from the contaminated soil and *Sutera fotida* can accumulate huge amount of chromium from the polluted areas. It has been reported that the naturally extracted or synthetically prepared humic/fulvic substances form complex with the toxicant (As) or with the micronutrient (Zn) to regulate the stability of the organo-metal-complex or release of the ions in the solution (Mukhopadhyay & Sanyal, 2004; Bhattacharya & Mukhopadhyay, 2006). The package of practice for crop cultivation (conservation or zero tillage) might have some impact (Pal et al., 2018; Thakur & Mukhopadhyay, 2018) on improving the soil structure that can also regulate the holding of nutrients in soil. Soil reaction at different pH influences the ionic sorption and desorption in the soil system (Bhabai & Mukhopadhyay, 2019).

1.11 CONCLUSION

The natural resources like soil, water, and air are the finest and life-saving components on earth for all the living systems, both under biotic and abiotic situations. A lump of soil, which requires billions of years to originate from the rocks, may act as a source of most of the nutrient elements for the plants and animals, and at the same time may act as a sink for the toxins. When the maximum bearing capacity of soil exceeds its limit, a kind of rejection of the solutes into the environment can initiate a natural disaster. The ecological balance may be disrupted. The availability of pure drinking water by 2050 will be of great concern as every drop of water will be counted for the stakeholders. The vision of "per crop per drop" will be focused on terms of conservation of water. A challenging situation will come to find out better ways for recharging the groundwater. Preserving the natural water bodies or surface water bodies through renovation/excavation will have to be prioritized. The water-harvesting structures to improve WUE in the agricultural sectors will be an alternative source of water during the lean period (January to May) when the groundwater recharge is at its minimum. More awareness among the stakeholders is to be emphasized to have pure water and air. The dearth of pure oxygen will be another challenge for survival. There should be an easy way out to extract oxygen from the air for subsequent use. The more there will be attention to nurturing natural resources (soil, air, and water), the more will be returned back to society for sustaining livelihood security. With the changing global scenario, the ecosystem is also changing, which can regulate the balance of the system. Immediate control measures should be prioritized to save the mother earth for future generations and also for progress in society as a whole. Approaches on simulation models and transfer function for predicting the relative contribution of the toxicants to the plants and animals are to be explored to establish standard protocols for the large-scale analysis of the impact on the environment. The implementing gaps are to be identified and remedial/corrective measures are to be taken at the earliest. In doing so, the traditional methods as well as the advanced scientific strategies to control pollution in natural resources may be advocated. Thus, we can save our environment by practicing the three R's, i.e., reduce, recycle, and reuse. All these issues are to be addressed by involving researchers, technologists, and planners to focus on the mitigation options on soil, water, and air pollution through a holistic approach.

KEYWORDS

- **air pollution**
- **biomagnification**
- **bioremediation**
- **conservation**
- **ecosystem**
- **natural resources**
- **nutrients**
- **pollutants**

REFERENCES

Babu, P. R., (**2015**). **Environmental Pollution. Available at:** https://slideplayer.com/slide/6188600 (accessed on 28 December 2023).

Balasubramanian, A., (2017). *Global Carbon Cycle Science*. Available at: https://www.slideshare.net/bala1957/global-carbon-cycle-78159379 (accessed on 28 December 2023).

Bhabai, B., & Mukhopadhyay, D., (2019). Adsorption envelope on phosphorus sorption-desorption in some soils of West Bengal. *Journal of Indian Society of Soil Science, 67*(4), 367–378. doi: 10.5958/0974-0228.2019.00040.9.

Bhattacharya, A., & Mukhopadhyay, D., (2006). Stability of the zinc-humic/fulvic complexation equilibria of soils of Terai agroecological region of West Bengal (India). *Extended Abstract of 18th World Congress of Soil Science*. Philadelphia, Pennsylvania, USA.

Ford, N., (2020). *Mechanism of Acid Deposition Caused by the Oxides of Nitrogen and Sulfur, E 11 Acid Deposition*. https://slideplayer.com/slide/7478594/ (accessed on 28 December 2023).

Hulburt, S., (2015). *Earth and the Terrestrial Worlds*. Available at: https://slideplayer.com/slide/2854876/ (accessed on 28 December 2023).

Janssen, A. J. H., Lettinga, G. D., & De Keizer, A., (1999). Removal of hydrogen sulfide from wastewater and waste gases by biological conversion to elemental Sulphur: Colloidal and interfacial aspects of biologically produced Sulphur particles. *Colloids Surf, A Physicochem. Eng. Asp., 151*(1, 2), 389–397. https://doi.org/10.1016/S0927-7757 (98)00507-X.

Mukhopadhyay, D., & Sanyal, S. K., (2004). Complexation, and release isotherm of arsenic in arsenic-humic/fulvic equilibrium study. *Soil Res., 42*(7), 815–824. https://doi.org/10.1071/SR03104.

Pal, D., Patra, P. K., & Mukhopadhyay, D., (2018). Effect of tillage and organic residues on yield of wheat (*Triticum aestivum* L.). *Research on Crops, 19*(3), 373–379. doi: 10.31830/2348-7542.2018.0001.2.

Reitsma, K. D., Clay, D. E., & Carlson, C. G., (2011). *Soil Fertility*. Available at: https://www.researchgate.net/publication/262798335 (accessed on 28 December 2023).

Sharp, T., (2017). *Earth's Atmosphere: Composition, Climate & Weather*. Available at: https://www.space.com/17683-earth-atmosphere.html (accessed on 28 December 2023).

Thakur, P., & Mukhopadhyay, D., (2018). Effect of boron and lime on yield assessment of wheat (*Triticum aestivum* L.). *International Journal of Current Microbiology and Applied Science, 7*, 1133–1140. https://doi.org/10.20546/ijcmas.2018.712.139.

World Bank, (1999). Environmentally & socially sustainable development network. Partnerships for global ecosystem management: Science, Economics, and law. *Proceedings and Reference Readings from the Fifth Annual World Bank Conference on Environmentally and Socially Sustainable Development*. Held at the World Bank and George Washington University, World Bank Publications, Washington, DC.

CHAPTER 2

Vegetation Dynamics and Climate Change: A Study of the Indian Himalayan Region

AMIT KUMAR,[1,2] RIMA KUMARI,[3] ALISHA PRASAD,[1] PREET LAL,[4] and PURABI SAIKIA[5]

[1]*Department of Geoinformatics, Central University of Jharkhand, Ranchi, India*

[2]*Department of Geography, Sikkim University, Tadong, India*

[3]*Department of Environmental Science and Engineering, Indian Institute of Technology (Indian School of Mines), Dhanbad, India*

[4]*Department of Civil and Environmental Engineering, Michigan State University, East Lansing, MI, United States*

[5]*Department of Environmental Sciences, Central University of Jharkhand, Ranchi, India*

ABSTRACT

Vegetation forms an integral component of an ecosystem wherein its responses to climate change (CC) act as a key indicator in global climate monitoring and also in maintaining ecosystem stability. Therefore, long-term analysis of vegetation and monitoring of inter-annual variations in vegetation are required to combat the effects of CC. The present study attempted to analyze the change in vegetation vigor on the basis of the normalized difference vegetation index (NDVI) from 2001 to 2017 in the Himalayan regions, where the NDVI has decreased in 2017 compared to 2001 in the

Sustainable Management and Conservation of Environmental Resources in India.
Sumit Chakravarty, Vineeta, Jahangeer A. Bhat, Munesh Kumar, Gopal Shukla (Eds.)

Western Himalayas while the same has increased in the Central Himalayas. Also, the long-term temperature and precipitation changes (1901 to 2017) in the region have been analyzed to correlate the effects of the climatic variables on the vegetation dynamics in the Himalayan region. The lowest mean temperatures have been recorded at higher altitudes of the Western Himalayas which continue to increase towards the east. Precipitation anomalies in the Western Himalayas were negligible while precipitation anomalies in the Central and Eastern Himalayas were highly variable. This rising temperature and precipitation anomaly in different Himalayan regions have been found to influence the vegetation dynamics of the region. The knowledge gained from this research will help with developing methods for adapting to climate change as well as mitigating the difficulties encountered in resolving the problems related to the shifting dynamics of the vegetation in the Himalayan region.

2.1 INTRODUCTION

Mountains have been globally recognized as one of the significant ecological systems possessing core areas of biodiversity that have been equipped with provisioning facilities of various natural resources and ecosystem services (ES) (Schickhoff et al., 2016). These mountains being located at higher altitudes and along the slopes form the most fragile ecosystems which are more vulnerable to climate extremes and various catastrophic events (Tiwari et al., 2017). They are most sensitive to the effects of changing climatic conditions and will continue to be highly vulnerable to the same in the future (IPCC, 2014a). This sensitivity of the mountainous ecosystem makes them even more suitable areas for comprehending and analyzing the repercussions of climate change (CC). Among the various mountain ecosystems of the world, the Himalayan regions being geodynamically young mountains with varied topography, altitude, species richness, and high species endemism make it even more unique among different mountain ecosystems of the world and thus grabs special attention of the researchers to study the effects of changing climatic conditions in the region (Ramachandran & Roy, 2018). Geographically, the Himalayan region encompasses Western Himalayas (Jammu and Kashmir, Himachal Pradesh, Uttarakhand), Central Himalayas (Nepal, Sikkim, Darjeeling hills of West Bengal), and Eastern Himalayas (Bhutan, Arunachal Pradesh, Nagaland, Assam, Meghalaya, Manipur, Mizoram, and Tripura).

Himalayan regions have a heterogeneous landscape along with one of the largest elevational and climatic gradients on Earth (Shooner et al., 2018) with a high level of anthropogenic disturbances and varied vegetation from tropical to subtropical, temperate forests, and alpine scrubs (Roy & Behera, 2005). The ecological integrity of this mountainous ecosystem is found to be at stake because of the increased vulnerability to the effects of CC that can be found to be more closely associated with human interference in the region. Although anthropogenic disturbances have played a significant role in shaping and setting the vegetation dynamics and landscapes of the mountainous ecosystems, however beyond a certain limit and degree of influence these have negatively affected the ecological integrity of the region (Garbarino et al., 2009; Diksha & Kumar, 2017). It has further aggravated the fragility of the Himalayan ecosystem, thereby making the region even more susceptible to the effects of CC. The low-intensity human interferences including lopping, grazing, litter collection, and soil removal and the high-intensity activities including forest fires, unplanned developmental activities, rapid demographic changes, and tourism, have played a significant role in altering the vegetation dynamics of the region (Malik et al., 2015; Kumari et al., 2019). Additionally, Himalayan forests have been severely affected by various natural disturbances of different intensities including landslides, earthquakes, floods, cloudbursts, heavy rainfalls, and different biotic interferences like insect damage, diseases, herbivory grazing, Jhum cultivation, hunting, timber felling, and collection of fuelwood, non-timber forest produces (NTFPs), medicinal herbs and wild edible plants (Saikia et al., 2017; Ahmad et al., 2021). The increased vulnerability due to CC in the Himalayan regions has endangered several taxa of the region more than any other mountainous ecosystem (Roy et al., 2012). Since vegetation forms the main component of any ecosystem which displays the effects of change in the surrounding environmental conditions, therefore the study of vegetation dynamics can be utilized as a measurable parameter to understand the degree of vulnerability to changing climatic conditions in the Himalayan regions.

Vegetation plays an utmost role in the ecosystem functioning, goods and services, as well as regulation and maintenance of the ecological integrity of the Himalayan regions. The increased risk of vulnerability also carries the maximum probability of changes in the composition, structure, function, and stability of the Himalayan ecosystem (Ravindranath et al., 2005). The changing climatic conditions in the Himalayan regions are likely to further intensify both the natural and anthropogenic disturbances, thereby threatening ecological resilience (Seidl et al., 2014). Different vegetation performs its function in equilibrium and accordance with the type of climatic conditions over specific sites (Yadava et al., 2017). The change in structure and composition of vegetation is generally triggered in response to anthropogenic influence and various climatic factors (Wang et al., 2015). The change in climatic conditions has also been found to induce changes in the phenology and the distribution pattern of different species (Jewitt et al., 2015; Trumbore et al., 2015). This change in vegetation dynamics acts as an indicator of change in local environmental and microclimatic conditions (Yue et al., 2009). The knowledge of the ecological processes and their interactions with the differing degree of biotic pressures may help in understanding and unveiling the complexity of the interaction between the vegetation dynamics and CC in the Himalayan regions. This knowledge becomes furthermore clear when coupled with the long-term analyses of the vegetation and their trends in the Himalayan regions (Malik et al., 2015). The long-term monitoring of the vegetation growth, in turn, will assist in predicting the changes in the same and in exploring the correlation between the vegetation dynamics and CC (Shao et al., 2017). Thus, the study of vegetation dynamics in the changing climatic scenario has become necessary to understand the relationship between the two and to identify the factors influencing the change in the vegetation structure and composition. This understanding is also required to identify and detect the early warning signals which will further serve in mitigating the effects of CC on the vegetation dynamics of the Himalayan regions.

2.2 CLIMATE CHANGE (CC) IMPACTS ON HIMALAYAN VEGETATION

The climatic and non-climatic stress-inducing parameters have been found to have a direct influence on the forest ecosystems resulting in large-scale degradation of forests and reduced ES (IPCC, 2014b). The repercussions of CC on vegetation are not immediate; rather it is a long-term process for vegetation to show any response against the changing climatic conditions. The stress created as a result of CC brings alteration in the vegetation dynamics which becomes visible only after the utilization of the adaptation capacity of the species to their maximum extent (Kumar et al., 2020). The changes in microclimatic conditions have significantly affected forest productivity, global biodiversity, and ecosystem stability and integrity in the Himalayan regions (Ravindranath et al., 2005). The plant species of the upper ecotone zones of the Himalayan regions are most sensitive to the changing climatic variables because of their limiting action in the growth and development of these species (Yadava et al., 2017). These species being narrowly adapted to temperature and precipitation change would certainly face repercussions of even a slight change in any of these climatic variables. The changes are generally observed in the form of spatial-shift, physiological, anatomical, or structural change, and variation in the phenology in response to CC in the Himalayan vegetation (Chaturvedi et al., 2011; Joshi et al., 2012; Chakraborty et al., 2013; Jewitt et al., 2015; Trumbore et al., 2015). The general response of the vegetations in terms of anatomical and structural changes to the effects of changing climatic scenarios is associated with growth retardation, decrement in total leaf area, and dry matter (Ahmad et al., 2016; Li et al., 2020). The effect of CC on the ecological integrity of the Himalayan ecosystem is also at stake because of the increased risk of invasions by exotic species (Priyanka & Joshi, 2013; Chakraborty et al., 2018; Lamsal et al., 2018). These species are believed to dominate over the native species because of their wide tolerance range against different climatic parameters, high growth and reproduction rate, broad physiological niche, short reproductive cycle, and better reproductive ability (Whitney & Gabler, 2008; Higgins & Richardson, 2014). The warming temperatures and loss of ice and snow in the Himalayan regions are further affecting the biodiversity of the region, prey-predator relationship, and water availability (Xu et al., 2009). Moreover, CCs have increased the incidences of forest fires, insect infestation and damage, variation in the frequency of strong winds, winter storms, and droughts bringing a huge loss to regional forestry sectors (Kirilenko & Sedjo, 2007; Seppälä, 2009). The increased incidences of forest fires due to warming in the Himalayan regions are mainly observed in altitudes ranging from 600 to 2,650 m (Negi, 2012). Enhanced CO_2 in the atmosphere and temperature changes have further affected various biogeochemical processes, including soil respiration, litter decomposition, nitrogen mineralization, nitrification, denitrification, methane emission, fine root dynamics, plant productivity, and nutrient uptake which in turn will alter forests dynamics (Norby et al., 2007; Mandal & Sharda, 2013). CC not only threatens biodiversity but also affects the socio-economic condition of the indigenous people of the Himalayan regions (Bharali & Khan, 2012).

2.3 LONG-TERM VEGETATION COVER CHANGES IN THE HIMALAYAN REGION

The monitoring of the inter-annual variations in the vegetation dynamics has become one of the important topics of research to establish the relationship between the ecological, environmental, and energy cycle with CC. In recent years, the NDVI has been widely used as one of the vital parameters to study the vegetation dynamics and their relationship with the different climatic factors (Zhang et al., 2013, 2017). Similarly, in the present study, NDVI has been used to assess the changes in the vegetation vigor in the Himalayan regions. The MODIS product (MOD13A1), having 500 m spatial and 16 days temporal resolution, has been used for vegetation change analysis during 2001–2017. The values of NDVI have been further classified into five different classes as very low (< 0.1), low (0.1 to 0.3), moderate (0.3 to 0.5), high (0.5 to 0.7), and very high (>0.7) (Figure 2.1).

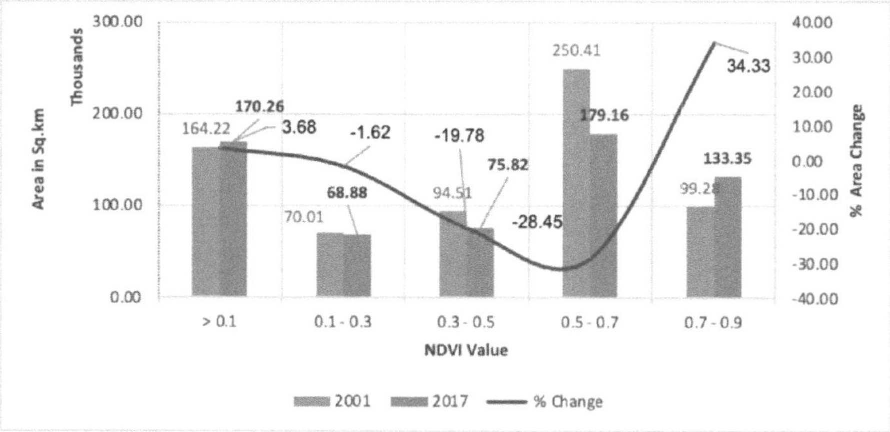

FIGURE 2.1 Area change in different NDVI classes between 2001 and 2017 for the Himalayan region.

The NDVI values for different classes demonstrated that the very high altitude of the Himalayas regions has very low NDVI, which insinuates the absence of vegetation due to the presence of snow and glaciers. In the Western Himalayas, the NDVI value has been found to decrease in 2017 compared to 2001 while, in the Central Himalayas, the NDVI value has increased. This can be visualized from the increased greenness in the region from the map (Figures 2.2(A) and (B)).

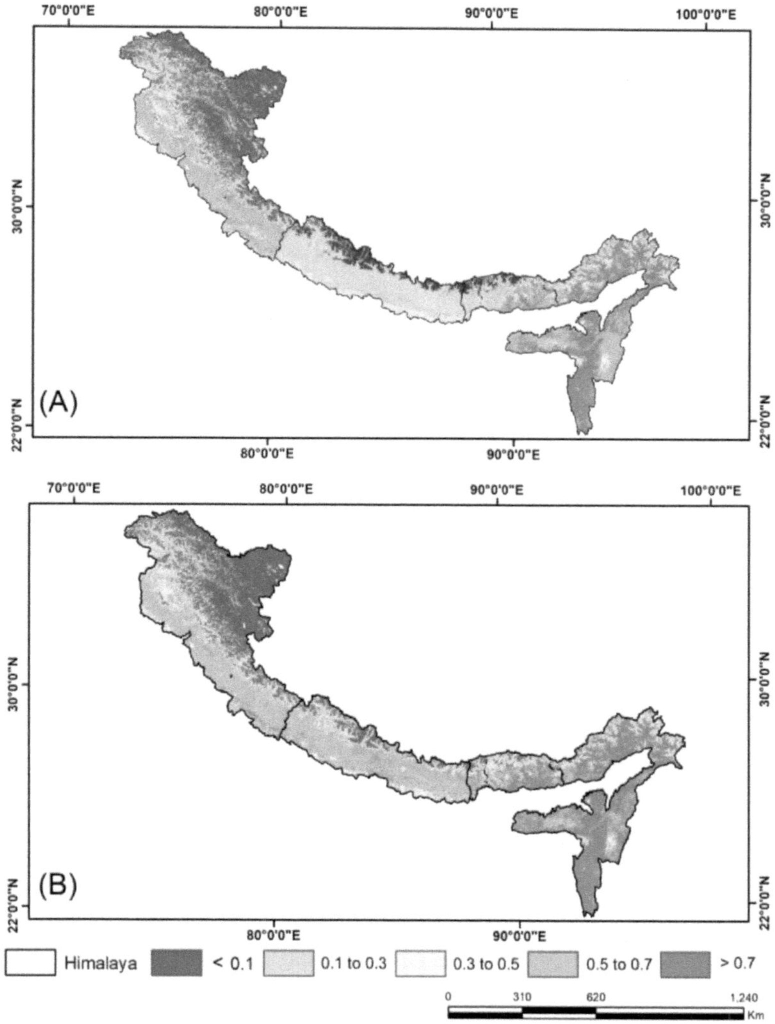

FIGURE 2.2 NDVI of the Himalayan region from MODIS MOD13A1 of (A) 2001; and (B) 2017.

In Eastern Himalayan Region (EHR), the lower parts especially near the proximity to major cities (Imphal and Aizwal) have shown increased greenness while, the eastern part of Assam has experienced a loss in greenness. The study exhibited the highest increase in area coverage was observed in the very high NDVI class (>0.7; 34.33% growth) followed by very low NDVI class (< 0.1; 3.68% growth). In contrast, a decrease can be observed in the high NDVI class (0.5 to 0.7; –28.45 change) followed by moderate NDVI class (0.3 to 0.5; –19.78% change) and low NDVI class (68.88% change). The decrement in the area coverage of high NDVI values in Western Himalayas conforms to the earlier reports (Shrestha et al., 2012; Zhang et al., 2013). A similar trend of greening at middle and lower elevations of Himalayan regions and browning at higher elevations (> 3,800 m) was also observed (Mishra & Mainali, 2017). The positive or negative trends of vegetation greenness over some times are generally linked with climatic driven factors (temperature and precipitation) and are also influenced by human interference particularly involving high intensity scaled activities (Kuenzer et al., 2015). In another study, an increment in the annual NDVI of Eastern Himalaya has been observed particularly at lower elevation areas dominated by broad-leaved and coniferous forests while, a significant decrease in the same with increased elevations (Saikia et al., 2016). The difference in the NDVI values at different elevations was linked with the rising temperature and precipitation with an insignificant effect on the same (Li et al., 2016). A similar negative correlation between the rising temperature and the NDVI and an insignificant positive correlation between the total precipitation and NDVI of the Central Himalaya was observed (Wu et al., 2020).

2.4 LONG-TERM TEMPERATURE CHANGES IN THE HIMALAYAN REGIONS

The long-term mean temperature from 1901 to 2017 for the Himalayan regions ranges from – 8°C to 25°C (Figure 2.3).

The lowest mean temperatures have been recorded in the higher altitude Western Himalayan region, which increases further on moving towards the east. The decadal temperature anomaly in the region ranged from – 0.82°C to 1.0°C. The highest anomalies were recorded in 1941, 2001, 2011, and 2017 which signifies that the mean temperatures during these years were more than the long-term mean. The lowest anomalies were recorded in 1911, 1961, 1981, and 1991. A similar trend of increment in the temperatures of the Himalayan regions was also found in the previous studies (Xu et al., 2009; Shrestha et al., 2012). This rapid warming in Himalayan regions in recent years reveals rising stress in vegetation cover due to increasing temperatures. Usually, the metabolic activities of these vegetations are adapted to lower temperatures; therefore, an increase in the temperature enhanced the chances of occurrence of water-stressed

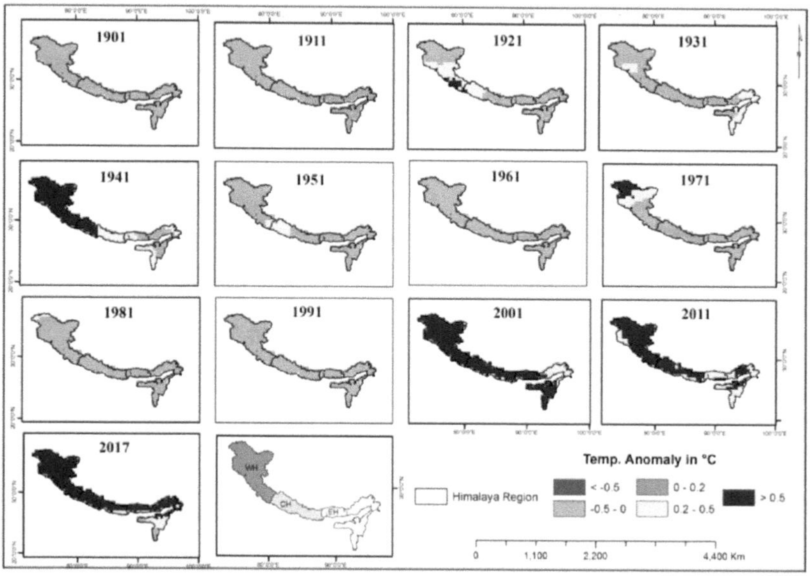

FIGURE 2.3 Temperature anomaly of the Himalayan region from 1901 to 2017.

conditions, thereby limiting the growth of the vegetation (Mishra et al., 2019). The germination of the seeds and seedlings particularly in the inner ranges of tree line ecotones has been affected because of the warming and soil moisture scarcity (Tiwari & Jha, 2018). The intensification of water-stressed conditions as a result of increasing temperatures in the Himalayan regions adversely affected the leaf phenology along with regeneration potential of several dominant species of that forest (Singh et al., 2011). This warming has further been shown to influence the spatial shift of species in the region by increasing their upper distributional limits (Chaturvedi et al., 2011; Joshi et al., 2012; Chakraborty et al., 2013; Dolezal et al., 2016).

2.5 LONG-TERM PRECIPITATION CHANGES IN THE HIMALAYAN REGION

The long-term cumulative precipitation ranges from 100 to 5,500 mm in the Himalayan regions from 1901 to 2017 (Figure 2.4) which usually increased from west to east.

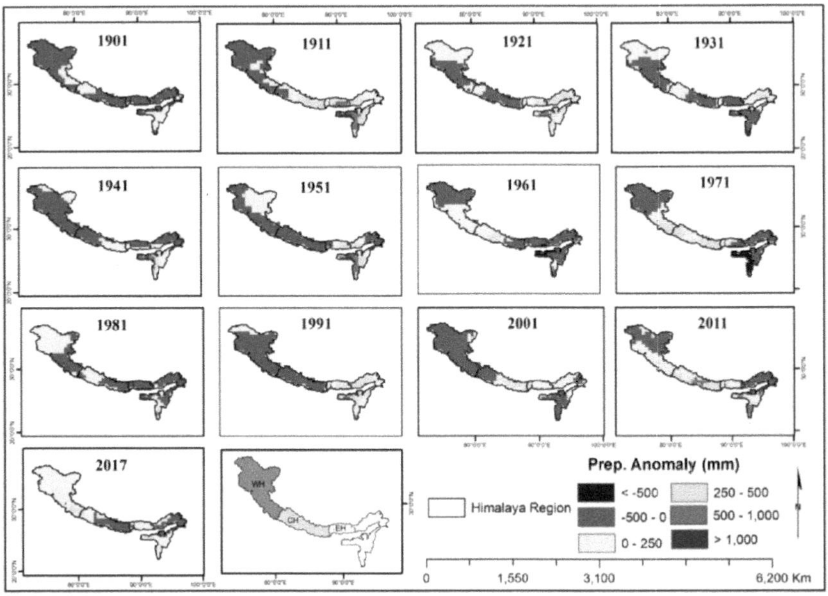

FIGURE 2.4 Precipitation anomaly of the Himalayan region from 1901 to 2017.

The decadal precipitation anomaly ranges from – 1,200 to 1,400 mm. There was negligible variation in precipitation anomaly in Western Himalayas while, the precipitation anomaly was highly variable in Central and Eastern Himalayas. The precipitation anomaly in Western Himalaya ranges from – 500 to 250 mm exhibiting a decrease in precipitation with respect to the long-term mean. However, the cumulative precipitation recorded was higher than the long-term mean cumulative precipitation during the years 1911, 1941, and 1951 as the precipitation anomaly measured 500 to 1,400 mm whereas, the precipitation anomaly in 1971 ranged from – 1,200 to 0 mm exhibiting less cumulative precipitation than the long-term mean. This precipitation anomaly in the region has highly influenced the green cover in the region. Generally, the response of vegetation against the changing climatic variables was mainly accorded to the changes in their phenological pattern, altitudinal shifts, and ultimately their population extinction that is further dependent on the tolerance level, adaptability, and degree of competition among species for their survival (Skyes, 2009; Ramachandran & Roy, 2018). Similar insights on the relationship between phenology and precipitation in the Himalayan regions have been reported by the earlier research where they revealed that the phenological events have changed significantly with the average advancement of 4.7 days at the start of the growing season for the local ecosystem of the Himalayan region (Shrestha et al., 2012). Moreover, the alteration in the precipitation regime has also been found to influence the upper limit of the tree line where the tree lines were normally higher in the eastern and central Himalayas with higher precipitation rate as compared to the western Himalayas where the precipitation was relatively lesser (Singh et al., 2017).

2.6 FUTURE RESEARCH PROSPECTS

CC impacts on vegetation and vice-versa are very sensitive in the Himalayan region due to the rapid global CC and ecosystem fragility. There is a need for further research in understanding various positive and negative impacts of greenness and brownness patterns over changing climate through ensembles of multiple datasets at different time periods which will help the researcher to understand the phenomenon in a better scientific way. Contribution of existing vegetation cover in controlling precipitation and temperature over the Himalayan region through regional climatic model simulation at different climate scenarios and sustainable pathways will also help to identify the actual phenomenon responsible for the vegetation change in the long term. Himalayan regions are mostly covered with a snow cover and due to global warming, it's melting at a higher rate, and it may impact water availability which is mostly considered to be unfavorable for the biota. Impacts of snow cover melting on vegetation change may also help in identifying the causes behind the vegetation change in Himalayan regions. Himalayan regions are mostly affected by climate extremes and there is a need to establish both climatic and ecological baselines to provide a context for determining the extreme ecological response of climate extremes. An in-depth study of CC impacts and responses of different endemic plant species in Himalayan forests at different altitudinal gradients will provide more valuable information toward its adaptation and mitigation.

2.7 CONCLUSION

The study of vegetation dynamics in the changing climatic scenario has become necessary to understand their relationship and also to identify the factors influencing the change in the vegetation structure and composition. The study exhibited an overall decrease in the green cover area while increasing only in vegetation with high vigor, i.e., the very high NDVI class as primarily observed in the EHR. The decreasing NDVI over a period from

2001 to 2017, particularly in the western Himalaya along with the anomalies observed in the rising temperature (>0.5°C) and decreasing precipitation pattern (–500 to 0 mm) for the entire Himalayan region, has influenced the dynamicity of the vegetation and thus the stability of the ecosystem. The understanding and anticipation of the impacts of these climatic variables on the Himalayan ecosystem are crucial in designing and developing effective policies and management strategies for reducing the fragility and vulnerability of the mountainous ecosystem. This anticipation can be achieved through long-term monitoring and modeling of the variability of the vegetation and the ecosystem. This, in turn, will help in improving the ecosystem structure and its function in the changing climatic conditions. The knowledge in the same will further contribute to addressing the challenges being faced while dealing with the issues associated with the change in vegetation dynamics of the Himalayan region and also in formulating adaptation strategies against the changing climatic conditions.

KEYWORDS

- **climate change**
- **climate change impacts**
- **forest cover change**
- **Himalayan region**
- **precipitation anomaly**
- **temperature anomaly**
- **vegetation dynamics**

REFERENCES

Ahmad, K. S., Hameed, M., Fatima, S., Ashraf, M., Ahmad, F., Naseer, M., & Akhtar, N., (2016). Morpho-anatomical and physiological adaptations to high altitude in some avenue grasses from Neelum Valley, Western Himalayan Kashmir. *Acta Physiologiae Plantarum, 38*(4), 1–14.

Ahmad, T., Pandey, A. C., & Kumar, A., (2021). Long-term precipitation monitoring and its linkage with flood scenario in changing climate conditions in Kashmir valley. *Geocarto International*. https://doi.org/10.1080/10106049.2021.1923829.

Bharali, S., & Khan, M. L., (2012). Climate change and its impact on biodiversity; some management options for mitigation in Arunachal Pradesh. *Current Science, 101*(7), 855–860.

Chakraborty, A., Joshi, P. K., Ghosh, A., & Areendran, G., (2013). Assessing biome boundary shifts under climate change scenarios in India. *Ecological Indicator, 34*, 536–547.

Chakraborty, A., Shukla, R., Sachdeva, K., Roy, P. S., & Joshi, P. K., (2018). The climate change conundrum and the Himalayan forests: The way forward into the future. *Proceedings of the National Academy of Sciences, India Section B: Biological Sciences, 88*, 837–847.

Chaturvedi, R. K., Gopalakrishnan, R., Jayaraman, M., Bala, G., Joshi, N. V., Sukumar, R., & Ravindranath, N. H., (2011). Impact of climate change on Indian forests: A dynamic vegetation modeling approach. *Mitigation and Adaptation Strategies for Global Change, 16*, 119–142.

Diksha, & Kumar, A., (2017). Analyzing urban sprawl and land consumption patterns in major capital cities in the Himalayan region using geoinformatics. *Applied Geography, 89C*, 112–123.

Dolezal, J., Dvorsky, M., Kopecky, M., Liancourt, P., Hiiesalu, I., Macek, M., Altman, J., et al., (2016). Vegetation dynamics at the upper elevational limit of vascular plants in Himalaya. *Scientific Reports, 6*, 1–13.

Garbarino, M., Weisberg, P. J., & Motta, R., (2009). Interacting effects of physical environment and anthropogenic disturbances on the structure of European larch (*Larix decidua* Mill.) forests. *Forest Ecology and Management, 257*(8), 1794–1802.

Higgins, S. I., & Richardson, D. M., (2014). Invasive plants have broader physiological niches. *Proceeding of the National Academy of Sciences, 111*, 10610–10614.

IPCC, (2014a). *Climate Change 2014: Impacts, Adaptation, and Vulnerability.* Part A: Global and sectoral aspects. Contribution of the working group II to the fifth assessment report of the intergovernmental panel on climate change. Cambridge University Press, Cambridge.

IPCC, (2014b). Summary for policymakers. In: Field, C. B., Barros, V. R., Dokken, D. J., Mach, K. J., Mastrandrea, M. D., Bilir, T. E., Chatterjee, M., et al., (eds.), *Climate Change 2014: Impacts, Adaptation, and Vulnerability* (pp. 1–32). Part A: Global and sectoral aspects. Contribution of working group II to the fifth assessment report of the intergovernmental panel on climate change, Cambridge University Press, Cambridge, United Kingdom and New York, USA.

Jewitt, D., Erasmus, B. F. N., Goodman, P. S., O'Connor, T. G., Hargrove, W. W., Maddalena, D. M., & Witkowski, E. T. F., (2015). Climate-induced change of environmentally defined floristic domains: A conservation based vulnerability framework. *Applied Geography, 63*, 33–42.

Joshi, P. K., Rawat, A., Narula, S., & Sinha, V., (2012). Assessing impact of climate change on forest cover type shifts in Western Himalayan Eco-region. *Journal of Forestry Research, 23*, 75–80.

Kirilenko, A. P., & Sedjo, R. A., (2007). Climate change impacts on forestry. *Proceeding of the National Academy of Sciences, 104*(50), 19697–19702.

Kuenzer, C., Dech, S., & Wagner, W., (2015). Remote sensing time series revealing land surface dynamics: Status quo and the pathway ahead. In: Künzer, C., Dech, S., & Wagner, W., (eds.), *Remote Sensing Time Series* (pp. 1–24). Springer International Publishing Switzerland.

Kumar, G., Kumari, R., Kishore, B. S. P. C., Saikia, P., Kumar, A., & Khan, M. L., (2020). Climate change impacts and implications: An Indian perspective. In: Roy, N.,

Roychoudhury, S., Nautiyal, S., Agarwal, S., & Bakshi, S., (eds.), *Socio-economic and Eco-biological Dimensions in Resource use and Conservation* (pp. 11–30). Environmental Science and Engineering, Springer, Cham.

Kumari, R., Banerjee, A., Kumar, R., Kumar, A., Saikia, P., & Khan, M. L., (2019). Deforestation in India: Consequences and sustainable solutions. In: Suratman, M. N., Latif, Z. A., Oliveira, G. D., Brunsell, N., Shimabukuro, Y., & Santos, C. A. C. D., (eds.), *Forest Degradation Around the World* (pp. 1–18). Intech Open, United Kingdom.

Lamsal, P., Kumar, L., Aryal, A., & Atreya, K., (2018). Invasive alien plant species dynamics in the Himalayan region under climate change. *Ambio, 47*, 697–710.

Li, H., Jiang, J., Chen, B., Li, Y., Xu, Y., & Shen, W., (2016). Pattern of NDVI-based vegetation greening along an altitudinal gradient in the eastern Himalayas and its response to global warming. *Environmental Monitoring and Assessment, 188*(3), 1–10.

Li, Y., Zou, D., Shrestha, N., Xu, X., Wang, Q., Jia, W., & Wang, Z., (2020). Spatiotemporal variation in leaf size and shape in response to climate. *Journal of Plant Ecology, 13*(1), 87–96.

Malik, Z. A., Pandey, R., & Bhatt, A. B., (2015). Anthropogenic disturbances and their impact on vegetation in Western Himalaya. *Journal of Mountain Science, 13*(1), 69–82.

Mandal, D., & Sharda, V. N., (2013). Appraisal of soil erosion risk in the Eastern Himalayan Region of India for soil conservation planning. *Land Degradation and Development, 24*, 430–437.

Mishra, A., Appadurai, A. N., Choudhury, D., Regmi, B. R., Kelkar, U., Alam, M., Chaudhary, P., et al., (2019). Adaptation to climate change in the Hindu Kush Himalaya: Stronger action urgently needed. In: Wester, P., Mishra, A., Mukherji, A., & Shrestha, A., (eds.), *The Hindu Kush Himalaya Assessment* (pp. 457–490). Springer, Cham.

Mishra, N. B., & Mainali, K., (2017). Greening and browning of the Himalaya: Spatial patterns and the role of climatic change and human drivers. *Science of the Total Environment, 587, 588*, 326–339.

Negi, P., (2012). Climate change, alpine tree line dynamics and associated terminology: Focus on north-western Indian Himalaya. *Tropical Ecology, 53*, 371–374.

Norby, R. J., Rustad, L. E., Dukes, J. S., Ojima, D. S., Parton, W. J., Del, G. S. J., McMurtrie, R. E., & Pepper, D. A., (2007). Ecosystem responses to warming and interacting global change factors. In: Canadell, J. G., Pataki, D., & Pitelka, L., (eds.), *Terrestrial Ecosystems in a Changing World* (pp. 23–36). The IGBP Series, Springer, Heidelberg.

Priyanka, N., & Joshi, P. K., (2013). Effects of climate change on invasion potential distribution of *Lantana camara*. *Earth Science and Climatic Change, 4*, 1–9.

Ramachandran, R. M., & Roy, P. S., (2018). Vegetation response to climate change in Himalayan hill ranges: A remote sensing perspective. In: Das, A. P., & Bera, S., (eds.), *Plant Diversity in the Himalaya Hotspot Region* (Vol. I, pp. 369–392). Bishen Singh Mahendra Pal Singh publisher, DehraDun.

Ravindranath, N. H., Joshi, N. V., Sukumar, R., & Saxena, A., (2005). Impact of climate change on forests in India. *Current Science, 90*, 354–361.

Roy, P. S., & Behera, M. D., (2005). Assessment of biological richness in different altitudinal zones in the Eastern Himalayas, Arunachal Pradesh, India. *Current Science, 88*, 250–257.

Roy, P. S., Kushwaha, S. P. S., Murthy, M. S. R., Roy, A., Kushwaha, D., Reddy, C. S., Behera, M. D., et al., (2012). *Biodiversity Characterization at Landscape Level: National Assessment* (p. 140). Indian Institute of Remote Sensing, Dehra Dun, India.

Saikia, P., Deka, J., Bharali, S., Kumar, A., Tripathi, O. P., Singha, L. B., Dayanand, S., & Khan, M. L., (2017). Plant diversity patterns and conservation status of eastern Himalayan forests in Arunachal Pradesh, Northeast India. *Forest Ecosystems, 4,* 28.

Saikia, P., Kumar, A., & Khan, M. L., (2016). Biodiversity status and climate change scenario in Northeast India. In: Nautiyal, S., Schaldach, R., Raju, K. V., Kaechele, H., Pritchard, B. I., & Rao, K. S., (eds.), *Climate Change Challenge (3C) and Social-Economic-Ecological Interface-Building* (pp. 107–120). Springer International Publishing, Switzerland.

Schickhoff, U., Singh, R. B., & Mal, S., (2016). Climate change and dynamics of glaciers and vegetation in the Himalaya: An overview. In: Singh, R., Schickhoff, U., & Mal, S., (eds.), *Climate Change, Glacier Response, and Vegetation Dynamics in the Himalaya* (pp. 1–26). Springer International Publishing Switzerland.

Seidl, R., Rammer, W., & Spies, T. A., (2014). Disturbance legacies increase the resilience of forest ecosystem structure, composition, and functioning. *Ecological Applications, 24,* 2063–2077.

Seppälä, R., (2009). A global assessment on adaptation of forests to climate change. *Scandinavian Journal of Forest Research, 24*(6), 469–472.

Shao, Z., Zhang, L., & Wang, L., (2017). Stacked sparse auto encoder modeling using the synergy of airborne LiDAR and satellite optical and SAR data to map forest above-ground biomass. *Journal of Selected Topics in Applied Earth Observations and Remote Sensing, 10,* 5569–5582.

Shooner, S., Davies, T. J., Saikia, P., Deka, J., Bharali, S., Tripathi, O. P., Singha, L. B., Khan, M. L., & Dayanandan, S., (2018). Phylogenetic diversity patterns in Himalayan forests reveal evidence for environmental filtering of distinct lineages. *Ecosphere, 9,* e02157.

Shrestha, U. B., Gautam, S., & Bawa, K. S., (2012). Widespread climate change in the Himalayas and associated changes in local ecosystems. *PLoS One, 7,* 1–10.

Singh, S. P., Singh, R. D., Gumber, S., & Bhatt, S., (2017). Two principal precipitation regimes in Himalayas and their influence on tree distribution. *Tropical Ecology, 58*(4), 679–691.

Singh, S. P., Singh, V., & Skutsch, M., (2011). Rapid warming in the Himalayas: Ecosystem responses and development options. *Climate and Development, 2*(3), 221–232.

Skyes, M. T., (2009). Climate change impacts: Vegetation. In: *Encyclopaedia of Life Sciences (ELS)*. John Wiley & Sons, Ltd, Chichester.

Tiwari, A., & Jha, P. K., (2018). An overview of tree line response to environmental changes in Nepal Himalaya. *Tropical Ecology, 59*(2), 273–285.

Tiwari, V. P., Verma, R. K., & Gadow, K. V., (2017). Climate change effects in the Western Himalayan ecosystems of India: Evidence and strategies. *Forest Ecosystems, 4*(13), 1–9.

Trumbore, S., Brando, P., & Hartmann, H., (2015). Forest health and global change. *Science, 349,* 814–818.

Wang, J., Wang, K., Zhang, M., & Zhang, C., (2015). Impacts of climate change and human activities on vegetation cover in hilly southern China. *Ecological Engineering, 81,* 451–461.

Whitney, K. D., & Gabler, C. A., (2008). Rapid evolution in invasive species, 'invasive traits' and recipient communities: Challenges for predicting invasive potential. *Diversity and Distribution, 14,* 569–580.

Wu, X., Sun, X., Wang, Z., Zhang, Y., Liu, Q., Zhang, B., Paudel, B., & Xie, F., (2020). Vegetation changes and their response to global change based on NDVI in the Koshi River Basin of Central Himalayas since 2000. *Sustainability, 12,* 1–15.

Xu, J., Grumbine, R. E., Shrestha, A., Eriksson, M., Yang, X., Wang, Y., & Wilkes, A., (2009). The melting Himalayas: Cascading effects of climate change on water, biodiversity, and livelihoods. *Conservation Biology, 23*(3), 520–530.

Yadava, A. K., Sharma, Y. K., Dubey, B., Singh, J., Singh, V., Bhutiyani, M. R., Yadav, R. R., & Misra, K. G., (2017). Altitudinal tree line dynamics of Himalayan pine in western Himalaya, India. *Quaternary International, 444*, 44–52.

Yue, Y., Wang, K., Bing, Z., Chen, Z., Jiao, Q., Bo, L., & Chen, H., (2009). Exploring the relationship between vegetation spectra and eco-geo-environmental conditions in Karst region, Southwest China. *Environmental Monitoring and Assessment, 160*, 157–168.

Zhang, B., Cui, L., Shi, J., & Wei, P., (2017). Vegetation Dynamics and their response to climatic variability in China. *Advances in Meteorology*, 1–10.

Zhang, Y., Gao, J., Liu, L., Wang, Z., Ding, M., & Yang, X., (2013). NDVI-based vegetation changes and their responses to climate change from 1982 to 2011: A case study in the Koshi River Basin in the middle Himalayas. *Global and Planetary Change, 108*, 139–148.

CHAPTER 3

Carbon Stock Potential and Ecosystem Services Provisioned by Major Land-Use Systems in Southern Assam, India: A Meta-Analysis

DEMSAI REANG,[1] AMITABHA NATH,[2] ARUN JYOTI NATH,[1] and ASHESH KUMAR DAS[1]

[1]*Department of Ecology and Environmental Science, Assam University, Silchar, Assam, India*

[2]*Department of Information Technology, North-Eastern Hill University, Shillong, Meghalaya, India*

ABSTRACT

Tropical ecosystems are global centers of carbon (C) storage both in vegetation and soils. Storage of C in these pools plays a significant role in maintaining the atmospheric C balance. But anthropogenic land use (LU) conversions alter biodiversity and C balance, responsible for global warming and climate change (CC). In return, CC threatens our ecosystems and their flow of services. Therefore, recognizing C stock potentials under different LU systems can offer vital facts for predicting effects of various LU on C pools. This information on C stocks is crucial for implementation of the emission reduction mechanisms (e.g., REDD+). The present study focuses on the C stock stored in different pools under the dominant LU systems and their ecosystem service production at a landscape level in Barak valley, southern Assam, northeast India. Barak valley covers 6,922 km^2 with representations of diverse major LU systems. Altogether six major LU have been studied viz. natural forest (NF), agricultural land (AG), plantation

Sustainable Management and Conservation of Environmental Resources in India.
Sumit Chakravarty, Vineeta, Jahangeer A. Bhat, Munesh Kumar, Gopal Shukla (Eds.)

forest (PF), agroforestry systems (AF), herbaceous systems (HV) and grasslands (GL). Among these LU, NF, AG, and AF constitute the predominant systems in terms of spatial coverage, i.e., 3,364 km^2, 2,164 km^2, and 799 km^2, respectively. Our meta-analysis revealed a greater stem density under NF (2,000 ha^{-1} to 4,000 ha^{-1}) compared to PF and AF (1,000 ha^{-1} to 2,000 ha^{-1}), implying more biomass in the NF. The scale of different major LU in terms of the biomass C and SOC stock exhibit more or less similar trends. The ecosystem services (ES) provided by the tree-based ecosystems, i.e., NF and AF were found higher compared to other LU for majority aspects under the four primary ES. We show the importance of tree-based ecosystems in carbon management, and higher levels of biodiversity under forest and agroforestry systems support higher ecosystem properties and provision of ecosystem services.

3.1 INTRODUCTION

Global warming and climate change (CC) due to CO_2 emissions and other greenhouse gases (GHGs) in the environment resulting from anthropogenic activities have resulted in an imbalance in our ecosystem, turning into a priority issue on the global political agenda (IPCC, 2014; Olorunfemi et al., 2018). Land-use (LU) change, particularly converting natural forests (NFs) into a managed system, alters the C balance (Yihenew & Getachew, 2013). Studying the role of different LU in the global carbon (GC) cycle has been critically important to reduce the imbalance of such a budget (Tian et al., 1998; Prentice et al., 2001). Next to fossil fuels combustion, change in LU practices constitutes the major contributor to CO_2 emissions responsible for distorting the C cycle (Lal, 2009). It is regarded as one of the most critical anthropogenic driven sources of atmospheric CO_2 (IPCC, 2014). Most of these emissions are attributed to losses from the aboveground terrestrial pools. Thus, understanding the changes in pools and fluxes of C in vegetation and soil in the terrestrial ecosystems has enticed the scientific community's attention (Sarkar et al., 2015) to progress the understanding of CC adaptation and mitigation.

The LU change from a native land may act either as a C sink or source (Zhang et al., 2015). For instance, change in the LU from primary to secondary forests results in depletion of C stocks both in vegetation and soil (Ogle et al., 2005; Ngo et al., 2013) while, degrading the soil properties too (Saha et al., 2011; Abera & Wolde-Meskel, 2013). Also, biodiversity loss through

LU conversion from NFs to plantation systems or grasslands (GL) results in the degradation of the natural vegetative structure, ultimately lowering the groundwater table (Ahrends et al., 2015), further affecting ecosystem C dynamics (De Camargo et al., 1999). Hence, it is imperative to assess the C stored in different pools under different LU systems to advance the scientific understanding of LU change and the terrestrial C cycle. Additionally, LU management could play a vital role in soothing the rising atmospheric CO_2 concentration (Post & Kwon, 2000). Therefore, it is crucial to characterize and quantify LU at local, regional and global levels and understand their influence on the terrestrial ecosystem processes controlling the C cycle (Prentice et al., 2001).

The tropical and sub-tropical ecosystems are reported to store more C in vegetation and soils than any other ecosystem (Eglin et al., 2010). Identifying the relationships between C stocks under different LU components can provide vital information for predicting the effects of various LU on C pools at a regional scale. The information can be utilized for developing policy and management guidelines for mitigating anthropogenic GHG emissions. Therefore, understanding the relationship between different LU systems and C stock is essential, as every LU system has a positive or negative impact on the C balance (Toru & Kibret, 2019). Besides, different LU systems also provide a varying degree of ecosystem goods and services that are essential and beneficial for human society (Tilman et al., 2002; Gilliam, 2007; Jose, 2009; Bauhus et al., 2010; Reed et al., 2017). Benefits are received either from natural or managed ecosystems. The ability of ecosystems to provide these services and goods depends on the less obvious ecosystem services (ES) or processes through which the goods are created and maintained (Krieger, 2001). But CC due to alteration in C balance threatens the flow of ES (Aznar-Sánchez et al., 2018).

A limited study of organic C stock under different LU systems is available for southern Assam, Northeast India (NEI). The available literature in the study region is limited or restricted to a particular ecosystem or LU system (Nath & Das, 2011; Borah et al., 2013; Rabha et al., 2014; Brahma et al., 2016, 2017; Nath et al., 2016; Kalita et al., 2017, 2020). In light of the above backdrop, the present study was attempted to assess the C stock stored in different pools under the major LU systems and assess the impact of the dominant LU systems on various ES at a landscape level in the Barak valley, southern Assam, north-east India.

3.2 MATERIALS AND METHOD

3.2.1 ABOUT THE STUDY REGION

The study was carried out for Barak valley region, i.e., the southernmost part of Assam, India (Figure 3.1).

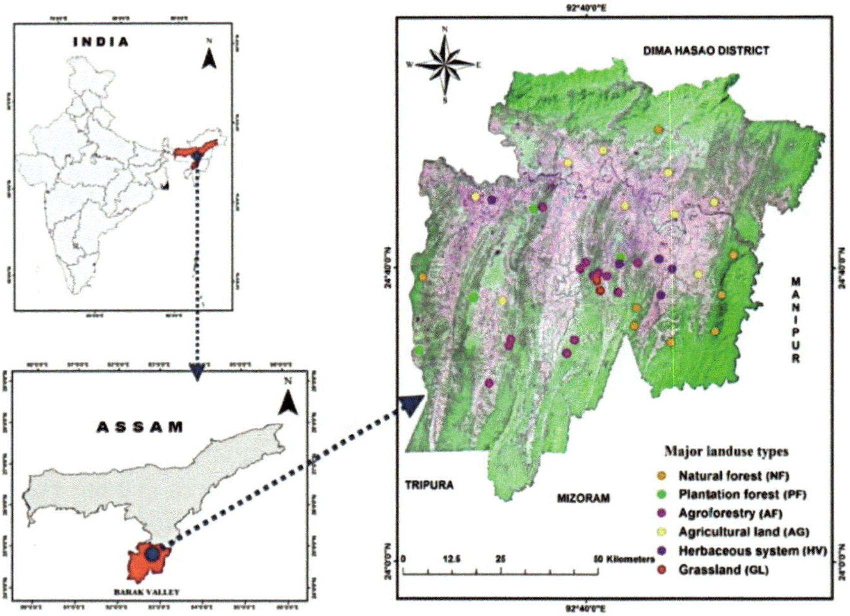

FIGURE 3.1 Map of the study area highlighting the different study location reported for major land-use system in Barak valley, north-east India.

The region is situated between 24°8′ and 25°8′ north latitude and 92°15′–93°15′ east longitude with a total geographical area (TGA) of 6,922 km^2. The area has an undulating topography characterized by hills, hillocks, vast plains and low-lying waterlogged areas. The Barak is the principal river with numerous tributaries, i.e., Dhaleswari, Longai, Jiri, Chiri, Sonai, Rukni, etc. Besides the rivers, the region is endowed with numerous wetlands, floodplain lakes, oxbow lakes, and other temporary reservoirs (Reang et al., 2018). The Barak valley has very gentle slopes of average height 22 m asl towards the east to 1,657 m asl in the north (Gupta, 2000). The vegetation in this region is dominated by Cachar tropical evergreen forest and Cachar semi-evergreen forest (Champion & Seth, 1968). The climate of the study area is sub-tropical warm and humid with an average annual rainfall of 2,390 mm, received mainly during the southwest monsoon season (May–September). The average maximum ranges from 25.4°C (January) to 33.5°C (August) and minimum temperature of 11.2°C (January) to 25.3°C (August). The soil of the region is a mixture of alluvial, sandy loam, silt clay loam.

3.2.2 MAJOR LAND-USE SYSTEMS (LUS) OF BARAK VALLEY

The landscape of Barak valley, north-east India is characterized by very small land holdings, making the landscape very heterogeneous (Reang et al., 2020). Despite the landscape heterogeneity, the dominant LU systems in the region comprised of NF, agricultural land (AG), agroforestry systems (AF), plantation forest (PF), herbaceous systems (HV) and GL (Devi et al., 2012; Das & Das, 2014; Reang et al., 2020). The corresponding spatial extent of the different LU systems in the region is presented in Figure 3.2.

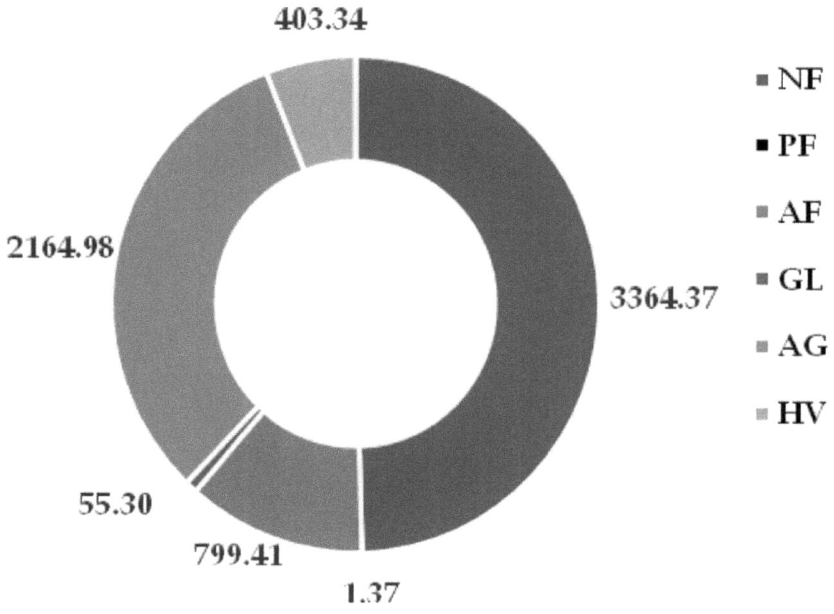

FIGURE 3.2 Area under major land-use systems in Barak valley (in sq. Km).

A brief description of the different LU systems and their components is given below. NF – this LU class includes naturally generated areas consisting of naturally immigrant and indigenous tree species. Vegetated areas primarily characterize NFs in the region under legal protection (reserve forests and wildlife sanctuary) and some areas under community and flood plain forests. The forest types include tropical evergreen and semi-evergreen forests. Some common growing tree species in the NFs of Barak valley include *Artocarpus chama, Palaquium polyanthum, Cynometra polyandra, Ficus benghalensis, Gmelina arborea, Mesua ferrea, Pterygota alata, Tetrameles nudiflora, Toona ciliata*, etc.

1. **Agricultural Land (AG):** This LU class mainly includes low-lying areas under wet-rice, irrigation, cultivated land, or land being prepared to raise crops. The AG is primarily concentrated in the central regions of the study area. In Barak valley, agriculture practice is dominated by relatively small-scale farmers growing mainly rice. The diverse physiographic conditions in the region have evolved diverse rice farming management practices, e.g., upland or dry rice (mountains and plateaus), rainfed lowland rice, flood-prone semi-deep-water rice and Boro rice (summer rice grown in shallow areas of wetlands). Paddy and vegetables are the principal crops grown in AG in the region.

2. **Agroforestry Systems (AF):** It is a collective name for LU systems where woody perennials are deliberately used on the same land-management units as crops. The AFs in Barak valley includes ancient tea-based systems and traditional homegardens and piper-based systems. The traditional piper-based AFs are established in NF comprising mainly of forests trees intercropped with *Piper betel*. In contrast, in the traditional homegardens, multipurpose trees (MPTs) domesticated includes *Mangifera indica, Artocarpus heterophyllus, Oroxylum indicum, Citrus grandis, Parkia timoriana, Syzygium cumini, Psidium guajava, Areca catechu, Bombax ceiba, Tamarindus indica, Litchi chinensis,* Bamboo sp. (*Bambusa cacharensis* and *B. vulgaris*), etc. The different AFs plays a crucial role in providing livelihoods, employment opportunities and the economy of the local people living in the region. In particular, AFs are crucial to smallholder farmers and other rural people because they can enhance their food supply, income, and health.

3. **Plantation Forest (PF):** This LU class comprises managed trees brought under plantation systems through human interventions. The PF concept is generally applied to single species, uniform planting densities and even stand age class distributions, unlike the NF stands comprising indigenous species in a more heterogeneous spatial

manner and age. The most commonly occurring PFs in Barak valley include mono plantation of trees such as *Hevea brasiliensis*, *Areca catechu*, *Tectona grandis*, and *Aquilaria malaccensis*.

4. **Herbaceous Systems (HV):** Under this LU class, we considered information pertaining mainly for two plants, i.e., wild *Musa* and pure bamboo stands or bush. *Musa* plants are regarded as herbaceous because their stems do not contain actual woody tissue while, bamboo plants are grass, they are not regarded as trees. The wild *Musa balbisiana* is the most widely occurring *Musa* species in Barak valley's tropical wet evergreen forests. At the same time, bamboo stands in the region are comprised of *Bambusa cacharensis, B. vulgaris, B. balcooa, B. assamica, B. nutans, Schizostachyum dullooa*, etc.
5. **Grasslands (GL):** This LU class comprises the traditionally managed *Imperata* GL in Barak valley. The GL are managed for harvesting of thatching material. In particular, majority of the resource-poor farmers utilizes the leaves of *Imperata* grass as roofing materials.

3.2.3 LITERATURE SEARCH AND DATA GENERATION

A literature survey was undertaken to identify the carbon (C) storage and sequestration potential under the major LU's in Barak valley. Journal articles, book chapters, and scientific reports were identified through a comprehensive literature search carried out using Web of Science, Scopus, Google, Google Scholar, and individual journal databases, using permutations of keywords. These include "major land-uses," "native forests," "agroforestry systems," "plantation," "grassland," "herbaceous vegetation," "bamboo," "management," "uses," "distribution," "ecosystem services," "soil health," "south Assam," "Barak valley." A total of 39 research articles were screened and studied thoroughly to generate the desired data set for analysis.

3.2.4 EVALUATING ECOSYSTEM SERVICES (ES)

We adopted a non-monetary approach to ES evaluation. Non-monetary approaches can be applied at various stages of ecosystem planning and management, e.g., problem framing, mapping, valuation, and decision-making

(Chan et al., 2012). The different indicators of ES were ranked on a scale of 1–5, 5 being the highest and 1 as the least.

3.3 RESULTS AND DISCUSSION

3.3.1 VARIATIONS IN STAND-LEVEL ATTRIBUTES

The stand density, basal area, aboveground biomass carbon (AGBC), and SOC stocks for the primary LUs are presented using a boxplot (Figure 3.3).

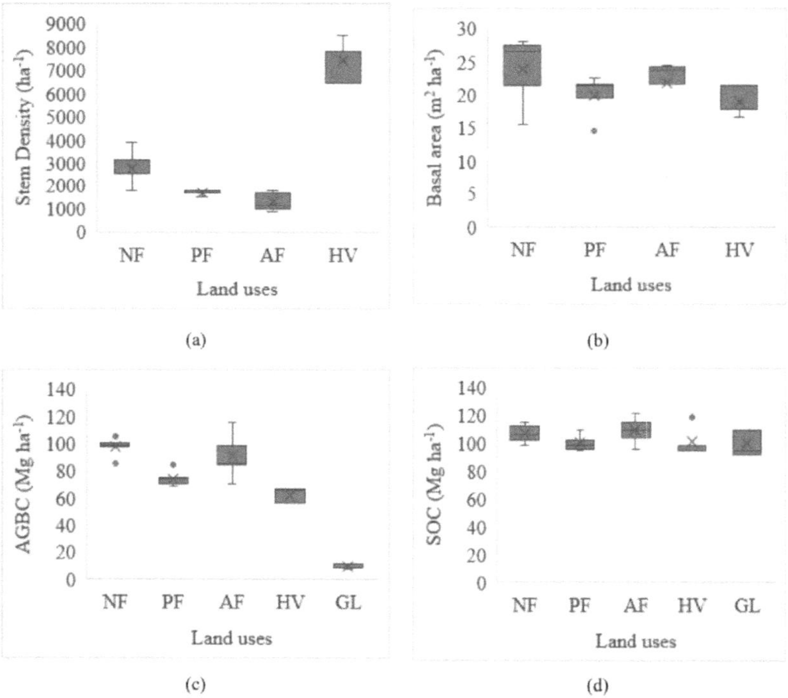

FIGURE 3.3 (a) Stem density; (b) basal area; (c) above ground biomass carbon; and (d) soil organic carbon stock under the different land-use system in Barak valley, north-east India. [NF: Natural forest; PF: plantation forest; AF: agroforestry; HV: herbaceous vegetation; and GL: grassland].

A box plot typically partitions the given data distribution into four equal-sized quartiles. The box indicates the positions of the upper and lower quartiles. This box's interior represents the *inner quartile range* which is 50% of the total data distribution. Thin lines, also known as *whiskers,* extend to the two extremes of the distribution, i.e., the minimum and maximum values in the data set. In addition to these, dots or star symbols are commonly used to represent outliers. It is visible from the plot that, for stand density, the data distributions were not similar for the different LU systems (Figure 3.3(a)). The stem density of PF and AF ranges from 1,000 ha^{-1} to 2,000 ha^{-1} whereas, for NF it extends from 2,000 ha^{-1} to 4,000 ha^{-1}. The density value of HV ranged between 7,000 ha^{-1} and 9,000 ha^{-1}. The data distribution was also not equal for different LU. PF and AF had comparatively smaller whiskers and *inner quartile* range in comparison to NF and HV. This essentially signifies proximity among the measured values as opposed to NF and HV. A similar trend was also observed with AGBC (Figure 3.3(b)).

In the case of AGBC, GL has distinctively different data distribution compared to other associated LU systems. The absence or tiny whiskers indicate belongingness of total data distribution within the *inner quartile* range. However, in the biomass C and SOC stock case, the scale of different LU exhibits more or less similar trends (Figure 3.3(c) and (d)). The size of the *inner quartile* range and the median for each LU were very much alike in these cases. Median or the average basal area value lies between 20 m^2 ha^{-1} and 27 m^2 ha^{-1} whereas, SOC stock hovers around 100 Mg ha^{-1} to 110 Mg ha^{-1}. This trend was uniform, unlike stem density and AGBC. Standard scaling was used to counter the variations in the inter-LU data distributions and bring uniformity. The standard scaling shifts the distribution to have a mean of zero and a standard deviation of one. Thereby, helps in reducing the inter-class variance up to a certain extent. In addition to this, as marked using dots in Figure 3(b)–(d), outliers were also removed to limit biases present in the input data.

3.3.2 IMPACT OF LAND-USE MANAGEMENT ON ECOSYSTEM SERVICES (ES)

The impact of major LU systems on various ES is shown in a heatmap in Figure 3.4.

Heat map provides a subtle way of visualizing the relation between two ordinal variables, as in this case. Each row of the heat map corresponds to different aspects of all four primary ES, and the columns correspond to various LU. A scale of 1 to 5 has been used to quantify the potential impact of different LU on these ES. Green and red symbolize two extreme results, whereas; other lighter shade colors represent the intermediate values. For

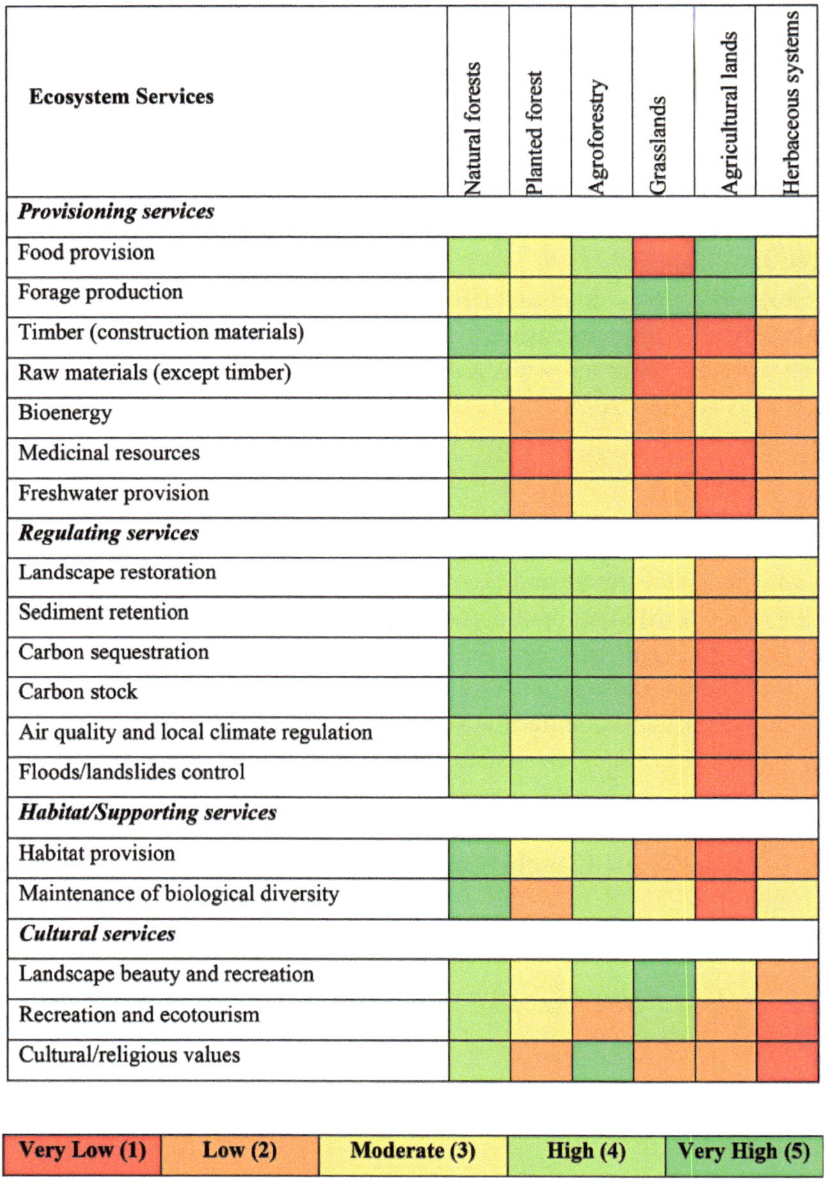

FIGURE 3.4 The major land-use system and their ecosystem services potential in Barak valley landscape.

example, in provisioning services, GL contributes very low whereas; AG contributes very high in provisioning food services. The impact of all these major LU on various other ES can be interpreted similarly.

3.3.2.1 PROVISIONING SERVICES

The material benefits people obtain from ecosystems are called provisioning services. The Barak valley region of NEI is primarily rural, and a larger section of the local populace primarily depends on the agricultural outputs for both food and economy (Altaf et al., 2014; Reang et al., 2020). Area put under agricultural practice constituted a dominant LU system in the region after NF, and much of the food production is derived from this agriculture LU system (Reang et al., 2020). Hence, AG has a very high impact on provisioning food services, followed by NF and AF. The GL and PF covered the least spatial area, contributing very minimal impact to food provisioning services. The tree species under PFs in Barak valley comprises *Areca catechu*, *Acacia* species, *Hevea brasiliensis*, *Aquilaria malaccensis, Tectona grandis*, etc. Of the two LU systems, i.e., PF and GL, the former to a certain degree contribute higher impact in food provisioning services through its fruit trees components and contributes economic security received from the plantation systems (Nath et al., 2020). GL and AG contributed very high services for forage production compared to other ecosystems. AF systems also showed moderately higher services for forage production since many livestock reared in the region were also supported through homegarden practices, a form of traditional AF system (Das & Das, 2014). Provisioning services for timber and construction materials was very high for all the tree-based ecosystems, i.e., NF, AF and PF. This was obvious since all these LU systems are primarily dominated by different tree species in Barak valley region (Devi & Das, 2013; Nandy & Das, 2013; Borah et al., 2016; Kalita et al., 2020; Nath et al., 2020). Likewise, the overall impact of the different LU management for majority aspects of provisioning ES was observed to be highest for NF and AF systems and the least under the PF.

3.3.2.2 REGULATING SERVICES

The different regulating services includes; landscape restoration, sediment retention, C sequestration, C stock, quality of local climate and flood/ landslides control demand natural or well-managed terrestrial ecosystems. Regulating ES was recorded to be very high in NF followed by AF and PF compared to other LU systems in Barak valley. Higher regulating services provided by these LU systems attribute to trees and other vegetation composition. Trees and shrubs contribute to better microclimate, creating a whole complex of environmental changes, affecting air temperature, soil temperature, soil moisture content (SMC), humidity, wind movement, pest and disease complexes (Sileshi et al., 2007). For instance, in regulating air and local climate quality, tree species provide shade while, rainfall and water availability are influenced by forest vegetation regionally and locally. Also, trees and other plants help remove pollutants from the atmosphere, thereby regulating the local air quality. The tree-based ecosystems effectively remove CO_2 from the atmosphere and act as a vital C sink. The AGBC stock and BGBC stock are known to be comparatively higher in these LU systems with a higher rate of sequestration in Barak valley landscape (Nath & Das, 2011; Borah et al., 2013; Rabha et al., 2014; Brahma et al., 2016, 2018; Nath et al., 2016). Other factors for higher regulating services by the tree-based ecosystems are preventing soil erosion and maintaining soil fertility through natural biological processes. In contrast, the least regulation service was observed for AG system. This is because most of the AG lands available for cultivation practice in this region are predominantly flood plains or low-lying areas, resulting in floods for most periods annually, thereby contributing very little to regulating services.

3.3.2.3 HABITAT SERVICES

Habitat or supporting services means providing living spaces and maintaining the diversity of plants and animals. Habitat services were higher in the case of NF followed by AF in Barak valley. Majority of the studies reporting higher plants and animal diversity were for NF and AF ecosystems of Barak valley (Das, 2008; Barbhuiya et al., 2009; Devi & Das, 2013; Dattagupta et al., 2014; Dey et al., 2015; Borah et al., 2016; Gogoi et al., 2016; Talukdar et al., 2018).

3.3.2.4 CULTURAL SERVICES

Cultural services are the non-material benefits people obtain from ecosystems, such as cultural identity, esthetic inspiration, sense of home, and spiritual experience related to the natural environment. The LU systems, i.e., NF, AF, and GL contribute a higher level of cultural services in the region by providing scope for mental peace through reduced pollution ecosystems, beautiful landscapes, rich biodiversity, etc. Animals and plants have always been a source of inspiration for arts, design, and culture to many ethnic communities. Some forms of recreation like hunting, picnic, and/or tourism, research, and adventures are other contributing factors. Also, the traditional AF systems is firmly bound to cultural and traditional rituals and practices among many tribal communities (Pnar, Rupini, Reang, Dimasa, Ranglong, Metei, Halam, Chorei, Hmar, etc.) inhabiting in Barak valley. The GL ecosystems provide spaces for picnic, walk, and play in its green spaces. AG and HV ecosystems have the most negligible impact on ecosystem cultural services.

The overall importance level or magnitude of impact represented by the varied LU on the ES is presented in the form of a radar chart (Figure 3.5).

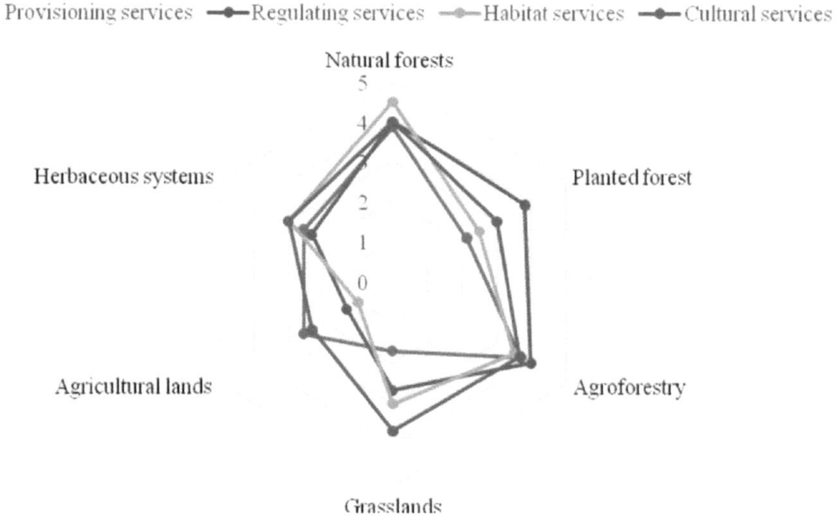

FIGURE 3.5 Radar chart showing the impact of various land-use classes on different ecosystem services.

Radar charts are particularly useful in analyzing multivariate observations with an arbitrary number of variables. Figure 3.5 indicates that almost every other LU class has a high impact on cultural services except for the PF. The PF has a moderate impact on the cultural services, but it has a high impact on regulating services. AF, HV, and NF have equal influence on all the ES. The HV has a moderate impact whereas; AF and NF have a high impact on all four services. Another noticeable insight is that AG contributes low or very low towards habitat and regulating services. It is the most negligible contributing factor among all the LU classes towards any of the ES as of the Barak valley landscape.

3.4 CONCLUSIONS

The tropical and subtropical ecosystems store more C in vegetation and soils than any other terrestrial ecosystem. The storage of C in these pools plays a significant role in maintaining the atmospheric C balance. However, converting natural into managed land use systems results in alteration of C balance responsible for global warming and climate change. In return, climate change is known to threaten ecosystems and their service flow. Therefore, recognizing the relationships between C stocks under different land use systems can offer vital facts for predicting the effects of various land uses on C pools as every land use positively or negatively impacts C balance. Barak valley of Assam covers 6,922 km^2 with representations of diverse major land use systems. The sustenance of humankind and social progress depends on the ecosystem services provided by the natural ecosystems. The ecosystem services provided by tree-based systems, i.e., natural forests and agroforestry, were observed to be very high for most of all four primary ecosystem services. Our study highlights the importance of a tree-based system amongst all major land use systems for C pools and storage potential and the varying ecosystem services provided by these land use systems. The data synthesis suggests that the human-managed agroforestry stores biomass C and SOC comparable to those under the natural forests. In the face of rapid depletion of natural forests through overexploitation, the promotion of agroforestry may continue the provision of essential goods and services required for human wellbeing. We suggest future studies to explore the ecosystem services of terrestrial ecosystems of southern Assam through a more holistic approach.

KEYWORDS

- **Barak valley**
- **carbon stock**
- **ecosystem services**
- **land-use management**
- **major land-use**
- **terrestrial ecosystems**
- **vegetation**

REFERENCES

Abera, G., & Wolde-Meskel, E., (2013). Soil properties, and soil organic carbon stocks of tropical andosol under different land uses. *Open Journal of Soil Science, 3*, 153–162.

Ahrends, A., Hollingsworth, P. M., Ziegler, A. D., Fox, J. M., Chen, H., Su, Y., & Xu, J., (2015). Current trends of rubber plantation expansions may threaten biodiversity and livelihoods. *Global Environmental Change, 34*, 48–58.

Altaf, H., Niranjan, R., & Avijit, D., (2014). Agricultural farm level efficiency in Barak valley region of Assam. *Scholars Journal of Arts, Humanities and Social Sciences, 2*(6B), 949–953.

Aznar-Sánchez, J. A., Belmonte-Ureña, L. J., López-Serrano, M. J., & Velasco-Muñoz, J. F., (2018). Forest ecosystem services: An analysis of worldwide research. *Forests, 9*(8), 453.

Barbhuiya, A. R., Sharma, G. D., Arunachalam, A., & Deb, S., (2009). Diversity and conservation of medicinal plants in Barak valley, Northeast India. *Indian Journal of Traditional Knowledge, 8*(2), 169–175.

Bauhus, J., Van, D. M. P., & Kanninen, M., (2010). Ecosystem goods and services from plantation forests. Routledge.

Borah, N., Nath, A. J., & Das, A. K., (2013). Aboveground biomass and carbon stocks of tree species in tropical forests of Cachar District, Assam, Northeast India. *International Journal of Ecology and Environmental Sciences, 39*(2), 97–106.

Borah, N., Rabha, D., & Athokpam, F. D., (2016). Tree species diversity in tropical forests of Barak valley in Assam, India. *Tropical Plant Research, 3*(1), 1–9.

Brahma, B., Nath, A. J., & Das, A. K., (2016). Managing rubber plantations for advancing climate change mitigation strategy. *Current Science*, 2015–2019.

Brahma, B., Pathak, K., Lal, R., Kurmi, B., Das, M., Nath, P. C., Nath, A. J., & Das, A. K., (2018). Ecosystem carbon sequestration through restoration of degraded lands in Northeast India. *Land Degradation & Development, 29*(1), 15–25.

Brahma, B., Sileshi, G. W., Nath, A. J., & Das, A. K., (2017). Development and evaluation of robust tree biomass equations for rubber tree (*Hevea Brasiliensis*) plantations in India. *Forest Ecosystems, 4*(1), 14.

Champion, H. G., & Seth, S. K., (1968). *A Revised Survey of the Forest Types of India*. Manager of Publications, Government of India press, Nasik.

Chan, K. M. A., Guerry, A. D., Balvanera, P., et al., (2012). Where are cultural and social in ecosystem services? A framework for constructive engagement. *BioScience, 62*(8), 744–756.

Das, A., (2008). *Diversity and Distribution of Herpetofauna and Evaluation of Their Conservation Status in the Barail Hill Range (Including the Barail Wildlife Sanctuary) Assam*. Final Report. Guwahati, India: Division of Herpetology, Aaranyak.

Das, T., & Das, A. K., (2014). Mapping and identification of homegardens as a component of the trees outside forests using remote sensing and geographic information system. *Journal of the Indian Society of Remote Sensing, 42*(1), 233–242.

Dattagupta, S., Gupta, A., & Ghose, M., (2014). Diversity of non-timber forest products in Cachar District, Assam, India. *Journal of Forestry Research, 25*(2), 463–470.

De Camargo, P., Trumbore, S., Martinelli, L., Davidson, E., Nepstad, D., & Victoria, R., (1999). Soil carbon dynamics in regrowing forest of eastern Amazonia. *Global Change Biology, 5*, 693–702.

Devi, N. L., & Das, A. K., (2013). Diversity and utilization of tree species in Meitei homegardens of Barak Valley, Assam. *Journal of Environmental Biology, 34*(2), 211.

Devi, N. P., Reddy, C. S., De, A., & Dutta, B. K., (2012). Spatial assessment of land use in Barak valley, Assam using satellite remote sensing data. *National Academy Science Letters, 35*(5), 493–495.

Dey, M., Roy, J., Tamuly, D., & Singh, P., (2015). Anuran Biodiversity of Barak Valley, Assam, and Impact of Human Interference. *Zoo's Print, 30*(4), 20–26.

Eglin, T., Ciais, P., Piao, S. L., Barré, P., Bellassen, V., Cadule, P., Chenu, C., et al., (2010). Historical and future perspectives of global soil carbon response to climate and land-use changes. *Tellus B: Chemical and Physical Meteorology, 62*(5), 700–718.

Gilliam, F. S., (2007). The ecological significance of the herbaceous layer in temperate forest ecosystems. *BioScience, 57*(10), 845–858.

Gogoi, M. J., Singha, S. H., & Deb, P., (2016). Butterfly (*Lepidoptera*) diversity in Barail Wildlife Sanctuary, Assam, India. *Journal of Entomology and Zoology Studies, 4*(4), 547–560.

Gupta, A., (2000). Forests of Barak Valley, Assam: History, management, and status of wildlife habitats. In: Arunachalam, A., & Khan, M. L., (eds.), *Sustainable Management of Forests-India*. International Book Distributors, Dehra Dun.

IPCC, (2014). *Climate Change 2014: Synthesis Report. Contribution of Working Groups I, II and III to the Fifth Assessment Report of the Intergovernmental Panel on Climate Change* [Core Writing Team, R. K. Pachauri and L. A. Meyer (eds.)]. IPCC, Geneva, Switzerland, 151 pp.

Jose, S., (2009). Agroforestry for ecosystem services and environmental benefits: An overview. *Agroforestry Systems, 76*(1), 1–10.

Kalita, R. M., Das, A. K., & Nath, A. J., (2017). Carbon stock and sequestration potential in biomass of tea agroforestry system in Barak Valley, Assam, North East India. *International Journal of Ecology and Environmental Sciences, 42*(5), 107–114.

Kalita, R. M., Das, A. K., Sileshi, G. W., & Nath, A. J., (2020). Ecosystem carbon stocks in different aged tea agroforestry systems: Implications for regional ecosystem management. *Tropical Ecology*, 1–12.

Krieger, D. J., (2001). *The Economic Values of Forest Ecosystem Services: A Review*. The Wilderness Society. Washington, DC, USA.

Lal, R., (2009). Challenges and opportunities in soil organic matter research. *European Journal of Soil Science, 60*(2), 158–169.

Nandy, S., & Das, A. K., (2013). Comparing tree diversity and population structure between a traditional agroforestry system and natural forests of Barak valley, Northeast India. *International Journal of Biodiversity Science, Ecosystem Services & Management, 9*(2), 104–113.

Nath, A. J., & Das, A. K., (2011). Carbon storage and sequestration in bamboo-based smallholder homegardens of Barak Valley, Assam. *Current Science*, 229–233.

Nath, A. J., Bhattacharyya, T., Ray, S. K., Deka, J., Das, A. K., & Devi, H., (2016). Assessment of rice farming management practices based on soil organic carbon pool analysis. *Tropical Ecology, 57*(3), 607–611.

Nath, P. C., Nath, A. J., Sileshi, G. W., & Das, A. K., (2020). Stand structure and functional attributes of agarwood (*Aquilaria malaccensis* Lam.) based smallholder farms in Northeast India. *Trees, Forests and People, 2,* 100027.

Ngo, K. M., Turner, B. L., Muller-Landau, H. C., Davis, S. J., Larjavaara, M., Hassan, N. F. B. N., & Lum, S., (2013). Carbon stocks in primary and secondary tropical forests in Singapore. *Forest Ecology & Management, 296*, 81–89.

Ogle, S. M., Breidt, F. J., & Paustian, K., (2005). Agricultural management impacts on soil organic carbon storage under moist and dry climatic conditions of temperate and tropical regions. *Biogeochemistry, 72*, 87–121.

Olorunfemi, I. E., Fasinmirin, J. T., & Akinola, F. F., (2018). Soil physico-chemical properties and fertility status of long-term land use and cover changes: A case study in forest vegetative zone of Nigeria. *Eurasian Journal of Soil Science, 7*(2), 133–150.

Post, W. M., & Kwon, K. C., (2000). Soil carbon sequestration and land-use change: Processes and potential. *Global Change Biology, 6*, 317–327.

Prentice, I. C., Farquhar, G. D., Fasham, M. J. R., et al., (2001). *The Carbon Cycle and Atmospheric CO_2, Chapter 3, Climate Change 2001.* The Third Assessment Report of Intergovernmental Panel on Climate Change (IPCC), Cambridge: Cambridge University Press.

Rabha, D., Borah, N., & Das, A. K., (2014). Assessment of aboveground and soil organic carbon stocks in Dipterocarpus forests of Barak Valley, Assam, Northeast India. *International Journal of Ecology and Environmental Sciences, 40*(1), 29–40.

Reang, D., De, A., & Das, A. K., (2018). Water resources of Barak valley, India: Spatial assessment of lentic and lotic system using remote sensing and GIS at 5.8 m resolution. *International Journal of Advanced Remote Sensing and GIS, 7*(1), 2633–2642.

Reang, D., De, A., & Das, A. K., (2020). Heterogeneous landscape assessment using multi-temporal LISS4 and hybrid approach: A case study in Barak Valley, India. *Indian Journal of Ecology, 47*(3), 842–848.

Reed, J., Van, V. J., Foli, S., Clendenning, J., Yang, K., MacDonald, M., Petrokofsky, G., Padoch, C., & Sunderland, T., (2017). Trees for life: The ecosystem service contribution of trees to food production and livelihoods in the tropics. *Forest Policy and Economics, 84*, 62–71.

Saha, D., Kukul, S. S., & Sharma, S., (2011). Land use impacts on SOC fractions and aggregate stability in typic ustochrepts of Northwest India. *Plant Soil, 339*, 457–470.

Sarkar, D. C., Meitei, B., Baisya, L. K., Das, A., Ghosh, S., Khumlo, L. C., & Rajkhowa, D., (2015). Potential of fallow chronosequence in shifting cultivation to conserve soil organic carbon in northeast India. *Catena, 135*, 321–327.

Sileshi, G., Schroth, G., Rao, M. R., & Girma, H., (2007). Weeds, diseases, insect pests and tri-trophic interactions in tropical agroforestry. In: Batish, D. R., Kohli, R. K., Jose, S., & Singh, H. P., (eds.), *Ecological Basis of Agroforestry* (pp. 73–94). CRC Press.

Talukdar, N. R., Singh, B., & Choudhury, P., (2018). Conservation status of some endangered mammals in Barak Valley, Northeast India. *Journal of Asia-Pacific Biodiversity, 11*(2), 167–172.

Tian, H., Melillo, J., Kicklighter, D. W., McGuire, A. D., et al., (1998). Effect of interannual climate variability on carbon storage in Amazonian ecosystems. *Nature, 396*, 664–667.

Tilman, D., Cassman, K. G., Matson, P. A., Naylor, R., & Polasky, S., (2002). Agricultural sustainability and intensive production practices. *Nature, 418*, 671–677.

Toru, T., & Kibret, K., (2019). Carbon stock under major land use/land cover types of Hades sub-watershed, eastern Ethiopia. *Carbon Balance and Management, 14*(1), 7.

Yihenew, G. S., & Getachew, A., (2013). Effects of different land use systems on selected physicochemical properties of soils in north western Ethiopia. *The Journal of Agricultural Science, 5*(4), 112–120.

Zhang, M., Huang, X., Chuai, X., Yang, H., Lai, L., & Tan, J., (2015). Impact of land use type conversion on carbon storage in terrestrial ecosystems of China: A spatial-temporal perspective. *Scientific Report, 5,* 10233.

CHAPTER 4

Floral Diversity in Various Land Use Types in Jodhpur District of Arid Rajasthan, India

KAUSHAL SINGH,[1] DEEPAK MISHRA,[2] and G. SINGH[2]

[1]*Department of Silviculture and Agroforestry, College of Forestry, Banda University of Agriculture and Technology, Banda, Uttar Pradesh, India*

[2]*Division of Forest Ecology, Arid Forest Research Institute, Jodhpur, Rajasthan, India*

ABSTRACT

Land use (LU) changes influence the physical, chemical, and biological characteristics of the land resulting in variations in land-cover types and consequently vegetation diversity. A study was conducted to assess the effect of five LU types on plant diversity in 45 villages of Jodhpur district of Western Rajasthan. The five LU types were forest, agriculture, *oran* (sacred groves), pasture and roadside. Diversity variables like species richness, Shannon-Weiner (H'), Simpson reciprocal index (1/D) were calculated. A total 277 plant species were recorded belong to 52 families and 174 genera in the studied area. Tree species dominated in different LUs were *Prosopis cineraria* (IVI-133.6) in agriculture, *Zizyphus rotundifolia* (IVI-54.9) in Orans, *Acacia tortilis* (IVI-98.6) in forests and roadside, and *Capparis decidua* (IVI-70.2) in pastureland. In shrubs, *Calotropis procera* dominated in agriculture (IVI-121), Oran (IVI-113), pasture (IVI-120) and roadside (IVI-124.1) and *Leptadenia pyrotechnica* (IVI-50.8) dominated in forest lands. For herbaceous species, agriculture land was dominated by *Eragrostis ciliaris* grass (IVI-20.7), whereas *Tephrosia purpurea* dominated in other LUs. Highest diversity variables of all plant life forms were in Orans except

Sustainable Management and Conservation of Environmental Resources in India.
Sumit Chakravarty, Vineeta, Jahangeer A. Bhat, Munesh Kumar, Gopal Shukla (Eds.)

for herbaceous species richness and population which were highest in forest lands. The diversity variables of shrubs were highest in agriculture whereas, roadside was found mostly planted with *Prosopis juliflora.* Because of sacredness and involvement of community in protection and conservation of *orans*; species diversity, richness, and evenness were generally high as compared to other LU types. Spatially considering forest ranges, highest tree, tree sapling population, species richness of tree, shrub, and herbaceous species, RSI of trees and RSI and species evenness of herbs were recorded in Mandor. All the diversity variables of tree saplings were highest in Bilara, shrub population in Balesar and herbaceous species population, RSI, and evenness in Osian, Bhopalgarh, and Shergarhforest ranges, respectively.

4.1 INTRODUCTION

Land use (LU) is characterized by arrangements, activities, and inputs people undertake in a certain land cover type to produce, change or maintain it (FAO, 1999). Over 70% of the world's poor are living in rural areas with LU as a major source of subsistence. It is also a low cost and rapidly implementable means of CC mitigation. Forests, agriculture, pastureland, human habitations, and various economic activities are different LUs. Agriculture is the prime one and most important for the survival option for mankind. At present, India has 23% forest cover, 3% pastures and grazing land, 46% land under agricultural use, 14% barren land, 6% cultivable wasteland and remaining 8% as fallow land (GoI, 2012). Out of the total geographical area (TGA) (328.73 million ha), LU statistics are available for roughly 305 million ha, contributing 93% of the total. Common property land resources are pastureland called '*gochar*' and natural woodland called '*oran.*' There are numerous studies worldwide examining LU and its effects on plant diversity, but the evidence for declining species diversity under LU change is a mix and not available separately for selected five LUs (Gerstner et al., 2014). Most of the evidence indicate that future climate and LU change is going to impact plant diversity negatively, but the response in drylands may differ (Newbold et al., 2015). The community pasture and *orans* are becoming over-crowded by increasing livestock population whereas, the productivity of forests and agriculture are at decline. All these LU practices are influencing biological diversity. To develop appropriate management strategies to preserve biodiversity, a more comprehensive understanding is needed concerning the effect of LUs on floral diversity. In view of this, the study on floral diversity in different LU types was conducted.

4.2 MATERIALS AND METHODS

4.2.1 STUDY AREA AND METHODS

The study was conducted in 45 villages (45 villages were taken as 5 replications of 9 forest ranges in jodhpur district) comprising nine forest ranges of Jodhpur district of Rajasthan, India (Figures 4.1(a)–(e)).

FIGURE 4.1 Details of land use under study. (a) Agriculture land at Lohawat (Bordi + Moongfali); (b) Pastureland at Gopa (Khejri + Bawali); (c) Forest land at Dechu (Israeli Babool plantation); (d) *Oran* land Kolu (Bordi + Ker + Khejri); (e) Roadside at Bhopalgarh (Israeli Babool).

These are situated at the latitude 26° 00′–27° 34′ north and 72° 00′–73° 33′ east longitude and altitude varies from 171–334 m from sea level. Jodhpur is the largest district of Rajasthan with geographical area of 22,901 km^2. Soil of the district is generally light yellowish brown in color, structureless to very weakly blocky friable, sandy-to-sandy loam and alkaline in nature. The mean maximum annual temperature varies from 24.6–41.6°C during peak summer months of April–June, and sometimes touches an extreme of 51°C. Cumulative annual rainfall of 12 years in the district Jodhpur was 461 mm. Total numbers of sampling points were 9 (forest range) × 5 (Landuse types) × 5 (village in each forest range as replication) = 225 observation points.

4.2.2 FLORISTIC ANALYSIS

A total of 225 plots each with size 32.5 m × 32.5 m (based on the species area relationship) were laid in the forests, *Oran* and pasture lands while, 100 m × 100 m plots were laid out in agriculture land (the selected size of quadrate was appropriate because of very less vegetation density in desert region). For recording observations on trees along the roadside, a 100 running meter stretch (width depended upon types of roads) along both the sides of the road were considered. Two sub-plots of 5 m × 5 m size were laid in two directions (N-E and S-W) in the plots of trees for shrub species, four sub-plots of 1 m × 1 m size were laid out at four corners of the main plots (laid for tree observations) in all land uses (LUs) for herbaceous vegetation. This is called nested stratified random quadrate sampling method. All trees of different tree species inside the plot having diameter at breast height (dbh) > 10 cm were counted for their population and measured for dbh. In case of shrubs, all the tillers/stems at 15 cm above the ground level were measured by using digital tree caliper. Diameter of the herbaceous vegetation was recorded for five representative plants of each species by using digital tree caliper after harvesting of all vegetations of the 1 m x 1 m quadrate. Vegetation study was conducted after monsoon period, i.e., during September to November months, when chances of availability of herbaceous species are highest in the region. Herbaceous vegetations were studied from 900 temporary quadrates following quadrate method of Misra (1968). Diversity variables like species richness (R), Shannon-Wiener diversity index (H'), species evenness (E) and reciprocal of Simpson's diversity index (RSI) and importance value index (IVI) were calculated following standard procedure (Whitford, 1949; Shannon & Wiener, 1963). All data were analyzed statistically using IBM

SPSS version 22 statistical package. Data on floral diversity were obtained from the five LUs and nine forest ranges as the main factors, these data were analyzed using two-way ANOVA considering the site as the replication in each village.

4.3 RESULTS AND DISCUSSION

A total 257 plant species represented by 52 families and 174 genera were recorded across the 5 LUs (Figures 4.1(a)–(e)) and 9 forest ranges in Jodhpur district, Rajasthan. Most dominant family was Poaceae with 53 species. It was followed by Papilionaceae with 34 species and Asteraceae with 16 species. There were 31 trees, 25 shrubs, 7 under-shrubs, 2 lianas (or woody climbers), 9 climbers, 5 twiners and 178 herbaceous species. In addition, 10 species of cultivated crops were also recorded in croplands. Highest IVI of tree species dominated in different LUs were *Prosopis cineraria* (IVI-133.6) in agriculture, *Zizyphus rotundifolia* (IVI-54.9) in *Orans*, *Acacia tortilis* (IVI-98.6) in forests and *Capparis decidua* (IVI-70.2) in pasture-land. In shrubs, *Calotropis procera* dominated in agriculture (IVI-121), roadside (IVI-124.1), oran (IVI-113) and pasture (IVI-120) and *Leptadenia pyrotechnica* (IVI-50.8) dominated in forest lands. For herbaceous species, *Eragrostis ciliaris* grass (IVI-20.7) dominated the agriculture land whereas, *Tephrosia purpurea* dominated in forest (IVI-28.4), *oran* (IVI-34.8), pasture (IVI-49.2), and roadside (IVI-36.2) LUs.

Diversity variable of trees, saplings, shrubs, and herbaceous species varied significantly ($P < 0.05$) between different land-use types. Tree species richness, H,' RSI and E of tree, tree saplings, population of shrubs, herbaceous diversity and RSI were highest in *oran* whereas, tree population, herbaceous species and herbs richness were highest in forest. Highest values of shrub richness, H,' RSI and evenness were found in agriculture. Herbaceous species evenness and tree sapling population were recorded maximum in pastureland and along the roadsides, respectively. The lowest population of the tree, and H,' RSI and E of tree sapling and shrubs were recorded in agriculture whereas, species richness of all the plant habits and diversity of tree, tree saplings and shrubs were in pasture lands. The lowest values of RSI and evenness of tree species and only evenness of tree sapling was in forest whereas, roadside showed lesser tree sapling concentration and even distribution of shrubs (Table 4.1).

TABLE 4.1 Diversity Variables of Various Plant Habits Influenced by Land Use Types in Jodhpur Thar Desert of India (Values Are Mean ± SE Replications)

Diversity Variables	Plant Habit	Land Use Types					F-Value	P-Value
		Agriculture	Forest	Oran	Pasture	Roadside		
Species population	Tree	11.96±0.7	137.56±12.48	120.22±7.13	76.22±6.58	89.63±10.12	41.320	0.000
	Sapling	21.84±4.80	118.67±20.6	114.67±15.6	96.11±14.44	156.04±32.87	9.156	0.000
	Shrub	116.8±16.3	198.44±30.92	442.44±84.61	242.0±36.80	384.73±70.68	8.446	0.000
	Herb	88.71±5.79	109.6±4.81	66.04±1.78	52.89±1.83	61.22±1.65	54.702	0.000
Species richness	Tree	3.18±0.23	2.58±0.22	3.24±0.20	2.33±0.16	2.62±0.22	3.867	0.005
	Sapling	1.44±0.14	1.53±0.16	1.82±0.17	1.28±0.11	1.29±0.11	2.166	0.075
	Shrub	2.53±0.18	2.51±0.28	2.38±0.25	2.09±0.23	2.33±0.22	0.644	0.632
	Herb	13.02±0.42	15.49±0.52	15.00±0.39	11.89±0.28	13.33±0.43	15.837	0.000
Shannon-Wiener diversity	Tree	0.85±0.07	0.62±0.07	0.93±0.06	0.61±0.07	0.66±0.07	4.453	0.002
	Sapling	0.25±0.06	0.25±0.06	0.39±0.07	0.21±0.0	0.22±0.00	1.795	0.132
	Shrub	0.60±0.06	0.57±0.08	0.54±0.0	0.50±0.07	0.52±0.08	0.338	0.852
	Herb	1.92±0.05	2.11±0.05	2.13±0.03	2.02±0.03	2.02±0.03	5.891	0.000
Simpson's reciprocal index	Tree	2.34±0.16	1.81±0.10	2.50±0.16	1.87±0.12	1.94±0.15	3.887	0.005
	Sapling	1.28±0.11	1.23±0.10	1.58±0.10	1.25±0.08	1.17±0.09	2.435	0.049
	Shrub	1.79±0.11	1.70±0.15	1.73±0.14	1.38±0.16	1.74±0.14	1.471	0.215
	Herb	5.23±0.25	6.40±0.31	6.54±0.24	6.13±0.20	6.02±0.22	4.947	0.001
Species evenness	Tree	0.68±0.05	0.56±0.05	0.79±0.04	0.61±0.06	0.59±0.06	3.080	0.017
	Sapling	0.28±0.06	0.24±0.06	0.42±0.07	0.25±0.06	0.28±0.06	1.457	0.217
	Shrub	0.59±0.05	0.50±0.06	0.47±0.06	0.48±0.06	0.44±0.06	0.930	0.448
	Herb	0.75±0.01	0.78±0.01	0.79±0.01	0.82±0.01	0.79±0.01	5.738	0.000

Among forest ranges, highest tree, tree sapling population, species richness of tree, shrub, and herbaceous species, RSI of trees and RSI and species evenness of herbs were recorded in Mandor. All the diversity variables of tree saplings were highest in Bilara, shrub population in Balesar and herbaceous species population, RSI, and evenness in Osian, Bhopalgarh, and Shergarhforest ranges, respectively. The lowest population, species richness, diversity, and RSI of tree, herbs, and evenness of trees were in Baap whereas, all the diversity variables of tree saplings and herbaceous species in Phalodi range. Shrub population and species richness was the lowest in Bilara and species diversity, RSI, and evenness of shrubs was in Luni (Table 4.1).

Greater number of tree species as compared to the shrub species in the region was due to their introduction under various plantations programs, like Desert Development Program, Desert Prone Area Program and Combating desertification program. Dominance of *Prosopis cineraria* and *Tecomella undulata* trees and *Pennisetum glaucum* crop in agriculture land, *Ziziphus rotundifolia* in *Orans* and *Capparis decidua* in pasture lands indicates their ecological amplitude and tolerance to drought stress and other edaphic factors (Singh, 2009). Because of dependency on these important multipurpose trees (MPTs) for food, fodder, fuel, and timber; people of the region protect and promote these species on their farmlands. Wide adaptability of *Prosopis cineraria* and *Ziziphus nummularia* has also been reported because of their availability throughout the desert with varying soil conditions (Singh et al., 2012). Use of *Acacia tortilis* as the dominant plantation species has resulted in the dominance of this species in forest land (Singh & Singh, 2015). Occurrence of *Calotropis procera* in all LUs is because of its ability to endure extreme temperatures and soil water stress under arid region (Karakilcik & Kalyar, 2014). Relatively high population of tree and herbaceous species and herbaceous species richness in forest land was due to tree plantation and protection of this area from grazing and trampling by the livestock that showed positive effects on tree population, and regeneration and species richness of herbaceous species. As compared to forest land, other LUs were greatly affected by grazing pressure of both domestic and wild animals which utilized maximum ground flora and reduces herbaceous population and species richness (Fensham et al., 2011). Relatively a smaller number of saplings was due to reduced seed production under browsing and canopy removal followed by reduced seed germination because of variable rainfall as well as weak capacity of recolonization of some of the species and seedling recruitment under human and livestock disturbance (Ceccon et al., 2006; Bradshaw & Waller, 2016).

LU changes influences the physical, chemical, and biological characteristics of the land resulting in variations in land-cover types and consequently vegetation diversity (Jurisch et al., 2013). It is shown by significantly high ($P < 0.05$) values of all diversity variables for tree, tree sapling, shrub population (Table 4.1) and herbaceous species in *Orans* (sacred groves). This was because of sacredness and community-controlled protection. Because of sacredness and involvement of community in protection and conservation of such area; species diversity, richness, and evenness were generally high in sacred groves as compared to cultivated land or non-sacred forests (Singh, 2016). However, overgrazing and trampling under increased livestock pressure coupled with environmental harshness in the region appears to be the real cause of low ($P < 0.01$) values of diversity indices of all plant habit in pastureland (Metera et al., 2010). Grazing puts an impact on the density and biomass of the individual species, alters species composition and reduces species richness under limited rainfall in dry areas (Skornik et al., 2010). Though plants have evolved different mechanisms to cope with the aridity, low and erratic rainfall and extremely high temperature regimes but heavy grazing generally modifies the magnitude of change in species richness under limited rainfall affecting diversity and productivity (Rutherford & Powrie, 2013). It is also supported by a positive correlation ($r = 0.195$, $P < 0.01$) between rainfall and herbaceous diversity. Low population and diversity in agriculture land was certainly due to farmer's preferences towards a particular species (i.e., *Prosopis cineraria, Tecomella undulata, Ziziphus nummularia*) and cultivation practices affecting vegetation diversity negatively. The agriculture lands of this region are a real example of traditional agroforestry system where some naturally growing MPTs and shrubs like *Prosopis cineraria, Tecomella undulata, Acacia senegal, Ziziphus mauritiana* and *Capparis decidua* are protected and promoted by the local people to cope with environmental adversities during which agriculture crop fails to provide sustenance (Bhandari et al., 2014). However, current expansion of the area under annual and cash crops together with mechanized farming is leading to decline in the population of these trees and shrubs species from agriculture lands affecting livestock and fuelwood production and the existing copping mechanism to droughts (Tanwar et al., 2018).

There were significant spatial variations ($P < 0.01$) in diversity variables of all plant life forms where species richness, diversity, Simpson's reciprocal index, population, and richness of herbaceous species were highest in Mandor range and lowest values in Baap range (Table 4.2).

TABLE 4.2 Diversity Variables of Various Plant Habits Influenced by Spatial Variation in Jodhpur Thar Desert of India (Values are Mean ± SE Replications)

DV	F	Forest Ranges									F Value	P Value
		Baap	Balesar	Bhopalgarh	Bilara	Luni	Mandor	Osian	Phalodi	Shergarh		
P	T	43.1±8.8	101.0±16.0	71.6±9.2	97.5±12.29	94.9±12.62	103.6±18.23	82.4±15.19	97.7±15.30	91.3±15.19	3.523	0.001
	Sl	68.15±21.2	69.9±15.4	82.6±15.3	143.99±40.	145.4±35.9	147.14±22.5	91.34±25.6	51.80±8.30	112.77±24.0	3.759	0.000
	S	327.9±57.4	525.5±139.7	153.6±32.4	86.25±31.8	140.7±46.5	478.5±101.5	189.2±35.8	174.3±34.3	415.90±73.1	6.928	0.000
	H	65.4±4.6	81.0±7.8	80.7±6.9	76.87±5.79	73.14±5.04	75.12±6.90	90.07±6.13	70.35±4.52	68.54±7.94	3.219	0.002
R	T	1.6±0.2	3.04±0.3	2.84±0.27	2.92±0.31	3.04±0.27	3.40±0.38	2.72±0.27	2.56±0.20	2.96±0.25	3.239	0.002
	Sl	1.1±0.1	1.44±0.1	1.56±0.17	1.88±0.31	1.68±0.17	1.52±0.16	1.72±0.23	1.08±0.11	1.48±0.17	2.050	0.043
	S	2.5±0.3	2.96±0.3	1.80±0.31	1.76±0.28	1.80±0.28	3.12±0.23	2.60±0.40	1.92±0.27	2.88±0.23	3.466	0.001
	H	11.6±0.4	14.88±0.5	14.88±0.64	15.12±0.73	12.88±0.60	15.28±0.59	13.44±0.47	13.08±0.56	12.60±0.58	7.133	0.000
H'	T	0.39±0.1	0.80±0.09	0.78±0.10	0.74±0.10	0.82±0.09	0.91±0.11	0.70±0.10	0.67±0.08	0.79±0.09	2.446	0.016
	Sl	0.17±0.06	0.23±0.07	0.27±0.07	0.41±0.11	0.31±0.07	0.24±0.08	0.37±0.10	0.08±0.04	0.29±0.08	1.784	0.083
	S	0.59±0.10	0.67±0.11	0.37±0.10	0.46±0.09	0.34±0.09	0.74±0.08	0.60±0.11	0.41±0.09	0.71±0.08	2.623	0.010
	H	1.91±0.05	2.12±0.03	2.11±0.07	2.11±0.06	1.94±0.05	2.14±0.04	2.03±0.04	1.93±0.06	2.07±0.04	3.468	0.001
RSI	T	1.44±0.18	2.14±0.19	2.24±0.19	2.09±0.19	2.25±0.20	2.57±0.30	2.00±0.22	1.89±0.15	2.19±0.19	2.216	0.028
	Sl	1.06±0.1	1.29±0.09	1.30±0.12	1.59±0.24	1.39±0.11	1.31±0.11	1.45±0.16	1.01±0.08	1.31±0.13	1.644	0.155
	S	1.83±0.15	1.96±0.20	1.44±0.20	1.41±0.20	1.34±0.17	2.00±0.15	1.82±0.24	1.52±0.16	1.71±0.18	1.906	0.062
	H	5.21±0.27	6.35±0.24	6.63±0.43	6.35±0.38	5.45±0.29	6.38±0.36	6.22±0.35	5.37±0.34	6.60±0.27	3.315	0.001
E	T	0.43±0.09	0.65±0.06	0.65±0.08	0.58±0.07	0.75±0.05	0.71±0.06	0.61±0.07	0.68±0.06	0.74±0.05	2.053	0.043
	Sl	0.24±0.08	0.28±0.08	0.30±0.08	0.39±0.09	0.37±0.08	0.27±0.08	0.34±0.09	0.10±0.06	0.35±0.09	1.270	0.262
	S	0.54±0.08	0.52±0.07	0.36±0.09	0.47±0.09	0.33±0.08	0.66±0.06	0.52±0.09	0.43±0.09	0.63±0.07	2.043	0.044
	H	0.78±0.01	0.79±0.01	0.79±0.02	0.78±0.02	0.77±0.01	0.79±0.01	0.79±0.01	0.76±0.02	0.83±0.01	2.039	0.044

Abbreviations: F: Form (T: tree; Sl: sapling; S: shrub; and H: herb); DV: diversity variables (R: species richness; H': Shannon-Wiener diversity index; E: species evenness; RSI: Simpson's reciprocal index; and P: species population).

It was due to varying physiographic and soil condition favoring habitat heterogeneity and biotic diversity in the region (CGWB, 2008). Thus, a combined effect of spatially structured environmental conditions like rainfall, soil conditions and plantation of different tree species might have favored vegetation diversity in Mandor as compared to the others forest ranges. This was also supported by positive correlation of herbaceous diversity with rainfall ($r = 0.195$, $P < 0.01$). Plant diversity at a particular site is influenced by environmental variability and spatial configuration of the habitats and increased environmental heterogeneity increases plant species richness (Redon et al., 2014). Number of species and the diversity value are low in the present study as compared to the diversity value of for young (5.06) and old (5.40) stands in tropical forests (Knight, 1975). The diversity values for tree ranging from 0.39 to 0.91 was lower and evenness index (0.43–0.75) was observed higher than the reported values of 0.67–0.79 and 0.02–0.05 in tropical dry deciduous forest of southern Rajasthan (Kumar et al., 2010). Similar pattern was observed for shrub species when compared with the values of species diversity and evenness value recorded for Aravalli forests in Pali district of Rajasthan (Singh & Singh, 2013). Variations in values of Simpson reciprocal index for trees, shrubs, and herbaceous species from 1.81, 1.38, 5.23 to 2.50, 1.79, 6.96, respectively and values of diversity index from 0.61, 0.50, 1.92 to 0.93, 0.60, 2.24, respectively falls in the lower range of the existing literature (Simpson, 1949; Margalef, 1972) indicating low floral diversity in the studied area. However, species evenness was little higher for tree, shrub, and herbaceous species (0.56, 0.44, 0.75 to 0.79, 0.59, 0.82, respectively) indicating abundance of different species available in patches under varying heterogeneity (Pandit, 2003; Harikrishna et al., 2014). Greater ($P < 0.01$) value of herbaceous species evenness (Table 4.2) in Shergarh range as compared to the other ranges was due to less fertile soil coupled with scattered availability of resources (soil, water, and nutrients). This lead to patchiness becoming more abundant among the species assemblage and increased the species evenness and decreased the species dominance (Larson & Belovsky, 2013). However, lowest values of most of the diversity variables of all plant habit in Baap range was explained by harsh edaphic and climatic conditions as diversity patterns of the plant communities are very much related to the local environment, including productivity and patchiness of the habitats (Adamidis et al., 2014). Sandy soils with relatively high aridity in Balesar range appeared favorable for shrub species indicated by highest number of individuals of shrubs in this range (Singh, 2008; Eldridge et al., 2011). It shows that shrubs are more adaptable to arid region compared to the tree species. But lesser ($P < 0.05$) population, species richness, diversity, Simpson's reciprocal index and species evenness for tree sapling in Phalodi range was associated with reduced seed production (tree lopping and debranching to use them as fodder) and less favorable edaphic (sandy and saline soils) and climatic conditions (Shrivastava & Kumar, 2015). Number of tree saplings are positively correlated with amount of annual rainfall ($r = 0.192$, $P < 0.01$) indicating low saplings in the forest ranges with relatively low rainfall. The diversity of herbaceous species ranged from 1.92–2.13 is quite lower than the values (2.59–3.18) reported by Krishna et al. (2014) for whole of Rajasthan. This difference appears to be due to relatively more adverse edaphic and climatic condition in Jodhpur.

4.4 CONCLUSION

Tree species dominated in different LUs were *Prosopis cineraria* in agriculture, *Zizyphus rotundifolia* in *Orans*, *Acacia tortilis* in forests, and *Capparis decidua* in pastureland. In shrubs, *Calotropis procera* dominated in agriculture, *oran* and pastures, and *Leptadenia pyrotechnica* dominated in forest lands. For herbaceous species, agriculture land was dominated by *Pennisetum glaucum*, co-dominated by *Ricinus communis.* Among the natural vegetation, *Eragrostis ciliaris* grass dominated in agriculture whereas, *Tephrosia purpurea* dominated in other LUs. Regeneration (21.8 to 156.0 individuals ha^{-1}) of trees appears low and was dominated by invasive species *Prosopis juliflora*. Highest diversity variables of all plant habit were in *Orans*, except for herbaceous species richness and population which were highest in forest land. Sacredness and community's role in retaining woody perennial and conservation enhanced diversity in *Oran*. The diversity variables of shrubs were highest in agriculture. Among forest ranges, indices of tree, shrubs, and herbaceous diversity were highest in Mandor range. Patches of varying heterogeneity enhanced herbaceous species evenness in Shergarh range whereas, shrub population was highest (525.51 population ha^{-1}) in Balesar forest range. However, lowest values of most of the diversity variables of all plant habit in Baap range was explained by harsh edaphic and climatic conditions.

ACKNOWLEDGMENT

We would like to express our sincere thanks to the director, Arid Forest Research Institute New Pali Road, Jodhpur for providing necessary facilities.

KEYWORDS

- **diversity**
- **floristic analysis**
- **Jodhpur**
- **land use**
- ***Prosopis juliflora***
- **shrub population**
- **species richness**

REFERENCES

Adamidis, G. C., Kazakou, E., Baker, A. J. M., Reeves, R. D., & Dimitrakopoulos, P. G., (2014). The effect of harsh abiotic conditions on the diversity of serpentine plant communities on Lesbos, an eastern Mediterranean Island. *Plant Ecol. Divers., 7*, 433–444.

Bhandari, D. C., Meghwal, P. R., & Lodha, S., (2014). Horticulture based production systems in Indian arid regions. In: Nandwani, D., (eds.), *Sustainable Horticultural Systems* (pp. 3–17). Springer Cham.

Bradshaw, L., & Waller, D. M., (2016). Impacts of white-tailed deer on regional patterns of forest tree recruitment. *Forest Ecol. Manag., 375*, 1–11.

Ceccon, E., Huante, P., & Rincon, E., (2006). Abiotic factors influencing tropical dry forests regeneration. *Braz. Arch. Biol. Technol., 49*, 305–312.

CGWB, (2008). *District Ground Water Brochure of Jodhpur District*. Central Ground Water Board, Ministry of Water Resources, Government of India. http://www.cgwb.gov.in/District_Profile/Rajasthan/Jodhpur.pdf (accessed on 28 December 2023).

Eldridge, D. J., Bowker, M. A., Maestre, F. T., Roger, E., Reynolds, J. F., & Whitford, W. G., (2011). Impacts of shrub encroachment on ecosystem structure and functioning: Towards a global synthesis. *Ecol. Lett., 14*(7), 709–722.

FAO, (1999). *Land Cover Classification System*. FAO, Rome. http://www.fao.org/nr/land/use/en (accessed on 28 December 2023).

Fensham, R. J., Silcock, J. L., & Dwyer, J. M., (2011). Plant species richness responses to grazing protection and degradation history in a low productivity landscape. *J. Veg. Sci., 22*, 997–1008.

Gerstner, K., Dormann, C. F., Stein, A., Manceur, A., & Seppelt, R., (2014). Editor's choice: Effects of land use on plant diversity-A global meta-analysis. *J. of Applied Eco., 51*, 10.1111/1365-2664.12329.

GoI, (2012). Special data dissemination standard division, Directorate of Economics & Statistics, Ministry of Agriculture, Govt. of India, New Delhi. http://lus.dacnet.nic.in (accessed on 28 December 2023).

Harikrishna, P., Reddy, C. S., Meena, S. L., & Katewa, S. S., (2014). Pattern of plant species diversity in grasslands of Rajasthan, India. *Taiwania, 59*(2), 111–118.

Jurisch, K., Hahn, K., Wittig, R., & Bernhardt, R. M., (2013). Land-use impact on the growth and survival of seedlings and saplings in West African savannas. *J. Veg. Sci., 24*(1), 101–112.

Karakilcik, Y., & Kalyar, M. N., (2014). The unexplored jewel of desert: Prospective role of Thar Desert in ecological and socio economic development of Pakistan. *JMEST, 1*(5), 411–420.

Knight, D. H. A., (1975). Phytosociological analysis of species-rich tropical forest on Barro Colorado Island, Panama. *Ecol., 45*(3), 259–284.

Krishna, P. H., Reddy, C. S., Meena, S. L., & Katewa, S. S., (2014). Pattern of plant species diversity in grasslands of Rajasthan, India. *Taiwania, 59*(2), 111–118.

Kumar, N. J., Patel, K., Kumar, R. N., & Bhoil, R. K., (2010). An assessment of carbon stock for various land use systems in Aravali Mountains, Western India. *Int. J. Environ. Sci., 5*, 600–608.

Larson, C. A., & Belovsky, G. E., (2013). Salinity and nutrients influence species richness and evenness of phytoplankton communities in microcosm experiments from Great Salt Lake, Utah, USA. *J. Plankton Res., 35*(5), 1154–1166.

Margalef, R., (*1972*). Homage to E. Hutchinson, or why is there an upper limit to diversity. *Trans. Conn. Acad., 44*, 211–235.

Metera, E., Sakowski, T., Sloniewski, K., & Romanowicz, B., (2010). Grazing as a tool to maintain biodiversity of grassland: A review. *Anim. Sci. Pap Rep., 28*(4), 315–334.

Misra, R., (1968). *Ecology Work Book* (p. 244). Oxford and IBH Publishing Co, New Delhi.

Newbold, T., Lawrence, N. H., Hill, S. L. L., Contu, S., Lysenko, I., Senior, R. A., Borger, L., et al., (2015). Global effects of land use on local terrestrial biodiversity. *Nature, 520*, 45–50.

Pandit, B. R., (2003). *Biodiversity of Tropical Dry Deciduous Forest Ecosystem* (Vol. 12, No. 699, pp. 1–9). XII World Forestry Congress, Quebec City, Canada.

Redon, M., Berge, L., Cordonnier, T., & Luque, S., (2014). Effects of increasing landscape heterogeneity on local plant species richness: How much is enough. *Landsc. Ecol., 29*(5), 773–787.

Rutherford, M. C., & Powrie, L. W., (2013). Impacts of heavy grazing on plant species richness: A comparison across rangeland biomes of South Africa. *S. Afr. J. Bot., 87*, 146–156.

Shannon, C. E., & Wiener, W., (1963). *The Mathematical Theory of Communication* (p. 177). University of Illinois press, Urbana, Illinois.

Shrivastava, P., & Kumar, R., (2015). Soil salinity: A serious environmental issue and plant growth promoting bacteria as one of the tools for its alleviation. Saudi J. Biol. Sci., 22*(2), 123–131.*

Simpson, E. H., (1949). Measurement of diversity. *Nature, 163,* 688.

Singh, G., & Singh, K., (2013). Aspect effect on diversity and soil carbon stock in a degraded forest of Rajasthan. *Indian Forester, 139*(12), 1061–1069.

Singh, G., (2008). *Study on Biological Diversity in Mangala, Saraswati and Rageshwari Areas of Rajasthan Hydrocarbon Project*. Final Report, Cairns Energy Pvt. Ltd. India.

Singh, G., (2009). Comparative productivity of *Prosopis cineraria* and *Tecomella undulata* based agroforestry systems in degraded lands of Indian Desert. *J. For. Res., 20*, 144–150.

Singh, G., (2016). *Sacred Groves of Rajasthan* (p. 292). Scientific Publisher.

Singh, G., Singh, K., Mishra, D., & Shukla, S., (2012). Vegetation diversity and role of *Leptadenia pyrotechnica* in biomass contribution and carbon storage in arid zone of India. *Arid Ecosyst., 2*(4), 264–272.

Singh, K., & Singh, G., (2015). Roadside vegetation diversity of Jodhpur district and its role in carbon sequestration and climate change mitigation. *Adv. For. Sci., 2*(2), 23–33.

Skornik, S., Vidrih, M., & Kaligaric, M., (2010). The effect of grazing pressure on species richness, composition, and productivity in North Adriatic Karst pastures. *Plant Biosyst., 144*(2), 355–364.

Tanwar, S. P. S., Bhati, T. K., Singh, A., Patidar, M., Mathur, B. K., Kumar, P., & Yadav, O. P., (2018). Rainfed integrated farming systems in arid zone of India: Resilience unmatched. *Indian J. Agron., 63*(4), 403–414.

Whitford, P. B., (1949). Distribution of woodland plants in relation to succession and clonal growth. *Ecology, 30*, 199–208.

CHAPTER 5

Forest Fire Detection and Monitoring for Its Management: Remote Sensing and GIS/Geospatial Technology Approach

DEBASISH CHAKRABORTY,[1] ARPAN BHOWMIK,[2] NILIMESH MRIDHA,[3] and KITDOR PYNGROPE[1]

[1]*ICAR–Research Complex for NEH Region, Umiam, Meghalaya, India*

[2]*Indian Agricultural Research Institute, Gogamukh, Assam, India*

[3]*National Institute of Natural Fiber Engineering and Technology, Kolkata, West Bengal, India*

ABSTRACT

Forest fires are one of the common causes of forest degradation and challenges for forest management. Forest fires are uncontrolled and non-prescribed combustion or burning of plants in a forest, or grassland and spread based on environmental factors (e.g., wind, topography). Human actions like land clearing, urbanization in nearby forest areas, etc., along with natural calamities like extreme drought or even in rare cases lightning may also cause forest fires. Forest fires are threat for both the forest wealth as well as the entire flora and fauna within the forests and thereby also affecting the biodiversity, ecology, and environment of the forest region. During the summers, when there is no rainfall for months, the forest becomes littered with dry senescent leaves and twigs that could burst into flames which may be ignited by a single spark. The number of forest fires and the area burned has increased considerably during the last decade. The frequency and intensity of wildfires increased due to climate change (CC), has been exacerbated in some nations

Sustainable Management and Conservation of Environmental Resources in India.
Sumit Chakravarty, Vineeta, Jahangeer A. Bhat, Munesh Kumar, Gopal Shukla (Eds.)

by fuel management regulations. Many of the burned places are in far-flung corners of the globe, where gathering information on fires is difficult or impossible on the ground. Satellite remote sensing (RS) has been used to solve this challenge, with impressive results in finding big fires, mapping burned areas, and tracking smoke plumes. RS, geographic information systems (GIS), and computer models will help firefighters manage flames more effectively.

5.1 INTRODUCTION

The world's total forest acreage is currently a little over 4 billion hectares, equating to 0.6 hectares per human (FAO, 2010). The forest cover is not distributed evenly across the landmass. According to the state of the world's forest report 2011, Europe has the most forest cover, accounting for about 45% of the world's total forest cover. South America ranks first in terms of land covered by forest, with about half of its landmass covered by forest, and Oceania ranks first in terms of per capita forest area. Changes in forest cover were slow and natural in the early history of the globe, but with the arrival of homo sapiens on the planet and its intervention, the situation changed dramatically. The current state of the world's forests is the result of these natural and manmade changes.

The expansion of forest resources in a country or within a region can be broadly signified by Forest cover. Forest Survey of India (FSI) has started the current assessment of forest cover in 1987 using remote sensing (RS) techniques. The current assessment is the 16th in the series providing continuous monitoring of forests and their changes across the country in the last 32 years. The Forest cover assessment published in the India State of Forest Report (ISFR) is a very important source of primary information on forests in the country and is widely used across the country by central and state governments, organizations, and other stakeholders (FSI, 2019). Forest cover reported in ISFR includes all lands having trees of more than one hectare in an area with more than 10% tree canopy density irrespective of legal status and ownership of the land. LISS III data with a resolution of 23.5 m allows mapping at the maximum scale of 1:50,000 at which the minimum mappable unit (MMU) becomes one hectare. The Forest cover in India is provided in Table 5.1.

TABLE 5.1 Forest Cover of India in Terms of Area

Class	Area (sq. km)	Percentage of Geographical Area
Very dense forest	99,278	3.02
Moderately dense forest	3,08,472	9.39
Open forest	3.04,499	9.26
Total Forest Cover	7,12,249	21.67
Total Geographical Area	3,287,469	100.00

Source: Adapted from FSI (2019).

The distribution of forest cover over the globe is influenced by a wide range of factors, including height, geology/soil type, and climate. Climate conditions, such as temperature and rainfall, have an important role in shaping forest dispersal. Globally, four different types of forests have been recognized based on these two factors as temperate, tropical, subtropical, and boreal forests. Further, these four types of forests are divided into 13 subtypes. Wildland fire ignition potential, fire behavior, and fire severity have been profoundly influenced by Weather and climate. Large-scale wind patterns across the hemispheres influence local weather and temperature, predisposing natural land fuels to fire. Changing climate and meteorological factors create both manmade and lightning-caused fires. After ignition, ambient atmospheric elements like air stability, fuel moisture, wind, and topography have a significant impact on fire behavior.

The mean and variability of weather characterize climate, which is the outcome of multiple nonlinear processes and interactions between the hydrosphere, atmosphere, biosphere, and geosphere. Although it is uncertain how expected climate change (CC) will affect future forest fire behavior, it is clear that extreme weather events happening over long periods of time will affect the moisture content of wildland fuels and hence fire regime features. According to most studies, future global temperatures will be warmer than current levels, and the region impacted by drought will expand, increasing the likelihood of wildfires (IPCC, 2007). The climate varies from one place to the next due to climate variations and variations in vegetation conditions. Fire seasons begin in lower latitude locations in both the Southern and Northern Hemispheres and proceed pole ward as the warm season progresses and the earth continues its journey around the sun. Annual fluctuations in a region's climatic variables, such as snowfall and spring melting, as well as the lengths and duration of growing seasons, are all essential factors in determining when fire seasons begin and end (Westerling et al., 2006).

5.2 FACTORS AFFECTING FOREST FIRE

5.2.1 CLIMATE AND ITS ABERRATIONS

Large wildfires requiring major suppression resources are usually believed to be linked to meteorological anomalies that impact the climate record. The bulk of the land burned by wildfires, which may remain underestimated in some areas (Soja et al., 2002), is generally related to a small number of documented fires each year and hence has been attributed to climatic change and growing fuel dangers (Stephens, 2005). Drought can occur when there is a lack of moisture for an extended period of time. Droughts are linked to long-term deviations from the regular pattern of large-scale weather patterns. It's thought that persistent weather patterns linked to drought are influenced by ocean surface temperatures (Ting & Wang, 1997; Herweijer et al., 2007). Droughts are also frequently linked to increased wildfire occurrence (Swetnam & Betancourt, 1998; Girardin et al., 2006) and to increased fire size (Maingi & Henry, 2007).

5.2.2 TELECONNECTIONS

The phrase "teleconnection" refers to the tendency for atmospheric circulation patterns from one region on the globe to be linked, either directly or indirectly, to another that spans over a broad area. In the study of air-sea interactions and global climate dynamics, teleconnections are critical. They are frequently valuable in deciphering global climatic patterns. The El-Nino Southern Oscillation is probably the most well-known teleconnection (ENSO). It's a global teleconnections phenomenon that's seen in the tropical Pacific Ocean, where the average sea surface temperature is either cooling or warming (Bridgman & Oliver, 2006). According to Notaro et al. (2007), future temperatures will be warmer due to increased carbon dioxide (CO_2) levels. According to certain research, global warming causes universal increases in fire frequency, as well as major regional differences in fire activity and burned area (Flannigan et al., 2005; Tymstra et al., 2007).

5.2.3 WEATHER PATTERN

The meteorological variables are relevant to affecting fire behavior result from synoptic scale forcing of weather which occurs at the micro-scale where fire, where fuels and topography interact. Circadian changes in temperature, relative humidity, wind speed and wind direction may significantly influence fire behavior (Flannigan & Harrington, 1987; Pyne et al., 1996). Use of geographic analysis showed that that the most relevant weather factors affecting fire occurrence in Alaska were growing season precipitation, temperature, and lightning frequency (Kasischke et al., 2002). Short-term local weather, particularly low relative humidity, unusual dry spells and windy weather generally kindred with cold fronts predispose wildland fuels to fire (Johnson & Miyanishi, 2001).

5.2.3.1 TEMPERATURE

The solar radiation that is not reflected by the Earth's surface is absorbed and transformed into heat. Heat which is defined as the energy transported from a higher temperature object to a lower temperature object, is an important part of the fire triangle. This heat energy is essential for the preheating or evaporative phase of combustion to commence (Johnson & Miyanishi, 2001). As a result, greater temperatures heat forest fuels, making them more susceptible to fire if an appropriate ignition source is available. The vertical and horizontal distribution of temperature is an important feature of temperature that relates to fire behavior. The degree of atmospheric stability describes vertical temperature disparities in the atmosphere. An unstable atmosphere is one that cools to some extent as it rises above the Earth's surface, which can lead to thunderstorms or cloud formation as air moves towards the surface from upward direction. Unstable air contributes to increased fire behavior in two ways: first, it allows for the formation of a well-defined convective plume or column, which can result in fire whirls and/or spotting; and second, it allows for stronger winds to mix down to the surface, which results in horizontal roll vortices and higher fire spread rates (Haines, 1982).

5.2.3.2 LIGHTNING

Wildfires are caused by lightning all around the world, but only when the fuel type and moisture are favorable. Positive cloud-to-ground (CG) lightning strikes are thought to cause more ignitions than their negatively charged counterparts. Most lightning-related wildfires are thought to be caused by more energy lightning strikes. CG lightning flashes send a negative electric charge to the earth in almost 90% of cases. Negative flashes are more likely to be multi-stroked than the 10% of CG flashes which transfer positive charge to the ground in a single stroke. Single-stroke flashes are supposed to allow for a longer continuous current, making them more likely to trigger fire start. Continuing currents are found in only about half of the negative flashes (Uman, 1969). Wildfire occurrence is also influenced by the type and status of the fuel. Based on a seven-year study that used geographic information system (GIS) layers of fuel types with the lightning occurrence, Latham & Williams (2001) discovered that some fuel types are more efficient in lightning-caused ignition.

5.2.3.3 MOISTURE

Atmospheric moisture in the form of water vapor and precipitation also plays a limiting factor in the occurrence of forest fire and fire behavior that affects fuel moisture in plants both dead and living. There is also evidence that the increase in amounts of fuel moisture also acts as a retardant to decrease combustion rate, ease of ignition and preheating of fuels. In case of saturated air, an equilibrium between evaporation and condensation exits. In areas that are prone to forest fires, condensation is lesser than evaporation causing wild land fuels to lose their moisture to the ambient air through evapotranspiration (ET). Ambient air evaporation rates are known to change due to the differences among the vapor pressure between the adjacent air and a water surface as shown by Johnson & Miyanishi (2001). Indirectly, ease of temperature changes is due to existing moisture on Earth's surface which is in contact with the air. Available energy from incoming solar radiation is used in evaporation during the times of moisture ponding or pooling. Otherwise, the energy from incoming solar radiation is converted to heat particularly when the surface moisture is scarce.

5.2.3.4 WIND

Air moves due to the pressure difference in the earth's atmosphere and as a result of frictional effects near the earth's surface. Wind affects the fire occurrence and the fire behavior at regional and microclimatic scales. Mountain winds can have devastating effects on forest fires specially air is forced to plunge through narrow mountain passes. Not only does the wind speed increase, but the falling air warms adiabatically, creating hot, dry, and fire-prone circumstances. The relative diurnal heating of valleys and mountain summits is also affected by mountain and valley breezes.

5.3 FOREST FIRES

5.3.1 GLOBAL SCENARIO AND MAJOR DRIVERS

Wildfire is a worldwide phenomenon that causes havoc with all types of plants. Forest fires are becoming more severe every day as a result of a variety of natural and largely human-caused elements. The dangers of rising forest fires have now been added to the list of other natural calamities that include floods, droughts, and earthquakes. Forest fires are very common in almost all kinds of vegetation, and they occur frequently in temperate and northern boreal forests during the summer months. Fire threats cause 5–20 million hectares of forest to be burned annually in Russia and North America. The situation is similar in Mediterranean woodlands, with roughly 0.6 million hectares of forest destroyed, in South America, cyclic climate variability caused by the El Nino Southern Oscillation (ENSO) phenomenon which is responsible for extreme forest fire events. Forest Fire is a common phenomenon in tropical forest, occurring every one to five years. Tropical savannas, which cover around 2.6 billion hectares and consist of continuous spread of grass interspersed with trees and plants, are prone to fire and burn on a frequent basis. Because of human pressure, the intensity of savannas has increased in recent years. Though all countries with forests are threatened by fire at some point, vulnerabilities are more for some countries. The FAO (2010) prepared the report "Global Forest Resources Assessment (2010)," and gathered some information on fire events in forest from many countries across the world (Table 5.2). Anthropogenic activities cause more than 90% of forest fires around the world, with the bulk being tied to land use (LU) patterns. The primary noteworthy factors for such occurrences are the conversion of forestland to agriculture, pastureland, and planting.

5.3.1.1 SHIFTING CULTIVATION

Shifting cultivation is an agricultural practice in which forest land is cleared and returned to the forest after a limited period of use for crop production. The forest vegetation is cut towards the conclusion of the rainy season, allowed to dry for a period of time, and then burned. The vegetation ash remains in place, providing enough nutrients for the agricultural crop. For a few years, this cleared area was used for agriculture until crop output declined. After that, the land is left to regenerate naturally, and new forest areas are sought for agricultural purposes. Millions of individuals around the world engage in the practice, resulting in massive forest destruction.

5.3.1.2 PERMANENT CONVERSION

With the growing population pressure, the forest areas in many countries are being permanently converted to non-forestry purposes, such as pastures, agricultural land, orchards, and resettlement. Such conversion of the forest also follows after clear-felling and setting the permanent forest vegetation on fire. As the world's population grows, many countries' forests are

TABLE 5.2 Average Area of Forest Annually Affected by Fire (2005)

Region/Sub-Region	Information Availability		Area of Forest Affected by Fire	
	Number of Countries	Percentage of Total Forest Area	1,000 ha	Percentage of Forest Area
Eastern and Southern Africa	8	29.3	452	0.6
Northern Africa	5	10	17	0.2
Western and Central Africa	8	19.7	7,849	11.9
Total Africa	21	22.4	8,318	5.4
East Asia	5	100	549	0.2
South and Southeast Asia	8	83.3	1,859	0.7
Western and Central Asia	16	51.7	50	0.2
Total Asia	29	87.9	2,457	0.5
Europe excl. Russian Federation	41	96.6	270	0.1
Total Europe	42	99.4	1,262	0.1
Caribbean	7	74.1	15	0.3
Central America	4	72.6	107	0.7
North America	4	100	3,437	0.5
Total North and Central America	15	98.9	3,558	0.5
Total Oceania	6	82.5	3,903	2.4
Total South America	5	14	333	0.3

Source: Adapted from FAO (2010)

permanently transformed into non-forestry uses such as pastures, farmland, resettlement, and orchards.

5.3.1.3 OTHER CAUSES

Forest fire issues are linked to other economic and social aspects in addition to agricultural use. In many developing nations, the collection of non-timber forest produce (NTFP) such as mahua, Sal-seeds, Kendu leaves, and other NTFPs is another major source of the forest fire. Aside from economic concerns, some tribal clans' light fires to commemorate traditional customs. This sparked fire frequently spreads to the nearby forest, resulting in disaster. Aggressive fire suppression has also contributed to extreme fire outbreaks by changing the composition of fire-prone tree species. As a result, forests, particularly dry forests, are dominated by high tree densities and heavy ground fuels, making major fires far more likely. Increased fire risk in the forests is partly due to the conversion of fire-resistant ancient forests to plantations.

5.3.2 INDIAN SCENARIO

India has one of the biggest biodiversity zones in the entire world which is abundant with unique and diversified flora and fauna. With 6,92,027 km^2 of forest cover, India is one of the richest areas of biodiversity in the world. India has about 1.7 M ha of productive conifer forest which includes various valuable timber species, i.e., spruce *(Picea), fir (Abies), kail (Pinus wallichiana), deodar (Cedrus deodara), teak (Tectona grandis), Sal (Shorea robusta)* and *Chir pine (Pinus roxburghii).* These woods have an estimated increasing stock of more than 200 million cubic meters, with a monitory value of more than INR 40,000–60,000 million (Bahuguna, 1999). Forests provide about 40% of the energy and 30% of the fodder demands in a country like India, which has around 17% of the world's human population and 18% of the world's cow population. Approximately 270 million tons of fuelwood, 280 million tons of feed, more than 12 million cubic meters of lumber, and a considerable amount of NTFPs are projected to be extracted from forests each year. Forest fire incidences in India have also been tremendously increased by the increase of human interference in the natural forest (NF) ecosystem. Forest fire periodically covers large forest areas which destroy timber, wildlife, and other properties. Due to recurrent fire events which are attributed to forest degradation, soil erosion, and reduced productivity, Indian ecosystem is under severe threat.

5.3.2.1 FIRE REGIME SPREAD IN INDIAN FOREST

The Forest fire spread over areas in India causes great loss to the environment and to the property. As far as annual economic loss is concerned, it is estimated that losses from forest fires come to be approximately INR ₹ 4,400 million. The FSI (1995) estimated that about 1.45 million hectares of forest area are affected by fire annually in India. Table 5.3 provides the forest fire incidents in selected states of India.

5.3.2.2 FIRE PRONE AREAS

Forest Survey of India (FSI), a pioneer government body, has been conducting field surveys since 1965 in different parts of the country to keep records of forest fires. The State of Forest Report (1995) data on forest fire ascribes around 50% of the forest areas as fire-prone (Table 5.4).

TABLE 5.3 Incidents of Forest Fire in Selected States of India

Fire Incidents in Forest Areas of India (area in km2)									
State/District	**Forest Area**	**Sample Plots**	**Extent of Fire Incidence (area in km2)**						
			Very Heavy	**Heavy**	**Frequent**	**Occasional**	**No Fire**	**Un-recorded**	**Total**
Andhra Pradesh	14,826.71	2,037	60.58	5.75	521.99	3,335.27	10,016.34	886.78	14,826.71
Assam	15,427.88	2,482	70.91	0.00	590.25	4,551.13	10,176.68	38.91	15,826.88
Bihar	5,347.01	296	57.72	0.00	452.62	3,330.74	1,505.93	0.00	5,347.01
Himachal Pradesh	10,269.40	4,878	163.70	0.00	671.45	3,811.38	5,054.92	567.98	10,269.43
Jammu and Kashmir	3,331.75	428	7.50	0.00	60.98	1,089.58	2,088.05	85.64	3,331.75
Haryana and Punjab	1,180.72	145	0.00	0.00	41.54	332.48	806.70	0.00	1,180.72
Karnataka	13,223.30	1,780	59.71	30.33	470.64	3,342.94	9,309.79	9.89	13,223.30
Manipur	15,154	1,880	0.00	151.5	454.62	5,758.52	8,789.32	0.00	15,154.00
Madhya Pradesh	19,625.91	1,947	136.53	23.07	1838.83	10,644.29	6,983.19	0.00	19,625.91
Maharashtra	8,165.54	1,355	0.00	0,00	186.83	4,222.57	3,756.94	0.00	8,166.34
Meghalaya	9,905.66	1,659	26.75	0.00	269.12	3,347.25	5,230.91	1,031.60	9,905.63
Nagaland	14,954.91	1,128	0.00	0.00	1084.23	12,038.70	1,831.98	0.00	14,954.91
Orissa	20,143.38	2,972	204.42	78.50	923.19	11,345.34	7,258.18	333.52	20,143.16
Rajasthan	20,178.79	2,446	71.39	0.00	99.03	4,348.12	14,763.26	896.99	20,178.79
Sikkim	1,707.77	401	47.12	0.00	18.14	544.84	1,097.67	0.00	1,707.77
Tripura	6,445.36	555	34.59	0.00	361.75	5,293.65	755.37	0.00	6,445.36
Uttar Pradesh	23,164.09	2,825	871.43	0.00	2092.51	11,124.10	9,076.05	0.00	23,164.09
West Bengal	5,764.81	1,471	4.78	0.00	656.43	1,356.52	3,444.32	302.76	5,764.81
Dadra and Nagar Haveli	186.49	62	0.00	0.00	0.00	180.89	5.5947	0.00	186.49
Total	209,003.4	30,747	1,817.13	289.19	10,794.2	89,998.33	101,951.19	4,154.07	209,004.10
Percentage (%)			0.87	0.14	5.16	43.06	48.78	1.99	100.00

Source: Adapted from India Disaster Report, Lok Sabha, Un-starred Question No. 1314 (1998).

TABLE 5.4 Data on Forest Fire Prone Areas in India

Type of Fire Damage	Area of Forest in km²
Very high	5426.664
High	949.6662
Frequent	35001.9828
Occasional	292090.1898
No fire	317188.5108
Total	650657.0136

Source: Adapted from State of Forest Report of 1995 (FSI, 1995).

5.3.2.3 DRIVING FORCES (NATURAL AND ANTHROPOGENIC)

More than 95% of forest fires are in India are caused by negligence or unknowingly by human beings, the rest are caused by natural reasons such as lightning, and extremely high temperature, which in some cases are rare. In general, around the world, the primary causes of forest fires events are anthropogenic. There are basically three components that are needed for the right combustion to produce fire: fuel, heat, and oxygen. Oxygen and fuel are naturally available in forest areas, it is heat that really initiates a forest fire event. Heat may be either through natural reasons or human artificial reasons. Natural reasons for forest fires are lightning, the friction of rolling stones, rubbing of dry bamboo clumps, and volcanic explosion. Anthropogenic reasons are shifting cultivation, clearing of forest to make pathways, encroaching upon forest land, burning farm residues, throwing of burning *bidi*/cigarettes, making charcoal in forest, and fire camps from picnickers.

5.4 FOREST FIRE DETECTION AND MONITORING

5.4.1 IDENTIFICATION OF FOREST FIRE

The major challenge in controlling forest fires is early detection of wildland fires and taking appropriate rapid action on new fires from spreading to new areas. There are four stages of assessment and analysis for an effective response to wildland fires, i.e., potential risk estimation, detection of fire starts, active fire monitoring, and assessment of post-fire degradation.

5.4.1.1 FIRE WATCHTOWERS

Fire lookout towers or watchtowers are generally located at high elevations inside the forest. A person has been given the responsibility to look for fires using binoculars or an Osborne fire finder (Fleming & Robertson, 2003). This system is very much uncertain and unreliable.

5.4.1.2 OPTICAL SMOKE DETECTION

Sensors capable of measuring radiation heat can be used for detecting fire or smoke inside the forest area with a smaller periphery. It is very difficult to install and maintain sensors over a large area as most of the forest is located in remote areas (Nagaraj & Vijayakumar, 2012). The other limitation includes the ability of the sensor would detect heat or smoke only when it reaches nearer.

5.4.1.3 AUTOMATIC VIDEO SURVEILLANCE SYSTEMS

Automatic video surveillance systems use charged coupled device (CCD) cameras and infrared (IR) image sensors that contain an array of light-sensitive capacitors or photodiodes and detect with a range of 10 km scanning diameter in less than 8 minutes, which is otherwise very difficult for a fire lookout personnel (Fire Watch Web Page; Khrt et al., 2001). This system is prevalent in Germany, Canada, and Russia. Clouds, light reflection, and smoke from industrial activities influenced the detection of these systems and limited their accuracy.

5.4.1.4 AERIAL PATROLLING

Unmanned aerial vehicles (UAV) or Aeroplane are used to monitor large forest areas as Automatic video surveillance systems is generally useful for small forest fields (up to 10 km). During aerial patrolling in an airplane over forests, the pilot warns the base station in case of fire or smoke activity. UAVs can fly atop the forest for several hours and transmit the collected data using both video and IR cameras to a base station on the ground that could be up to 50 km away (Aerovision Web Page).

5.4.1.5 SATELLITE REMOTE SENSING (RS) IN FIRE DETECTION

Conventional ground-based methods have severe limitations to monitor forest fires in remote and inaccessible forest areas. Near real-time RS data can provide reliable and impartial observations of fire activity over large regions (Badarinath et al., 2007). Several satellite sensors used for forest fire studies are listed in Table 5.5.

Satellite systems mainly used thermal sensors like AVHRR for fire detection. The Defense Meteorological Satellite Program (DMSP), optical line-scan system (OLS) sensor with daily revisit overflights and regularly collects images in visible bands during its night-time pass. NOAA Geostationary Operational Environmental Satellite (GOES) sensor provides imageries in visible and thermal band every 15 minutes of the USA and every 30 minutes elsewhere. But the most extensively used sensor for monitoring and detecting wildfires is AVHRR.

There are several research studies on fire detection through capturing camera segments and processing and image processing and classification. Roy (2003) used satellite images from forests and detected fire possibility. The researcher offered a fully automated method of forest fire detection from TIR sensor data using the random field theory involving pre-processing of the images to model the Gaussian field. moderate resolution imaging spectroradiometer (MODIS) based multi-year daily active fire product having a good detection rate with four daily overpasses has been developed (Giglio et al., 2003). Some global efforts for assessing fire dynamics are MODIS Rapid Response System Global Fire Maps, GLOBSCAR, whereas some regional and national initiatives include European Forest Fire Information System, WEB FIRE MAPPER, GOFC/GOLD Fire Monitoring and Mapping Implementation Team, Canadian Wildland Fire Information System and Indian Forest Fire Response and Assessment System (INFFRAS). For early warning of forest fire, a Bio-Ad hoc sensors network system has been developed by the researchers for mountain areas, where animals are used as wireless Ad hoc nodes.

Among many techniques for forest fire detection used by authorities, satellite detection is the most common and widely used technique across the globe. Table 5.6 highlights the various techniques and their comparison amongst them.

5.4.1.6 NEAR REAL-TIME FOREST FIRE DETECTION SYSTEM

FSI gets near real-time geospatial data of forest fires from the National Remote Sensing Center (NRSC). MODIS and SNPP-VIIRS satellite sensor data are used to generate the active fire hotspots. FSI prepares the newest

TABLE 5.5 Major Satellite Systems Used for Fire Assessment from Global to Local Scale

Sensor	Major Application	Spatial Resolution	Swath Width	Bands (μm)	Major Advantages
BIRD	Active fire detection and burnt area assessment.	185 m	190 km	MIR 3.4–4.2 TIR 8.5–9.3	Specialized satellite, nighttime data.
MODIS	Active fire detection and burnt area assessment.	250 m 500 m 1 km	2,330 km	36 bands including 3.9 and 11 μm.	Easily accessible, high spatial and temporal resolution.
AVHRR	Active fire detection and burnt area assessment.	1.1 km 0.56 km	2,400 km 3,000 km	0.58–0.91 10.3–12.9 0.58–0.91	High temporal frequency.
DMSP-OLS	Active fire detection	–	–	–	High temporal resolution, night time data.
GOES-10 and 12	Active fire detection.	1 km 4 km 8 km	Hemisphere	0.55–0.75 6.50–7.00 10.20–11.20 11.5–12.5	Very high temporal resolution.
ATSR	Active fire detection and burnt area assessment.	1 km	500 km	3.51–3.89 1.57–1.63 10.4–11.3 11.5–12.5	Good spectral configuration.
IRS-AWiFS	Active fire detection, burnt area assessment and validation.	56 m	370 km	0.52–0.59 0.62–0.68 0.77–0.86 1.55–1.70	SWIR band, high spatial resolution.
Landsat TM and ETM	Active fire detection, burnt area assessment and validation.	30 m	185 km	0.45–0.52 0.52–0.60 0.63–0.69 0.76–0.90 1.55–1.75 10.4–12.5 2.08–2.35	MIR band, high spatial resolution

Source: Adapted from Joseph et al. (2009).

forest cover map of India where the active fire coordinates from NRSC are superimposed to exclude any fire hotspots coming from agricultural, industrial or any sources other than forest. The active fire coordinates are then joined with different attributes like state, district, and Survey of India 1:50000 toposheet's numbers as forest fire information. The refined forest fire information is then converted into *.KML (keyhole mark-up language) format and is sent to the selected nodal officers of each state. During the fire season, i.e., December to next year June, forest fire alerts of detection and dissemination are done twice daily through SMS messages to every registered user in the country. FSI started this activity in 2004 (FSI Annual Report, 2019).

5.5 FOREST FIRE STATISTICS IN INDIA

Global burnt area (BA) information is increasingly used in climate and dynamic vegetation models for appropriate representation of the growing interaction between fire and climate (Mouillot et al., 2014). The extended use of these models emphasizes the requirement of consistent long-term time-series data of Burned Area, with actual error and uncertainty characterization. Chirici & Corona (2005) showed the mixed signature in BA pixels with average reflectance similar to unburnt ground coverage for imagery with a coarse spatial resolution (pixel size between 180 m and 1 km). According to Lentile et al. (2006), IRS AWiFS sensor is capable of estimating the BA map of Mediterranean Europe with improved BA demarcations and small burned scar numbers, which are otherwise not detected in low-resolution sensor data like MODIS. The study on forest fires of Uttarakhand hills shows a BA assessment for April 2016 to the forest vegetation with a BA of 3774.14 km^2, comprising 15.28% of the total forest area of the state (Gupta et al., 2018). The study conducted by the European Space Agency (ESA) on global forest BA using SPOT-VGT (spatial resolution: 1 km) reported BA as 9% of the total forest area of India in 2000 (IFFN, 2002). The study of Harikrishna et al. (2012) reconfirmed that coarse-scale detection does not represent the actual area figures of small-BAs. Different studies on Indian forest fires have utilized either coarse resolution data (Giriraj et al., 2010; Badarinath et al., 2011) or moderate to high-resolution data, focused to certain states or districts (Reddy et al., 2009, 2012, 2014; Sharma et al., 2012).

Reddy et al. (2017) assessed countrywide forest BA using ResourceSat-2 AWiFS data for the year 2014 (5.1). They found that the total BA under

TABLE 5.6 Comparison of Different Forest Fire Detection Techniques

Comparison	SBT	CBT	NBT	SaBT	ABT	FBT	AMS	RBT
Cost	Low	High	Low	Very high	High	Low	High	Medium
Practicality	High	Medium	Low	Very high	High	Low	Very low	Very low
Frequency of use in literature.	High	Medium	Rare	Very high	Medium	Rare	Very rare	Very rare
Detection device mobility.	Fixed	Fixed	Fixed or mobile	Mobile	Mobile	Fixed or mobile	Mobile	Fixed
Battery power provided through.	RD	RD	RD	Solar	RD	RD	RD	RD
Information on fire behavior.	Yes	No	Yes	Yes	Yes	Yes	Yes	No
Detection to notification delay.	Small	Long	Small	Very long	Long	Small	Long	Small
False alarm	Low	Medium	Medium	Very low	Medium	Medium	High	High
Fire localization error.	Low	High	Low	Very high	High	Low	High	High

Abbreviations: SBT: Sensor based techniques (wireless sensor networks); CBT: camera based technique (image and video processing); NBT: neural networks based technique; SaBT: satellite based technique; ABT: air borne techniques or UAV; FBT: fuzzy logic based techniques; AMS: animals as mobile sensors; RBT: radio-acoustic based techniques; and RD: rechargeable devices.

Source: Adapted from Chowdary et al. (2018).

the state of Odisha had a maximum BA (8186.46 sq. km). However, out of the total forest cover, the maximum BA was observed in Andhra Pradesh (25.43%) followed by Bihar, Tamil Nadu, Odisha, and Telangana (Figure 5.1 and Table 5.7).

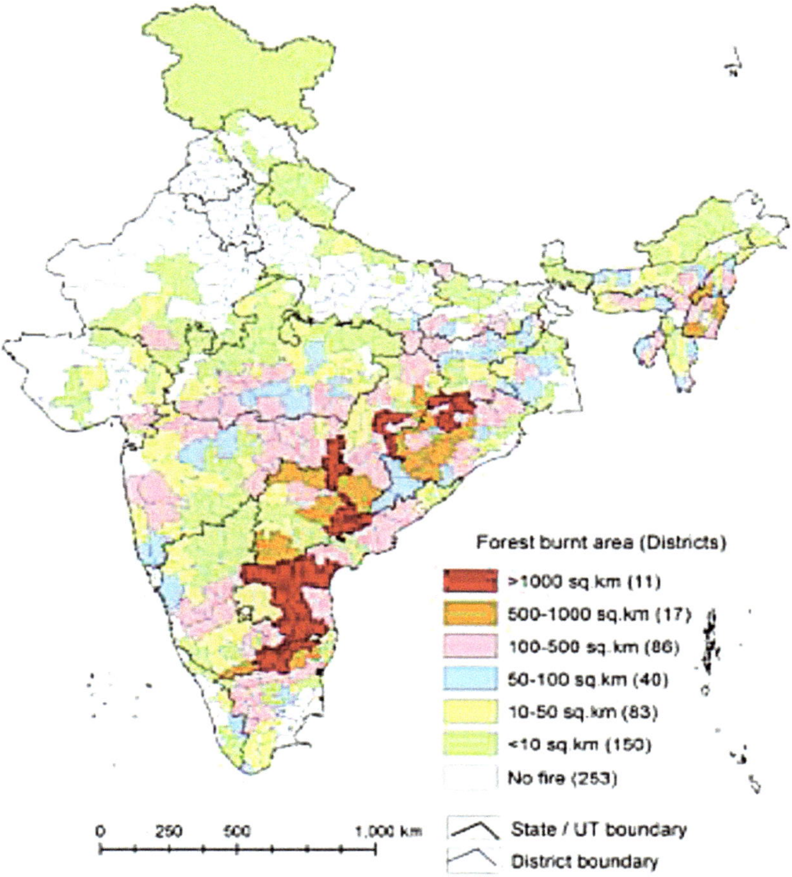

FIGURE 5.1 Map of districts wise total forest burnt area for the year 2014.

Source: Modified from Reddy et al. (2017).

Trans-Himalayas and Islands were lacking forest fires due to prevalent climatic conditions and moist vegetation. Maximum numbers of forest fires observed in North-Eastern parts of India are due to shifting cultivation. The classification accuracy and the kappa statistics of the produced maps were calculated as 91.46% and 0.89, respectively based on 500 ground control points.

TABLE 5.7 State-Wise Total Forest Burnt Area in India

State/UT	Forest Burnt Area (sq. km)
Odisha	8186.46
Andhra Pradesh	6611.86
Maharashtra	5066.66
Chhattisgarh	4606.69
Tamil Nadu	4275.65
Madhya Pradesh	3342.66
Telangana	2955.23
Jharkhand	2587.40
Manipur	1974.23
Karnataka	1920.35
Bihar	1773.22
Nagaland	975.79
Assam	941.11
Tripura	739.00
Gujarat	487.81
Uttar Pradesh	459.07
Meghalaya	457.50
Mizoram	421.03
West Bengal	386.37
Rajasthan	364.17
Arunachal Pradesh	102.7
Kerala	82.01
Uttarakhand	42.01
Total	48,758.98

5.6 FOREST FIRE ALERT SYSTEM

5.6.1 OUTREACH

The National Forest Inventory program of FSI showed that the heavily affected and mildly affected area due to forest fires comprises 9.89% and 54.40% of forest areas which indicates almost two-thirds of our forest areas are vulnerable to forest fires. There has been significant improvement in the user base of a forest fire alerts, which has increased significantly from around 2000 in 2017 to more than 66,000 in 2019 depicting the usefulness of the service particularly in those states where customized alerts are being circulated to the administrative jurisdiction of field officials. The total number of alerts based on MODIS and SNPP-VIIRS sensors are 29,547 and 2,10,286, respectively from November 2018 to June 2019.

5.6.2 EARLY WARNING AND DANGER RATING

According to Merrill & Alexander (1987), fire danger rating quantifies the capability or potential of a fire to start, spread, and do destruct and thus has long been used to monitor a variety of fire management problems. There are different formal fire danger rating systems (FDRS) that existed in Canada, Australia, and the United States which are constantly upgrading themselves for about 75 years (Deeming et al., 1977; Luke & McArthur, 1978; Stocks et al., 1989).

5.6.3 FOREST FIRE DANGER RATING SYSTEMS (FDRS)

Prominent forest fire danger rating systems (FDRS) in the world employed in the world is listed below:

1. **Canadian Forest Fire Danger Rating System (CFFDRS):** Originally, CFFDRS was developed for Canada. However, a set of customizable constituents forms this System the only one in the world capable of being adapted to any scale, from local to global. Thus, this rating system adopted by several other countries most notably New Zealand, Fiji, Mexico, Alaska, Florida, Southeast Asia nations, Croatia, Russia, Chile, and the State of Michigan (USA).
2. **National Fire Danger Rating System (NFDRS) in USA:** It includes three different types of processors like weather information management system (WIMS), Fire Weather Plus and Weather Pro, and Fire Family Plus used to produce the desired outputs of the NFDR.
3. **Australian Forest Fire Danger Rating System:** This system was developed based on the projected rate of spread for a fire burning on level ground in a standard dry forest and the difficulty level to control fires under different weather conditions. This Fire Danger Rating System has been used as the standard forest fire danger rating system in eastern Australia since the late 1950s.
4. **Other Forest Fire Danger Rating Systems (FDRS):** Several countries trying to develop their own FDRS from the three systems mentioned above. With slight modification, New Zealand has adopted the CFFDRS.

5.6.4 FOREST FIRE DANGER RATING SYSTEMS (FDRS) IN INDIA

5.6.4.1 EVOLUTION OF NEAR REAL-TIME FOREST FIRE ALERTS

The FSI Forest Fire Alerts system through periodic changes has upgraded itself to assist foresters as well as common people in a superior way. FSI Fire Alerts System (known as FAST 3.0) was launched on 16th January 2019 through the Large Forest Fire Monitoring Program using near real-time SNPP-VIIRS data. The major objectives were to point out, trace, and delineate serious forest fire events and at the same time to help monitor and control such fires at the senior level in the State Forest Department.

5.6.4.2 FEATURES OF THE FSI FIRE ALERT SYSTEM

Features in forest fire alert system 3.0 (FAST 3.0) or FSI fire alert system is given below:

- Large forest fire monitoring program using SNPP-VIIRS sensor data to automatically identify and track large forest fire events;
- FSI forest fire geoportal;
- Customized alerts for 20 states (beat level) and 2 states (range level);
- System for improved feedback;
- Improved accuracy through customization in filter, masking with forest cover, rationalization of trees outside forest layer;
- Improved and updated database through Map server-based web portal (open source) for a dynamic alert display to state forest departments.

5.6.4.3 GENERATION AND DISSEMINATION OF FOREST FIRE ALERTS

The process of generation and diffusion of forest fire alerts are described below:

- NRSC provides the active fire spots or hotspots in their ground station at Shadnagar, Telangana that are shared with FSI through email.
- NRSC provides fire alerts comprising all thermal anomalies identified by the sensors both within and outside forests. FSI then filters active fire spots outside the forest using a custom filter. Different

attributes like State, District, Division, Range, Beat, Compartment boundaries, etc., are added to the active forest fire spots to enrich forest fire information.

- The processed information is then sent to State Nodal Officers and registered users and uploaded on the FSI website.
- The registered users who have specified their areas of interest get alerts of forest fires through SMS.

5.6.4.4 ADVANTAGES OF REMOTE SENSING (RS) AND GEOSPATIAL TECHNOLOGY

- Remote sensors can measure the detectable changes in electromagnetic (EMR) spectrum before (standing vegetation) and after combustion (ash and char) due to changes in energy in the form of light and heat during the burning process. Since the 1970s, different Earth-observing satellite sensors have been deployed to detect these variations, compute the BAs, and determine the composition and distribution of by-products after the fire (Robinson, 1991; Fuller, 2000).
- Satellite remote sensing can detect spatial and temporal forest fire and burnt scars of earlier burning events up to three weeks after fire incidence (Eva & Lambin, 1998).
- Remote sensing being a cost-effective and timesaving method can identify the active fire location, fire events, and the extent of the fire-affected area (Chuvieco et al., 1999).
- Satellite imagery integrated with ground truth points provide valuable information regarding pre-fire vegetation conditions and thus greatly help in fire damage assessment. Pre-fire condition is essential to assess probable fire risk and to make logical judgments on the preventive measures required to keep away potential soil erosion, which topographic and meteorological conditions may encourage on newly burnt land (Chuvieco & Martin, 1994).

5.6.4.5 DISADVANTAGES OF REMOTE SENSING (RS) AND GEOSPATIAL TECHNOLOGY

- Weather conditions like clouds and rain mostly absorb parts of the frequency spectrum and affect the accuracy and reliability of satellite-based systems.

- Low resolution of satellite imagery cannot detect small fire, so these satellite-based systems can monitor a large area only when forest fire has grown large.
- More importantly, the long scan period (e.g., 2 days) cannot provide timely detection.
- FSI generates the forest fire alerts based on only six satellite overpasses in a day and thus cannot detect events of fire activity that do not overlap with the satellite overpasses.
- The time delay (1–1.5 hr.) between fire detection by the satellite and dissemination of the alert restricts the usefulness of satellite detection for tactical fire operations.
- It is very difficult to determine the actual size of the active burning spots from satellite imagery as a pixel size of 1-km^2 may symbolize a fire as small as 100 m^2.

Hence, the exhortation to use the FSI forest fire alerts should be for strategic and not for tactical firefighting purposes. An integrated forest fire detection approach by combining all other sources of detection such as ground-based sensors, watchtower observations, local information, etc., should be developed.

5.7 CONCLUSION

The use of RS and GIS for wildland fire detection, monitoring, and management is growing rapidly. There have been many successes in detecting fires and mapping fire areas, especially for large fires and over inaccessible areas. When combined with field data, RS can be very helpful in mapping and analyzing both active fire characteristics and post-fire effects. RS has a great potential for scholars and managers seeking to map, understand, predict, and assess the ecological effects of the forest fire. Remotely sensed data can also be useful for quantifying carbon released by fire as well as estimating increases in vegetation growth and carbon sequestration post-fire. These technologies can be useful for understanding the fire dispersion by identifying the fire smoke, its location, and emissions parameters being tied to weather and dispersion models. Overall, the use of geospatial technologies provides an ecosystem for lot of applications regarding the management of forest fires as a whole.

KEYWORDS

- **AVHRR**
- **burned areas**
- **forest degradation**
- **geographical information systems**
- **near real-time forest fire alerts**
- **optical line scan system**
- **remote sensing**

REFERENCES

Badarinath, K. V. S., Kharol, S. K., & Kiran, C. T. R., (2007). Use of satellite data to study the impact of forest fires over the northeast region of India. *IEEE Geoscience and Remote Sensing Letters, 4*, 485–489.

Badarinath, K. V. S., Sharma, A. R., & Kharol, S. K., (2011). Forest fire monitoring and burnt area mapping using satellite data: A study over the forest region of Kerala State, *India. Int. J. Remote Sens., 32*, 85–102.

Bahuguna, V. K., (1999). Forest fire prevention and control strategies in India (No. 20, pp. 5–9). International Forest Fire News.

Bridgman, H. A., & Oliver, J. E., (2006). *The Global Climate System*. Cambridge University Press, Cambridge.

Chand, K. T. R., Badarinath, K. V. S., Prasad, K. V., Murthy, M. S. R., Elvidge, C. D., & Tuttle, B. T., (2006). Monitoring forest fires over the Indian region using DMSP-OLS night-time satellite data. *Remote Sens. Environ., 103*, 165–178.

Chirici, G., & Corona, P., (2005). An overview of passive remote sensing for post-fire monitoring. *Forest, 2*(3), 282–289.

Chowdary, V., Gupta, M. K., & Singh, R., (2018). A review on forest fire detection techniques: A decadal perspective. *Int. J. Engg. Tech., 7*(3), 1312–1316.

Chuvieco, E., & Martin, M. P., (1994). A simple method for fire growth mapping using AVHRR channel 3 data. *Int. J. Remote Sens., 15*, 3141–3146.

Chuvieco, E., Deshayes, M., Stach, N., Cocero, D., & Riano, D., (1999). Short-term fire risk: Foliage moisture content estimation from satellite data. In: Chuvieco, E., (ed.), *Remote Sensing of Large Wildfires in the European Mediterranean Basin* (pp. 17–38). Berlin, Springer-Verlag.

Deeming, J. E., Burgan, R. E., & Cohen, J. D., (1977). *The National Fire-Danger Rating System.* USDA For. Serv. Intermt. For. and Range Exp. Stn., Ogden, Utah. Gen. Tech. Rep. INT-39.

Eva, H., & Lambin, E. F., (1998). Burnt area mapping in Central Africa using ATSR data. *Int. J. Remote Sens., 19*, 3473–3497.

FAO, (2010). *Global Forest Resources Assessment, Main Report*. FAO Forestry Paper 163, Rome, Italy. Available at: www.fao.org/forestry/fra/fra2010/en/ (accessed on 28 December 2023).

Flannigan, M. D., & Harrington, J. B., (1987). Synoptic conditions during the Porter Lake burning experiment. *Climatol. Bull., 21*, 19–40.

Flannigan, M. D., Logan, K. A., Amiro, B. D., Skinner, W. R., & Stocks, B. J., (2005). Future area burned in Canada. *Climate Change, 72*, 1–16.

Fleming, J., & Robertson, R. G., (2003). *Fire Management Tech Tips: The Osborne Fire Finder*. Tech. Rep. 0351 1311-SDTDC, USDA Forest Service, October.

FSI, (1995). *India State of Forest Report*. Ministry of Environment, Forests and Climate Change, Government of India.

FSI, (2017). *India State of Forest Report*. Ministry of Environment and Forests, Government of India.

FSI, (2019). *Annual Report, Technical Information Series* (Vol. I, No. 2), Ministry of Environment and Forests, Government of India.

FSI, (2019). *India State of Forest Report*. Ministry of Environment and Forests, Government of India.

Fuller, D. O., (2000). Satellite remote sensing of biomass burning with optical and thermal sensors. *Prog. Phys. Geogr., 24*, 543–561.

Giglio, L., Descloitres, J., Justice, C. O., & Kaufman, Y., (2003). An enhanced contextual fire detection algorithm for MODIS. *Remote Sens. Environ., 87*, 273–282.

Girardin, M. P., Bergeron, Y., Tardif, J. C., Gautheir, S., Flannigan, M. D., & Mudelsee, M., (2006). A 229-year dendroclimatic-inferred record of forest fire activity for the Boreal Shield of Canada. *Int. J. Wildland Fire, 15*(3), 375–388.

Giriraj, A., Shilpa, B., Jentsch, A., Sudhakar, S., & Murthy, M. S. R., (2010). Tracking fires in India using advanced along track scanning radiometer (A) ATSR data. *Remote Sens., 2*, 591–610.

Gupta, S., Roy, A., Bhavsar, D., Kala, R., Singh, S., & Kumar, A. S., (2018). Forest fire burnt area assessment in the biodiversity rich regions using geospatial technology: Uttarakhand Forest fire event 2016. *J. of the Ind. Soc. of Remote Sen., 46*, 945–955.

Haines, D. A., (1982). Horizontal roll vortices and crown fires. *J. Appl. Meteorol., 21*, 751–763.

Harikrishna, P., & Reddy, C. S., (2012). Assessment of increasing threat of forest fires in Rajasthan, India using multi-temporal remote sensing data (2005–2010). *Curr. Sci., 102*(9), 1288–1297.

Herweijer, C., Seager, R., Cook, E. R., & Emile-Geay, J., (2007). North American droughts of the last millennium from a gridded network of tree-ring data. *J. Clim., 20*(7), 1353–1376.

IFFN, (2002). *Fire Situation in India* (pp. 23–27). https://www2.fire.uni-freiburg.de/iffn/country/in/in_5.htm (accessed on 28 December 2023).

India Disaster Report, Lok Sabha, Unstarred Question No. 1314, (1998).

IPCC, (2007). *Summary for Policymakers, in Climate Change 2007: Impacts, Adaptation and Vulnerability* (p. 17). Contribution of Working Group II to the Fourth Assessment Report of the Intergovernmental Panel on Climate Change. Cambridge University Press, Cambridge, UK.

Johnson, E. A., & Miyanishi, K., (2001). *Forest Fires: Behavior and Ecological Effects* (p. 594). Academic Press, San Diego, CA.

Joseph, S., Anitha, K., & Murthy, M. S. R., (2009). Forest fire in India: A review of the knowledge base. *J. For Res., 14*, 127–134.

Kasischke, E. S., Williams, D., & Barry, D., (2002). Analysis of the patterns of large fires in the boreal forest region of Alaska. *Int. J. Wildland Fire, 11*(2), 131–144. doi: 10.1071/WF02023.

Khrt, E., Knollenberg, J., & Mertens, V., (2001). An automatic early warning system for forest fires. *Annals of Burns and Fire Disasters, 14,* 3.

Latham, D., & Williams, E., (2001). Lightning and forest fires. In: Johnson, E. A., & Miyanishi, K., (eds.), *Forest Fires: Behavior and Ecological Effects* (pp. 375–418). Academic Press, San Diego, CA.

Lentile, L. B., Holden, Z. A., Smith, A. M., Falkowski, M. J., Hudak, A. T., Morgan, P., & Benson, N. C., (2006). Remote sensing techniques to assess active fire characteristics and post-fire effects. *Int. J. Wildland Fire, 15*(3), 319–345.

Luke, R. H., & McArthur, A. G., (1978). *Bushfires in Australia* (p. 359). CSIRO Div. For. Res. Australian Gov. Publ. Serv., Canberra.

Maingi, J. K., & Henry, M. C., (2007). Factors influencing wildlife occurrence and distribution in eastern Kentucky, USA. *Int. J. Wildland Fire, 16*, 23–33.

Merrill, D. F., & Alexander, M. E., (1987). *Glossary of Forest Fire Management Terms* (4th edn.). National Research Council of Canada, Ottawa. NRCC No. 26516.

Nagaraj, B., & Vijayakumar, P., (2012). Controller tuning for industrial process-a soft computing approach. *Int. J. Advance. Soft Comput. Appl., 4*(2).

Notaro, M., Vavrus, S., & Liu, Z., (2007). Global vegetation and climate change due to future increases in CO_2 as projected by a fully coupled model with dynamic vegetation. *Ontario. Can. J. For. Res., 21*, 277–287.

Pyne, S. J., Andrews, P. L., & Laven, R. D., (1996). *Introduction to Wild Land Fire* (2nd edn., p. 769). Wiley.

Reddy, C. S., Harikrishna, P., Anitha, K., & Joseph, S., (2012). Mapping and inventory of forest fires in Andhra Pradesh, India: Current status and conservation needs. *ISRN Forestry*.

Reddy, C. S., Jha, C. S., Manaswini, G., Alekhya, V. V. L. P., Pasha, S. V., Satish, K. V., Diwakar, P. G., & Dadhwal, V. K., (2017). Nationwide assessment of forest burnt area in India using resourcesat-2 AWiFS data. *Curr. Sci., 112*(7), 1521–1532.

Reddy, C. S., Khuroo, A. A., Hari, K. P., Saranya, K. R. L., Jha, C. S., & Dadhwal, V. K., (2014). Threat evaluation for biodiversity conservation of forest ecosystems using geospatial techniques: A case study of Odisha, India. *Ecol. Eng., 69*, 287–303.

Reddy, C. S., Navatha, K., Rachel, B., Murthy, M. S. R., & Manikya, R. P., (2009). Forest fire monitoring in Sirohi district, Rajasthan using remote sensing data. *Curr. Sci., 97*(9), 1287–1290.

Robinson, J. M., (1991). Fire from space: Global fire evaluation using infrared remote sensing. *Int. J. Remote Sens., 12*, 3–24.

Roy, P. S., (2003). Forest fire and degradation assessment using satellite remote sensing and geographic information system, satellite remote sensing and GIS application in agricultural meteorology. *Proceedings of a Training Workshop* (pp. 361–400). Dehra Dun, India.

Satendra, & Kaushik, A. D., (2014). *Forest Fire Disaster Management*. National Institute of Disaster Management, Ministry of Home Affairs, New Delhi. Available at: https://nidm.gov.in/pdf/pubs/forest%20fire.pdf (accessed on 28 December 2023).

Sedano, F., Kempeneers, P., Strobl, P., Mclnerney, D., & San, M. J., (2012). Increasing spatial detail of burned scar maps using IRS–AWiFS data for Mediterranean Europe. *Remote Sens., 4*(3), 726–744.

Sharma, R. K., Sharma, N., Shrestha, D. G., Luitel, K. K., Arrawatia, M. L., & Pradhan, S., (2012). Study of forest fires in Sikkim Himalayas, India using remote sensing and GIS techniques. *Climate Change in Sikkim- Patterns, Impacts, and Initiatives*, 233–244.

Soja, A. J., Sukhinin, A., Cahoon, D. R., Shugart, H. H., Stackhouse, P. W., & Conard, S. G., (2002). *Fire Frequency, Distribution, and Area Burned in Siberia Described Using AVHRR-derived Products*. American Geophysical Union, Spring Meeting abstract A22E-09.

Stephens, S. L., (2005). Forest fire causes and extent on United States Forest Service lands. *Int. J. Wildland Fire, 14*, 213–222.

Stocks, B. J., Lawson, B. D., Alexander, M. E., Van, W. C. E., McAlpine, R. S., Lynham, T. J., & Dube, D. E., (1989). The Canadian Forest fire danger rating system: An overview. *For Chron., 65*, 450–457.

Swetnam, T. W., & Betancourt, J. L., (1998). Mesoscale disturbance and ecological response to decadal climatic variability in the American Southwest. *J. Clim., 11*(12), 3128–3147.

Ting, M., & Wang, H., (1997). Summertime U.S. precipitation variability and its relation to Pacific Sea surface temperature. *J. Clim., 10*(8), 1853–1873.

Tymstra, C., Flannigan, M. D., Armitage, O. B., & Logan, K., (2007). Impact of climate change on area burned in Alberta's boreal forest. *Int. J. Wildland Fire, 16*, 153–160.

Uman, M. A., (1969). *Lightning*. McGraw-Hill, New York.

Westerling, A. L., Hidalgo, H. G., Cayan, D. R., & Swetnam, T. W., (2006). Warming and earlier spring increase western U.S. Forest wildfire activity. *Science, 313*, 940–943.

CHAPTER 6

Agroforestry Systems for Ecosystem Services in India

S. B. CHAVAN,[1] A. R. UTHAPPA,[2] A. R. CHICHAGHARE,[3] DHIRAJ KUMAR,[4] CHHAVI SIROHI,[5] and VIJAYSINH D. KAKADE[1]

[1]*ICAR–National Institute of Abiotic Stress Management, Malegaon, Baramati, Pune, Maharashtra, India*

[2]*ICAR–Central Coastal Agricultural Institute, Old Goa, Goa, India*

[3]*Department of Silviculture and Agroforestry, Kerala Agricultural University, Thrissur, Kerala, India*

[4]*ICAR–Indian Institute of Soil Science, Bhopal, Madhya Pradesh, India*

[5]*Department of Forestry, CCS Haryana Agricultural University, Hisar, Haryana, India*

ABSTRACT

Ecosystem services (ES) are the potential instrument for providing an incentive to farmers in terms of tangible benefits (timber, fodder, fuel, food, NTFP, and crop productivity) and intangible benefits (soil fertility, moisture conservation, microclimate moderation, and check soil erosion). Existing research studies indicate that the incorporation of agroforestry in a land-use can prevent degradation of the environment, enhance agricultural productivity, sequester carbon, enhance the quality of water, alleviate soil health and maintain health and quality of ecosystems along with generating steady incomes and additional advantages for wellbeing of human. The quantification of different services (goods and functions) is a complex task due to the absence of standard methodologies. The research in the area of ES is lacking,

Sustainable Management and Conservation of Environmental Resources in India.
Sumit Chakravarty, Vineeta, Jahangeer A. Bhat, Munesh Kumar, Gopal Shukla (Eds.)

but the scope of the study is very wide. The present review emphasizes the basic understanding of ES rendered by different tree-based systems and an overview of their various types of valuation methods. Despite the tremendous potential value of these services to society, farmers rarely compensate for them. When landowners are compensated for providing these non-market services, agroforestry systems will become more economically viable. For stimulating farmers and landowners to adopt these tree-based systems, which is more socially advantageous, new approaches towards accounting for externalities such as payment of ecosystem services (PES) or other incentives of land use (LU) should be strengthened. Additionally, agroforestry may become a financially luring option among farmers, policymakers, and other stakeholders if their ecosystem services are better valued, accounted, and incorporated into formal decision-making procedure.

6.1 INTRODUCTION

Existence of human life is solemnly dependent on nature and its processes. A. G. Tansley coined the term ecosystem in the year 1935 and became a myriad in ecological studies. Ecosystems serve humanity in a variety of ways from cradle to grave. The value of ecosystem is dimensionless and cannot be counted. The services and goods offered by the natural ecosystem are commonly known as ecosystem services (ES). Various biological properties, ecological processes, and habitat are all referred to as ecosystem functions. Ecosystem goods and services are the advantages that humans gain from ecosystem functions, either directly or indirectly. This Ecological services and the natural capital stocks that provide them are crucial to the sustaining life-support system on earth. They contribute to human well-being in both direct and indirect ways, and hence form part of the planet's overall economic worth. The excellent and popular book of Daily (1997) on "*Nature's Services: Societal Dependence on Natural Ecosystems*" was based on ecosystem functions and services, where "The circumstances and processes by which natural ecosystems, and the species that make them up, maintain, and satisfy human existence" is called ESs. Natural ecosystems provide numerous necessary items to human societies, such as seafood, game animals, feed, fuelwood, lumber, and pharmaceuticals. These goods are crucial and forms well-known economic components. Costanza et al. (1997) published very key research on the value of the world's ESs and natural capital in The Nature Publishing house. They estimated that the economic value of 17 ESs for 16 biomes was ranged between US $16 trillion to US $54 trillion (1,012) annually, with mean value of US $33 trillion annually, almost twice the gross world product (sum of the gross national product of all nations – that is, US $18 trillion).

According to the FAO's Global Forest Resources Assessment–2015 report (FAO, 2018), the world's forests are shrinking as human populations expand and demand for food and land rises. However, the pace of net forest loss has decreased by more than half. In 1990, the world's forest acreage was 4,128 million ha; by 2015, it had shrunk to 3,999 million ha. This is a change from 31.6% in 1990 to 30.6% in 2015. From 2010 to 2015, reported natural forest (NF) area decreased by a net 6.5 million ha per year. This reduction in the world forest has directly reduced the ESs of forest. This reduction of forest alarmed various natural calamities such as devastation of land by the cyclone, floods, and drought. In such a difficult climatic situation, role of planted trees provides a viable solution for the conservation of natural resources. In this context, agroforestry is being considered as a possibility for enhancing tree-crop interactions. Agroforestry holds a lot of promise for minimizing deforestation and forest degradation, supporting rural livelihoods and habitat for species outside forest, and relieving resource demand on conservation areas. However, a variety of issues, including design aspects of prospective agroforestry innovations, perceived requirements, regulations, availability, and distribution of components of production, and risk perception, continue to limit broad adoption of agroforestry innovations. The acceptance and adoption of agroforestry systems have increased over the last decade. For promoting ESs, these tree-based systems are tested ways as they are environment-friendly and ecologically sustainable land-use practice. Hence, the review is presented to understand the role of agroforestry in ESs.

6.2 AGROFORESTRY AND ESS

Since time immemorial trees are the basis of life and they are forming a dominant component in agroforestry. From the arctic to the south temperate area, agroforestry has been practiced for centuries, although it is most prevalent in the tropics. It is a land-use management system that blends trees on farms along with animals. Agroforestry goods and services directly support about 1.2 billion people (almost 20% of the world's population) in urban and rural regions of developing nations (Dagar & Tewari, 2016).

Agroforestry is a contemporary strategy for developing sustainable land use (LU) and increasing food production by combining agricultural crops and/or animals with woody species (bamboos, shrubs, palms, trees, etc.) in some type of spatial arrangement or temporal sequence. The science behind the evolution of agroforestry practices is the 'biomimicry of nature' and 'ecological engineering.' Jones et al. (1997) explained the science and role of trees in the ecosystem, which affect the availability of resources to other organisms directly or indirectly by generating changes in physical state in biotic or abiotic materials. The nature of agroforestry systems helps to generates various ESs as generated by forest ecosystem. As agroforestry is the systematic science of growing trees for six Fs, i.e., Fodder, fruit, fiber, fuel, fertilizer, and food (Chavan et al., 2015a, 2016a). The positive interactions among different components can generate spillway of numerous benefits. The quantification of ESs from agroforestry is still in the infant stage due to the unavailability of standard methodologies and complexity.

Human reliance on Earth's ecosystems dates back to the dawn of *Homo sapiens*. The phrase "environmental services" was first used in a report of the Study of Critical Environmental Problems published in 1970, which identified services such as insect pollination, fisheries, climate management, and flood control. Variations of the name were used in subsequent years, but 'ESs' finally established the norm in scientific literature. But credit goes to Ehrlich & Ehrlich (1981) to coin the term "ESs." The benefits that the natural environment offers to humans are referred to as ESs. As recalled by Mooney & Ehrlich (1997), Plato (in his book Critias) mentioned the services delivered by forest ecosystems; he hypothesized that the deforestation of Attica in ancient times had led to current (i.e., around 400 BC) drying out of springs and erosion of soil. Since the dawn of agriculture, humans have been fully aware of the connections between nature and their livelihoods. However, the increasing dominance of urban lifestyles has weakened this awareness of our dependence on nature; the formerly obvious links have to be recalled and reemphasized. The Millennium Ecosystem Assessment (2005) was contributed essentially to firmly establishing ESs on the policy agenda, and the research on ESs has developed tremendously since its publication. The millennium environment assessment (MEA) emphasized the state of the ecosystem and ESs, classifying them into four categories: provisioning, regulating, cultural, and supporting services (Millennium Ecosystem Assessment, 2005; Figure 6.1).

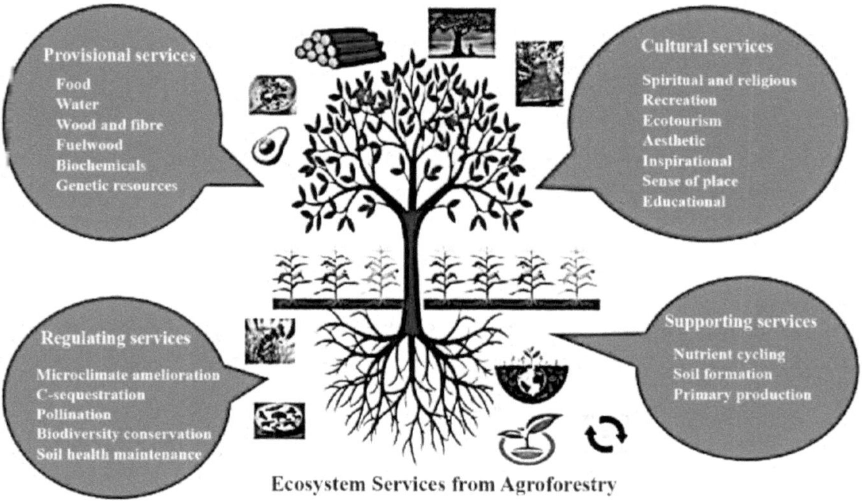

FIGURE 6.1 ESs from agroforestry.

The scientific community's rising interest in the study of ES prompted the launch of a unique research journal on ES in 2012, with a solid foundation in reports and publications from the Millennium Ecosystem Assessment and TEEB programs. Cerda et al. (2020) identified six coffee-based agroforestry which includes a simple agroforestry system dominated by service trees, three medium diversified and two highly diversified system including service trees, timber trees, fruit trees in Costa Rica which were promising for reducing losses while, simultaneously providing other ESs. Most of these systems use less fungicides indicating chemical inputs can be reduced under agroforestry. The survey by Seruni et al. (2020) in Central Java, Indonesia found that farmers who engage in parcelized cut and carry agroforestry earn on an average 11.1% higher than those adopting monocropping. According to Braat & Groot (2012), definitions of the term have developed over time, with differing focus paid to the ecological base or the economic use (paraphrased and given bellow):

- The circumstances and processes by which natural ecosystems and the organisms that make them up maintain and fulfill human life are known as ecosystem services (Daily, 1997).
- Direct and indirect benefits human obtained from various ecosystem functions are called ecosystem services (Costanza et al., 1997).
- The benefits that individuals receive from ecosystems are known as ecosystem services (World Resources Institute, 2005).

- Nature's components that are enjoyed, consumed, and employed to benefits humans directly called ecosystem services (Boyd & Spencer, 2007).
- Ecosystem services are the that features of ecosystems which is used to promote human well-being either actively or passively (Fisher et al., 2009).
- Ecological services are ecosystem's contribution to betterment of human society that are both direct and indirect (TEEB, 2010).

6.2.1 PROVISIONAL SERVICES

Provisioning services are the tangible benefits that humans receive from ecosystems such as water, food, timber, and other items. Rural communities in many areas of world rely on provisioning services for their livelihoods. In this context, the value of the services may be far greater than the rates at which they are sold on local marketplaces, although many provisioning services are sold on the open market (FAO, 2018). The provisional services of trees the direct benefits obtained from trees in terms of products like food, fodder, fuelwood, fruits, non-wood forest products, biofuel, and bioenergy. The services can be easily quantified by using standard technologies. Bado et al. (2021) compared the performance of millet, cowpea with *Ziziphus* trees and without *Ziziphus* in Sahel and study concluded that the inclusion of *Ziziphus* trees in low input cropping systems of smallholder farmers enhances agricultural productivity and incomes along while ensuring sustainable management of soils, land resources, and ESs.

6.2.1.1 TIMBER PRODUCTION

Timber is a strong economic choice and one of the key sources of long-term revenue for farmers in agroforestry system, because it can provide larger profits and lower expenses than traditional agricultural systems (Beer et al., 1998; Bertomeu, 2006). Due to its high durability, various hues, textures, and application tropical timber (hardwood) has a unique commercial value (De'Sousa et al., 2016). The event of crop failures, such as droughts, insect outbreaks, and market failures, timber also serves as a crisis-prevention tool (Ramírez et al., 2001). Traditionally farmers use these trees for home constructions (beams) or agricultural implements (bullock drawn implements). The commercial agroforestry systems are a prominent source of industrial raw material such as plywood, pencil, and paper (Table 6.1).

TABLE 6.1 Provisional Services (Wood Production) by Different Tree Species

Tree Species	Wood Production	References
Acacia nilotica	3 to 4 m^3/ha/yr.	Nicodemus (2014)
Ailanthus excelsa	50–75 t/ha at a rotation of 5–6 years.	Rajasugunasekar (2014)
Casuarina equisetifolia	Irrigated: 100–150 t/ha of dry biomass in 4 years.	Nicodemus (2014)
	Rainfed: 75–100 t/ha in 6 years.	
Dalbergia sissoo	4 to 6 m^3/ha/yr.	Pandey (2000)
Eucalyptus	60–70 t/ha in six years.	Sivakumar (2014)
–	8 to 12 m^3/ha/yr.	Pandey (2000)
Gmelina arborea	250–300 t/ha in 10 years	Mayavel et al. (2014)
–	10 to 15 m^3/ha/yr.	Vijayaraghavan (2014)
Melia dubia	0.4 to 0.5 m^3 of timber in 5 years.	Warrier (2014)
Neolamarckia cadamba	70–100 t/ha at 6–7 years rotation.	Vijayaraghavan (2014)
Populus deltoides	20–25 m^3/ha/yr (8–10 years).	Pandey (2000)
Tectona grandis	2.85 m^3/ha/yr.	Palanisamy (2014)

The wood production varies with the species, site quality, spacing, and management practices. The poplar-based agroforestry system of Indo-Gangetic Plains (IGP) helps to get highest income. The economic analysis of bund system of Poplar plantation with eight-year rotation (146 trees ha^{-1}) has provided ₹ 60,000 INR (Dwivedi et al., 2007). Tropical agroforestry has the ability to provide a sustainable hardwood supply to world. Agroforestry contributed about 9–10% of green fodder for livestock, 60% raw material for paper pulp, 65% of small timber, 70% to 80% plywood and 50% of fuelwood in addition to its other environmental benefits (Ram et al., 2018). It was estimated that the homegardens of Kerala, contain about 6.6–50.8 m^3 ha^{-1} standing stock of commercial timber on an average (Kumar et al., 1994). Dhanda & Verma (1995) reported productivity of poplar grown in agroforestry as 444.1 m^3 at nine years with MAI of 49.3 m^3 ha^{-1} y^{-1} and 292.0 m^3 at five years rotation with MAI of 58.4 m^3 ha^{-1} y^{-1} in India. In case of bamboo-based agroforestry, the biomass production was varied 49 t ha^{-1} (Patil et al., 2004) to 241.2 t ha^{-1} (Kittur et al., 2016) in India. Nath et al. (2009) estimated 121.51 t ha^{-1} of total biomass in village bamboo grove, with culm components accounting for 86% of the growing stock, followed by branch (10%) and leaf (4%). The high differences were observed in volume production of *Tectona grandis* ranging from 8 to 12 m^3 ha^{-1} y^{-1} and 25–30 m^3 ha^{-1} y^{-1} in different sites and different zones of India (Katwal, 2003).

6.2.1.2 FODDER

Pan-tropically agroforestry, livestock-keeping, and the interactions between these practices are crucial for the livelihoods of rural communities. In a wide range of farming systems across the world, fodder trees are major feed sources for cattle (Table 6.2).

A study conducted at NRCAF, Jhansi to compare the biomass yield from silvipastoral system consisting of three components (*Albizia amara* or *Leucaena leucocephala*), shrub (*Dichrostachys cinerea*), grass (*Chrysopogon fulvus*), *legumes* (*Stylosanthes hamate* and *S. scabra*) and natural grassland for eight years reported that, the mean total biomass yield of 12.62 t ha^{-1} y^{-1} (4.55 t from grasses + 0.51 t from lopped leaves and pods from trees + 1.66 t from fuel wood through pruning + 5.90 t from small timber, fuel, lopped fodder leaves of trees) was recorded under silvopasture that was approximately four times greater than natural grassland yield (3.16 t ha^{-1} $year^{-1}$). These results showed that well established silvipasture can produce above 12 t ha^{-1} y^{-1} biomass on a same area that produced only 3 t ha^{-1} $year^{-1}$ biomass through natural vegetation. Similarly, *Dalbergia sissoo* based silvipastoral system at 10 years rotation produced total biomass of 9.62 t ha^{-1} $year^{-1}$ (5.40 t from forage + 0.63 t from tree leaves + 1.71 t fuel wood + 1.88 t from timber) followed by *Hardwickia binata* based system (7.79 t ha^{-1} yr^{-1}). The dry forage yield recorded with pasture alone was 6.91 t ha^{-1} yr^{-1} (Palsaniya et al., 2011). Two-tier intensive system with hybrid Napier + mulberry (11,111 trees ha^{-1}) was found promising silvo-pasture system in terms of fodder yield (24 Mg ha^{-1} $year^{-1}$ dry matter) and crude protein yield (3.15 Mg ha^{-1} $year^{-1}$ for fulfilling fodder deficit in India (Varsha et al., 2019). Study in Zimbabwe reported introduction of fodder bank trees (*Acacia angustissima, Leucaena pallida, L. diversifolia*, and *L. leucocephala*) in smallholder's dairy farm provided gross margins of US $13 to US $334 and B:C ratios ranged 1.12–3.03. Study further reported that he margins and ratios changed based on the amount of tree fodder supplied; more utilization resulted in better returns (Moyo & Ayuk, 2001). In the semi-arid region nearby Segou in Mali, *Gliricidia sepium* produces about 2 tons ha^{-1} y^{-1} and *Pterocarpus* spp. produces around 0.5 tons ha^{-1}

TABLE 6.2 Potentials of Various Regions for Silvipastoral Systems in India Text

Region	Accents and Components	Results
Himalaya	Pastures in alpine area and intensive silvipasture and hortipasture in temperate areas. • *Alnus nepalensis-Stylosanthus guyensis* + *Panicum maximum* • *Ficus hookerii* + *Thysanolaena maxima* • *Dalbergia sissoo* + *Chrysopogon fulvus* • *Albizia lebbek*+ *Euliopsis binata* • *Bauhinia purpurea*+ *Eulaliopsis binata* • *Leucaena leucocephala*+ *Hybrid napier* • *Acacia catechu*+ *Eulaliopsis binata*	• 5–13.5 t/ha/yr. dry forage yield. • 8–10 kg tree^{-1} dry matter yield. • 64–71 t/ha fuel wood and 5.2–5.5 t/ha/yr. Fodder grass in a short rotation of 4–5 years.
Indo-gangetic plains	Tree-grass combinations as silvipasture and hortipasture – *A. nilotica/Gmelina arborea/Madhuca latifolia/Mangifera indica/Leucaena leucocephala/Pithecellobium dulce* + *Phragmites australis/Saccharum spontaneum/Saccharum arundinaceum/Imperata cylindrical* combinations.	• 8–12.3 t/ha/yr. biomass production.
Arid and semi-arid	Silvipasture and hortipasture and grazing lands – *Prosopis cineraria/Zizyphus nummularia/Acacia nilotica/Ailanthus excelsa/Acacia. leucophloa/Albizia procera/L. leucocephala* + *Dichanthium/Cenchrus/Lasiurus/Panicum/Heteropogon/Chrysopogon.*	• *A. excelsa* – *Cenchrus setigerus*+ *Vigna auguiculata* produced dry biomass of 10.0 t ha^{-1} (3.93 t forage + 3.13 t leaf fodder + 2.91 t fuel wood). • 10.2 t/ha DM from *Leucaena leucocephala-Cenchrus ciliaris* + *Clitoria. ternatea.*
Humid and sub-humid	Quick growing leguminous trees in crop rotations, plantations, and silvipasture – *L. leucocephala/Lyricidia/Sesbania/Ficus hookeri* + fodder grasses like *Pennisetum pedicellatum* and *Stylo*.	• Have high forage yield potential.
Coastal plains	Silvipasture and hortipasture – *Sesbania grandiflora/L. leucocephala/* Desmenthes + Congo signal/Stylosanthes/Centrosema.	• 22–29 t ha fodder yield can be obtained from coconut based silvipasture.

Source: Adapted from: Palsaniya et al. (2011).

y^{-1} of fodder (Hamer & Franzel, 2007). In Himalaya, *Grewia optiva* is heavily lopped for fodder production and a mature tree can produce 25–75 kg $tree^{-1}$ y^{-1} of green fodder.

6.2.1.3 ENERGY

About 66% of households (including 20% of urban and 75% of rural households) in world were fuelwood dependent as source energy for cooking and heating especially in undeveloped nations (World Health Organization, 2018). More than 56% of rural families in most of the Indian states used fuelwood for cooking purposes. Total consumption of fuelwood in India was 106 million tons in 1993–1994 which increased to 130 million tons in 2011–2012 due to rise in population. During 2011–2012 period, agroforestry contributed 64% of total fuelwood for cooking in India followed by forest resources (24%), and 12% from common's land (Singh et al., 2021). Biofuels and bioenergy have been raised as an alternate source of energy, but obtaining a sustainable supply of biomass remains a significant problem. Agroforestry may be a viable option for integrating perennial woody bioenergy crops with traditional agricultural crops to meet energy demands while maintaining food supply. Biodiesel has a productivity of about 20 million tons per year. Only a few million tons (20–25% of estimated potential of 0.1 to 20 million tons) of have been utilized (due to lack of demand) (Dhyani, 2014). In tropical woody species, more than 100 species of tree borne oilseeds (TBOs) have been identified as a source of fatty oils with estimated potential of 11.3 lakh tons (Dhyani et al., 2015). The potential sources include *Pongamia pinnata, Madhuca longifolia, Azardirachta indica, Calophyllum innophyllum* and *Simarouba glauca* (Chavan et al., 2014, 2015b). These biofuel species can be grown on wasteland and degraded land and add to the farmer's income. As per the National Biofuel Policy, 2008 of India, the biofuel-based agroforestry systems could help to reduce the gap between supply and demand (Government of India, 2008; Table 6.3). Torreiro et al. (2020) studied potential of selected agroforestry biomasses in Europe as biofuels in energy recovery tests through micro-cogeneration and found that biomasses from agroforest reaches about 97% of cogenerating efficiency including 9% net electric yield and 88% thermal yield indicating low valued biomass from agroforest including pruned vines, branches, and scrub can be good alternative to obtain clean energy.

TABLE 6.3 Production of Biofuels in India

Species	Oil Content (%)	Estimated Seed Production (106 t yr–1)	Oil Production (t ha–1yr–1)
Ricinus cuminis	45–50	0.25	0.5–1.0
Jatropha curcus	50–60	0.20	2.0–3.0
Madhuca indica	35–40	0.20	1.0–4.0
Shorea robusta	10–12	0.20	1.0–2.0
Linseed	35–45	0.15	0.5–1.0
Azadiractchta indica	20–30	0.10	2.0–3.0
Pongamia pinnata	30–40	0.06	2.0–4.0
Simaruba glauca	70	–	40–40 kg/tree
Garcinia indica	22–25	–	30–35 kg/tree
Olea europae	60–70	–	2–3 kg/tree

Source: Dhyani et al. (2015); Singh & Singh (2010).

Trees may supply feedstocks for many kinds of bioenergy production, such as solid biomass, which can be utilized as firewood, charcoal, and power generation. Annual net financial benefit of biomass production in the agroforestry landscapes in Europe was estimated around € 29 ha^{-1} y^{-1} (Kay et al., 2019a). Around 1,000 hectares of seven-year-old poplar plantation is predicted to produce enough biomass to run a 5-MW power plant for a year (Ranjan, 2020). Liquid biofuels such as biodiesel produced from oilseeds, whereas ethanol can be made from lingo-cellulosic biomass (Sharma et al., 2016). The need for fuelwood is a major issue for most policies since it is directly linked to deforestation. Around 70% of the rural population still uses trees as a source of energy (Chauhan et al., 2015). Agroforestry energy plantation can be successful strategy to fulfill the demand of fuelwood. For providing sustainable and efficient biomass production for local power generation without compromising food production, there is need to design proper site specific and suitable agroforestry. Fast growth, drought resistance, good coppicing ability, high calorific value, and low ash content are all desirable characteristics for ideal fuelwood tree species. Many species such as such as *Prosopis juliflora, Leucaena leucocephala, Melia dubia, Eucalyptus* spp., *Casuarina equisetifolia, C. junghuhniana* and *several* bamboo species meet these requirements (Shankar, 2012). The yield depends on age of tree, pruning resistance and growth rate. On an average tropical tree provides 100–150 kg of fuel wood per year per tree. Popular culture in India is providing 5 Mt of firewood every year. Singh et al. (2021) estimated that if total calorific energy of fuelwood obtained from agroforestry was valued it would cost US $4,053 million accounting 1297.4 PJ energy. In terms of self-reliance, conserving limited foreign exchange, and ensuring renewable and almost carbon-neutral fuel and energy security for the poor, especially in developing nations, agroforestry is crucial for fuelwood production.

6.2.2 REGULATING SERVICES

Ecosystems provide a variety of regulating services, including as maintaining air and soil quality, flood and disease management, and crop pollination. They are frequently unseen and hence taken for granted. Damage to them might result in significant losses that are difficult to recover. Agroforestry systems are mimicking the NF to regulate the environment and obtain a variety of benefits. These include climate moderation to reduce evaporation and stress in animals to increase productivity; microclimate changes (CCs) for shade and windbreak, pest and disease management; and flood control by boosting soil infiltration characteristics.

6.2.2.1 MICROCLIMATE AMELIORATION

Trees on the farm influence understory crop saturation deficit, wind velocity, air temperature and radiation flux, all of which have a substantial impact on adjusting the photosynthetic rate and duration of photosynthesis and consequent plant development, transpiration, and soil water usage (Monteith et al., 1991). Rao et al. (2011) has quoted some of the examples from different researchers which were used to explain science behind the role of agroforestry in microclimate enrichment. Review by Beer et al. (1998) found that shade trees may buffer temperature extremes by up to 5°C in coffee and cocoa plantation. Sida et al. (2018) studied wheat yield under *Faidherbia albida* trees and found that yield and growth attributes of wheat under trees is higher that of sole wheat due to buffering action of trees against heat stress during sensitive stage that enhances grain yield. The removal of shade trees raised soil surface temperature by around 4°C and lowered relative air humidity at 2 m above ground by about 12%, according to Steffan-Dewenter et al. (2007). Faidherbia trees dropped soil temperature at 2-cm depth by 5–10C depending on the movement of shadow in the Sahel, where soil temperatures frequently exceed 50–60°C, a key limitation to establishing a successful crop (Vandenbeldt & Williums, 1992). The air around a maize crop growing beneath a *Grevillea robusta* stand was likewise more humid than the free environment above the trees, according to data from an agroforestry study in Kenya (Ong & Leaky, 1999). Larger available nitrogen (35–55%) under canopy of trees than open wheat and 6°C less midday air temperature and was observed under tree canopy than in open (Sida et al., 2018). Tea yield increased by 53% due to integration of shade trees as compared to mono-cropping (Barrios et al., 2018).

The main finding of Ujah & Adeoye (1984) showed the influence of windbreaks on crop productivity and microclimate (as give bellow):

- On the leeward side, the shelterbelt lowered wind velocity (2 m above the ground). At distances between 20 and 150 meters from the belt, the drop in wind velocity varied from 20% to 10%.
- At 20 m from the belt on the leeward side, maximum air temperatures were 0.8–1.5°C higher than in the open field.
- The belt's wind-moderating impact is expected to extend 15 times the height of the trees, while the belt's effect on air temperature is up to 10 times the height of the trees.
- Close to the belt, maximum soil temperatures at 5 cm deep were 0.5–1.0°C higher than in the open.
- However, moisture depletion from the topsoil was faster on the exposed side than on the leeward side immediately afterward the end of the rains, due to increased wind speeds on the unprotected side hastening the evaporation process.
- Higher yield of millet obtained in sheltered area as compared to open and with a peak in yield is reached at a distance of four times the tree height.

In scenario of CC, where heat and moisture stress becoming more widespread as a result of CC related extremities, trees and their influence on microclimate amelioration might be a good starting point for designing more robust and climate-smart agriculture (CSA).

6.2.2.2 IMPROVEMENT OF WATER STATUS AND THEIR QUALITY

Water is an extremely valuable national resource. Human activities frequently impair the water quality in the streams, lakes, estuaries, wetlands, and aquifers on which we rely. In both rural and urban regions, trees minimize and impede runoff and trap pollutants. By lowering runoff, chemical leaching into groundwater, and groundwater contamination, agroforestry is helpful in minimizing non-point source (NPS) pollution from cropland (Udawatta et al., 2002). Roots of trees take nutrients from subsurface plumes with high nitrate contents and assimilated in tree biomass, over several years helps in cleaning the aquifer (Ranjan, 2020). Pavlidis & Tsihrintzis (2018) reviewed the contribution of agroforestry in pollution reduction and affirmed that around 20–100% of soil nitrogen and 90% of pesticides removed and prevent from leaching into runoffs by planting trees along with crops. Doty et al. (2017) studied the effect of poplar on groundwater plumes contaminated with *Trichloroethylene* in region of Silicon Valley, USA, and suggested that industrial pollutants were effectively removed by planting poplar. Most of the researchers has confirmed the perennial vegetative buffers improve water quality in agricultural watersheds and have nitrogen removal efficiencies ranging from 20% to 90% (Lee et al., 2003). Trees can theoretically improve soil water content beneath their canopies if water 'retained' by reducing soil evaporation and funneling of captured rainfall as stem flow surpasses that absorption by root systems under their canopies (Ong & Leaky, 1999). Study by Bargues-Tobella et al. (2020) in Africa conformed found that soil and groundwater recharge can be improved by maintaining or promoting an appropriate tree cover in tropical African dryland thus agroforestry holds key for water security in such drier areas. Agroforestry improves water quality by reducing runoff and leaching of nutrients and soil and water remediation. By forming safety-network of deep and fine roots, agroforestry blocks percolation of chemicals in soil, agroforestry helps in reducing soil and groundwater contamination (Pavlidis et al., 2021). Rhoades (1997) found 4% to 53% higher soil water in the agricultural root zone beneath *Faidherbia albida* canopies than in open Malawi.

The perennial woody component of agroforestry modify water cycling by lowering runoff, and enhancing infiltration, boosting water retention in the soil, transpiration, rain and cloud interception (with both positive and negative effects). Some of these relationships have been demonstrated in Costa Rica through a number of research. Rain interception was 16% in coffee (*Coffea arabica*) plantations with regularly pruned *Erythrina poeppigiana* (555 trees/ha) and 7.5% in coffee (*Coffea arabica*) farms with unpruned *Cordia alliodora* (135 trees/ha), respectively (Jimnez, 1986). During dry season, trees are commonly described as "sponges," collecting rainwater and gently releasing it to keep groundwater and streams alive (Calder et al., 2004). Pavlidis et al. (2021) investigate nutrient and pesticide uptake of wheat-poplar agroforestry systems and they observed 80–100% reductions for nutrients and 75–100% decrease for herbicides near tree row than the control points away from the trees. This exhibited the role of agroforestry in pollution abatement through uptake the excess of agrochemicals that would otherwise remain in soil and contaminate soil and water. Similar reductions were observed in maize-olive system (Pavlidis et al., 2018), maize-poplar system and a potato-poplar system (Pavlidis et al., 2020). Poplar trees were studied as a buffer for nitrogen leaching from animal farms by Manevski et al. (2019), and nitrate leaching was expected to be decreased by 75% to 80%, affirming potential of agroforestry in pollutant reduction. To make payments for ecosystem services (PES)-based agroforestry for nitrate remediation of severely polluted aquifers is financially appealing to farmers, requires long term conversion of croplands into agroforest on wide scale. A PES-based agroforestry related bioremediation might be a win-win situation for farmers and the environment, particularly in areas where solutions are limited or expensive, and water quality is quickly deteriorating owing to increasing population and related urbanization (Ranjan, 2020).

Woody perennial buffers precipitate dust from exhaust fans by decreasing air speed, absorb gaseous ammonia, and divert the odor plume into the atmosphere above the buffer, all at a minimal cost. Plants have the potential to absorb gaseous ammonia, which is one of the serious issues of the poultry sector (Lin et al., 2006). This leads to faster development rates since plants in front of exhaust fans were found to have greater nitrogen and dry matter weights than control plants (Patterson et al., 2006). Malone et al. (2006) reported 46%, 49%, and 99% reduction in ammonia, dust, and wind speed, respectively, downwind of trees, during summer season. Reduction in potato-poplar system exhibited 86% potassium, 90% nitrate 92% for NH_4^+, 85% for NO_2^- and 100% for PO_4^{3-} reduction while, maize-poplar system exhibited reductions of 73% K^+, 77% NO_3^-, 97% NH_4^+, and up to 100% PO_4^{3-} (Pavlidis et al., 2018, 2020). Integration of trees on farms helps in reducing recalcitrant pollutants from groundwater plumes at much lower cost than expensive alternatives like reverse osmosis or any other chemical and biological treatment (Manevski et al., 2019).

6.2.2.3 *PREVENTION OF SOIL EROSION*

The system metrics such as nutrient efficiency, biomass, and soil loss, and cardamom output have been determined as a baseline for evaluating conventional cardamom-based agroforestry ESs. The efficiency of soil loss (kg/ha) in diverse land-use systems (LUS), including cardamom-based agroforestry (30 kg/ha), forest-based agroforestry (74 kg/ha), horticulture-based agroforestry (145 kg/ha), and rainfed agriculture (477 kg/ha), shows that cardamom agroforestry has the least soil loss. According to one research, a single bamboo plant may bind up to 6 meters of soil (Anonymous, 1997). In Jiulongjang and Dayingjang rivers in China it was found that each clump could protect 12 m of river embankment also yielding shoots and bamboo timber (Maoyi & Banik, 1996). *Bambusa tulda*'s root was seen to stretch horizontally for 5.2 meters (White & Childers, 1945). *Phyllostachys heterocycla*, *P. viridis*, and *P. nigra* had total lengths of surviving rhizome per hectare of 50 to 170, 90 to 250, and 200 to 320 cm, respectively, for monopodial species (Zhou et al., 2005). Zero-till and conventional plowing without hedgerows were compared to *L. leucocephala* and *Gliricidia sepium* based intercropping system at 2 and 4 m spacing. Mean rates of soil loss over two years were 8.75 t ha^{-1} y^{-1} under plowing, 0.95 t ha^{-1} y^{-1} under hedgerow intercropping (HI) (Young, 1989). According to Wiersum (1984), as compared to other practices, different agroforestry practices cause least soil erosion (Table 6.4).

TABLE 6.4 Soil Loss from Various Agroforestry Systems Across World

Agroforestry System	Soil Loss in Agroforestry System	Soil Loss in Control/ Monocrop	References
Gliricidia hedgerow	6.1 Mg ha^{-1} y^{-1}	10.2 Mg ha^{-1} y^{-1}	Adhikary et al. (2017)
Leucaena hedgerow.	6.5 Mg ha^{-1} y^{-1}	10.2 Mg ha^{-1} y^{-1}	Adhikary et al. (2017)
Gliricidia hedgerow.	6.3 Mg ha^{-1} y^{-1}	12.5 Mg ha^{-1} y^{-1}	Adhikary et al. (2017)
Leucaena hedgerow.	8.0 Mg ha^{-1} y^{-1}	12.5 Mg ha^{-1} y^{-1}	Adhikary et al. (2017)
Young mixed tree system (2-year-old).	2.0 g m^{-2} in one month.	32.00 g m^{-2} in one month.	Beliveau et al. (2017)
Cardamom under forest trees.	0.65 Mg ha^{-1} y^{-1}	120 Mg ha^{-1} y^{-1}	Moench (1991)
Pepper based agroforestry.	3.5 Mg ha^{-1} y^{-1}	120 Mg ha^{-1} y^{-1}	Moench (1991)
Mixed species garden.	1.45 Mg ha^{-1} y^{-1}	120 Mg ha^{-1} y^{-1}	Moench (1991)
Eucalyptus + Cassava	0.78 Mg ha^{-1} y^{-1}	1.79 Mg ha^{-1} y^{-1}	Ghosh et al. (1989)
Leucaena + Cassava	0.834 Mg ha^{-1} y^{-1}	1.79 Mg ha^{-1} y^{-1}	Ghosh et al. (1989)
Rubber and tea (*C. sinensis*) system.	1,401 g m^{-2} (sand loss by splash for six month).	1,611 g m^{-2} (sand loss by splash for six month).	Liu et al. (2016)
Sapota + cowpea + castor	2.69 Mg ha^{-1} y^{-1}	4.30 Mg ha^{-1} y^{-1}	Jinger et al. (2022)
Cardamom-based agroforestry.	3.23 Mg ha^{-1} y^{-1}	7.73 Mg ha^{-1} y^{-1}	Mishra et al. (2022)
Live hedges	2 Mg ha^{-1} y^{-1}	27 Mg ha^{-1} y^{-1}	Roose & Ndayizigiye (1997)

Under agroforestry as compared to pure cropping systems, 50% reduction in erosion, 75% higher infiltration, 57% lower runoff, 22% higher macro-aggregation and 30% higher mean weight diameter was observed (Muchane et al., 2020). Béliveau et al. (2017) compared erosion in forest system, 2-year-old agroforest and cropping system with short cycle in Brazil and found that soil particle loss, total mercury and cation mobilized in agroforestry system was similar to forest but significantly lower as compared to short cycle cropping system. Density of eroded soil particles in runoff was 1.9 g/L in agroforestry system, 2.2 g/L in forest and 54 g/L in short cycle cropping system. This study highlighted role of agroforestry in erosion control resemble to indigenous forest even at young stage. Soil erosion control property of agroforestry systems depends on types and its management aspect. Not all existing agroforestry systems control erosion, study in Indonesia found that only some of agroforestry systems have ability to protect farmland from erosion, indicating need of improved agroforestry for reducing soil erosion to maintain watershed quality and support sustainable agriculture (Rachman et al., 2020). Comparison study of different perennial systems with cropping in a clay-pan soil in USA during 2017 and 2018 by Alagele et al. (2020). The study found that establishment of perennial vegetation systems like agroforestry system in erosion-prone locations within watershed helps to reduce runoff sediment transport from agricultural land. Purwaningsih et al. (2020) recommended use agroforestry with proper species and arrangement as technique to control landslide reactivation on volcanic foot slopes in Java. Agroforestry function in two ways to control erosion of soil; supplementary and directly, by supplementary function, plants stabilize soil structure and improves texture and enhances soil productivity while vegetation directly by itself reduces erosion by reducing impact of rainfall (Rachman et al., 2020).

6.2.2.4 *IMPROVEMENT OF SOIL HEALTH*

Integration of trees in the crop lands can assist in preserving the physico-biochemical properties both under sequential and simultaneous agroforestry. By contributing to organic matter to the soil as well as releasing nutrients through mineralization, litterfall, and prunings helps in soil fertility improvement. The quality and amount of litterfall or prunings, the soil type, climatic conditions of the location, and the rising activity of microorganisms in the root zone all are deciding factor in soil improvement. Tree roots are usually found deep below the crop zone, and they feed on nutrients from the lower soil layers, resulting in increased nutrients and water use efficiency (WUE) with least competition with crops. Shade trees for crops and Silvopastural systems provided improved environmental conditions, which boosted the activity of soil engineers (termites, ants, and earthworms), resulting in micro- and macro-faunal aggregates being formed (Rodríguez et al., 2021). The soil fertility enhancement varies with the species and the quantity of litter added to the soil (Table 6.5).

TABLE 6.5 Soil fertility enhancement through trees

Species	Soil Fertility Enhancement	References
Populus deltoides	SOC: 2.9–4.8 Mg ha^{-1}	Gupta et al. (2009)
Eucalyptus tereticornis	NPK: 62.03–75.6, 4.1–4.3, 27.9–30.6 kg/ha.	Bargali (1995)
Leucaena leucocephala	SOC: 12.3 g/kg soil; SOM: 4.4 g/kg soil or 1.78%; N:150–160 kg N/ha; P, K: 0.25, 163 ppm.	Kang & Mulongoy (1992); Gupta (1995); Kang (1997); Youkhana & Idol (2009)
Prosopis cineraria	SOM: 0.2%; N: 139 kg/ha; P: 18.73 kg/ha.	Aggarwal et al. (1976); Sharma & Gupta (1989)
Dalbergia sissoo	SOM: 5.1 g/kg soil.	Gupta (1995)
Acacia nilotica	SOM: 1.4 g/kg soil.	Dagar et al. (2001)
Sesbania rostrata	307 kg N ha^{-1}	Patel et al. (1996)
S. cannabina	209 kg N ha^{-1}	Pandey et al. (2010)
Azadirachta indica	NPK: 98, 2.25, 32 kg ha^{-1}; Ca:131 kg ha^{-1}	Pandey et al. (2010)
Alnus nepalensis	N: 155 kg ha^{-1}	Sharma et al. (2002)
Alnus nepalensis	NPK: 32.7, 4.77, 37.2 kg ha^{-1}	Danso et al. (1995)
Dendrocalamus strictus	SOC: 0.21%; SOM: 0.36%.	Patil et al. (2004)
Gliricidia sepium	NPK: 21, 2.5, 18 kg ha^{-1}; Zn, MN, Cu, Fe: 85, 64, 365, 728 g ha^{-1}	Rao et al. (2011)
Robinia pseudoacacia	SOM: 1.13%; N: 35–150 kg ha^{-1}, PK: 0.60, 92 ppm; NH_4^+: 2.63 ppm.	Youkhana & Idol (2009)
Erythrina indica	NPK: 46.3, 6.11, 36.2 kg ha^{-1}	Danso et al. (1995)
Parkia roxburghii	NPK: 37.1, 5.9, 24.2 kg ha^{-1}	Das et al. (2010)
Acacia auriculiformis	NPK: 30.5, 5.36, 20.2 kg ha^{-1}	
Cassia siamea	NPK: 45.5, 6.25, 40.2 kg ha^{-1}	

Abbreviations: SOC: Soil organic carbon; and SOM: soil organic matter.

Various workers have reported that, trees directly or indirectly influence the soil physic-chemical-biochemical properties. Rivest et al. (2020) found that windbreaks increased the C:N ratio, total nitrogen and soil organic carbon (SOC) while, reducing the soil pH relative to adjacent crop fields. Meta-analysis by Muchane et al. (2020) revealed that agroforestry can reduce soil erosion rates by 50%, enhances SOC by 21%, rises nitrogen by 13%, enhances available phosphorous by 11% and alleviates pH of soil by 2% compared to pure cropping. This might be due to reduced runoff, high infiltration, increased macro-aggregation and more stable soil structure in agroforest. Raveendra et al. (2021) conducted three-year study on soil physical and chemical properties under coconut monocropping, five-year-old and 20-year-old Gliricidia-coconut-based mixed systems and found that 20-year-old gliricidia-coconut-based mixed systems have 22% higher organic matter content and 20% higher available phosphorus content, and 69% higher total exchangeable potassium content as compared to coconut monocrop. This study demonstrated the potential of coconut-gliricidia mixed system for rehabilitation of degraded coconut growing soils in Sri Lanka.

By fixing atmospheric nitrogen in plants, biological nitrogen fixation (BNF) is an important source of nutrients for agro-ecosystems. Leguminous trees' role in building up nitrogen in degraded soils via BNF is well acknowledged as a major component of the ES of nutrient cycling in many studies (Ram et al., 2017). BNF potential varies with species, ranging from high rates of up to 472 kg N_2 ha^{-1} $year^{-1}$ in *Calliandra calothyrsus, Gliricidia sepium, Leucaena leucocephala*, and to low rates of less than 50 kg N_2 ha^{-1} yr^{-1} in *Acacia melanoxylon* and *Acacia holoserica* (Giller, 2001), whereas roots of *Casuarina* species fix up to 350 kg N $ha^{-1}yr^{-1}$. Improvement of SOC by 36.5% and total nitrogen stocks by 33.2% at 0–10 cm soil layer observed in alley cropping of rubber trees with *Coffea liberica* as compared to rubber monoculture (Li et al., 2020). In agroforestry system, BNF trees contribute towards complementary interaction which makes agroforestry system sustainable for nutrient requirements.

Leguminous trees help in enrichment of nitrogen in soil by the process of N-fixation due to accumulation of nitrogen rich biomass into soil by litter fall, prunings, and root decay. Not only leguminous but inclusion of some non-leguminous shade trees such as rambutan and durian in cocoa agroforest had positively influence soil nitrogen and soil carbon (Wartenberg et al., 2020). Mycorrhiza associated with most of the species helps in increase phosphorous mobilization in the rhizosphere and increased labile soil organic phosphorus and soil inorganic phosphorus. By synchronizing nutrient release, demand of crop and its use, agroforestry can play crucial role in increasing nutrients availability in soil and improving nutrient use efficiency. As humid and sub-humid tropical areas facing soil degradation, risk of loss of biodiversity and majority of soils are deficient in nutrients, low in nutrient reserve, erosion-prone, and affected with acidity and associated toxicities. Agroforestry has a high potential to relieve most of these constraints while ensuring food production, health and nutrition improvement and conserving natural resources (Stocker et al., 2020). Stocker et al. (2020) also demonstrated that agroforestry improves several soil physical attributes over a short time by diversification of tree root systems and accumulation of plant residues. Barrios et al. (2018) found that soil macro-porosity was 76% lower in coffee without shade trees due to loss of native earthworm populations compared to shade trees. Piza et al. (2021) compared decomposition and release of nutrients in various land-uses practices in Columbia and reported similar litter decomposition and nutrient mineralization in coffee agroforests as compared to nearby forest. Froufe et al. (2019) found at least 100, 5, and 10 kg ha^{-1} of N, P, and K returns, respectively through litter fall in agroforestry.

6.2.2.5 CARBON SEQUESTRATION

Carbon sequestration is the process of removing extra carbon from the atmosphere and storing it in another reservoirs. Practically it is mostly accomplished through bringing more areas under tree plantations. Agroforestry is significantly more intensely managed than traditional forests and generate far more biomass. It has received much interest as a method for biological carbon sequestration as a result of its recent identification as a GHG-mitigation approach under the Kyoto Protocol. Ajit et al. (2017) reported that the mean carbon sequestration capacity of the agroforestry system on farmer's field at national level was 0.21 Mg C $ha^{-1}yr^{-1}$ based on varied edapho-climatic variables. At the national level, existing AFS are estimated to mitigate 109.34 million tons CO_2 annually which may offset one-third (33%) of the total GHG emissions from the agriculture sector. Recently the rate of one ton of carbon is US $11 as per the European market (https://greenliving.lovetoknow.com). So, on this basis, agroforestry in India on total agricultural area (142 m ha) can contribute to receive $328 million USD from Global Environmental Fund. Amadi et al. (2017) studied GHG emissions along a 31-year-old shelterbelt-cropped field transect in Canada and found that soil N_2O emissions were raised with growing distance from the shelterbelt. Kay et al. (2019b) estimated that integration of trees over 8.9%

of agricultural land in Europe has the capacity to store 1.4% to 43.4% of all agricultural GHG emissions in Europe, which will help to achieve future zero-emission agriculture. Two-tier intensive system with hybrid Napier + mulberry (11,111 trees ha^{-1}) is found promising silvopasture system fixes about 11 Mg C ha^{-1} y^{-1} for fulfilling Intended Nationally Determined Contribution (INDC) commitment of India along with the production of quality fodder (Varsha et al., 2019).

According to Zomer et al. (2016), agroforest and trees on farm contribute significantly to CC mitigation. They projected 45.3 Pg C on agricultural land globally, with trees providing more than 75% contribution, by combining regionally and bioclimatically stratified Intergovernmental Panel on Climate Change (IPCC) Tier 1 baseline projections of carbon sequestration with this tree cover study. Tree cover rose by 3.7% during 2000–2010, resulting in a biomass carbon gain of nearly 2 Pg C (or 4.6%). Biomass carbon grew from 20.4 to 21.4 t C ha^{-1} on average over the world. A transition from intensively cultivated and degraded land to smallholder AFS led to substantial gains in the carbon stock (Chiemela et al., 2018). Rodríguez et al. (2021) investigated capacity of different systems to store SOC in Amazon valley of Columbia. Silvipasture (131 Mg ha^{-1}) had the highest SOC content, followed by pasture (112.2 Mg ha^{-1}), home gardens (105.8 Mg ha^{-1}), natural regeneration (100.6 Mg ha^{-1}) and secondary forest (100.6 Mg ha^{-1}). They observed that SOC storage in some agroforestry systems is even higher than forest. This study indicated that adopting silvipasture might be sound strategy to increase soil carbon storage deforested Amazonia soils of Columbia (Figure 6.2). Another meta-analysis by Feliciano et al. (2018) on soil C-sequestration potential among various agroforestry systems concluded that silvipasture and mixed homegardens have the greatest average rates of carbon sequestration in the soil among all.

6.2.2.6 *BIODIVERSITY CONSERVATION*

Agroforestry is a sophisticated and integrated strategy that allows trees, crops, pastures, and animals to coexist in a controlled manner, while also providing habitat for soil fauna and flora, insects, birds, and mammals. Many traditional agroforestry practices are finest examples of conservation agrobiodiversity. By sheltering and providing habitat for a wide range of flora and wildlife, inclusion of tree adds to the diversity in system. It also aids in the preservation of the genetic variety of wild cultivars or landraces, as

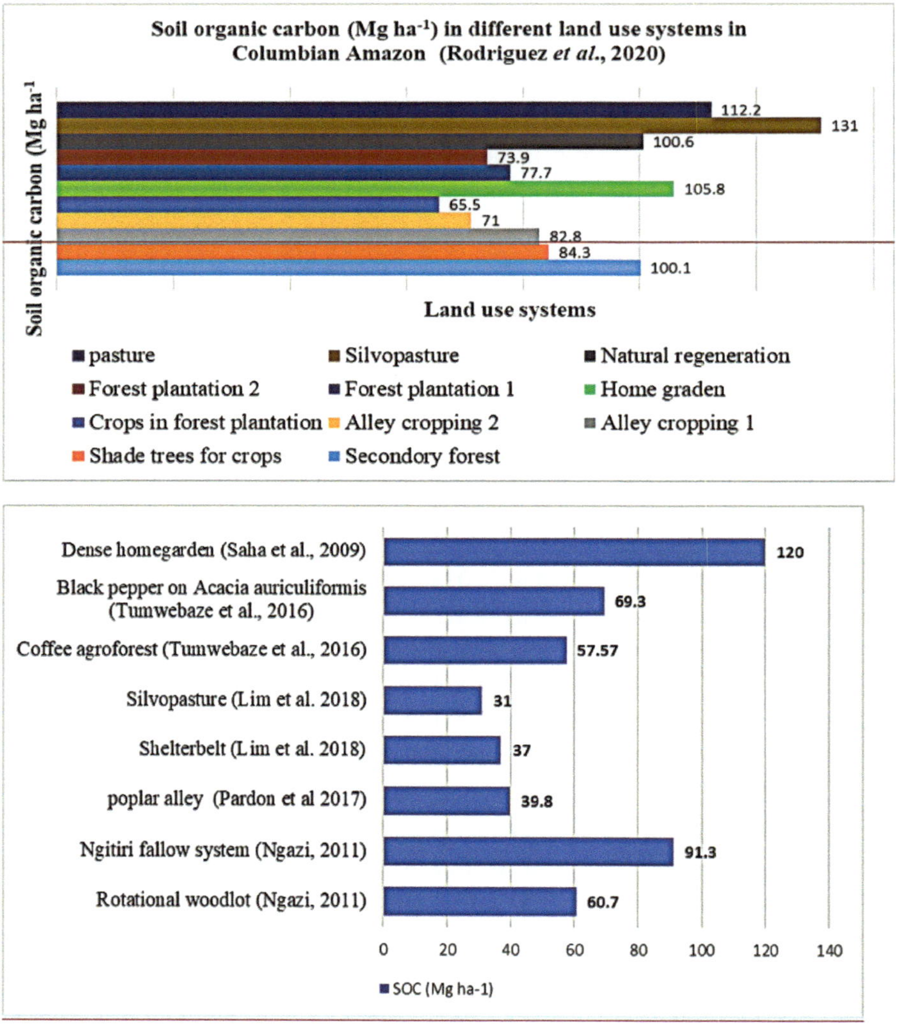

FIGURE 6.2 SOC storage (Mg ha^{-1}) in various land-use systems in agroforestry systems.

well as trees that are endangered and require immediate protection (Chavan et al., 2014). A review by Udawatta et al. (2019) concluded that agroforestry systems across Canada and America having higher faunal, floral, and microbial diversity than many mono-cropping systems. Many studies even found that shaded coffee systems having higher numbers of omnivorous, frugivorous, nectarivorous birds than nearby forest but with fewer insectivorous birds (Komar, 2015). The intermediate level of biodiversity between forest and agricultural land is supported by agroforestry systems was confirmed by various studies (Sagastuy & Krause, 2019). Philpott & Bichier (2012) observed 113 bird species with 7,718 of individuals in shaded coffee system in Mexico. Pardon et al. (2019) studied diversity of ground-dwelling arthropods in agroforestry systems of Belgium and found that rows of trees was having higher abundance and diversity of woodlice, millipedes, and macro-detritivores as compared to arable zone. Like numerous studies, this study also indicated the role of agroforestry in the preservation of arthropod biodiversity while, improving decomposition processes and nutrient cycling in agricultural fields. However, further study on management approaches to fully exploit these ecological services is needed. Study by Cardinael et al. (2019) in temperate hedgerow cropping systems indicated higher abundance, richness, and biomass of earthworm beneath tree windbreaks as compared to adjacent arable farm. Holden et al. (2019) also found that the richness of arbuscular mycorrhizal fungi (AMF) was higher under hedgerows and at their margins as compared to nearby cropped areas. Agroforestry systems are acting as *in-situ* conservation strategy to protect and sustainable utilization. Shastri et al. (2002) reported 952 individuals belonging to 93 species in just 1.7 hectares of agroecosystem in Karnataka.

Tree based farms act as an important wintering site for migratory birds and support bats, small mammals, arthropods, and amphibians as well. These birds play important role in pest control by preying upon coffee berry borer and coffee leaf miner (Philpott & Bichier, 2012). Several studies found that these birds in coffee-based agroforestry system reduce the insect population in coffee and shade trees (Philpott et al., 2004) while, limiting their outbreaks. Tea yield was increased by 53% due to the integration of shade trees as compared to monocropping. Mixed trees agroforestry maintained about 93% of crop pollinators from a nearby forest and 86% lower infestation of coffee berry borer observed in coffee with shade trees than open system (Barrios et al., 2018). Agroforests connect various fragmented forest patches and act as a corridor for wildlife (Haggar et al., 2019). The predation rate of arthropods in agroforestry plots was significantly higher as compared to control plots, i.e., with no trees (Martin-Chave et al., 2019). Understory with herbaceous vegetation other than crops develops on the rows of trees were generally considered as reservoirs for weeds but study by Boinot et al. (2019) disproved this hypothesis and found that very few species managed to disperse in interspaces while, it restricted to a short distance in alley cropping systems. Rather understory vegetation strips act as refugia for detrivorous microbes and more beneficial arthropods such as pollinators and natural enemies of pests thus enhances cycling of nutrients, pollination, and biological control of the pest. It also helps in the improvement of structure and porosity and reducing soil erosion (Boinot et al., 2019). Numerous agroforestry systems like alley cropping, shelterbelts, and scattered trees on farmland resemble to semi-natural landscape which act as nesting sites, habitat, reproduction sites, breeding areas, shelter, and corridors for many florae and faunas as these systems provide them with food, protection against predators and links the fragmented forest areas. Integration of trees in agricultural land enhances complexity and heterogeneity of landscapes thus helps in the integration of be pollinating insects, water body and beneficial species together in landscape (Amadi et al., 2017). Udawatta et al. (2019) highlighted that knowledge regarding selection of species and their management for different ESs requires under various socio-ecological contexts needs to be shared with farmers and translated into viable options with farmers.

6.2.3 CULTURAL AND RECREATIONAL SERVICES

Cultural services are non-material benefits people derive from ecosystems, including ecotourism, recreation, cultural heritage values, sense of place, customs, social relations, esthetic values, inspirations, educational values, knowledge systems, spiritual, and religious values and cultural diversity (Masiero et al., 2019). Cultural services have long been acknowledged as valuable, but in the few cases when they are, they are frequently described as "intangible," "subjective," and impossible to measure in biological or economic terms. Esthetics is frequently mentioned as an example of cultural ES, although more detailed practical criteria to help evaluations are rarely given. Beauty or esthetic value in different areas of ecosystems, as indicated in support for parks, scenic drives, and the selection of residential areas, according to the Millennium Assessment. De Groot (2014) defined esthetic services as "appreciation of natural scenery," while Chan (2011) associates rural esthetic values with "the amount or layout of open space in agricultural or wooded (land use/land cover) types." Visual landscape esthetics, particularly scenic beauty, are emphasized in these concepts.

Agroforestry system in *Prosopis cineraria*-based agroforestry of arid zone provides the recreation and cultural value. The system has been considered as religiously sacred of the desert people and harvesting or felling is considered as a taboo. During cropping season, the system imparts scenic and natural values. Most of the natural based systems like *Grevellia robusta* of Assam and Nilgiris, *Alnus nepalensis* of NE India, *Grewia optiva* of Himalaya and Bamboo and *Tectona grandis* based systems of Western Ghat are the excellent examples of recreational and cultural values of agroforestry. The quantification on the basis of economics is still unclear and need further specific research.

6.3 VALUATION OF AGROFORESTRY-BASED ESS

Existing research studies indicate that the incorporation of agroforestry in LU can prevent degradation of the environment, enhance agricultural productivity, sequester carbon, enhance the quality of water, alleviate soil health and maintaining health and quality of ecosystems along with generating steady incomes and additional advantages for wellbeing of human (Brown et al., 2018). For stimulating farmers and landowners' agroforestry, a more socially beneficial LU, new approaches towards accounting for externalities such as PES or other incentives of LU should be strengthened. Many existing approaches are used for valuation of agroforestry derived ESs.

6.3.1 CAPITAL BUDGETING

The is most frequent approach for analyzing the profitability of alternatives with varying inputs and outputs, also referred as cost-benefit analysis or cashflow analysis. Net present value (NPV) and soil expectation value (SEV) are primary measures of the financial viability of an agroforestry system (Cary & Frey, 2020). If NPV is higher for agroforestry system as compared to other alternatives then agroforestry, it is considered as adoptable. As market analysis is best suited for valuing direct use items such as timber and non-wood products, the SEV is utilized due to the availability of a functioning market for these products (Chiputwa et al., 2020). Baron et al. (2018) used cumulative net margin, NPV and an equivalent annual value (EAV) for comparing the profitability of arable cropping, olive monoculture and agroforestry systems in Greece using the Farm-safe model. Lehmann et al. (2020) evaluated the agronomic productivity through application of land equivalent ratio (LER) and economic viability of diverse agroforestry systems in Europe by using gross margin as an indicator. By using Monte Carlo model of financial risk and returns, Cary & Frey (2020) estimated that alley cropping has a mean SEV of US $3,955. With the inclusion of environmental externalities and Yield-SAFE model, De'Jalon et al. (2018) estimated that economic performance of the silvo-arable system with economic EAVs. Jing et al. (2021) used ecological service value (ESV) as measure for valuation of ES and studied impact of change in LU in period of 1991–2015 on ecological services in Nansi Lake Basin in China. Dev et al. (2020) analyzed profitability of bamboo-based agroforestry in Bundelkhand, India by using LER.

6.3.1.1 MARKET METHODS

If ES is sold in a market, the economic value may be calculated directly from the price that customers are willing to pay (WTP) for the related services or commodity. According to their ongoing prices in the market, provisioning services such as wood, food, fuel, and non-wood products can easily be valued. In Ethiopia, the total value of all provisioning services from agroforestry systems for coffee was assessed to be $1,100 per hectare for semi-forest coffee and $2,450 per hectare for garden coffee (Namirembe et al., 2015). Based on the market price of carbon, Namirembe et al. (2015) estimated the above ground carbon stock in Ethiopia's coffee agroforestry system at US $3,370 ha^{-1} and US $21,000 ha^{-1} according to the social price of carbon. Campos et al. (2020) studied valuation of the environmental assets, ESs, and incomes at basic and social prices.

6.3.1.2 PRODUCTION FUNCTION APPROACHES

In case, ES might help to boost the production of products or services traded on marketplaces, the production function approaches are applied. Based on the contribution of a given ES to the increased production of a commodity, it is valued. Considering ES as a component in a market good's manufacturing process and it is valued by considering Es-related variation in production. Inadequate data and lack of understanding about the cause-effect linkages between the ESs are limiting the application of production function approaches to value these services (Aisbett & Kragt, 2010).

6.3.1.3 DOSE RESPONSE FUNCTION APPROACHES

This method is used to estimate the impact of a change in quality of environment and the subsequent impact on the target output. This method may also be applied for value the favorable health effects of improved air quality owing to agroforestry (Mercer et al., 2017).

6.3.1.4 COST-BASED APPROACHES

Regulating services such as regulation of water quality, erosion reduction, and air quality improvements are valued using cost-based approaches. Only if replacement is possible, the replacement or restoration cost technique examines the value of ES in terms of the cost of replacing them through alternative or artificial ways. Value of pollination estimated by valuing of artificial pollination and the valuation of soil fertility improvement could be estimated in terms of the cost of replacing with fertilizer purchases. The cost of avoiding or replacing an ES loss is used in the avoidance cost principle. Health costs associated with drinking low-quality water owing to the loss of watershed trees that safeguard water quality can be used to value improvement in water quality service of agroforestry (Aisbett & Kragt, 2010).

6.3.2 VALUATION OF ES AND EXTERNALITIES FROM AGROFORESTRY

6.3.2.1 TRAVEL COST METHOD

If accurate market data is not readily accessible or markets do not exist, site-based facilities are valued based on the expenses individuals pay to use them. This method first used by Hoteling in 1949 to estimate the value of recreational sites (Grilli, 2018). This approach is based on real expenditures and other costs, such as the value of time spent by visitors to a certain site to visit. It is used to value the recreational site for valuation of the services and benefits derived from that specific location (Christie et al., 2012).

6.3.2.2 CONTINGENT VALUATION

Contingent valuation is commonly used to assess the non-consumptive value of ESs like biodiversity conservation and cultural assets. Customers are asked about their willingness to pay for a shift in an environmentally friendly service like better air and water quality in hypothetical market (Chiputva et al., 2020). It is a questionnaire-based approach to assessing non-usage and passive use values in

ES under alternative land-use scenarios (Mercer et al., 2017). It may also use to predict how likely farmers are ready to accept payments in return for supplying various ESs. Otter & Langenberg (2020) analyzed and quantified the level of willingness to pay for environmental benefits of agroforestry systems from 1,714 German taxpayers and its determining factors from the perspective of German taxpayers by applying contingent evaluation method and observed that on an average German taxpayer willing to pay (WTP) US $38.79 yearly for complimentary benefits of agroforestry. Another study done by Okwu-Abolo et al. (2020) to analyze public willingness to pay for perceived ESs in rubber-based agroforestry systems in Edo and Delta states of Nigeria with contingent valuation method using the payment card method found that nearly 80% of respondents were WTP 100–1,000 per month monthly perceived ecosystem services (ES) provided by rubber-based agro-forestry system. Intensive data requirement, cost of implementation and careful design of survey are some limitations. It possible that somebody overestimate or underestimate willingness to pay based on their false belief (TEEB, 2018). Okwu-Abolo et al. (2020) observed that WTP values by male were higher than female for perceived ESs in rubber-based agroforestry systems in Edo and Delta states of Nigeria. WTP value for ESs can be useful for evaluating the feasibility of PES scheme to guarantee the quality of ES.

6.3.2.3 CHOICE EXPERIMENTS

This method is also applied for valuation the non-consumptive ESs, like biodiversity conservation or cultural values (Mercer et al., 2017). Instead of single scenario, it requires respondents must pick from a set of two or more choice sets, each with possibilities containing different combinations of characteristics (Otter & Langenberg, 2020). The influence of each feature on willingness to pay for an ES may be assessed through statistical analysis of the respondents' sets of options (Chiputwa et al., 2020). The willingness to pay for a certain condition is the sum of the WTP for all its features (Petcharat et al., 2020). Choice experiments are particularly very useful for assessing policies with a wide range of attributes (Mercer et al., 2017; Bartkowski et al., 2020). Shrestha & Alavalapati (2004) examined WTP of households in south-central Florida by using choice method for good and ES produced by establishment of silvipasture in Florida. By considering C-sequestration, wildlife conservation and regulation of water quality three features, they found that households would be WTP US $30.24–$71.17 yearly for five years for silvipasture rendered ESs. Petcharat et al. (2020) studied valuation of ESs by choice method in the Bang Ka Chao Green Area, Thailand, and found that Bangkok residents were WTP US $42 yearly to improve Bang Kachao's ESs.

6.3.2.4 HEDONIC PRICE METHOD

The hedonic technique is used to assess the esthetic and recreational aspects of agroforestry. It is based on revealed preference approach assuming that the prices of land or property is associated with two attributes land or property feature including environmental amenities. Even though two properties appear to be identical, their qualities and environmental features may differ, affecting the price a potential buyer is ready to pay (Aisbett & Kragt, 2010). It used to estimate value of nearby recreational or scenic opportunity, economic benefits of esthetic views and economic damages associated with air pollution or water pollution. Its use in valuation of most of the agroforestry services is limiting as agroforestry markets are rarely as dynamic as properties (TEEB, 2018; Bartkowski et al., 2020).

6.3.2.5 BENEFIT TRANSFER TECHNIQUES

When conducting study on site is not feasible due to time, large area and resources, to estimate values of ES indirectly benefit transfer techniques is used. Thus, value of ES that is obtained from one location or at a given time from earlier studies were used for valuation of ES at different location or at different time (Mercer et al., 2017; Temesgen et al., 2018). Temesgen et al. (2018) studied ES values of different LU systems in Ethiopia, landscape dominated by agroforestry by benefit transfer method and estimated that agroforestry has total economic value of \$604.4 ha^{-1} y^{-1} based on standardized values of USD in 2007.

6.3.2.6 RISK AND UNCERTAINTY

As most of the agroforestry models having long gestation period and requires three to six years for realization of first benefits as compared to arable cropping, risk and uncertainty playing crucial role in consideration of investment in agroforestry. Freedom to amend or postpone decisions in the face of uncertainty and change in return are crucial decision criteria for farmers in the adoption process. Mean-variance analysis also called E-V for "expected value/variance" is traditional risk management approach that identifies a set of "efficient" alternatives that either reduce risk for a given level of return or maximize returns for a given level of risk (Chiputwa et al., 2020). Flexible assumptions and limited information render E-V inaccurate when comparing various distributions of outcomes across options, thus stochastic dominance method is used. Stochastic dominance is commonly used to estimate a partial ordering based on partial information for initial screenings of alternatives (Mercer et al., 2017). For estimating, the value of flexibility for the alternatives for uncertain future real options approach is used. This method provides flexibility of abandon or exercise options at different points in time (Aisbett & Kragt, 2010).

6.3.2.7 PES IN AGROFORESTRY

Payment for ES is rapidly gaining attention as a potential mechanism to integrate protection of natural resources within production landscapes. The value world's ESs is re-evaluated by Costanza et al. (2014) by using updated data. The worldwide value of ESs was predicted to be US $33 trillion per year in 1997, but the entire global ESs in 2011 is US $125 trillion per year (considering updated unit values and variation to biomes). According to Costanza et al. (2014), the annual loss of eco-services owing to LU change from 1997 to 2011 was US $4.3–20.2 trillion. The financial value of Indian forest was an astounding amount of US $1.7 trillion by the expert group of Government of India (Hindustan Times, 2016). As per Bahuguna & Bist (2013) the overall economic value of Indian forests is estimated to be 6.96 lakh crore per year, or 6.86% of GDP, and reflects the basic minimum approximate values of products and services obtained from forests. In India, the overall net value of benefits (after subtracting disserves) offered by the protected area ranges from US $13–148 million per year or US $203–2,294 per hectare per year, depending on the valuation method (Ninan & Andreas, 2016). Singh et al. (2021) studied valuation of fuelwood from agroforestry in India and estimated that 196.4 million additional domestic LPG connections costing US $36,487.5 would be required to replaces energy from the agroforestry-derived fuelwood for 2011–2012. US $1116.6 million saved yearly by substituting agroforestry-sourced fuelwood for cow dung and dung cake can be used for manure instead (Singh et al., 2021).

These tree-based system have tremendous potential to play important role in environmental security and livelihood of the rural people (Chavan et al., 2015c, 2016b; Handa et al., 2016). Contribution of agroforestry in ESs has widely recognized for provisional services but underdeveloped in regulating and cultural services, due to less accurate quantifying methodologies. Globally, some of the research quantified ESs in financial values (Costanza et al., 1997, 2014; Sileshi et al., 2007; Sridhar et al., 2016). Around 10 ESs of the tree-based intercropping (TBI) system of Canada were documented, including agriculture provisioning, timber provisioning, windbreak, air quality regulation, biological control, pollination services, soil improvement, water regulation, nutrient cycling and climate regulation, with the results revealing a total annual margin of US \$2,645 ha^{-1} y^{-1} (averaged over 40 years). Non-market services had a total economic value of US \$1,634 ha^{-1} $year^{-1}$, which was higher than the value of marketable items (i.e., timber and agricultural products). Although, for the rotation period, the total economic value of all ESs was high as US \$54,782 ha^{-1}, but farmers only profited from agricultural and timber goods (Mahbubul et al., 2014). According to Milder et al. (2010), biodiversity conservation markets could advantage 10–15 million low earning families in underdeveloped nations by 2030, carbon markets could aid 25–50 million people, watershed protection markets might advantage 80–100 million population, and landscape beauty and recreation markets could advantage 5–8 million population around world.

6.4 CONCLUSION

Agroforestry, a tree-based system has been considered as a climate-smart agricultural strategy and it provides various products and economic and ecological benefits. Many of these ESs are valued and markets are readily available while, on the other side hundreds of beneficial services generated by trees are neither documented nor valued. In the last decade, ESs, its quantification and valuation has been the buzz word. In agroforestry systems, different ES *viz.*, provisional, regulating, supporting, and cultural services are quantified. The type and quantum of ES produced varies with agroforestry systems and practices. The effect of ES generated may be at local, regional or in the global scale. Payment for ESs has further increased the scope of agroforestry and its adoption in a large scale. Proper designing and choice of species in agroforestry will enhance the ESs. Agroforestry may compete with traditional farming and livestock systems when incentives or rewards for ESs are incorporated. Further research is needed on the financial and economic valuation of agroforestry derived ES, risk analysis, their influence on the adoption of agroforestry among farmers. As available studies on non-market valuation studies of ESs produced by agroforestry were very few, many studies are needed particularly for services, which are benefitting society such as maintenance of biodiversity and, water quality enhancement to compare with alternative land-use practices. To avoid double-counting to

ES when two or more ES are interconnected or overlapped each other, there is a need to better understand interdependencies between the ESs. Despite the potential worth of these services to society, landowners are rarely compensated for them. Agroforestry systems will be commercially feasible in many circumstances only if landowners are compensated for supplying non-market services. Additionally, agroforestry may become a financially luring option among farmers, policymakers, and other stakeholders if their ecosystem values widely recognized and incorporated into formal policy-making processes.

KEYWORDS

- **agroforestry**
- **carbon sequestration**
- **contingent valuation**
- **ecosystem services**
- **formal policy**
- **microclimate**
- **soil and moisture conservation**
- **stakeholders**

REFERENCES

Aggarwal, R. K., Gupta, J. P., Saxena, S. K., & Muthan, K, D., (1976). Studies on physic-chemical and ecological changes under twelve years old five desert tree species of Western Rajasthan. *Indian For., 100*, 863–872.

Aisbett, E., & Kragt, M. E., (2010). *Valuing Ecosystem Services to Agricultural Production to Inform Policy Design: An Introduction* (No. 96385). Environmental Economics Research Hub.

Ajit, H. A. K., Dhyani, S. K., Bhat, G. M., Malik, A. R., Dutt, V., & Jain, A., (2017). Quantification of carbon stocks and sequestration potential through existing agroforestry systems in the hilly Kupwara district of Kashmir valley in India. *Curr. Sci., 113*, 781–785.

Alagele, S. M., Anderson, S. H., & Udawatta, R. P., (2020). Agroforestry, grass, biofuel crop, and row-crop management effects on soil water dynamics for claypan landscapes. *Soil Sci. Soc. A. J., 84*(1), 203–219.

Amadi, C. C., Farrell, R. E., & Van, R. K. C., (2017). Greenhouse gas emissions along a shelterbelt-cropped field transect. *Agric. Ecosyst. Environ., 241*, 110–120.

Anonymous (Healing degraded land), (1997). *INBAR Magazine, 5*(3), 40–45.
Bado, B. V., Whitbread, A., & Manzo, M. L. S., (2021). Improving agricultural productivity using agroforestry systems: Performance of millet, cowpea, and ziziphus-based cropping systems in West Africa Sahel. *Agric. Ecosyst. Environ., 305,* 107175.
Bahuguna, V. K., & Bisht, N. S., (2013). Valuation of ecosystem goods and services from forests in India. *Indian For., 139*(1), 1–13.
Bargali, S. S., (1995). Litter fall, nutrient, and leaf litter decomposition in an age series of eucalyptus plantations in Central Himalaya. *Oecology, 4*, 31–38.
Bargues-Tobella, A., Hasselquist, N. J., Bazie, H. R., Bayala, J., Laudon, H., & Ilstedt, U., (2020). Trees in African drylands can promote deep soil and groundwater recharge in a future climate with more intense rainfall. *Land Degrad. Dev., 31*(1), 81–95.
Baron, G., Giannitsopoulos, M., Pantera, A., Graves, A., & Burgess, P. J., (2018). Comparison of the profitability of an arable rotation, a monoculture olive system and a silvoarable system in Greece using the farm-SAFE model. In: *EURAF, 4th European Agroforestry Conference- Agroforestry as Sustainable Land Use.* Nijmegen, Netherland.
Barrios, E., Valencia, V., Jonsson, M., Brauman, A., Hairiah, K., Mortimer, P. E., & Okubo, S., (2018). Contribution of trees to the conservation of biodiversity and ecosystem services in agricultural landscapes. *Int. J. Biodivers. Sci. Ecosyst. Serv. Manag., 14*(1), 1–16.
Bartkowski, B., Bartke, S., Helming, K., Paul, C., Techen, A. K., & Hansjurgens, B., (2020). Potential of the economic valuation of soil-based ecosystem services to inform sustainable soil management and policy. *Peer J.* https://doi.org/10.7717/peerj.8749.
Beer, J., Muschler, R., Kass, D., & Somarriba, E., (1998). Shade management in coffee and cacao plantations. *Agrofor. Syst., 38*, 139–164.
Béliveau, A., Lucotte, M., Davidson, R., Paquet, S., Mertens, F., Passos, C. J., & Romana, C. A., (2017). Reduction of soil erosion and mercury losses in agroforestry systems compared to forests and cultivated fields in the Brazilian Amazon. *Journal of Environmental Management, 203*, 522–532.
Bertomeu, M., (2006). Financial evaluation of smallholder timber-based agroforestry systems in Claveria, Northern Mindanao, the Philippines. *Small-scale For., 5*, 57–81.
Boinot, S., Fried, G., Storkey, J., Metcalfe, H., Barkaoui, K., Lauri, P. E., & Meziere, D., (2019). Alley cropping agroforestry systems: Reservoirs for weeds or refugia for plant diversity? *Agric. Ecosyst. Environ., 284,* 106584.
Boyd, J., & Spencer, B., (2007). What are ecosystem services? The need for standardized environmental accounting units. *Ecol. Econ., 63*, 616–626.
Braat, L. C., & Groot, R. D., (2012). The ecosystem services agenda: Bridging the worlds of natural science and economics, conservation and development, and public and private policy. *Ecosyst. Services, 1*(1), 4–15.
Brown, S. E., Miller, D. C., Ordonez, P. J., & Baylis, K., (2018). Evidence for the impacts of agroforestry on agricultural productivity, ecosystem services, and human well-being in high-income countries: A systematic map protocol. *Environ. Evidence, 7*(1), 24.
Calder, I. R., Amezaga, J. M., Aylward, B., Bosch, J., Fuller, L., Gallop, K., Gosain, A., et al., (2004). Forest and water policies- The need to reconcile public and science perceptions. *Geol. Acta, 2*, 157–166.
Campos, P., Álvarez, A., Oviedo, J. L., Ovando, P., Mesa, B., & Caparrós, A., (2020). Income and ecosystem service comparisons of refined national and agroforestry accounting frameworks: Application to holm oak open woodlands in Andalusia, Spain. *Forests, 11*(2), 185.

Cardinael, R., Hoeffner, K., Chenu, C., Chevallier, T., Be´ral, C., Dewisme, A., & Cluzeau, D., (2019). Spatial variation of earthworm communities and soil organic carbon in temperate agroforestry. *Biol. Fertil. Soils, 55*, 171–183.

Cary, M. A., & Frey, G. E., (2020). Alley cropping as an alternative under changing climate and risk scenarios: A Monte-Carlo simulation approach. *Agric. Syst., 185,* 102938.

Cerda, R., Avelino, J., Harvey, C. A., Gary, C., Tixier, P., & Allinne, C., (2020). Coffee agroforestry systems capable of reducing disease-induced yield and economic losses while providing multiple ecosystem services. *Crop Prot., 105*149.

Chan, K. M. A., (2011). Cultural services and non-use values. In: Kareiva, P., Daily, G., Ricketts, T., Tallis, H., & Polasky, S., (eds.), *The Theory and Practice of Ecosystem Service Valuation in Conservation* (pp. 206–228). Oxford University Press, Oxford.

Chauhan, D., Chavan, S. B., Keerthika, A., Jha, A., Dwivedi, R. P., Newaj, R., Anil, K. A., (2015). Socio-economic fuelwood survey in Bundelkhand region of Central India. In: Mishra, G. C., Yadav, S. K., (eds.), *Energy Research and Environmental Management: An Innovative Approach* (pp. 124–129). Krishi Sankriti Publications: New Delhi.

Chavan, S. B., Keerthika, A., Dhyani, S. K., Handa, A. K., Newaj, R., & Rajarajan, K., (2015a). National agroforestry policy in India: A low hanging fruit. *Curr. Sci., 108*, 1826–1834.

Chavan, S. B., Keerthika, A., Gunaga, R. P., Uthappa, A. R., Handa, A. K., Jha, A., Newaj, R., et al., (2015c). Status of polyembryony in tree borne oilseeds: A review. *Int. J. Bioresource Stress Manag., 6*, 509–512.

Chavan, S. B., Keerthika, A., Jha, A., & Ranjan, R., (2015b). Biodiesel production processes. In: Devi, S. V., Sridhar, K. B., Chavan, S. B., Handa, A. K., & Dhyani, S. K., (eds.), *Agroforestry for Biofuels and Bioenergy* (pp. 75–77). National Research Center for Agroforestry Institute: Jhansi, Uttar Pradesh, India.

Chavan, S. B., Toky, O. P., Handa, A. K., & Dhyani, S. K., (2016b). Innovative agroforestry for environmental security in India. *World Agric., 6*(8), 51–58.

Chavan, S. B., Uthappa, A. R., Sridhar, K. B., Keerthika, A., Handa, A. K., Newaj, R., Kumar, N., et al., (2016a). Trees for life: Creating sustainable livelihood in Bundelkhand region of Central India. *Curr. Sci., 11*(6), 994–1002.

Chavan, S., Newaj, R., Keerthika, A., Ram, A., Jha, A., & Kumar, A., (2014). Agroforestry for adaptation and mitigation of climate change. *Popular Kheti, 2*, 214–220.

Chiemela, S. N., Noulèkoun, F., Chiemela, C. J., Zenebe, A., Abadi, N., & Birhane, E., (2018). Conversion of degraded agricultural landscapes to a smallholder agroforestry system and carbon sequestration in drylands. *Int. J. Clim. Change Strategies Manag., 10*, 472–487.

Chiputwa, B., Ihli, H. J., Wainaina, P., & Gassner, A., (2020). Accounting for the invisible value of trees on farms through valuation of ecosystem services. In: Rusinamhodzi, L., (ed.), *The Role of Ecosystem Services in Sustainable Food Systems* (pp. 229–261). Academic Press.

Christie, M., Fazey, I., Cooper, R., Hyde, T., & Kenter, J. O., (2012). An evaluation of monetary and non-monetary techniques for assessing the importance of biodiversity and ecosystem services to people in countries with developing economies. *Ecol. Econ., 83*, 67–78.

Costanza, R., D'Arge, R., De Groot, R., Farber, S., Grasso, M., Hannon, B., Limburg, K., et al., (1997). The value of the world's ecosystem services and natural capital. *Nature, 387*(6630), 253–260.

Costanza, R., De'Groot, R., Sutton, P., Phoeg, S. V., Anderson, S. J., Kubiszewskia, I., Farbere, S., Turner, R. K., (2014). Changes in the global value of ecosystem services. *Global Environ. Change, 26*, 152–158.

Dagar, J. C., & Tewari, J. C., (2016). *Agroforestry Research Developments: Anecdotal to Modern Science*. Nova Publishers, New York.

Dagar, J. C., Singh, G., & Singh, N. T., (2001). Evaluation of forest and forest trees used for rehabilitation of semiarid alkali soils in India. *Arid Land Res. Manag., 15*, 115–133.

Daily, G., (1997). *Nature's Services: Societal Dependence on Natural Ecosystems Island*, Washington DC.

Danso, S. K. A., Zapata, F., & Awonaike, K. O., (1995). Measurement of biological N_2 fixation of field-grown *Robinia pseudoacacia* L. *Soil Bio. Biochem., 27*, 415–419.

Das, A., Tomar, J. M. S., Munda, G. C., & Ramesh, T., (2010). Productivity and economics of lowland rice as influenced by incorporation of N fixing tree leaves. *Nutr. Cycl. Agroecosyst., 87*, 9–19.

De Groot, S. A., (2014). *In Search of Beauty.* Developing Beautiful Organizations. Eindhoven University of Technology, *Eindhoven*.

De' Jalón, S. G., Graves, A., Palma, J. H., Williams, A., Upson, M., & Burgess, P. J., (2018). Modelling and valuing the environmental impacts of arable, forestry and agroforestry systems: A case study. *Agrofor. Syst., 92*, 1059–1073.

De'Sousa, K. F. D., Detlefsen, G., & De'Melo V. F. E., (2016). Timber yield from smallholder agroforestry systems in Nicaragua and Honduras. *Agrofor. Syst., 90*, 207–218.

Dev, I., Ram, A., Ahlawat, S. P., Palsaniya, D. R., Singh, R., Dhyani, S. K., Kumar, N., et al., (2020). Bamboo-based agroforestry system (*Dendrocalamus strictus* + sesame-chickpea) for enhancing productivity in semi-arid tropics of central India. *Agrofor. Syst., 94*(5), 1725–1739.

Dhanda, R. S., & Verma, R. K., (1995). *Growth Performance of Populus deltoides Bartr. in Agroforestry Plantations in Punjab*. In: Khurana, D. K., (ed.), *Poplars in India: Recent Research Trends.* IDRC-UHF: Solan.

Dhyani, S. K., (2014). Introduction to biofuels- a renewable source. In: Devi, V. S., Sridhar, K. B., Chavan, S., Handa, A. K., & Dhyani, S. K., (eds.), *Training Manual on 'Agroforestry for Biofuels and Bioenergy* (p. 229). ICAR-NRC for Agroforestry, Jhansi, U. P.

Dhyani, S. K., Devi, V. S., & Handa, A. K., (2015). *Tree Borne Oilseeds for Oil and Biofuel.* Technical Bulletin, ICAR-CAFRI, Jhansi.

Doty, S. L., Freeman, J. L., Cohu, C. M., Burken, J. G., Firrincieli, A., & Simon, A., (2017). Enhanced degradation of TCE on a superfund site using endophyte-assisted poplar tree phytoremediation. *Environ. Sci. Technol., 51*(17), 10050–10058.

Dwivedi, R. P., Kareemulla, K., Singh, R., Rizvi, R. H., & Chauhan, J., (2007). Socio-economic analysis of agroforestry systems in Western Uttar Pradesh. *Indian Res. J. Ext. Edu., 7*, 18–23.

FAO, (2018). *The State of the World's Forests: Forests Pathways to Sustainable Development.* Food and Agriculture Organization of the United Nations, Rome.

Feliciano, D., Ledo, A., Hillier, J., & Nayak, D. R., (2018). Which agroforestry options give the greatest soil and above ground carbon benefits in different world regions? *Agric. Ecosyst. Environ., 254*, 117–129.

Fisher, B., Turner, R. K., & Morling, M., (2009). Defining and classifying ecosystem services for decision making. *Ecol. Econ., 68*, 643–653.

Froufe, L. C. M., Schwiderke, D. K., Castilhano, A. C., Cezar, R. M., Steenbock, W., Seoane, C. E. S., & Vezzani, F. M., (2019). Nutrient cycling from leaf litter in multistrata successional agroforestry systems and natural regeneration at Brazilian Atlantic rainforest biome. *Agrofor. Syst., 94*, 1–13.

Giller, K. E., (2001). *Nitrogen Fixation in Tropical Cropping Systems* (2nd edn.) CAB International Publishing, Wallingford.

Government of India, (2008). *National Policy on Biofuels, 2008*. Ministry of New and Renewable Energy, Government of India, New Delhi. Available at: http://mnre.gov.in/file-manager/UserFiles/biofuel-policy.pdf (accessed on 28 December 2023).

Grilli, G., Gavin, L., John, C., & Stephen, H., (2018). A travel cost evaluation of the benefits of two destination salmon rivers in Ireland. *Journal of Outdoor Recreation and Tourism, 23*, 1–7.

Gupta, N., Kukal, S. S., Bawa, S. S., & Dhaliwal, G. S., (2009). Soil organic carbon and aggregation under poplar-based agroforestry system in relation to tree age and soil type. *Agrofor. Syst., 76*, 27–35.

Gupta, R. K., (1995). Multipurpose trees. In: Gupta, R. K., (ed.), *Multipurpose Trees for Agroforestry and Wasteland Utilization* (pp. 331–335). Center for Research on Environmental Application: Training and Education, Dehradun.

Haggar, J., Pons, D., Saenz, L., & Vides, M., (2019). Contribution of agroforestry systems to sustaining biodiversity in fragmented forest landscapes. *Agric. Ecosyst. Environ., 283*, 1–8, 106567.

Hamer, A. S., & Franzel, M. B., (2007). Assessing profitability of fodder banks using farmers' criteria in the desert margins of West Africa. *Land Degrad. Dev., 18*, 670–679.

Handa, A. K., Toky, O. P., Dhyani, S. K., & Chavan, S. B., (2016). Innovative agroforestry for livelihood security in India. *Teeb Agric., 6*(2), 7–16.

Hindustan Times, (2016). Home Page. Available at: https://www.hindustantimes.com/india-news/india-s-forests-valued-at-rs-115-trillion-but-tribals-unlikely-to-get-a-share/story-06QqOb6PGrkcxZ7l1ACDI.html accessed on 28 December 2023).

Holden, J., Grayson, R., Berdeni, D., Bird, S., Chapman, P. J., Edmondson, J. L., Firbank, L. G., et al., (2019). The role of hedgerows in soil functioning within agricultural landscapes. *Agric. Ecosyst. Environ., 273*, 1–12.

Jimnez, F., (1986). *Balance hídrico de dos Sistemasagroforestales: Café-poro y Café Laurel en Turrialba, Costa Rica* (p. 104). M. Sc. Thesis. UCR-CATIE, Turrialba, C. R.

Jing, Y., Chang, Y., Cheng, X., & Wang, D., (2021). Land-use changes and ecosystem services under different scenarios in Nansi Lake Basin, China. *Environ. Monit. Assess., 193*(1), 1–14.

Jones, C. G., John, L. H., & Shachak, M., (1997). Positive and negative effects of organisms as physical ecosystem engineers. *Ecology, 78*(7), 1946–1957.

Kang, B. T., & Mulongoy, K., (1992). Nitrogen contribution of woody legumes in alley cropping systems. In: Mulonogy, K., Gueye, M., Spencer, D. S. C., (eds.), *Biological Nitrogen Fixation and Sustainability of Tropical Agriculture* (pp. 367–375). Willey.

Kang, B. T., (1997). Alley cropping-soil productivity and nutrient recycling. *For. Ecol. Manag., 91*, 75–82.

Katwal, R. P. S., (2003). Teak in India: Status, prospects, and perspectives. In: *Proceedings of International Conference on Quality Timber Products of Teak from Sustainable Forest Management*. Peechi, Kerala, India.

Kay, S., Graves, A., Palma, J. H., Moreno, G., Roces-Díaz, J. V., Aviron, S., Chouvardas, D., et al., (2019a). Agroforestry is paying off–economic evaluation of ecosystem services in European landscapes with and without agroforestry systems. *Ecosys. Services, 36*, 100896.

Kay, S., Rega, C., Moreno, G., Den'Herder, M., Palma, J. H., Borek, R., Crous-Duran, J., et al., (2019b). Agroforestry creates carbon sinks whilst enhancing the environment in agricultural landscapes in Europe. *Land Use Policy, 83*, 581–593.

Kittur, B. H., Sudhakara, K., Kumar, M. B., Kunhamu, T. K., & Sureshkumar, (2016). Bamboo based agroforestry systems in Kerala, India: Performance of turmeric (*Curcuma longa* L.) in the subcanopy of differentially spaced seven-year-old bamboo stand. *Agrofor. Syst., 90*, 237–250.

Komar, O., (2015). Priority contribution. Ecology and conservation of birds in coffee plantations: A critical review. *Bird Conserv. Int., 16*, 1–23.

Kumar, M. B., George, S., & Chinnamani, S., (1994). Diversity, structure and standing stock of wood in the homegardens of Kerala in Peninsular India. *Agrofor. Syst., 25*, 243–262.

Lee, K. H., Isenhart, T. M., & Schultz, R. C., (2003). Sediment and nutrient removal in an established multi-species riparian buffer. *J. Soil Water Conserv., 58*(1), 1–8.

Lehmann, L. M., Smith, J., Westaway, S., Pisanelli, A., Russo, G., Borek, R., Sandor, M., et al., (2020). Productivity and economic evaluation of agroforestry systems for sustainable production of food and non-food products. *Sustainability, 12*(13), 5429.

Li, X. A., Ge, T. D., Chen, Z., Wang, S. M., Ou, X. K., Wu, Y., Chen, H., & Wu, J. P., (2020). Enhancement of soil carbon and nitrogen stocks by abiotic and microbial pathways in three rubber-based agroforestry systems in Southwest China. *Land Degrad. Dev., 31*, 2507–2515. https://doi.org/10.1002/ldr.3625.

Lin, X. J., Barrington, S., Nicell, J., Choiniere, D., & Vezina, A., (2006). Influence of windbreaks on livestock odor dispersion plume in the field. *Agric. Ecosyst. Environ., 116*, 263–272.

Mahbubul, A., Olivier, A., Paquette, A., Dupras, J., Revéret, J. P., & Messier, C., (2014). A general framework for the quantification and valuation of ecosystem services of tree-based intercropping systems. *Agrofor. Syst., 88*(4), 679–691.

Malone, G., VanWicklen, G., Collier, S., & Hansen, D., (2006). Efficacy of vegetative environmental buffers to capture emissions from tunnel ventilated poultry houses. In: *Proceedings of the Workshop on Agricultural Air Quality: State of the Science* (pp. 875–878). Potomac, Maryland.

Manevski, K., Jakobsen, M., Kongsted, A. G., Georgiadis, P., Labouriau, R., Hermansen, J. E., & Jorgensen, U., (2019). Effect of poplar trees on nitrogen and water balance in outdoor pig production- a case study in Denmark. *Sci. Total Environ., 646*, 1448–1458.

Maoyi, F., & Banik, R. L., (1996). Bamboo production system and their management. In: People and the environment: Propagation and management. In: Rao, V. R., Sastry, C. B., & Widjaja, E., (eds.), *Proceedings of 5th International Bamboo Workshop and 4th International Bamboo Congress, Bali* (Vol. 1, pp. 18–33).

Martin-Chave, A., Béral, C., & Capowiez, Y., (2019). Agroforestry has an impact on nocturnal predation by ground beetles and Opiliones in a temperate organic alley cropping system. *Biol. Control., 129*, 128–135.

Masiero, M., Pettenella, D., Boscolo, M., Barua, S. K., Animon, I., & Rao, M., (2019). *Valuing Forest Ecosystem Services: A Training Manual for Planners and Project Developers*. FAO, Rome.

Mayavel, A., Soosai, R. J., Singh, G. B., & Krishnakumar, N., (2014). *Cultivation Techniques for Gmelina Arborea* (pp. 23–27). In Transfer of tree cultivation technologies to Krishi Vigyan Kendras (KVKs) of Tamil Nadu and Pondicherry; Institute of Forest Genetics and Tree Breeding. Coimbatore.

Mercer, D. E., Li, X., Stainback, A., & Alavalapati, J., (2017). Valuation of agroforestry services. In: Schoeneberger, M. M., Bentrup, G., & Patel-Weynand, T., (eds.), *Agroforestry:*

Enhancing Resiliency in US Agricultural Landscapes Under Changing Conditions. Department of Agriculture, Forest Service: Washington, DC, Gen. Tech. Report WO-96.

Milder, J. C., Scherr, S. J., & Bracer, C., (2010). Trends and future potential of payment for ecosystem services to alleviate rural poverty in developing countries. *Ecol. Soc., 15*(2), 1–4.

Millennium Ecosystem Assessment, (2005). *Ecosystems and Human Well-Being: Synthesis.* Island Press, Washington, DC.

Monteith, J. L., Ong, C. K., & Corlett, J. E., (1991). Microclimatic interactions in agroforestry systems. *For. Ecol. Manage., 45*, 31–44.

Mooney, H. A., & Ehrlich, P. R., (1997). Ecosystem services: A fragmentary history. In: Daily, G. C., (ed.), *Nature's Services: Societal Dependence on Natural Ecosystems* (pp. 11–19). Island Press, Washington, DC.

Moyo, S., & Ayuk, E. T., (2001). Costs and benefits of tree fodder with small-scale dairy farmers in Zimbabwe. *Living Trees S. Afr., 6*, 1–3.

Muchane, M. N., Sileshi, G. W., Gripenberg, S., Jonsson, M., Pumariño, L., Barrios, E., (2020). Agroforestry boosts soil health in the humid and sub-humid tropics: A meta-analysis. *Agric. Ecosys. Environ., 295,* 106899.

Namirembe, S., Mcfatridge, S., Duguma, L., Bernard, F., & Minag, P., (2015). *Agroforestry: An Attractive REDD Policy Option?* World Agroforestry Center (ICRAF), Nairobi, Kenya.

Nath, A. J., Das, G., & Das, A. K., (2009). Above ground standing biomass and carbon storage in village bamboos in North East India. *Biomass Bioenergy, 33*(9), 1188–1196.

Nicodemus, A., (2014). *Cultivation Techniques for Casuarina* (pp. 6–11). In Transfer of Tree Cultivation Technologies to Krishi Vigyan Kendras (KVKs) of Tamil Nadu and Puducherry. Institute of Forest Genetics and Tree Breeding, Coimbatore.

Ninan, K. N., & Andreas, K., (2016). Valuing forest ecosystem services and disservices- Case study of a protected area in India. *Ecosyst. Serv., 20*, 1–14.

Okwu-Abolo, C., Adekunle, M. F., Mesike, C. S., & Omoigberale, J. A., (2020). Public willingness to pay for perceived ecosystem services in rubber agroforestry systems in Edo and Delta States, Nigeria. *J. Res. For. Wildlife Environ., 12*(3), 50–57.

Ong, C. K., & Leakey, R. R. B., (1999). Why tree-crop interactions in agroforestry appear at odds with tree-grass interactions in tropical savannahs. *Agrofor. Syst., 45*, 109–129.

Otter, V., & Langenberg, J., (2020). Willingness to pay for environmental effects of agroforestry systems: A PLS-model of the contingent evaluation from German taxpayer's perspective. *Agrofor. Syst., 94*(3), 811–829.

Palanisamy, K., (2014). *Cultivation Techniques for Teak* (PP. 1–5). Transfer of Tree Cultivation Technologies to Krishi Vigyan Kendra (KVKs) of Tamil Nadu and Puducherry. Institute of Forest Genetics and Tree Breeding. Coimbatore.

Palsaniya, D. R., Dhyani, S. K., & Rai, P., (2011). *Silvipasture in India: Present Perspectives and Challenges Ahead.* Scientific Publishers.

Pandey, A. K., Gupta, V. K., & Solanki, K. R., (2010). Productivity of neem-based agroforestry system in semi-arid region of India. *Range Manage. Agrofor., 31*(2), 144–149.

Pandey, D., (2000). Development of plantation forestry and joint forest management in India. In: *Proceedings of the International Conference on Timber Plantation Development.* Corporate Document Repository, Food and Agriculture Organization of the United Nations, Rome.

Pardon, P., Reheul, D., Mertens, J., Reubens, B., De Frenne, P., De Smedt, P., Proesmans, W., et al., (2019). Gradients in abundance and diversity of ground dwelling arthropods as a function of distance to tree-rows in temperate arable agroforestry systems. *Agriculture, Ecosystems & Environment, 270*, 114–128.

Patel, L. B., Sidhu, B. S., & Beri, V., (1996). Symbiotic efficiency of Sesbania rostrata and S. cannabina as affected by agronomic practices. *Biol. FerTL. Soils, 21*, 149–151.

Patil, V. D., Sornikar, P. N., Adsul, P. B., & Thengal, P. D., (2004). Profile studies, organic matter build-up and nutritional status of soil under Bamboo (*Dendrocalamus strictus*) based agroforestry system. *J. Soil Crop., 34*, 48–54.

Patterson, P., Adrizal, B. R., Myers, C., Martin, G., Shockey, R., & Van, D. G. M., (2006). Plant foliar nitrogen and temperature on commercial poultry farms in Pennsylvania. In: *Proceedings of the Workshop on Agricultural Air Quality: State of Science* (pp. 453–457).

Pavlidis, G., & Tsihrintzis, V. A., (2018). Environmental benefits and control of pollution to surface water and groundwater by agroforestry systems: A review. *Water Resour. Manag., 32*, 1–29.

Pavlidis, G., Karasali, H., & Tsihrintzis, V. A., (2020). Pesticide and fertilizer pollution reduction in two alley-cropping agroforestry cultivating systems. *Water Air Soil Pollut., 231*(5), 241. https://doi.org/10.1007/s11270-020-04590-2.

Pavlidis, G., Karasali, H., & Tsihrintzis, V. A., (2021). Dynamics of changes in the concentrations of herbicides and nutrients in the soils of a combined wheat-poplar tree cultivation: A field experimental model during the growing season. *Agrofor. Syst., 10*, 1–18.

Pavlidis, G., Tsihrintzis, V. A., Karasali, H., & Alexakis, D., (2018). Tree uptake of excess nutrients and herbicides in a maize-olive tree cultivation system. *J. Environ. Sci. Health., 53*, 1–12.

Petcharat, A., Lee, Y., Chang, J. B., (2020). Choice experiments for estimating the non-market value of ecosystem services in the Bang Kachao Green Area, Thailand. *Sustainability, 12*(18), 7637.

Philpott, S. M., & Bichier, P., (2012). Effects of shade tree removal on birds in coffee agroecosystems in Chiapas, Mexico. *Agric., Ecosys. Environ., 149*, 171–180.

Philpott, S., Greenberg, R., Bichier, P., & Perfecto, I., (2004). Impacts of major predators on tropical agroforest arthropods: Comparisons within and across taxa. *Oecologia, 140*, 140–149.

Piza, P. A., Suárez, J. C., Andrade, H. J., (2021). Litter decomposition and nutrient release in different land use located in Valle del Cauca (Colombia). *Agrofor. Syst.* https://doi.org/10.1007/s10457-020-00583-6.

Purwaningsih, R., Sartohadi, J., Setiawan, M. A., (2020). Trees and crops arrangement in the agroforestry system based on slope units to control landslide reactivation on volcanic foot slopes in Java, Indonesia. *Land, 9*(9), 327.

Rachman, L. M., Hidayat, Y., Tarigan, S. D., Sitorus, S. P., Fitri, R., & Ain, A. Q., (2020). The effect of agroforestry system on reducing soil erosion in upstream Ciliwung watershed. In: *IOP Conference Series: Earth and Environmental Science* (Vol. 556). IOP Publishing.

Rajasugunasekar, D., (2014). *Cultivation Techniques for Ailanthus* (pp. 19–22). Transfer of Tree Cultivation Technologies to Krishi Vigyan Kendras (KVKs) of Tamil Nadu and Puducherry. Institute of Forest Genetics and Tree Breeding, Coimbatore, India.

Ranjan, R., (2020). Payments for ecosystems services-based agroforestry and groundwater nitrate remediation: The case of *Poplar deltoides* in Uttar Pradesh, India. *J. Cleaner Prod.*, 125059, 1–15. https://doi.org/10.1016/j.jclepro.2020.125059.

Ram, A., Dev, I., Uthappa, A. R., Kumar, D., Kumar, N., Chaturvedi, O. P., Dotaniya, M. L., & Meena, B. P., (2017). Reactive nitrogen in agroforestry systems of India. In: *The Indian Nitrogen Assessment* (pp. 207–218). Elsevier.

Ramírez, O. A., Somarriba, E., Ludewigs, T., & Ferreira, P., (2001). Financial returns, stability, and risk of cacao-plantain-timber agroforestry systems in Central America. *Agrofor. Syst., 51*(2), 141–154.

Ranjan, R., (2020). Payments for ecosystems services-based agroforestry and groundwater nitrate remediation: The case of Poplar deltoides in Uttar Pradesh, India. *J. Cleaner Prod.,* 125059.

Rao, S., Venkateswarlu, C., Dinesh, B. B., Wani, M., Dixit, S. P., Sahrawat, S., & Sumanta, K. L., (2011). *Soil Health Improvement with Gliricidia Green Leaf Manuring in Rainfed Agriculture, On Farm Experiences.* Central Research Institute for Dryland Agriculture, Hyderabad.

Raveendra, S. A. S. T., Nissanka, S. P., Somasundaram, D., Atapattu, A. J., & Mensah, S., (2021). Coconut-*Gliricidia* mixed cropping systems improve soil nutrients in dry and wet regions of Sri Lanka. *Agrofor. Syst.*, 1–13.

Rhoades, C. C., (1997). Single-tree influences on soil properties in agroforestry: Lessons from natural forest and savanna ecosystems. *Agrofor. Syst., 35*, 71–94.

Rivest, M., Whalen, J. K., & Rivest, D., (2020). Variation of soil microbial and earthworm communities along an agricultural transect with tree windbreak. *Agrofor. Syst., 94*(5), 1639–1649.

Rodríguez, L., Suárez, J. C., Rodríguez, W., Artunduaga, K. J., & Lavelle, P., (2021). Agroforestry systems impact soil macroaggregation and enhance carbon storage in Colombian deforested Amazonia. *Geoderma, 384*, 114810.

Sagastuy, M., & Krause, T., (2019). Agroforestry as a biodiversity conservation tool in the Atlantic Forest? motivations and limitations for small-scale farmers to implement agroforestry systems in North-Eastern Brazil. *Sustainability., 11*(24), 6932.

Seruni, A. P., Aguilar, F. X., Cai, Z., Gold, M. A., & Roshetko, J. M., (2020). Parcelized cut-and-carry agroforestry systems for confined livestock. *Small-scale For.*, 1–25. https://doi.org/10.1007/s11842-020-09460-7.

Shankar, A. V., (2012). *Agroforestry – A Blog on Agrihortisilviculture: Melia Dubia for Biomass Power Production.* Available at: http://agrowmania.blogspot.in/2012/05/melia-dubia-for-biomass-power.html (accessed on 28 December 2023).

Sharma, B. D., & Gupta, I. C., (1989). Effect of trees cover on soil fertility in western Rajasthan. *Indian For., 115*, 348–354.

Sharma, G., Sharma, R., Sharma, E., & Singh, K. K., (2002). Performance of an age series of *Alnus*- Cardamom plantations in the Sikkim Himalaya: Nutrient dynamics. *Annals of Botany, 89*, 273–282.

Sharma, N., Bohra, B., Pragya, N., Ciannella, R., Dobie, P., & Lehmann, S., (2016). Bioenergy from agroforestry can lead to improved food security, climate change, soil quality, and rural development. *Food Energy Secur., 3*(5), 165–183.

Shastri, C. M., Bhat, D. M., Nagaraja, B. C., Murali, K. S., & Ravindranath, N. H., (2002). Tree species diversity in a village ecosystem in Uttara Kannada district in Western Ghats, Karnataka. *Curr. Sci., 82*, 1080–1084.

Shrestha, R. K., & Alavalapati, J., (2004). Valuing environmental benefits of silvopasture practice: A case study of the Lake Okeechobee watershed in Florida. *Ecol. Econ., 49*(3), 349–359.

Sida, T. S., Baudron, F., Kim, H., & Giller, K. E., (2018). Climate-smart agroforestry: *Faidherbia albida* trees buffer wheat against climatic extremes in the Central Rift Valley of Ethiopia. *Agric. For. Meteor., 248*, 339–347.

Sileshi, G., Akinnifesi, F. K., Ajayi, O. C., Chakeredza, S., Kaonga, M., & Matakala, P. W., (2007). Contributions of agroforestry to ecosystem services in the Miombo eco-region of eastern and southern Africa. *Afr. J. Environ. Sci. Technol., 1*(4), 68–80.

Singh, M., Babanna, S. K., Kumar, D., Dwivedi, R. P., Dev, I., Kumar, A., Tewari, R. K., et al., (2021). Valuation of fuelwood from agroforestry systems: A methodological perspective. *Agrofor. Syst.*, 1–17.

Singh, S. P., & Singh, D., (2010). Biodiesel production through the use of different sources and characterization of oils and their esters as the substitute of diesel: A review. *Renew. Sustain. Energ. Rev., 14*, 200–216.

Sivakumar, V., (2014). *Cultivation Techniques for Eucalyptus*. In Transfer of Tree Cultivation Technologies to Krishi Vigyan Kendras (KVKs) of Tamil Nadu and Puducherry. Institute of Forest Genetics and Tree Breeding, Coimbatore, India.

Sridhar, K. B., Kumar, S., Chavan, S. B., Uthappa, A. R., Diwedi, R. P., Singh, M., & Singh, R., (2016). *Butea monosperma*: The Kulpvrksha of Bundelkhand Region. *Indian Farming, 65*(11), 1–4.

Steffan-Dewenter, I., Kessler, M., & Barkmann, J., (2007). Tradeoffs between income, biodiversity, and ecosystem functioning during tropical rainforest conversion and agroforestry intensification. *PNAS, 104*, 4973–4978.

Stöcker, C. M., Bamberg, A. L., Stumpf, L., et al., (2020). Short-term soil physical quality improvements promoted by an agroforestry system. *Agroforest Syst., 94*, 2053–2064.

TEEB, (2008). *TEEB for Agriculture and Food: Scientific and Economic Foundations*. Available at: https://teebweb.org/wp-content/uploads/2018/11/Foundations_Report_Final_October.pdf (accessed on 28 December 2023).

TEEB, (2010). In: Kumar, P., (ed.), *The Economics of Ecosystems and Biodiversity: Ecological and Economic Foundations*. TEEB, London, UK and Washington, DC.

Temesgen, H., Wu, W., Shi, X., Yirsaw, E., Bekele, B., & Kindu, M., (2018). Variation in ecosystem service values in an agroforestry dominated landscape in Ethiopia: Implications for land use and conservation policy. *Sustainability, 10*(4), 1126.

Torreiro, Y., Pérez, L., Pineiro, G., Pedras, F., Rodríguez-Abalde, A., (2020). The role of energy valuation of agroforestry biomass on the circular economy. *Energies, 13*(10), 2516.

Udawatta, P. R., Rankoth, L., & Jose, S., (2019). Agroforestry and biodiversity. *Sustainability, 11*(10), 2879.

Udawatta, R. P., Krstansky, J. J., Henderson, G. S., & Garrett, H. E., (2002). Agroforestry practices, runoff, and nutrient loss: A paired watershed comparison. *J. Environ. Qual., 31*, 1214–1225.

Ujah, J. E., & Adeoye, K. B., (1984). Effects of shelterbelts in the Sudan savanna zone of Nigeria on microclimate and yield of millet. *Agric. For. Meteor., 33*, 99–107.

Vandenbeldt, R. J., & Williams, J. H., (1992). The effect of soil surface temperature on the growth of millet in relation to the effect of *Faidherbia albida* trees. *Agric. For. Meteor., 60*, 93–100.

Varsha, K. M., Raj, A. K., Kurien, E. K., Bastin, B., Kunhamu, T. K., & Pradeep, K. P., (2019). High density silvopasture systems for quality forage production and carbon sequestration in humid tropics of Southern India. *Agrofo. Syst., 93*(1), 185–198.

Vijayaraghavan, A., (2014). *Cultivation Techniques for Neolamarckia Cadamba.* In Transfer of Tree Cultivation technologies to Krishi Vigyan Kendras (KVKs) of Tamil Nadu and Puducherry. Institute of Forest Genetics and Tree Breeding, Coimbatore.

Warrier, R. R., (2014). *Cultivation Techniques for Melia Dubia.* In Transfer of Tree Cultivation Technologies to Krishi Vigyan Kendras (KVKs) of Tamil Nadu and Puducherry. Institute of Forest Genetics and Tree Breeding, Coimbatore.

Wartenberg, A. C., Blaser, W. J., Roshetko, J., Noordwijk, M., & Six, J., (2020). Soil fertility and *Theobroma cacao* growth and productivity under commonly intercropped shade-tree species in Sulawesi, Indonesia. *Plant Soil., 453*, 87–104.

White, D. G., & Childers, N. F., (1945). Bamboo for controlling soil erosion. *J. Am. Soc. Agron., 37*(10), 839–847.

Wiersum, K. F., (1984). Surface erosion under various tropical agroforestry systems. In: O'Loughlin, C. L., & Pearce, A. J., (eds.), *Symposium on Effects of Forest Land Use on Erosion and Slope Stability* (pp. 231–239). East-West Center, Honolulu, Hawaii, USA.

World Health Organization, (2018). *Opportunities for Transition to Clean Household Energy: Application of the Household Energy Assessment Rapid Tool (HEART).* India.

World Resources Institute, (2005). *World Resources, 2005: The Wealth of the Poor: Managing Ecosystems to Fight Poverty.* World Resources Institute.

Youkhana, A., & Idol, T., (2008). First-year biomass production and soil improvement in *Leucaena* and *Robinia* stands under different pollarding systems. *J. Tropical For. Sci., 20*(3), 181–187.

Young, A., (1989). *Agroforestry for Soil Conservation.* CAB International, Wallingford, UK, in cooperation with ICRAF.

Zhou, B. Z., Maoyi, F., Xie, J. Z., Yang, X. S., & Li, Z. C., (2005). Ecological functions of bamboo forest: Research and application. *J. For. Res., 16*(2), 143–147.

Zomer, R. J., Neufeldt, H., Xu, J., Ahrends, A., Bossio, D., Trabucco, A., Noordwijk, M., & Wang, M., (2016). Global tree cover and biomass carbon on agricultural land: The contribution of agroforestry to global and national carbon budgets. *Nature Scientific Reports, 6,* 29987. doi: 10.1038/srep29987.

Additional References

Adhikary, P. P., Hombegowda, H. C., Barman, D., Jakhar, P., & Madhu, M., (2017). Soil erosion control and carbon sequestration in shifting cultivated degraded highlands of eastern India: performance of two contour hedgerow systems. *Agroforestry Systems, 91*(4), pp. 757–771.

Ghosh, S. P., Kumar, B. M., Kabeerathumma, S., & Nair, G. M., (1989). Productivity, soil fertility and soil erosion under cassava-based agroforestry systems. *Agroforestry Systems, 8*(1), pp. 67–82.

Jinger, D., Kumar, R., Kakade, V., Dinesh, D., Singh, G., Pande, V. C., Bhatnagar, P. R., Rao, B. K., Vishwakarma, A. K., Kumar, D., & Singhal, V., (2022). Agroforestry for controlling soil erosion and enhancing system productivity in ravine lands of Western India under climate change scenario. *Environmental Monitoring and Assessment, 194*(4), pp. 1–17.

Kunhamu, T. K., Aneesh, S., Mohan, K. B., Jamaludheen, V., Raj, A. K., & Niyas, P., (2018). Biomass production, carbon sequestration and nutrient characteristics of 22-year-old support trees in black pepper (*Piper nigrum* L) production systems in Kerala, India. *Agroforestry Systems, 92*(5), 1171–1183.

Lim, S. S., Baah-Acheamfour, M., Choi, W. J., Arshad, M. A., Fatemi, F., Banerjee, S., Carlyle, C. N., et al., (2018). Soil organic carbon stocks in three Canadian agroforestry systems: From surface organic to deeper mineral soils. *Forest Ecology and Management, 417*, 103–109.

Liu, W., Zhu, C., Wu, J., & Chen, C., (2016). Are rubber-based agroforestry systems effective in controlling rain splash erosion? *Catena, 147*, pp. 16–24.

Mishra, P. K., Rai, A., Abdelrahman, K., Rai, S. C., & Tiwari, A., (2022). Land degradation, overland flow, soil erosion, and nutrient loss in the Eastern Himalayas, India. *Land, 11*(2), p. 179.

Moench, M., (1991). Soil erosion under a successional agroforestry sequence: A case study from Idukki District, Kerala, India. *Agroforestry Systems, 15*(1), pp. 31–50.

Ngazi, H. S., (2011). *The Potential of Rotational Woodlot and Ngitiri Systems in Carbon Sequestration and Soil Fertility Improvement in Meatu District, Tanzania.* Doctoral dissertation, Sokoine University of Agriculture.

Roose, E., & Ndayizigiye, F., (1997). Agroforestry, water, and soil fertility management to fight erosion in tropical mountains of Rwanda. *Soil Technology, 11*(1), pp. 109–119.

Saha, S. K., Nair, P. K., Nair, V. D., & Kumar, B. M., (2009). Soil carbon stock in relation to plant diversity of homegardens in Kerala, India. *Agroforestry Systems, 76*(1), 53–65.

Tumwebaze, S. B., & Byakagaba, P., (2016). Soil organic carbon stocks under coffee agroforestry systems and coffee monoculture in Uganda. *Agriculture, Ecosystems & Environment, 216*, 188–193.

CHAPTER 7

Agroforestry Systems for Improved Above and Below Ground Biodiversity

PRABHU GOVINDASAMY,[1,2] K. RAJARAJAN,[3] MANOJ CHOUDHARY,[4] MUKESH CHAUDHARY,[1] MAHENDRA PRASAD,[1] SONU KUMAR MAHAWER,[1] HANAMANT M. HALLI,[1,5] M. C. KEERTHI,[1,6] DIBYENDU DEB,[7] V. C. TYAGI,[1] R. SRINIVASAN,[1] BHOLURAM GURJAR,[1] V. RAJAGOPAL,[5] and S. SURESH RAMANAN[2]

[1]*ICAR–Indian Grassland and Fodder Research Institute, Jhansi, India*

[2]*ICAR–National Research Centre for Banana, Tiruchirappalli, Tamil Nadu, India*

[3]*ICAR–Central Agroforestry Research Institute, Jhansi, India*

[4]*ICAR–Indian Agricultural Research Institute, Hazaribag, Jharkhand, India*

[5]*ICAR–National Institute of Abiotic Stress Management, Baramati, Maharashtra, India*

[6]*ICAR–Indian Institute of Horticulture Research, Bengaluru, Karnataka, India*

[7]*ICAR–Indian Agricultural Research Institute, Gogamukh, Assam, India*

ABSTRACT

Ecosystem health and biodiversity directly related to human life on earth. Natural resources and biodiversity are essential for human livelihood security. Biodiversity is also at the focal point of the United Nations sustainable development goal number 15 (Life on Earth) on the 2030 agenda.

Sustainable Management and Conservation of Environmental Resources in India.
Sumit Chakravarty, Vineeta, Jahangeer A. Bhat, Munesh Kumar, Gopal Shukla (Eds.)

Conservation and enrichment of biodiversity are very important for many economic activities, such as agriculture, forestry, fishing, and tourism. In agriculture, each management measures have a direct effect on above and below ground diversity. The agroforestry system, one of these management practices, can positively influence the diversity of flora and fauna by altering the physio-chemical properties and biological properties of soil, and the micro-climate of the site. This chapter portrays the role of trees, microclimate, and soil features on above and below-ground biodiversity growth and development. Adoption of agroforestry practices in areas close to the large natural habitat or between protected areas (PA) has been shown to improve flora and faunal diversity by linking landscapes. It is believed that high floristic diversity will also improves the soil microbial diversity, as different forms of plant life (trees, shrubs, and grasses) promote the growth of different kinds of soil microbes. The diversity of microbes enhances the state of soil fertility by accelerating the decomposition rate of crop residues, rhizo-deposition, and solubilization of nutrients. Agroforestry system is very much like a natural habitat because of perennial trees that influence micro-climate, nutrient cycling and other biotic processes. All of the above natural processes encourage the plant diversity in agroforestry systems. Rich in plant diversity offer multiple overarching opportunities for insect conservation because it provides foraging resources and site to construct net and lay egg; it enhances landscape connectivity besides mitigating pesticide exposure. Due to that, the entomofauna of agro-forestry systems are very diverse. Tree and shrub species utilized in the agroforestry system also provide abundant nectar that increases the variety and number of bees. All this information will serve as useful guidelines for achieving the United Nations Sustainable Development Goal 15 of 2030 agenda.

7.1 INTRODUCTION

The variability among the living entities in an ecosystem is referred as "biological diversity" (Secretariat of the Convention on Biological Diversity, 2005). It is the blanket for supporting life on planet earth. It plays a fundamental part in ensuring the prosperity of human life. Anthropogenic pressure is directly or indirectly causing the biodiversity loss (Bowler et al., 2020). Still, last decade, the predominant reasons listed out for decline in biodiversity were overexploitation of natural resources, ecosystem degradation and pollution, climate change (CC), loss of natural habitats and threats from invasive alien species (Rands et al., 2010). Efforts are being taken to conserve the biodiversity through *in situ* as well as *ex situ* measures. *In situ* conservation of the native biodiversity in protected areas (PA) like biosphere reserves, national parks, sanctuaries, and community reserves are very important and have contributed tremendously. However, there are lacunae and constraints in expanding *in situ* conservation measures for biodiversity conservation (Larsen et al., 2012; Naidoo et al., 2019). Especially, a highly populated country like India where 3.6% of the total land area is already under PA (NSO, 2020) and chances for further expanding the protected area network is implausible. PA throughout global are under intense pressure from human beings in human-dominated landscapes such as agricultural lands, roads, and housing (Jones et al., 2018). Human activities near PA further may hamper the biodiversity conservation (Jose, 2012).

Before getting into the functionality of agroforestry systems on conservation of biodiversity, one should have a good foundational understanding on: What is Agroforestry? There are numerous definitions for agroforestry given by many scientists and researchers in many literatures (Somarriba, 1992; Dagar & Tewari, 2017). Agroforestry means the integration of agricultural crops with trees in a piece of land. In a typical agroforestry model, different components – crops, trees, and/or animals are brought together in the same piece of land and these components must interact with each other aimed to achieve sustainable land use (LU) practice. Much scientific evidence has proved the limitless tangible and intangible benefits of agroforestry. Still last decade agroforestry researchers mainly focused on the economic benefits of agroforestry rather than ecosystem services (ES) like biodiversity conservation. There are several existing studies have been vigorously analyzed the functional role of agroforestry in conservation of above and below ground biodiversity (Schroth et al., 2004; Sridhar & Bagyaraj, 2017). With such voluminous literature, it will be a tedious job in compiling all the information on this chapter. However, we will delimit our discussion with the role of agroforestry towards biodiversity conservation in respect with plant, faunal, and insect diversity along with soil microbial properties. We will also discuss on the caveats and researchable areas in agroforestry related to biodiversity conservation. Biodiversity conservation is a one of the secondary roles of agroforest. The ES provided by the agroforestry can be grouped into four major categories: biodiversity conservation, carbon sequestration, air and water quality improvement and soil nutrient enrichment (Jose, 2012). Since 1980, the need for biodiversity conservation gained momentum and reached the global limelight with the formation of Convention on Biological Diversity in 1992. Its prominent conservation effort was the Aichi Biodiversity

Targets for 2010–2020. It was (Schroth et al., 2004) postulated with three hypotheses which provides a possible explanation for biodiversity conservation: agroforestry–deforestation hypothesis; agroforestry–habitat hypothesis, and agroforestry–matrix hypothesis (Atangana et al., 2014). In nutshell, the agroforestry practice aids in biodiversity conservation through (Figure 7.1):

i. It is a habitat of many species can tolerate frequent disturbance;
ii. Preserve endangered species;
iii. Reduce pressure on forest land and reduce deforestation levels;
iv. Preventing the degradation of the habitat;
v. Act as connecting corridors between fragmented protected areas.

This is possible owing to the certain features of the agroforestry like multi-strata and multispecies combination which ensures high floristic diversity; minimal agronomic activities (cropping done in the spacing between the trees) and retained over the years. These features aid in floral, faunal, and soil biodiversity conservation.

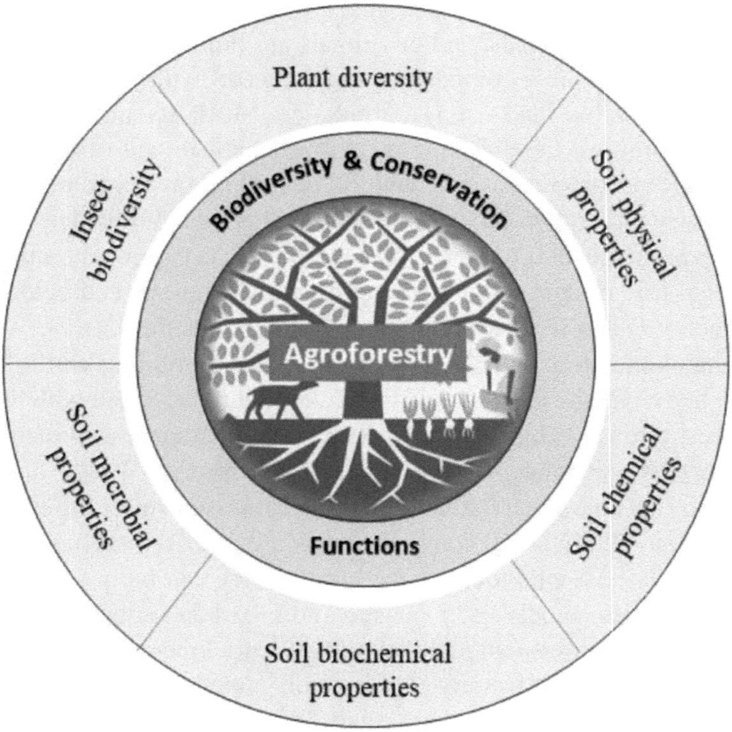

FIGURE 7.1 Agroforestry for biodiversity and conservation – mechanisms and role.

Generally, even commercial agroforestry systems like alley cropping and boundary plantations will be able to support native flora to a certain extent. In that case, intermediate and subsistence agroforestry practices like homegardens, shelterbelts, and windbreaks will have higher floristic diversity. Researchers use diversity indices like Shannon's diversity index, the Shannon–Wiener index, Simpson index, Margalef diversity index to assess the diversity. Most of the studies have reported that agroforestry has potential to maintain higher plant diversity compared to monoculture and agricultural land, irrespective of structural difference in agroforestry system adopted. More pertinently, the understory or crop management activities influence plant diversity levels. Homegardens in tropics have proven for high floral, avian, and soil microbial diversity. The species richness and evenness of homegardens was 18–30% similarity to native forests (Tynsong & Tiwari, 2010; Bardhan et al., 2012).

Regarding faunal diversity, a good number of literatures stating that agroforests act as refuge or connecting corridors for avian (Martin et al., 2021), insects (Peters & Carroll, 2012; Varah et al., 2020), bats and even bigger mammals like elephants (Krishnan et al., 2019; Ferreira et al., 2020). Adoption of agroforestry practices in areas close to the large natural habitat or between PA has proven to improve the faunal diversity by connecting the landscapes: the agroforestry–habitat hypothesis. Multistory agroforestry practice acts as an optimum roosting shelter for avian communities. It is also reported that even coffee grown under the shade of trees will be able to support birds if appropriate species are selected and planted (Bakermans et al., 2012). This implies there is a need to have a conceptual understanding of the functional role of perennial trees which will provide economic returns after a very long time. Depending upon the tree component, the birds and bats in the agroforestry system can be beneficial (Maas et al., 2016). Sometimes, the bird also causes trouble by predating on the fruit trees planted in agroforestry (Dadhwal et al., 1988). These are intricate issues where scientific expertise and common sense have to combine in designing appropriate agroforestry models accounting the geography, seasonality, vegetation strata and habitat complexity.

It is believed that high floristic diversity will also enhance soil microbial diversity with different plant life-forms (trees, shrubs, and herbs) being present in the same piece of land. The heterotrophic microbial community will be thriving there. These heterotrophic microbes enhance the nutrient cycling process and improve soil chemical properties too. In addition to nutrient pumping by trees from sub-soil layers, some trees can fix atmospheric nitrogen through *Rhizobium* symbiosis or actinomycetes like *Frankia.* This also enhances soil fertility status. A significant work using state-of-the-art technology has proven that agroforestry can also improve soil porosity, soil aggregate stability and even improve soil enzyme activity (Udawatta et al., 2008). Agroforestry is even stated best strategy as well as a tool for managing the natural resources sustainably. This multifunctional land use pattern forms a mosaic landscape supporting to flora and fauna, especially in areas where PA are inside the matrix of agricultural lands. Agroforestry has a close resemblance to natural habitat because of the perennial trees which influence microclimate, nutrient cycling and other biotic processes. This resemblance favors the conservation of native biodiversity to a certain extent. It has been cautiously stated that agroforestry or agroforests are better compared to monoculture and modern agriculture, and it mimics the natural forests (NFs) in the context of biodiversity conservation and sustainability (Kumar & Nair, 2006).

7.2 AGROFORESTRY SYSTEM AND PLANT DIVERSITY

Plant diversity provides a better substitute for costly agricultural inputs such as fertilizers, agro-chemicals, and pollinators (Isbell et al., 2017). Moreover, the production systems rich in biodiversity has often been associated with efficient use of applied inputs and encouraging mutual interaction between organisms and other ecosystem facilities (Smith et al., 2008; Cardinale et al., 2012). Numerous studies have been also reported the impact of plant diversity on the productivity of cultivated systems (Naeem et al., 1994; Loreau et al., 2001). Lehman & Tilman (2000) described that plant diversity promotes community productivity but may not individual species productivity. Over the period, the intensive agricultural systems being were responsible for the loss of fertile soils to erosion, declining soil fertility, and loss of biodiversity. In this scenario, the agroforestry (AF) system (AF, "those integrates perennial trees with crops"), is believed to be a possible solution for reverting biodiversity and ecosystem function (Sistla et al., 2016; Udawatta et al., 2019). AF has been acknowledged to conserve and enrich biodiversity in the biosphere (Méndez et al., 2009). Jose (2011) reported that even after with deliberate interventions of human on species selection for higher consumption in an AF system, they still hold much species diversity and richness compared to intensive production systems. The AF systems promotes plant diversity and ES through providing habitat to species that can tolerate regular disturbance and sensitive to disturbance, encouraging biodiversity by providing various ES, and reducing the conversion of natural habitat (Udawatta et al., 2019). Further, the AF system profoundly affects microclimate and soil properties which can also directly influences the plant diversity (Noble & Dirzo, 1997).

Hortipasture, silvopasture, windbreaks, riparian buffers, alley cropping, trees in cultivated lands, hedgerow intercropping (HI), home garden AF, contour strips in cropland, and forest farming are the predominant AF systems across the globe (Wood & Burley, 1991). Some of the suitable trees for AF systems are; Bihul (*Grewia optiva* J.R. Drumm. ex. Burret), Chinaberry tree (*Melia azedarach* L.), Indian Gooseberry (*Emblica officinalis* L.), north Indian rosewood (*Dalbergia sissoo* Roxb), physic nut (*Jatropha curcas* L.), purple bauhinia (*Bauhinia purpurea* L.), and orchid tree (*Bauhinia variegate* (L.) Bench), in India, ashes (*Fraxinus* spp.), basswood (*Tilia* spp.), chestnut (*Castanea* spp.), pecan [*Carya illinoensis* (Wanganh.) K. Koch], oaks (*Quercus* spp.), and walnut (*Juglan niegra* L.), in North America, cocoa (*Theobroma cacao* L.), white acacia [*Faidherbia albida* (Delile) A.Chev.], African oil-palm (*Elaeis guineensis* Jacq.), and prairie acacia [*Acaciella angustissima* (Mill.) Britton & Rose] in Africa, messmate (*Eucalyptus oblique* L.'Hér), Australian blackwood (*Acacia melanoxylon* R.Br.), river oak (*Casuarina cunninghamiana* Miq.) and native pepper [*Tasmannia lanceolate* (Poir.) A.C.Sm.] in Australia and holm oak (*Quercus rotundifolia* Lam.), cork oak (*Quercus suber* L.), holly oak (*Quercus ilex* L.), wild cherry (*Prunus avium* L.), and sycamore (*Acer pseudoplatanus* L.) in Europe (Garrett et al., 2009; Bijalwan, 2013). Fruit trees commonly suitable for AF systems are guava (*Psidium guajava* L.), lemon (*Citrus limon* (L.) Osbeck), mango (*Mangifera indica* L.), and papaya (*Carica papaya* L.) (Bijalwan, 2013).

Numerous studies have been reported that the adaptation of the AF system promoted plant diversity (Huang, 1998; Huang et al., 2002; Bhagwat et al., 2008). Jose (2012) reported a higher species richness, and diversity in the AF system compared to monoculture systems. Compared to forest, a higher mean richness of taxa (60% greater) was recorded for AF system (Bhagwat et al., 2008). A comparative study on home garden AF and forest system conducted in Bangladesh, found a 30% similarity in tree species between homegarden AF and forest systems, and the tree species richness was increased with an increase in the size of homegarden AF (Bardhan et al., 2012). Studies on plant diversity conducted in the Amazon Basin, the Congo Basin, and Southeast Asia, reported intermediary diversity in AF systems and mono-cropping with either perennial or annual crops (Tomich et al., 1998).

In a comparative study between traditional agroforestry systems and tropical dry forests in Tehiacian-Cuicatlan Valley, Mexico, Rendón-Sandoval et al. (2020) reported that traditional agroforestry system can be able to harbored 68% of the regional plant species, of which 95% were native to the region and remaining 5% were exotic. The same authors also noted that the total species richness (alpha, beta, and gamma) of traditional agroforestry systems and tropical dry forests were found similar, but the species abundance was significantly higher (90 species) in traditional dry forests systems (90 species greater). Likewise, Sistla et al. (2016) reported that among land-use systems (LUS) both agroforestry and secondary forest had a greater species richness (45 and 20 species, respectively) compared to pasture (16). Further, the similarity index (Jaccard similarity index) showed that agroforestry and secondary forestry system shared the highest species (38 species), followed by agroforestry systems and pasture (28 species), and secondary forestry and pasture had the lowest species share (15 species).

In a recent study, Salazar-Díaz & Tixier (2019) interestingly compared the income and species diversity in different agroforestry systems and found that increase in plant species diversity has increased the income of fruit trees (+3.70 estimate), firewood (+0.01), timber (+0.26) and total trees (+66.61) but decreased the income of banana (–0.12) and cacao (–0.23). This was mainly due to the complementary effect of the higher strata group (fruit trees, firewood, and wood) than the lower strata (banana and cocoa). Findings from different corners of the globe evidently suggest that plant diversity is highly associated with agroforestry systems compared to mono-cropping agricultural practices and sometimes even perhaps superior to the secondary forestry systems. However, the positive effect of plant diversity considerably varies with different AF systems.

7.3 AGROFORESTRY SYSTEM AND SOIL PHYSICAL PROPERTIES

The features of soil physical properties affect the crop growth and trees in any LU systems including agroforestry by altering soil aeration, movement of water, acquisition of nutrients, root penetration and anchorage. Agroforestry ecosphere has the capacity to sustain or preserve greater heights of biodiversity and higher biomass against the traditional farming. The biodiversity in agroforestry system effects the recycling ability of the soil because of improved soil structure, texture, aggregate stability, soil depth, porosity, reducing erosion, bulk density, water holding capacity (WHC) and rate of infiltration (Brito-Vega et al., 2018). Deforestation, unscientific agricultural practices, over-construction, alignment of roads and over grazing are major reason for soil degradation (Araujo et al., 2012). Plant litter addition by diverse plant biodiversity and its decomposition improves the soil structure, porosity, bulk density, pore density and WHC (Hillel, 1982; Singh & Sharma, 2007). Different plant species in agroforestry system adds residue litter in soil act as caring layer for refining soil physical characteristics against soil compaction, soil erosion and soil moisture conservation (Xiong et al., 2008).

7.3.1 SOIL AGGREGATE

Biodiversity in agroforestry systems may directly or indirectly affect the aggregate stability through leachates from weathering, substance secreted from stems, leaves, and roots, decaying plant materials, surface residues and canopies that shield aggregates against interruption due to rainfall, wind, and scattering through running water. With 0.84 to 1.00 mm aggregates in diameter are not eroded by water and wind stroke. Soil aggregate is a good physical parameter to evaluate the land degradation. The leaf litter adding the organic carbon by perennial plants improves the soil aggregate stability. Agroforestry and grassland have greater larger aggregate (> 5 mm) as compared to seasonal crop cultivated fields (Somasundaram et al., 2012). The diversity in grass buffer and agroforestry has higher water stable aggregates in surface soils than row crop management and grazed pasture (Paudel et al., 2012). Protected aggregates in the agroforestry system store more soil moisture and carbon which support various plant species than traditional intensive production systems.

7.3.2 BULK DENSITY AND POROSITY

The physical properties like porosity and bulk density are inversely related. Fallow land from long duration and continuous use of mechanization with heavy machinery cause soil compaction and reduce the soil porosity. The soil that is less in organic matter generally has more natural compaction. The biodiversity in agroforestry system is effective in loosening the soil. Leaf litter and root residue in soil lowers the bulk density and increases porosity. Bulk density is a soil health indicator which affects soil porosity, WHC, infiltration rate and plant rooting depth influence soil productivity and fertility (Ernest et al., 2015). Odewumi et al. (2013) observed lower bulk density and higher porosity in soil under teak canopy than in soil outside teak canopy. Similarly, lower soil bulk density was also found due to biodiversity conservation in agroforestry systems as compared to grass buffer, grazed pasture and row crop (Paudel et al., 2012). Non-significant increase in bulk density was found in grazed land while, slightly lower mean bulk density was found under agroforestry systems (Tornquista et al., 1999). Soil bulk density affects the plant available soil moisture, therefore the low bulk density under agroforestry systems encourages plant diversity.

7.3.3 WATER HOLDING CAPACITY (WHC)

Irrigation availability and use efficiency are extremely threatened for agriculture (Gregory & Ingram, 2000; Fischer et al., 2005). Traditional way of irrigation practices cannot completely exploit the rainwater because of more soil surface runoff, drainage, and evaporation from surface (Ong et al., 2007). The occurrence of biodiversity in agroforestry is a practicable choice to sustain productivity while, providing necessary tree products and environmental facilities. Different tree species-based cropping systems and good soil structure also assist soil moisture conservation by adding organic carbon and reducing evaporation and soil erosion. The conserved biodiversity in agroforestry systems and NF enhances soil moisture content (SMC) through soil profile depth because of organic matter and reduction in evaporation (Saha et al., 2004). The higher WHC in Alder (*Alnus nepalensis*) based agroforestry system had higher plant diversity as compared to Khasi mandarin (*Citrus reticulata*) was reported by Tripathi et al. (2009).

7.4 AGROFORESTRY SYSTEM AND SOIL MOISTURE, TEMPERATURE, AND GHG EMISSION

Increased net greenhouse gas (GHG) emissions have resulted into global CC threatening ecosystems and people's livelihoods. The atmospheric concentrations of the GHGs (CO_2, CH_4, and N_2O) have increased to 398 ppm, 1,833 ppb, and 327 ppb, respectively (WMO, 2015). The world is therefore looking for ways to halt further increases in GHGs. Agroforestry represents an opportunity to reduce CO_2 concentrations in the atmosphere by increasing carbon stocks in agricultural lands. Relative to conventional agriculture, it contributes significantly to carbon sequestration, increases a range of regulating ES, and enhances biodiversity. In addition, it has been highlighted as one of the agricultural practices with the greatest potential for CC mitigation and adaptation. It also improves water availability; protect crops, pastures, and livestock from harsh-climatic events (Aertsens et al., 2013).

Agroforestry can increase biomass carbon stocks as the biomass of trees consists of 46–51% carbon. Agroforestry can also enhance soil carbon sequestration through enhanced fine root production, rhizo-deposition, and litter fall (Lorenz & Lal, 2014). Kim et al. (2016) compiled and analyzed data from 56 peer-reviewed publications and conclude that agroforestry system (at an average age of 14 years) sequestered 7.2 t C ha^{-1} y^{-1}, with biomass carbon sequestration contributing about 70% and soil carbon sequestration contributing about 30% of the overall gain. Soils under agroforestry also oxidized 1.6 ± 1.0 kg CH_4 ha^{-1} y^{-1} and emitted 7.7 ± 3.3 kg N_2O ha^{-1} y^{-1}. Compared with adjacent agricultural fields, soils under agroforestry had lower bulk density, higher soil organic carbon (SOC) and soil nitrogen contents, but pH and net CH_4 and N_2O emissions from soils under agroforestry remained nearly the same as under agriculture. Overall, young agroforestry stands contributed to mitigating 27 ± 14 t CO_2 eq ha^{-1} y^{-1}. Kay et al. (2019) estimated that agroforestry on 8.9% of European agricultural land could potentially store between 1.4 and 43.4% of the total European agricultural GHG emissions. Earthworms were believed to stimulate carbon sequestration in soil aggregates, but it increases emissions of the main GHG such as carbon dioxide (CO_2) and nitrous oxide (N_2O) (Lubbers et al., 2013). Increase in biodiversity of other soil fauna in the presence of earthworms decrease faunal-induced N_2O emission despite enhanced carbon cycling (Lubbers et al., 2020).

Agroforestry has received increased attention during the last two decades for biodiversity conservation (Huang et al., 2002; McNeely & Schroth, 2006). Agroforestry tempted biodiversity improvements have been reported in both temperate and tropical regions (Udawatta et al., 2019). Some studies have indicated significantly greater biodiversity in agroforestry compared to forests and tree monoculture management (Huang et al., 2002; Sistla et al., 2016). In a meta-analysis, Bhagwat et al. (2008) reported 60% higher mean richness of species in agroforestry as compared to forests. Several studies have indicated a strong relationship between biodiversity and ES and strongly emphasize the importance of biodiversity conservation for enhanced ES (Hooper et al., 2005; Udawatta et al., 2019).

Tanzania is the richest country in terms of biodiversity (UNEP, 1998). In the past 2–3 decades, deforestation increased substantially, and the biodiversity is expected to change as different species try to adjust and cope with the impact of CC. Because invasive species tend to adapt better to changing climates than the desirable pasture species (Malcolm et al., 2002), pasture-lands also are likely to be rendered a no-go area for animal grazing (Mushi et al., 2015). In a study on 34 countries during the period of 1995–2014, Bhuiyan et al. (2018) found that GHGs emissions considerably affected the potential habitat area and biodiversity index.

The soil biodiversity depends on various physicochemical properties of soil such as porosity, soil temperature, soil moisture and the presence of inorganic elements and components, organic matters and the pH of the soil. Soil moisture is one of the main constraints on soil biota in many ecosystems and one of the key factors governing microbial activities. A meta-analysis found that reduced precipitation generally has negative impacts on fungal biomass, collembolans, and enchytraeids, and no consistent effects were observed on other taxa (Nielsen et al., 2015). In contrary to this, Pereira et al. (2013) believed that reduced precipitation can promote the formation of micro-aggregates in which microorganisms are protected from adverse effects of lower SMC, thereby providing sites for continued ecosystem function. Fluctuations in soil temperature also have significant effects on belowground communities, but the observed local responses are highly dependent on interactions between soil moisture and temperature. The effects of changes in precipitation may be ameliorated or amplified by simultaneous changes in temperature or may be moderated by the current temperature regime. Lalthanzara et al. (2011) found higher density of earthworm with SMC in agroforestry systems of Mizoram, India. Studies of the traditional agroforestry practices in the Pacific islands exhibited their diversity in both species and cultivars. In Papua New Guinea, there is wide diversity of cultivated plants in shifting cultivation system, on an average that ranges from 6 to 10 cultivated species per 25 m^2. In some parts of Indonesia (Eastern islands), the agroforestry system contains eight varieties of coconuts, 25 *Musa* spp., 13 *Colocasia taro*, 14 *Cyrtosperma taro*, and 25 breadfruits. In Hawaii nearly 200 varieties of sweet potato, 6 sugarcane, and 35 banana varieties were introduced and maintained in the garden (Manner, 1981).

7.5 AGROFORESTRY SYSTEM AND SOIL CHEMICAL PROPERTIES

Agroforestry system plays a crucial role in maintaining the soil fertility through several mechanisms. Agroforestry system enhances soil fertility, reduce erosion, improve water quality, enhance biodiversity, and sequester carbon (Nair et al., 2009; Young, 1991). Biodiversity contributes to enhanced ecosystems, such as ecosystem stability and productivity, and improved nutrition and human health (Wall et al., 2015). The higher the soil biodiversity, the better the soil fertility. Conservation of soil fertility is extremely important because fertile soil supports a large number flora and fauna. The activities of the soil biota are essential for the greater soil function. For example, both soil microorganisms and fauna can moderate vegetation succession and plant diversity (De Deyn et al., 2003; Philippot et al., 2013). Symbiotic N-fixing bacteria and mycorrhizal fungi associated with plant roots aid in nutrient uptake, in particular N and P, respectively. It has been estimated that symbiotic microorganisms are responsible for ~5–20% of all N (up to 80% in boreal and temperate forests) and up to 75% of P that is taken up by plants annually, and more than 20,000 plant species depend on microbial symbionts for growth and survival (Van der Heijden, 2008). Moreover, changes in the composition of symbiotic microorganisms may alter the plant community dynamics by modifying plant-herbivore interactions and competition between different plant species (Philippot et al., 2013). Similarly, pathogens strongly influence competition between plant species and can contribute to the persistence of rare species, maintain biodiversity by suppressing dominant species, and facilitate invasion of non-native species (Van der Heijden, 2008). However, non-symbiotic soil microbes also influence nutrient dynamics. For example, free-living N-fixing bacteria are common in most terrestrial ecosystems, contributing a significant amount of N to soils with rates of up to >3 kg N ha^{-1} $year^{-1}$ in some dryland and tundra ecosystem types (Cleveland et al., 1999).

Furthermore, rapid immobilization of N by soil microbes acts as a short-term N sink, preventing loss to groundwater or nearby ecosystems; later, this N can be released and act as a source of N for plant uptake (Bardgett et al., 2003). In addition, soil microbes contribute to plant growth via exudation of organic acids and solubilization of precipitated P (Landeweert et al., 2001). Conversely, microbes can have negative effects on plant growth by competing for nutrients particularly in nutrient limited ecosystems, by transforming N to more labile forms, or by reducing N pools through denitrification (Van der Heijden, 2008). As shown by Bélanger and Pilling (2019), avoiding losses of soil biodiversity requires a more holistic approach and interdisciplinary studies to advance our understanding of the main drivers for soil biodiversity gains and losses, and interlinking practical measures to recover, sustain, and enhance these drivers. Soil biodiversity is a valuable resource that is essential for sustainable food security. Global efforts are thus urgently needed to preserve it. Agroforestry systems is one such management practice can support the soil biodiversity.

7.6 AGROFORESTRY SYSTEM AND SOIL BIO-CHEMICAL PROPERTIES

The biological and biochemical properties related to cycling of biogenic elements and to the transformation of organic matter in the soil are being considered the most suitable properties for soil quality estimation (Visser & Parkinson, 1992). These properties include all variables directly associated to microbial activity (microbial biomass carbon and nitrogen, respiration, and nitrogen mineralization capacity) and activities of exocellular hydrolytic enzymes involved in the elemental cycles of the nutrients (C, N, S, and P) (Nannipieri et al., 1995). The general biochemical parameters which are most commonly used to estimate the alteration in soil quality includes soil microbial biomass carbon (SMBC), phosphatase (acid or alkaline), β-glucosidase, O-diphenol oxidase and urease activity.

Soil biochemical biodiversity is the result of several factors such as plant roots, fungi, mycorrhizal fungi and bacteria (Tarafdar, 1988; Tarafdar & Claassen, 1988; Dinkelaker & Marschner, 1992; Tarafdar & Marschner, 1994). Several tree species have been identified as a suitable component in the agroforestry (Niang et al., 1996). For instance, according to George et al. (2001) numerous potential agroforestry tree species found capable to attain large concentrations of phosphorus in their biomass in spite of being grown in soils having low available phosphorus. George et al. (2001) also suggested that agroforestry species have actively increased rhizospheric phosphatase activity either directly by exudation or indirectly by stimulation of soil microbial activities. Further, their data revealed that *Tithonia* species (137 μM) and *Tephrosia* species (108 μM) showed higher concentrations of citric acid in the rhizosphere than maize (96 μM).

Agroforestry system may duplicate several characteristics of undisturbed ecosystems and combine several soil-conserving attributes of tree crops. Trees may also improve above- and below-ground microclimate whereas; meso- and micro-fauna and micro-flora around plant roots may alter physical, chemical, biological, and biochemical properties of soil. Number of researchers reported that the higher biochemical properties of soil in agroforestry system as compared to grasslands (GL), mono-cropping or other sole pastures in different climatic conditions. Udawatta et al. (2009) reported that β-glucosidase enzyme activity was significantly lower in the crop area compared to grass buffer, agroforestry buffer, and grass waterway treatment areas. Similarly, in a study carried out under the Ginkgo-tea agroforestry the SMBC was low in the pure tea plantation in rows of hedges (G1) (264.2) as compared to tea intercropping with Ginkgo seedlings (grafted) at a spacing of 10 m × 10 m (G2) (287.5); and tea intercropping with Ginkgo seedlings (grafted) at a spacing of 6 m × 6 m (G3) (295.8); urease and dehydrogenase activity was also higher in treatment G2 as compare to others (G1 and G2) (Tian et al., 2013).

7.7 AGROFORESTRY AND MICROBIAL DIVERSITY

It was presumed that one square meter of belowground earth portion may contains about 20,000 species of bacteria, fungi, virus, protozoa, nematodes, and many other species (Sridhar, 2009). These microbes are important for most of the bio-geochemical processes, nutrient cycling, and degradation of chemicals. Whereas, in recent past, anthropogenic activities including intensive agriculture coupled with CC have resulted in unsustainable exploitation of biological diversity at an alarming rate (Udawatta et al., 2009). It was opined that deforestation due to agricultural intensification is one of the major contributors for loss of biological diversity, ES and increased load on natural resources (Steffan-Dewenter et al., 2007; Sridhar & Bagyaraj, 2017). Therefore, integrated agricultural practices favor the biological diversity. In this context, adaption of AFS has a global role in biodiversity and species richness, therefore receiving an increased attention in both temperate and tropical regions (Udawatta et al., 2009). Microbial diversity in AFS may be classified into belowground and aboveground diversity and they are dependent on the nature of AFS (composition), soil quality, rhizosphere, phyllosphere, and abiotic factors (Keesstra et al., 2016; Radhakrishnan & Varadharajan, 2016). The important microbes in AFS include nitrogen fixers (symbiotic and non-symbiotic), phosphate-mobilizing organisms (mainly mycorrhiza) and disease-preventing endophytic microbes. For instance, a study conducted at the Guelph, Canada revealed that tree-based cropping system has greater arbuscular mycorrhizal fungi (AMF) richness, composition, and several taxa in comparison to mono-cropping system (Bainard et al., 2011). Similarly, greater richness and activity of saprotrophic fungi and ectomycorrhizal fungi with higher carbon ratio were reported under silvopasture system in Germany (Beuschel et al., 2019). In AFS, higher density of tree roots, shrubs, grass, and annual crops add quality litter and rhizodeposition that in turn intensify the inevitable interactions, improved root colonization and higher spore densities in the crop's rhizosphere (Jose et al., 2004).

Unlike fungi, AFS was found with greater bacterial abundance (10^6 copies per gram of dry soil) and species richness in comparison to monocrop in Canada (Banarjee et al., 2016). This improvement in bacterial communities was mainly associated with the improved soil pH and carbon content. Likewise, a study using 16S rRNA gene copies revealed that adoption of hedgerows and woodlands had a greater bacterial abundance and species richness (Banarjee et al., 2016). Higher plant diversity and favorable rhizosphere conditions possibly improved the microbial biomass and nitrogen mineralization rates in temperate wind breaks compared to conventional practices (Kaur et al., 2000; Wojewoda & Russel, 2003). Trees also influence the above and belowground microclimate and may alter the soil microbial properties. Likewise, higher total soil microbial biomass and robust microbial community compositions (especially more gram-negative bacteria) were witnessed in alley-cropping with fruit trees and native grasses (Kremer & Hezel, 2013). The differential distribution of crops, pasture, shrubs, and trees in AFS systems influence the quantity and quality of organic matter, and microclimate thus influences the activity of soil enzymes (Jose, 2009; Banarjee et al., 2016). As per the meta-analysis estimation, AFS found to have 60% greater microbial species richness than common cropping systems. However, important changes in microbial activity due to different AFS practices are discussed in Table 7.1.

Greater enzyme activity was reported in a silvopasture system (wood trees; *Populus deltoides* + tall fescue grass; *Festuca arundinacea,* buffers), compared toa row-cropping (corn-soybean rotation) system (Meyers et al., 2001; Paudel et al., 2011, 2012). Therefore, AFS could play an important role in sustainable developments, and ES by influencing microbial diversity and soil health improvement. Therefore, diversity of tree species and food crops in AFS could decide and manage the composition of microbial diversity.

7.8 AGROFORESTRY SYSTEM AND INSECT DIVERSITY

Biological diversity, often referred to as biodiversity, is vital for the survival of human life and the health of our biosphere (Jose, 2012). Insects are the most diverse group of organisms, accounting for 70% of all biological

TABLE 7.1 Microbial Changes in AFS

Particulars	Observations	References
AFS and mono-cropping.	Improved the efficacy of bio-inoculants for *Glycine max*, *Phaseolus mungo*, and *Cicer arietinum*.	Shukla et al. (2018)
AFS with tree species.	The tree Ensete had greater AMF diversity and density than tree-coffee and multiple species cropping systems.	Torres-Gomez et al. (2018)
Native oak-pine forest and a *Cupressus lusitanica* plantation.	Native forest had greater edible mushroom biomass due to presence of ectomycorrhizal hosts with the double market value.	Nygren et al. (2012)
Determination of bacterial communities in AFS.	*Acidobacteria*, *Actinobacteria*, and *Alphaproteo bacteria* were the dominant bacterial phyla and subphyla across land uses.	Banarjee et al. (2016)
Microbial biomass and grasslands.	Soil under AFS harbors greater microbial biomass over soils with mono-crops or open grasslands.	Lakshmipathy et al. (2012)
Microbial diversity and plant species.	Mixtures of plant species in AFS allow a larger diversity of microbes than monocultures.	Kaur et al. (2000); Lagerlof et al. (2014)
Nitrogen fixation symbionts in AFS.	*Frankiaceae* has wider association with forest tree species much greater than *Proteobacteria.*	Freitas et al. (2010)
AMF colonization and land use systems.	Root colonization and species diversity were significantly higher in natural forests, grasslands, and were least in paddy fields.	Dobo et al. (2018)
Microbial biomass carbon and alley cropping system.	The microbial biomass carbon in soil was more in all the three alleyed crops (finger millet, peanut, and pigeon pea) grown in between *Gliricidia*.	Cardoso & Kuyper (2006)
Microbiological conditions and land use systems.	Continuously practicing of AFS practices for 20–30 years improved the soil micro-biological and chemical conditions over traditional agriculture fields.	Balakrishna et al. (2017)

Abbreviations: AFS: Agroforestry systems; and AMF: arbuscular mycorrhizal fungi.

diversity found on Earth's surface. Insects have significant functional significance because of their large number of individuals and considerable intra- and inter-specific variation. The insect-plant interaction is highly effective in maintaining the earth's bio-cycles. It is an important baseline in all aspects of biodiversity conservation (Samways, 1993). It is estimated that over 85% of the world's flowering plants rely on animals, primarily insects, for pollination (Ollerton et al., 2011). Animal pollination is responsible for around 35% of global agriculture production (Klein et al., 2007; Eilers et al., 2011). Therefore, the conservation of insect biodiversity is also essential for deriving sustainable benefits for present and future generations.

The global-scale alteration of natural ecosystems is a key driver of biodiversity loss and jeopardizes ecological functioning, sustainable LU, and economic viability (Hoekstra et al., 2005). Land-use intensification, in addition to natural ecosystem alteration, is a major cause of declining insect diversity. The efforts of environmental bodies to conserve biodiversity have been exceptional over the last two decades. Despite the effort, the pressure on biodiversity continues to increase with the loss of one endangered species every 20 minutes, and more species might be lost in the next 20 years than those lost in the past century (Conservation International, 2012). Each country defines a certain area as a protected area to protect biodiversity within its natural habitat. However, PA alone cannot conserve biodiversity, especially in human-dominated landscapes where natural ecosystems are under significant strain to supply food, wood, and fiber demands. Diversified agricultural systems, such as agroforestry, are proposed as key tools in nature conservation programs (McNeely, 2004; Schroth et al., 2004). Agroforestry is a traditional land-use method that is beneficial in preserving and improving biodiversity from the farm to the landscape level. Agroforestry approaches give several overarching options for insect conservation, such as habitats containing feeding resources and nesting or egg laying places; it improves landscape connectivity and limits pesticide exposure. As a result, the entomofauna of agroforestry systems is quite diversified (Campbell, 1984).

Current scientific evidence suggests that agroforestry practices can boost crop pollination and productivity by offering a high value for floral resources. The tree and shrub species utilized in the agroforestry system provide abundant nectar with relatively high sugar contents, such as maple (*Acer* spp.), basswood (*Tilia* spp.), willow (*Salix* spp.), horse chestnut (*Aesculus* spp.), cherry and plum (*Prunus* spp.), brambles (*Rubus* spp.), and service berry (*Amelanchier* spp.). Pollen is a high-protein resource used by native bees, honeybees, and some wasps to feed their brood or provision their eggs. Several woody species flowers early in the spring in temperate locations, providing pollinators with some of the season's earliest pollen and nectar resources (Stubbs et al., 1992; Loose et al., 2005).

The biodiversity observed in agroforestry systems is a close replica of the natural ecosystem; thus, the importance of agroforestry in biodiversity conservation is rapidly expanding (Shanker & Solanki, 2000). Furthermore, the significance of agroforests for biodiversity protection grows as human pressure on natural ecosystems increases. Although different insect groups respond differently, insect species generally grow in diverse agroforestry systems. According to Bos et al. (2007), cacao-dominated agroforestry systems maintain high insect diversity and species richness. Despite the loss of 24.5–87.5% of the species documented at forest sites due to conversion, the richness and diversity of species in managed agroforestry systems were comparable to or greater than that found at wild forest sites. Söderström et al. (2001) found more bird diversity in grassland areas with shrubs and trees in Sweden. This was partly attributed to a rise in the abundance and diversity of insects and other invertebrates. In addition to facilitating pollination, insects are an essential component of the food web, feeding on plant parts and being preyed upon while aiding in seed dissemination. Insects are a very important component of any ecosystem as they influence plant growth patterns, plant populations, plant communities, and community succession (Brown, 1985).

7.9 BIOMASS ESTIMATION AND BIODIVERSITY ASSESSMENT USING GIS AND REMOTE SENSING (RS)

By definition, remote sensing (RS) is a process where information about something is acquired without being in direct contact (Lillesand et al., 2015). It is by nature completely opposite of *in situ* data acquisition where data is gathered about an object by physical contact. RS technique was first ever deployed long back in 1839 in case of photography for topographic surveying (Gebelein, 2012). During World War-I aerial photography was prominently used for reconnaissance survey to identify usable versus damaged barriers and other facilities for troop movement (Gebelein, 2012; Lillesand et al., 2015). In 1972, after the first satellite Landsat with multi-spectral scanner (MSS) was launched, its capability of capturing earth surface image and its digital processing was came into use for crop yield forecasting through projects like crop identification technology assessment for RS (CI-TARS) (MacDonald, 1984), large area crop inventory experiment (LACIE) (MacDonald et al., 1975) and Agriculture and Resources Inventory Surveys through Aerospace Remote Sensing (AgRISTARS) (AgRISTARS, 1981). The CI-TARS program was a full fazed demonstration to show the capabilities of RS through analysis of aircraft multispectral data, spectral discriminability of soybean and corn inventory in the United States (Bauer et al., 1975). Whereas LACIE and AgRISTARS were initiatives to implement agricultural applications based on Landsat data after launching of Landsat 1 supported by funds provided by multiple agencies (Roy et al., 2014). The scientific agency of United States Department of Commerce namely National Oceanic and Atmospheric Administration (NOAA) is performing large scale crop monitoring as well as yield forecasting since 1980 using advanced very high-resolution radiometer (AVHRR) (Malingreau, 1986; Bolton & Friedl, 2013). In recent times, there have been significant progress made in respect of both radiometric calibration and spatial and spectral resolution with the launch of moderate resolution imaging spectroradiometer (MODIS) by National Aeronautics and Space Administration (NASA). Although the observation frequency is satisfactory for both AVHHR and MODIS and widely available across the world, the resolution is quite coarse; for MODIS these are 250, 500, and 1,000 m spatial resolution whereas, for AVHHR even less resolution of 1.1 km for local area coverage and 4 km for global coverage (Kidwell, 1988). The launch of Landsat Thematic Mapper with spatial resolution 30 m has significantly improved the spatial resolution. But the accessibility of this medium resolution data is mostly restricted to regions of USA as it is produced by United States Department of Agriculture (USDA) and National Agricultural Statistics Service (NASS) only at regional scale and for most of the world not available (Bolton et al., 2013). Another problem with Landsat data is its less frequency; the repeat period of 16 days for Landsat throws a challenge to collect frequent RS data during different crop growth stages. To overcome this challenge, researchers have come up with an innovative idea through which high resolution images obtained from Landsat are merged with higher frequency data of MODIS or AVHHR (Bolton et al., 2013).

In case of India, researchers are aggressively poised to use RS in crop yield studies. Space Application Center (SAC) of India along with other collaborators started an ambitious program to use space borne RS for crop production forecasting under IRS utilization program in 1983 (Dadhwal et al., 2002). Following which a project named large area crop acreage (LACA) was launched sponsored by Ministry of Agriculture for yield forecasting of few main crops like rice, wheat, groundnut, sorghum, mustard, sugarcane, and cotton; since 1988 it renamed as 'crop acreage and production estimation' (CAPE) project. Indian Space Research Organization (ISRO) implemented its crop monitoring program through an institute called Mahalanobis National Crop Forecast Center (MNCFC) under Department of Agriculture and Cooperation, Ministry of Agriculture whose primary aim is to use RS in crop forecasting. ISRO is taking the lead here and after the successful launch of IRS series in 1988, LISS-1 (linear imaging self-scanning system) was adopted widely for crop yield modeling (Sridhar et al., 1994; Panigrahy et al., 1999).

7.9.1 USE OF GIS AND RS IN BIODIVERSITY ASSESSMENT

The present pandemic condition due to devastating disease like coronavirus is a kind of wake-up call to the mankind to reconsider the relationship it has with the nature. Continuous biodiversity loss and a severe degradation of ecosystem are taking a huge toll on human wellbeing and its overall existence. Biodiversity loss has become a global concern since last few decades. The United Nations has mentioned its concern on biodiversity loss in its several reports. The convention of biological diversity in its Global Biodiversity report has critically mentioned that to address long term food security and fight CC, conserving the biodiversity is of immense importance. They concluded that it is the need of the hour to check biodiversity loss at any cost to check future pandemics.

For biodiversity assessment and monitoring also GIS and RS either separately or in combination are very important. To maintain the biodiversity the need of a database catering variety of information is the first priority. These databases should cater geographically based information on different species and population of endangered species, also an indication of any potential hot spot. In order to create such information rich database, satellite data are immensely helpful (Salem, 2003). Satellite data are also rich source for creating spatial layers for vegetation type and to create density maps. In case of biodiversity monitoring, providing a synoptic view with high temporal resolution of a forest and a real time basis thematic information is the key and RS techniques are very useful in this case (Prasad et al., 2015). Thus, these advanced techniques help us to overcome the difficulties we faced in case of field-based data collection. Application of RS and GIS also have added advantages of being inexpensive and less time-consuming process even in case of covering a large area in a consistent manner (Muldavin et al., 2001). On the other hand, GIS can be used to accommodate large amount of spatial and aspatial data catering information regarding target surveys and temporal distribution species habitat. Such a geo-database can be very helpful for change detection and assessment of biodiversity.

In the limited scope of this chapter, we will mention few of the significant studies on biodiversity deploying GIS and RS. Behera et al. (2005) conducted a study on rapid assessment of biological richness in a part of Eastern Himalaya with the help of geospatial tools. The study was unique in the sense that it suggested a way to manage conservation areas of biodiversity as well as methods to rapidly assess the biological richness where the vegetation and land cover types were classified using RS. The result was then validated using ground truthing that improved the accuracy of the study up to 89%. Another study conducted in 2003 on agricultural landscapes of Costa Rica (Daily et al., 2003), evaluated non-flying mammal species abundance and diversity in five habitat types of the study area. Satellite images (Landsat TM) of different time and seasons were used in the study to assess the change in forest cover and its proximity to other land cover types. Supervised classification of satellite images was conducted to identify different land cover types and the classification was further verified using aerial photographs of the same period as well as ground truthed data. The study revealed an interesting finding that the species richness of mammal's changed with change in habitat type but was independent on its distance from the extensive forest. After the forest clearance, they found to persist even more than five-kilometer distance from relatively extensive forest.

7.9.2 USE OF GIS AND RS IN BIOMASS ESTIMATION

Aboveground biomass (AGB) estimation is very essential to understand the types of LU and land cover, crop vigor and also to calculate the biosphere carbon share in global carbon (GC) cycle (Baccini et al., 2008). AGB estimation can be done using destructive method or by measuring physical parameters of vegetation like diameter at breast-height, canopy cover tree density, etc., and using them as predictors in allometric equations (Basuki et al., 2009). Volumetric method is very common for smaller plants like shrubs, herbs, etc. (Muldavin et al., 2008). Space borne RS is a good alternative to measure AGB with high accuracy as it is less labor and cost intensive also allows collection of data from places which are otherwise physically not accessible. Launching of satellite series like Landsat and NOAA in 1970s initiated this approach. With the advent of new and sophisticated satellites providing higher spatial and spectral resolution accuracy of the data improved significantly (Li et al., 2016). Using this satellite data as predictors, mathematical equations are developed to predict AGB of an area (Deb et al., 2017). Vegetation indices derived combining reflectance spectra of red and near infrared (IR) bands are generally used as inputs in such equations (Yang et al., 2012). There are numbers of vegetation indices with varying degree of simplicity, reliability, inclusion, and exclusion of factors and studies have used many of them comprehensively to estimate AGB of different land cover classes (Schulze-Brüninghoff et al., 2019).

In order to conserve biodiversity and check its further loss, we need to continuously keep an eye on it. But it is easier said than done keeping in mind the vast area we need to keep under our radar. Thankfully, the advancement made in RS and geospatial technologies in last few decades make it possible to monitor the real-time status and scanning for any potential risk in future times for timely adoption of precautionary measures.

7.10 CONCLUSION

Biodiversity protects the functional ecosystem that delivers clean water and air, oxygen, plant pollination, pest control and other ES. Food and nutritional security are also depending on the biodiversity. Biodiversity conservation is helpful for the protection and use of natural resources. To increase biodiversity, hedge row planting, agroforestry, the encouragement of biodiversity in arable land, conservation agriculture, mulching, and the wise use of pesticides are necessary. The agroforestry system is a practice that enables the ES by enhancing the diversity of microbes, insects, and other micro- and macro-fauna. The nitrogen fixing trees, beneficial insects, AM fungi that mobilize phosphates and endophytic microbes that prevent disease are the most important components of agroforestry Selecting agroforestry systems suited to soil, site, climate, social, and local needs can help improve biodiversity. A more rapid assessment and monitoring of biodiversity using RS and GIS techniques further helps to make meaningful mid-term corrections in long-term agroforestry systems. To conclude, agroforestry systems could become a viable option for improved biodiversity, maintenance of ES and the production system.

KEYWORDS

- **biochemical reaction**
- **biomass**
- **ecosystem**
- **geo-informatics**
- **insect diversity**
- **plant diversity**
- **remote sensing**

REFERENCES

Aertsens, J., De Nocker, L., & Gobin, A., (2013). Valuing the carbon sequestration potential for European agriculture. *Land Use Policy, 31*, 584–594.

AgRISTARS, (1981). *Program Support Staff; AgRISTARS. Agriculture, and Resources Inventory Surveys Through Aerospace Remote Sensing, Annual Report – Fiscal Year 1980, NASA, Lyndon B.* Johnson Space Center, Houston, TX.

Araujo, A., Leite, L., Iwata, B. F., Lira, M. A., Jr, Xavier, G. R., & Figueiredo, M. V. B., (2012). Microbiological process in agroforestry systems: A review. *Agron. Sus. Dev., 32*, 215–226.

Atangana, A., Khasa, D., Chang, S., & Degrande, A., (2014). Agroforestry and biodiversity conservation in tropical landscapes. In: *Tropical Agroforestry* (pp. 227–232). Dordrecht, Springer Netherlands.

Baccini, A., Laporte, N., Goetz, S. J., Sun, M., & Dong, H., (2008). A first map of tropical Africa's above-ground biomass derived from satellite imagery. *Environ. Res. Lett., 3,* 045011.

Bainard, L. D., Klironomos, J. N., & Gordon, A. M., (2011). Arbuscular mycorrhizal fungi in tree-based intercropping systems: A review of their abundance and diversity. *Pedobiologia, 54*, 57–61.

Bakermans, M. H., Rodewald, A. D., Vitz, A. C., & Rengifo, C., (2012). Migratory bird use of shade coffee: The role of structural and floristic features. *Agrofor. Sys., 85*, 85–94.

Balakrishna, A. N., Lakshmipathy, R., Bagyaraj, D. J., & Ashwin, R., (2017). Influence of alley copping system on AM fungi, microbial biomass C and yield of finger millet, peanut, and pigeon pea. *Agrofor. Syst., 91*, 487–493.

Banarjee, S., Baah-Acheamfour, M., Carlyle, C. M., Bissett, A., Richardon, A. E., Siddique, T., Bork, E. W., & Chang, S. X., (2016). Determinants of bacterial communities in Canadian agroforestry systems. *Environ. Microbiol., 18*, 1805–1816.

Bardgett, R. D., Streeter, T., & Bol, R., (2003). Soil microbes compete effectively with plants for organic nitrogen inputs to temperate grasslands. *Ecology, 84*, 1277–1287.

Bardhan, S., Jose, S., Biswas, S., Kabir, K., & Rogers, W., (2012). Homegarden agroforestry systems: An intermediary for biodiversity conservation in Bangladesh. *Agrofor. Sys., 85*, 29–34.

Basuki, T. M., Van, L. P. E., Skidmore, A. K., & Hussin, Y. A., (2009). Allometric equations for estimating the above-ground biomass in tropical lowland *Dipterocarp* forests. *For. Ecol. Manag., 257*, 1684–1694.

Bauer, M. E., Cary, T. K., Davis, B. J., & Swain, P. H., (1975). *Crop Identification Technology Assessment for Remote Sensing (CITARS): Data Processing at the Laboratory for Applications of Remote Sensing* (Vol. 6).

Behera, M. D., Kushwaha, S. P. S., & Roy, P. S., (2005). Rapid assessment of biological richness in a part of Eastern Himalaya: An integrated three-tier approach. *For. Ecol. Manag., 207*, 363–384.

Bélanger, J., & Pilling, D., (2019). *The State of the World's Biodiversity for Food and Agriculture*. Food and Agriculture Organization of the United Nations.

Beuschel, R., Piepho, H. P., Joergensen, R. G., & Wachendorf, C., (2019). Similar spatial patterns of soil quality indicators in three poplar-based silvo-arable alley cropping systems in Germany. *Biol. Fertil. Soils, 55*, 1–14.

Bhagwat, S. A., Willis, K. J., Briks, H. J. B., & Whittaker, R. J., (2008). Agroforestry: A refuge for tropical biodiversity? *Trends Ecol. Evol., 23*, 261–267.
Bhuiyan, M. A., Khan, H. U. R., Zaman, K., & Hishan, S. S., (2018). Measuring the impact of global tropospheric ozone, carbon dioxide and sulfur dioxide concentrations on biodiversity loss. *Environ. Res., 160*, 398–411.
Bijalwan, A., (2013). Vegetation status of agroforestry systems in Tehri district of Garhwal Himalaya, India. *Asian J. Sci. Technol., 4*, 11–14.
Bolton, D. K., & Friedl, M. A., (2013). Forecasting crop yield using remotely sensed vegetation indices and crop phenology metrics. *Agric. For. Meteorol., 173*, 74–84.
Bos, M. M., Höhn, P., Saleh, S., Büche, B., Buchori, D., Steffan-Dewenter, I., & Tscharntke, T., (2007). Insect diversity responses to forest conversion and agroforestry management. In: *Stability of Tropical Rainforest Margins* (pp. 277–294). Berlin, Heidelberg, Springer.
Bowler, D. E., Bjorkman, A. D., Dornelas, M., Myers-Smith, I. H., Navarro, L. M., Niamir, A. Supp, S. R., et al., (2020). Mapping human pressures on biodiversity across the planet uncovers anthropogenic threat complexes. *People and Nature, 2*, 380–394.
Brito-Vega, H., Salaya-Dominguez, J. M., Gomez-Mendez, E., Gomez-Vazquez, A., & Antele-Gomez, J. B., (2018). Physico-chemical properties of soil and pods (*Theobroma cacao* L.) in cocoa agroforestry systems. *J. Agron., 17*, 48–55.
Brown, V. K., (1985). Insect herbivores and plant succession. *Oikos, 44*, 17–22.
Campbell, C. A. M., (1984). The influence of overhead shade and fertilizers on the homoptera of mature upper-amazon cocoa trees in ghana. *Bull. Entomol. Res., 74*, 163–174.
Cardinale, B. J., Duffy, J. E., Gonzalez, A., Hooper, D. U., Perrings, C., Venail, P., Narwani, A., et al., (2012). Biodiversity loss and its impact on humanity. *Nature, 486*, 59–67.
Cardoso, M. I., & Kuyper, T. W., (2006). Mycorrhizas and tropical soil fertility. *Agric. Ecosys. Environ., 116*, 72–84.
Cleveland, C. C., Townsend, A. R., Schimel, D. S., Fisher, H., Howarth, R. W., Hedin, L. O., Perakis, S. S., et al., (1999). Global patterns of terrestrial biological nitrogen (N_2) fixation in natural ecosystems. *Glob. Biogeochem. Cycles, 13*, 623–645.
Conservation International, (2012). https://www.conservation.org/blog/despite-pandemic-there-is-still-work-to-do (accessed on 28 December 2023).
Dadhwal, K. S., Narain, P., & Dhyani, S. K., (1988). Agroforestry systems in the Garhwal Himalayas of India. *Agrofor. Sys., 7*, 213–225.
Dadhwal, V. K., Singh, R. P., Dutta, S., & Parihar, J. S., (2002). Remote sensing based crop inventory: A review of Indian experience. *Trop. Ecol., 43*, 107–122.
Dagar, J. C., & Tewari, V. P., (2017). Evolution of agroforestry as a modern science. In: *Agroforestry* (pp. 13–90). Springer Singapore.
Daily, G. C., Ceballos, G., Pancheco, J., Suzan, G., & Sanchez-Azofeifa, A., (2003). Countryside biogeography of neotropical mammals: Conservation opportunities in agricultural landscapes of costa Rica. *Biol. Conserv., 17*, 1814–1826.
De Deyn, G. B., Raaijmakers, C. E., Zoomer, H. R., Berg, M. P., De Ruiter, P. C., Verhoef, H. A., Bezemer, T. M., & Van, D. P. W. H., (2003). Soil invertebrate fauna enhances grassland succession and diversity. *Nature, 422*, 711–713.
Deb, D., Singh, J. P., Deb, S., Datta, D., Ghosh, A., & Chaurasia, R. S., (2017). An alternative approach for estimating above ground biomass using resourcesat-2 satellite data and artificial neural network in Bundelkhand region of India. *Environ. Monit. Assess., 189,* 576.
Dinkelaker, B., & Marschner, H., (1992). *In vivo* demonstration of acid phosphatase activity in the rhizosphere of soil-grown plants. *Plant Soil, 144*, 199–205.

Dobo, B., Asefa, F., & Asfaw, Z., (2018). Effect of tree-enset-coffee based agro-forestry practices on arbuscular mycorrhizal fungi (AMF) species diversity and spore density. *Agrofor. Syst., 92*, 525–540.

Eilers, E. J., Kremen, C., Greenleaf, S. S., Garber, A. K., & Klein, A. M., (2011). Contribution of pollinator-mediated crops to nutrients in the human food supply. *PLoS One, 6,* e21363.

Ernest, H., Kumar, A., Bhople, B. S., Srivastava, P. K., Singh, R. D., & Sahu, S. K., (2015). Soil quality and health management through farming practices: To ensure agricultural sustainability. *Res. J. Pharm. Biol. Chem. Sci., 6*, 1983–1992.

Ferreira, A. S., Peres, C. A., Dodonov, P., & Cassano, C. R., (2020). Multi-scale mammal responses to agroforestry landscapes in the Brazilian Atlantic Forest: The conservation value of forest and traditional shade plantations. *Agrofor. Sys., 94*, 2331–2341.

Fischer, G., Shah, M., Tubiello, F. N., & Van, V. H., (2005). Socio-economic and climate change impacts on agriculture: An integrated assessment, 1990–2080. *Phil. Trans. Roy. Soc. B., 360*, 2067–2083.

Freitas, A. D. S., Sampaio, E. V. S. B., Santos, C. E. R. S., & Fernandes, A. R., (2010). Biological nitrogen fixation in tree legumes of the Brazilian semi-arid Caatinga. *J. Arid Environ., 74*, 344–349.

Garrett, H. E., McGraw, R. L., & Walter, W., (2009). Alley cropping practices. In: Garrett, H. E., (ed.), *North American Agroforestry: An Integrated Science and Practice* (2nd edn., pp. 133–162). *American Society of Agronomy*, Madison, WI, USA.

Gebelein, J., (2012). History of remote sensing and GIS as it relates to assessment of land use and land cover changes over time. In: *A Geographic Perspective of Cuban Landscapes* (pp. 55–75). Springer, Dordrecht.

George, T. S., Gregory, P. J., Robinson, J. S., Buresh, R. J., & Jama, B. A., (2001). *Tithonia diversifolia*: Variations in leaf nutrient concentration and implications for biomass transfer. *Agrofor. Syst., 52*, 199–205.

Gregory, P. J., & Ingram, J. S. I., (2000). Global change and food and forest production: Future scientific challenges. *Agric. Ecosyst. Environ., 82*, 3–14.

Hillel, D., (1982). *Introduction to Soil Physics*. Academic Press, San Diego, CA.

Hoekstra, J. M., Boucher, T. M., Ricketts, T. H., & Roberts, C., (2005). Confronting a biome crisis: Global disparities of habitat loss and protection. *Ecol. Lett., 8*, 23–29.

Hooper, D. U., Chapin, F. S., Ewel, J. J., Hector, A., Inchausti, P., Lavorel, S., Lawton, J. H., et al., (2005). Effects of biodiversity on ecosystem functioning: A consensus of current knowledge. *Ecol. Monogr., 75*, 3–35.

Huang, W., (1998). Productive coexistence and gain in agroforestry systems. *Acta For. Fenn., 260*, 1–72.

Huang, W., Luukkanen, O., Johanson, S., Kaarakka, V., Räisänen, S., & Vihemäki, H., (2002). Agroforestry for biodiversity conservation of nature reserves: Functional group identification and analysis. *Agrofor. Syst., 55*, 65–72.

Isbell, F., Adler, P. R., Eisenhauer, N., Fornara, D., Kimmel, K., Kremen, C., Letourneau, D. K., et al., (2017). Benefits of increasing plant diversity in sustainable agroecosystems. *J. Ecol., 105*, 871–879.

Jones, K. R., Venter, O., Fuller, R. A., Allan, J. R., Maxwell, S. L., Negret, P. J., & Watson, J. E., (2018). One-third of global protected land is under intense human pressure. *Sci., 360*, 788–791.

Jose, S., (2009). Agroforestry for ecosystem services and environmental benefits: An overview. *Agrofor. Syst., 76*, 1–10.

Jose, S., (2011). Managing native and non-native plant species in agroforestry. *Agrofor. Syst., 83*, 101–105.

Jose, S., (2012). Agroforestry for conserving and enhancing biodiversity. *Agrofor. Sys., 85*, 1–8.

Jose, S., Gillespie, A. R., & Pallardy, S. G., (2004). Interspecific interactions in temperate agroforestry. *Agrofor. Syst., 61*, 237–255.

Kaur, B., Gupta, S. R., & Singh, G., (2000). Soil carbon, microbial activity, and nitrogen availability in an agroforestry systems on moderately alkaline soils in northern India. *Appl. Soil Ecol., 15*, 283–294.

Kay, S., Rega, C., Moreno, G., Den, H. M., Palma, J. H., Borek, R., Crous-Duran, J., et al., (2019). Agroforestry creates carbon sinks whilst enhancing the environment in agricultural landscapes in Europe. *Land Use Policy., 83*, 581–593.

Keesstra, S. D., Bouma, J., Wallinga, J., Tittonell, P., Smith, P., Cerdà, A., Montanarella, L., et al., (2016). The significance of soils and soil science towards realization of the United Nations sustainable development goals. *Soil, 2*, 111–128.

Kidwell, K. B., (1998). *Polar Orbiter Data Users' Guide* (TIROS-n, NOAA-6, NOAA-7, NOAA-8, NOAA-9, NOAA-10, NOAA-11, NOAA-12, and NOAA-14), Washington, DC.

Kim, D. G., Kirschbaum, M. U., & Beedy, T. L., (2016). Carbon sequestration and net emissions of CH_4 and N_2O under agroforestry: Synthesizing available data and suggestions for future studies. *Agric. Ecosyst. Environ., 226*, 65–78.

Klein, A. M., Vaissiere, B. E., Cane, J. H., Steffan-Dewenter, I., Cunningham, S. A., Kremen, C., & Tscharntke, T., (2007). Importance of pollinators in changing landscapes for world crops. *Proc. Biol. Sci., 274*, 303–313.

Kremer, R. J., & Hezel, L. F., (2013). Soil quality improvement under an ecologically based farming system in Northwest Missouri. *Renew. Agric. Food Syst., 28*, 245–254.

Krishnan, V., Kumar, M. A., Raghunathan, G., & Vijayakrishnan, S., (2019). Distribution and habitat use by Asian elephants (*Elephas maximus*) in a coffee-dominated landscape of Southern India. *Trop. Conserv. Sci., 12,* 194008291882259.

Kumar, B. M., & Nair, P. K. R., (2006). *Tropical Homegardens.* Springer, The Netherlands.

Lagerlof, J., Adolfsson, L., Börjesson, G., Ehlers, K., Vinyoles, G. P., & Sundh, I., (2014). Land-use intensification and agroforestry in the Kenyan highland: Impacts on soil microbial community composition and functional capacity. *Appl. Soil Ecol., 82*, 93–99.

Lakshmipathy, R., Balakrishna, A. N., & Bagyaraj, D. J., (2012). Abundance and diversity of AMF across a gradient of land use intensity and their seasonal variations in Nilgiri biosphere of Western Ghats, India. *J. Agric. Sci. Technol., 14*, 903–918.

Lalthanzara, H., Ramanujam, S. N., & Jha, L. K., (2011). Population dynamics of earthworms in relation to soil physico-chemical parameters in agroforestry systems of Mizoram, India. *J. Environ Biol., 32*, 599–605.

Landeweert, R., Hoffland, E., Finlay, R. D., Kuyper, T. W., & Van, B. N., (2001). Linking plants to rocks: Ectomycorrhizal fungi mobilize nutrients from minerals. *Trends Ecol. Evol., 16*, 248–254.

Larsen, F. W., Turner, W. R., & Brooks, T. M., (2012). Conserving critical sites for biodiversity provides disproportionate benefits to people. *PLoS One, 7*, e36971.

Lehman, C. L., & Tilman, D., (2000). Biodiversity, stability, and productivity in competitive communities. *Am. Nat., 156*, 534–552.

Li, F., Zeng, Y., Luo, J., Ma, R., & Wu, B., (2016). Modeling grassland aboveground biomass using a pure vegetation index. *Ecol. Indic., 62*, 279–288.

Lillesand, T., Kiefer, R. W., & Chipman, J., (2015). *Remote Sensing and Image Interpretation*. John Wiley & Sons.

Loose, J. L., Drummond, F. A., Stubbs, C., Woods, S., & Hoffman, S., (2005). *Conservation and Management of Native Bees in Cranberry* (p. 33). MAFES Technical Bulletin, USA.

Loreau, M., Naeem, S., Inchausti, P., Bengtsson, J., Grime, J., Hector, A., Hooper, D., et al., (2001). Biodiversity and ecosystem functioning: Current knowledge and future challenges. *Sci., 294*, 804–808.

Lorenz, K., & Lal, R., (2014). Soil organic carbon sequestration in agroforestry systems. A review. *Agron. Sustain. Dev., 34*, 443–454.

Lubbers, I. M., Berg, M. P., De Deyn, G. B., Van, D. P. W. H., & Van, G. J. W., (2020). Soil fauna diversity increases CO_2 but suppresses N_2O emissions from soil. *Global Change Biol., 26*, 1886–1898.

Lubbers, I. M., Van, G. K. J., Fonte, S. J., Six, J., Brussaard, L., & Van, G. J. W., (2013). Greenhouse-gas emissions from soils increased by earthworms. *Nat. Clim. Change, 3*, 187–194.

Maas, B., Karp, D. S., Bumrungsri, S., Darras, K., Gonthier, D., Huang, J. C. C., Lindell, C. A., et al., (2016). Bird and bat predation services in tropical forests and agroforestry landscapes, *Biol. Rev., 91*, 1081–1101.

MacDonald, R. B., (1984). A summary of the history of the development of automated remote sensing for agri- cultural applications. *IEEE T. Geosci. Remote, 22*, 473–481.

MacDonald, R. B., Hall, F. G., & Erb, R. B., (1976). The use of Landsat data in a large area crop inventory experiment (LACIE). In: *Proceedings of the 10th International Symposium of Remote Sensing of Environment* (Vol. I) Ann Arbor, Michigan.

Malcolm, J. R., Markham, A., Neilson, R. P., & Garaci, M., (2002). Estimated migration rates under scenarios of global climate change. *J. Biogeogr., 29*, 835–849.

Malingreau, J. P., (1986). Global vegetation dynamics: Satellite observations over Asia. *Int. J. Remote Sens., 7*, 1121–1146.

Manner, H. I., (1981). Ecological succession in new and old swiddens of montane Papua New Guinea. *Hum. Ecol., 9*, 359–377.

Martin, D. A., Andriafanomezantsoa, R., Dröge, S., Osen, K., Rakotomalala, E., Wurz, A., Andrianarimisa, A., & Kreft, H., (2021). Bird diversity and endemism along a land-use gradient in Madagascar: The conservation value of vanilla agroforests. *Biotropica, 53*, 179–190.

McNeely, J. A., & Schroth, G., (2006). Agroforestry and biodiversity conservation- traditional practices, present dynamics, and lessons for the future. *Biodivers. Conserv., 15*, 549–554.

McNeely, J. A., (2004). Nature vs. nurture: Managing relationships between forests, agroforestry, and wild biodiversity. In: *New Vistas in Agroforestry* (pp. 155–165). Dordrecht, Springer.

Méndez, V. E., Shapiro, E. N., & Gilbert, G. S., (2009). Cooperative management and its effects on shade tree diversity, soil properties and ecosystem services of coffee plantations in Western El Salvador. *Agrofor. Syst., 76*, 111–126.

Meyers, R. T., Zak, D. R., White, D. C., & Peacock, A., (2001). Landscape-level patterns of microbial community composition and substrate use in upland forest ecosystems. *Soil Sci. Soc. Am. J., 65*, 359–367.

Muldavin, E. H., Moore, D. I., Collins, S. L., Wetherill, K. R., & Lightfoot, D. C., (2008). Aboveground net primary production dynamics in a Northern Chihuahuan Desert ecosystem. *Oecologia, 155*, 123–132.

Muldavin, E. H., Neville, P., & Harper, G., (2001). Indices of grassland biodiversity in the Chihuahuan desert ecoregions derived from remote sensing. *Conserv. Biol., 15*, 844–855.
Mushi, D. E., Eik, L. O., Bernués, A., Ripoll-Bosch, R., Sundstøl, F., & Mo, M., (2015). Reducing GHG emissions from traditional livestock systems to mitigate changing climate and biodiversity. In: *Sustainable Intensification to Advance Food Security and Enhance Climate Resilience in Africa* (pp. 343–365). Springer, Cham.
Naeem, S., Thompson, L. J., Lawler, S. P., Lawton, J. H., & Woodfin, R. M., (1994). Declining biodiversity can alter the performance of ecosystems. *Nature, 368*, 734–737.
Naidoo, R., Gerkey, D., Hole, D., Pfaff, A., Ellis, A. M., Golden, C. D., Herrera, D., et al., (2019). Evaluating the impacts of protected areas on human well-being across the developing world. *Sci. Adv., 5,* eaav3006.
Nair, P. K. R., Kumar, B. M., & Nair, V. D., (2009). Agroforestry as a strategy for carbon sequestration. *J. Plant Nutr. Soil Sci., 172*, 10–23.
Nannipieri, P., Landi, L., & Badalucco, L., (1995). La capacitàmetabólica e la qualitàdelsuolo. *Agronomía, 29*, 312–316.
Niang, A., Amadalo, B., Gathumbi, S., & Obonyo, C., (1996). Maize yield response to green manure application from selected shrubs and tree species in western Kenya: A preliminary assessment. *East Afr. Agric. For. J., 62*, 199–207.
Nielsen, U. N., Wall, D. H., & Six, J., (2015). Soil biodiversity and the environment. *Ann. Rev. Environ. Resources, 40*, 63–90.
Noble, I. R., & Dirzo, R., (1997). Forest a human dominated ecosystems. *Sci., 277*, 522–525.
NSO, (2020). EnviStats-India 2020. *Environment Statistics* (Vol. I). New Delhi.
Nygren, P., Fernández, M. P., Harmand, J. M., & Leblanc, H. A., (2012). Symbiotic dinitrogen fixation by trees: An underestimated resource in agroforestry systems. *Nutr. Cycl. Agroecosyst., 94*, 123–160.
Odewumi, S. G., Iwara, A. I., & Ogundele, F. O., (2013). Effects of teak (*Tectona grandis*) cultivation on soil physical and chemical properties in Ajibode community, Ibadan, Oyo state. Wudpecker. *J. Agri. Res., 2*, 49–54.
Ollerton, J., Winfree, R., & Tarrant, S., (2011). How many flowering plants are pollinated by animals? *Oikos, 120*, 321–326.
Ong, C. K., Anyango, S., Muthuri, C. W., & Black, C. R., (2007). Water use and water productivity of agroforestry systems in the semiarid tropics. *Ann. Arid Zone, 46*, 244–255.
Panigrahy, S., Parihar, J. S., Patel, N. K., Dadhwal, V. K., Medhavy, T. T., Ghose, B. K., Ravi, N., et al., (1999). Rice acreage estimation for Orissa using remotely sensed data. *J. Indian Soc. Remote Sens., 19*, 16–26.
Paudel, B. R., Udawatta, R. P., Kremer, R. J., & Anderson, S. H., (2012). Soil quality indicator responses to row crop, grazed pasture, and agroforestry buffer management, *Agrofor. Syst., 84*, 311–323.
Paudel, B., Udawatta, R. P., Kremer, R. J., & Anderson, S. H., (2011). Agroforestry and grass buffer effects on soil quality parameters for grazed pasture and row-crop systems. *Appl. Soil Ecol., 48*, 125–132.
Pereira, E. I. P., Chung, H., Scow, K. M., & Six, J., (2013). Microbial communities and soil structure are affected by reduced precipitation, but not by elevated carbon dioxide. *Soil Sci. Soc. Am. J., 77*, 482–488.
Peters, V. E., & Carroll, C. R., (2012). Temporal variation in coffee flowering may influence the effects of bee species richness and abundance on coffee production. *Agrofor. Syst., 85*, 95–103.

Philippot, L., Raaijmakers, J. M., Lemanceau, P., & Van, D. P. W., (2013). Going back to the roots: The microbial ecology of the rhizosphere. *Nat. Rev. Microbiol., 11*, 789–799.

Prasad, N., Semwal, M., & Roy, P. S., (2015). Remote sensing and GIS for biodiversity conservation. In: *Recent Advances in Lichenology* (pp. 151–179). Springer, New Delhi.

Radhakrishnan, S., & Varadharajan, M., (2016). Status of microbial diversity in agroforestry systems in Tamil Nadu, India. *J. Basic Microbial., 56*, 662–669.

Ramesh, T., Manjaiah, K. M., Tomar, J. M. S., & Ngachan, S. V., (2013). Effect of multipurpose tree species on soil fertility and CO_2 efflux under hilly ecosystems of Northeast India. *Agrofor. Syst., 87*, 1377–1388.

Rands, M. R., Adams, W. M., Bennun, L., Butchart, S. H., Clements, A., Coomes, D., Entwistle, A., et al., (2010). Biodiversity conservation: Challenges beyond. *Sci., 329*, 1298–1303.

Rendón-Sandoval, F. J., Casas, A., Moreno-Calles, A. I., Torres-García, I., & García-Frapolli, E., (2020). Traditional agroforestry systems and conservation of native plant diversity of seasonally dry tropical forests. *Sustainability, 12,* 4600.

Roy, D. P., Wulder, M. A., Loveland, T. R., Woodcock, C. E., Allen, R. G., Anderson, M. C., Helder, D., et al., (2014). Landsat-8: Science and product vision for terrestrial global change research. *Remote Sens. Environ., 145*, 154–172.

Saha, R., Mishra, V. K., & Tomar, J. M. S., (2004). Soil water retention-transmission characteristics of various agro-forestry systems under hilly agriculture in Meghalaya. *Indian J. Hill Farming, 17*, 106–110.

Salazar-Díaz, R., & Tixier, P., (2019). Effect of plant diversity on income generated by agroforestry systems in Talamanca, Costa Rica. *Agrofor. Syst., 93*, 571–580.

Salem, B. B., (2003). Application of GIS to biodiversity monitoring. *J. Arid Environ., 54*, 91–114.

Samways, M. J., (1993). Insects in biodiversity conservation: Some perspectives and directives. *Biodivers. Conserv., 2*, 258–282.

Schroth, G., Izac, A. M. N., Vasconcelos, H. L., Gascon, C., Da Fonseca, G. A., & Harvey, C. A., (2004). *Agroforestry and Biodiversity Conservation in Tropical Landscapes.* Island Press, Washington, USA.

Schulze-Brüninghoff, D., Hensgen, F., Wachendorf, M., & Astor, T., (2019). Methods for LiDAR-based estimation of extensive grassland biomass. *Comput. Electron. Agr., 156*, 693–699.

Secretariat of the Convention on Biological Diversity, (2005). *Handbook of the Convention on Biological Diversity Including its Cartagena Protocol on Biosafety* (3rd edn.). Montreal, Canada.

Shanker, C., & Solanki, K. R., (2000). Agroforestry: An ecofriendly land-use system for insect management. *Outlook. Agric., 29*, 91–96.

Shukla, A., Kumar, A., Chaturvedi, O. P., Nagori, T., Kumar, N., & Gupta, A., (2018). Efficacy of rhizobial and phosphate-solubilizing bacteria and arbuscular mycorrhizal fungi to ameliorate shade response on six pulse crops. *Agrofor Syst., 92*, 499–509.

Singh, B., & Sharma, K. N., (2007). Tree growth and nutrient status of soil in a poplar (*Populus deltoides* Bartr.)-based agroforestry system in Punjab, India. *Agrofor. Syst., 70*, 125–134.

Sistla, S. A., Roddy, A. B., Williams, N. E., Kramer, D. B., Stevens, K., & Allison, S. D., (2016). Agroforestry practices promote biodiversity and natural resource diversity in Atlantic Nicaragua. *PLoS One, 11,* e0162529.

Smith, R. G., Gross, K. L., & Robertson, G. P., (2008). Effects of crop diversity on agroecosystem function: Crop yield response. *Ecosyst., 11*, 355–366.

Söderström, B. O., Svensson, B., Vessby, K., & Glimskär, A., (2001). Plants, insects, and birds in semi-natural pastures in relation to local habitat and landscape factors. *Biodivers. Conserv., 10*, 1839–1863.

Somarriba, E., (1992). Revisiting the past: An essay on agroforestry definition, *Agrofor. Sys., 19*, 233–240.

Somasundaram, J., Singh, R. K., Ali, S., Sethy, B. K., Singh, D., Lakaria, B. L., Chaudhary, R. S. et al., (2012). Soil aggregates and other properties as influenced by different long term land uses under table landscape topography of Chambal region, Rajasthan, India. *Indian J. Soil Conserv., 40*, 212–217.

Sridhar, K. R., & Bagyaraj, D. J., (2017). Microbial biodiversity in agroforestry systems. In: *Agroforestry* (pp. 645–667). Springer.

Sridhar, K. R., (2009). Fungi in the tree canopy- an appraisal. In: Rai, M., & Bridge, P., (eds.), *Applied Mycology* (pp. 73–91). CAB International, Wallingford.,.

Sridhar, V. N., Dadhwal, V. K., Chaudhari, K. N., Sharma, R., Bairagi, G. D., & Sharma, A. K., (1994). Wheat production forecasting for a predominantly unirrigated region in Madhya Pradesh, India. *Remote Sens., 15*, 1307–1316.

Steffan-Dewenter, I., Kessler, M., Barkmann, J., Bos, M. M., Buchori, D., Erasmi, S., Faust, H., et al., (2007). Tradeoffs between income, biodiversity, and ecosystem functioning during tropical rainforest conversion and agroforestry intensification. *Proc. Natl. Acad. Sci. USA, 104*, 4973–4978.

Stubbs, C. S., Jacobson, H. A., Osgood, E. A., & Drummond, F. A., (1992). Alternative forage plants for native (wild) bees associated with low bush blue berry, *Vaccinium* spp., in Maine. *Maine Agricultural Experiment Station Technical Bulletin, 148*, 63.

Tarafdar, J. C., & Claassen, N., (1988). Organic phosphorus compounds as a phosphorus source for higher plants through the activity of phosphatases produced by plant roots and microorganisms. *Biol. Fert. Soils, 5*, 308–312.

Tarafdar, J. C., & Marschner, H., (1994). Phosphatase activity in the rhizosphere and hyphosphere of VA mycorrhizal wheat supplied with inorganic and organic phosphorus. *Soil Biol. Biochem., 26*, 387–395.

Tarafdar, J. C., (1988). Activity of urease and phosphatases in the root-soil interface of vegetable crops. *J. Hortic. Sci., 63*, 605–608.

Tian, Y., Cao, F., & Wang, G., (2013). Soil microbiological properties and enzyme activity in Ginkgo–tea agroforestry compared with monoculture. *Agrofor. Syst., 87*, 1201–1210.

Tomich, T. P., Van, N. M., Vosti, S. A., & Witcover, J., (1998). Agricultural development with rainforest conservation: Methods for seeking best bet alternatives to slash-and-burn, with applications to Brazil and Indonesia. *Agric. Econ., 19*, 159–174.

Tornquista, C. G., Hons, F. M., Feagley, S. E., & Haggar, J., (1999). Agroforestry system effects on soil characteristics of the Sarapiqui region of Costa Rica. *Agric. Ecosyst. Environ., 73*, 19–28.

Torres-Gomez, M., Garibay-Orijel, R., Casas, A., & Perez-Salicrup, D. R., (2018). Ectomycorrhizal trees intermingled within *Cupressus lusitanica* plantations sustain the diversity and availability of edible mushrooms. *Agrofor. Syst., 92*, 575–588.

Tripathi, O. P., Pandey, H. N., & Tripathi, R. S., (2009). Litter production, decomposition, and physicochemical properties of soil in 3 developed agroforestry systems of Meghalaya, Northeast India. *Afr. J. Plant Sci., 3*, 160–167.

Tynsong, H., & Tiwari, B. K., (2010). Plant diversity in the homegardens and their significance in the livelihoods of War Khasi community of Meghalaya, North-east India. *J. Biodivers., 1*, 1–11.

Udawatta, P., Rankoth, R. L., & Jose, S., (2019). Agroforestry and biodiversity. *Sustainability, 11,* 2879.

Udawatta, R. P., Gantzer, C. J., Anderson, S. H., & Garrett, H. E., (2008). Agroforestry and grass buffer effects on pore characteristics measured by high-resolution x-ray computed tomography. *Soil Sci. Soc. Am. J., 72*, 295–304.

Udawatta, R. P., Kremer, R. J., Garrett, H. E., & Anderson, S. H., (2009). Soil enzyme activities and physical properties in a watershed managed under agroforestry and row-crop systems. *Agric. Ecosyst. Environ., 131*, 98–104.

UNEP, (1998). *Tanzania: Country Study on Biological Diversity*. Office of the Vice President, Dar es Salaam. United Nations Environment Program.

Van, D. H. M. G. A., Bardgett, R. D., & Van, S. N. M., (2008). The unseen majority: Soil microbes as drivers of plant diversity and productivity in terrestrial ecosystems. *Ecol. Lett., 11*, 296–310.

Varah, A., Jones, H., Smith, J., & Potts, S. G., (2020). Temperate agroforestry systems provide greater pollination service than monoculture. *Agric. Ecosyst. Environ., 301,* 107031.

Visser, S., & Parkinson, D., (1992). Soil biological criteria as indicators of soil quality: Soil microorganisms. *Am. J. Altern. Agric., 7*, 33–37.

Wall, D. H., Nielsen, U. N., & Six, J., (2015). Soil biodiversity and human health. *Nature, 528*, 69–76.

WMO, (2015). *WMO Greenhouse Gas Bulletin*. No. 11., World Meteorological Organization, Geneva. Switzerland. Available at: http://library.wmo.int/ (accessed on 28 December 2023).

Wojewoda, D., & Russel, S., (2003). The impact of a shelterbelt on soil properties a microbial activity in an adjacent crop field. *Pol. J. Ecol., 51*, 291–307.

Wood, P. J., & Burley, J., (1991). *A Tree for all Reasons: The Introduction and Evaluation of Multipurpose Trees for Agroforestry*. International Center for Research in Agroforestry.

Xiong, Y., Xia, H., Li, Z., Cai, X., & Fu, S., (2008). Impacts of litter and understory removal on soil properties in a subtropical *Acacia mangium* plantation in China. *Plant Soil, 304*, 179–188.

Yang, J., Weisberg, P. J., & Bristow, N. A., (2012). Landsat remote sensing approaches for monitoring long-term tree cover dynamics in semi-arid woodlands: Comparison of vegetation indices and spectral mixture analysis. *Remote Sens. Environ., 119*, 62–71.

Young, A., (1991). Soil fertility. In: Avery, M. E., Cannel, M. G. R., & Ong, C. K., (eds.), *Biophysical Research for Asian Agroforestry* (pp. 187–208). Winrock International USA and South Asia Books, USA.

CHAPTER 8

Agroforestry Systems: An Effective Tool for Carbon Sequestration

CHANDINI PRADHAN, A. K. GHOSH, PREETI SINGH, and RAJENDRA GADHWAL

Department of Soil Science and Agricultural Chemistry, Institute of Agricultural Science, Banaras Hindu University, Varanasi, Uttar Pradesh, India

ABSTRACT

The consequences of global warming on climate change (CC) are now evident to all. The escalating level of greenhouse gases is the root cause of global warming is the increase in greenhouse gases (GHGs). As per reports of the Intergovernmental Panel on Climate Change (IPCC) the levels of current GHG, especially CO_2 have increased drastically about 30% in the atmosphere as compared to the pre-industrial levels. Such drastic increase of CO_2 has occurred due to anthropogenic activities like vehicular emission due to combustion of fossil fuels, emission from industries and replacement of tropical forests by agricultural lands. In order to mitigate this hazard C-sequestration has emerged as an effective way wherein the atmospheric carbon is being converted and stored in a form that is unavailable. In terms of sequestration, plant and soil organic carbon (SOC) act as an excellent carbon sink (25% of global carbon (GC) stocks, i.e., 2,000 ± 500 Pg) and their benefits can be maximized if they are managed and manipulated in various biomes. One such option is agroforestry systems which judiciously combine the woody perennials or trees with crops, livestock, and pastures for deriving maximum ecological benefit from their interaction. The tree component in such systems serve as a long-term sink for carbon. Moreover, agroforestry systems have the ability to sequester carbon both in the aboveground and

Sustainable Management and Conservation of Environmental Resources in India.
Sumit Chakravarty, Vineeta, Jahangeer A. Bhat, Munesh Kumar, Gopal Shukla (Eds.)

belowground biomass portions as well as soil. It is estimated that the agroforestry systems can potentially store approximately up to 30–300 Mg C ha^{-1} in one meter soil depth. Thus, the chapter encapsulates brief definitions, extent, and types of agroforestry systems present across world, mechanism C-sequestration by agroforestry systems and ancillary benefits associated with them. In addition, some lights have been thrown on international policies for promoting agroforestry all of the world, their execution, loopholes, and recommendations.

8.1 INTRODUCTION

Climate change (CC) currently is the big elephant in the room. It is omnipresent throughout the globe, making its presence evident through the shift in rainfall pattern, rise in temperature, forest fires, and alteration of seasonal pattern and melting of polar ice caps. The main driving force for this alarming CC is the phenomenon of global warming, which is the phenomenon of rise of temperature of near-surface air and oceans of earth is caused due to the drastic rise in levels of greenhouse gasses (GHGs). Ever since industrial revolution carbon dioxide (CO_2) and other GHGs have been accumulating in earth's atmosphere and contributing to the greenhouse effect. The rise of concentration of these gasses in atmosphere since industrial revolution and their radiative forcing is given in Table 8.1.

TABLE 8.1 Rise in Levels of Green House Gasses Since Industrial Revolution

GHG	Pre-Industrial Atmospheric Concentration	Atmospheric Concentration (2017)	Radiative Forcing (%)
CO_2	280 ppm	405.5 ppm	62
CH_4	700 ppb	1,859 ppb	17
N_2O	270 ppb	329.9 ppb	6

The atmospheric concentration of the main culprit CO_2 which has a radiative forcing of 62% has risen up from 280 ppmv in 1,750 up to 409.8 ppmv in 2019 (NOAA, 2020). The rate of rise of CO_2 is also escalating with each coming year. During the 1960s the rate of rise in concentration of CO_2 in atmosphere was 0.6 ppm per year which has elevated to 2.3 ppm per year during 2009–2018 (NOAA, 2020). Such steep rise in levels of GHGs in atmosphere has led to several changes in climate and environment. For instance, the global surface temperatures have increased by 0.4 to 0.8°C during the 20th century while, precipitation enhanced by 18.5 mm d^{-1} or by 10% and sea level rose by

18–75 cm (Cheddadi et al., 2001; IPCC, 2001, 2007). Land-surface precipitation in majority of northern hemisphere mainly in mid- and high-latitude has increased at the rate of 0.5–1% per decade while, in sub-tropical land areas it has decrease at the rate of 0.3% per decade. Such change in temperature and moisture regimes altered the growing season and species composition (Buckland et al., 2001). Further, CC when coupled with faulty agricultural practices like excessive use of fertilizers, herbicides, and pesticides to boost the food production has resulted in decrease in the soil organic carbon (SOC) pool and structural stability, rise in susceptibility of soil to water runoff and erosion and disturbance in water and nutrient cycles causing an adverse impact on biomass productivity, biodiversity, and the environment. It is quite evident that the impact of global warming (Figure 8.1) is jeopardous enough to adopt all possible mitigation strategies in order to avert it.

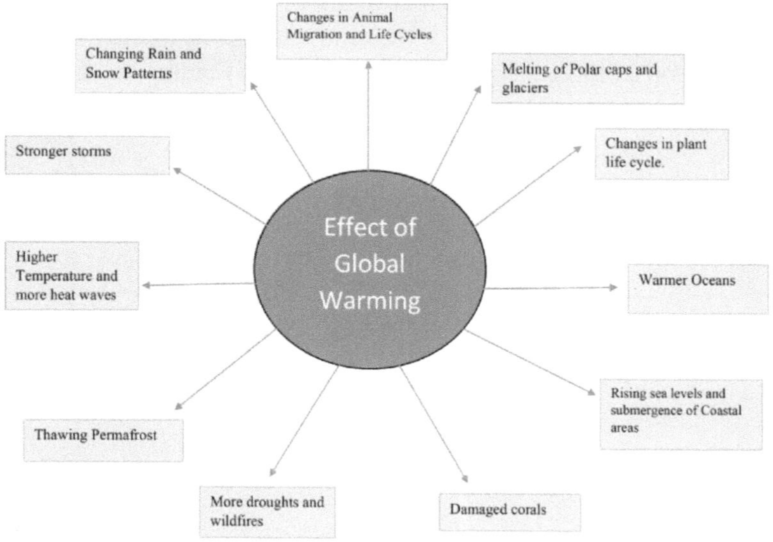

FIGURE 8.1 Adverse effect of global warming.

CO_2 being the most pivotal cause of global warming, any strategy involving reduction in CO_2 from atmosphere is a win-win strategy against mitigating CC *per se*. For this purpose, at first the sources and sinks of carbon are needed to be identified and then ways for curtailing the sources and means of enhancing the carbon sinks needs to be figured out. One such potential and effective strategy of doing so is the process of carbon sequestration. But just identification of carbon sequestration is not enough, the real trick lies in identification of the land use (LU) system which leads to effective carbon sequestration along with providing socio-economic benefit to the people. Keeping these criteria in consideration, the age-old practice of agroforestry system, in which crop plants and livestock are grown and reared along with the trees, emerges as a viable and efficient solution. This chapter highlights how carbon is cycled globally between its sources and sinks, what is carbon sequestration and how it plays a role in controlling global warming, in-depth study of role agroforestry function in carbon sequestration, role of international agencies like United Nation and current policies involving agroforestry and carbon trading, limitations of agroforestry and finally the measures to improve its efficiency.

8.2 GLOBAL CARBON (GC) CYCLE: SOURCES AND SINKS

The earth consists of five principal pools of carbon. These pools include the oceanic pool which has the greatest amount, followed by the geologic, pedologic (soil), biotic and the atmospheric pool (Figures 8.2 and 8.3).

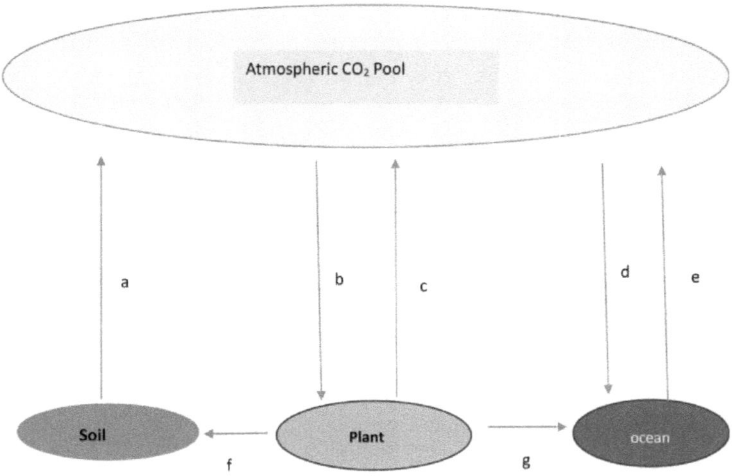

a: Soil respiration **e: Ocean respiration**
b: Photosynthesis **f: decomposition of plant**
residues
c: Plant respiration **g: organic residues flowing through**
river
d: ocean uptake

FIGURE 8.2 Global carbon cycle.

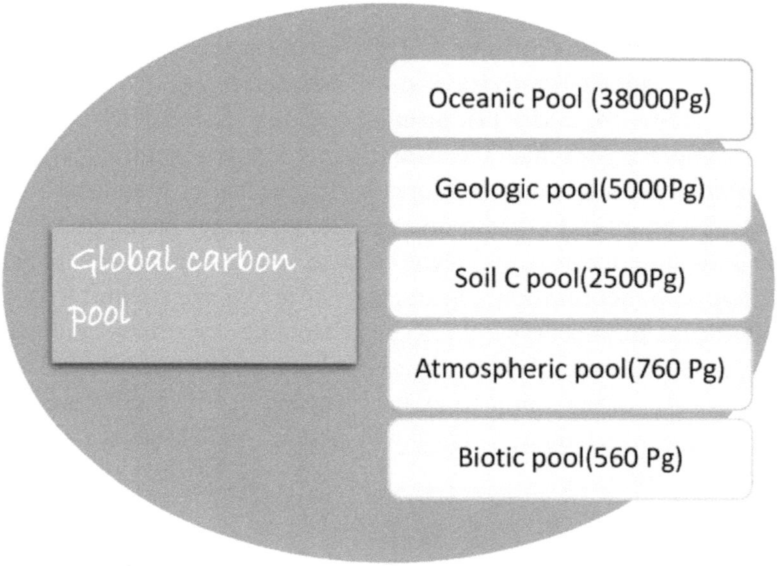

FIGURE 8.3 The components of global carbon pool.

Carbon is continuously cycled among these inter-connected pools (Figure 8.2). The oceanic pool is absorbing about 92 Gt C year^{-1} while, emitting 90 Gt C year^{-1} into the atmosphere, thereby sequestering two giga ton carbon per year. The biotic pool harnesses 120 Gt C year^{-1} from atmospheric pool through photosynthesis, out of which 60 Gt C year^{-1} is returned to atmosphere via plant respiration and remaining half is returned thorough soil respiration or organic matter mineralization. Among these pools, the one involved in fostering of the terrestrial ecosystems is the pedologic or soil C pool which is four times the biotic pool and three times of atmospheric pool. This pool encompasses of two components, i.e., SOC and the soil inorganic carbon (SIC) pool. These pools vary in there abundance depending upon the climate and vegetation. For instance, the SIC pool predominates in dry region whereas; the SOC pool predominates at peat or organic soils in temperate region but is low in arid regions. Additionally, different eco regions affect size of SOC pool for instance higher in cool and moist regions than warm and dry regions.

All these sources of carbon are also the sinks of carbon. Ever since the inception of earth carbon is balanced or cyclic throughout these pools. The problem arose when the concentration of carbon increased in atmospheric pool due to increase in sources in contrast to the non-expanded sinks. The major cause which has contributed to this can be broadly divided into two categories. The first is burning of fossil fuels and industrial emissions which accounted for emission of about 270 ± 30 Gt of carbon from AD 1850 to 2000 whereas, the second is alteration in LU and deforestation which accounted for emission of 136 ± 55 Gt C during the same period (IPCC, 2001). Out of these factors, the one which is linked to the realm of agriculture is the change in LU pattern which includes deforestation, burning of biomass, intensification of agriculture and mono-cropping. These changes have not just reduced the CO_2 fixation by trees and plant but also have enhanced the depletion of SOC due to mineralization and decomposition of vegetation. It is evident that shift of natural ecosystem into unsustainable agriculture promotes CO_2 emission the flip side of the coin is that the process can be reversed. This reversion will not just mitigate global warming but also will improve SOC, consequent of which there will be development of soil structure, fertility, and biomass productivity.

8.3 CARBON SEQUESTRATION

The dictionary meaning of the term sequester means 'to hold on to' or 'to keep separate' and are thus unavailable for other uses. As per this meaning, the term carbon sequestration simply means to capture the carbon from atmosphere and accumulate it in long-lived pool to separate it from other parts of ecosystem through change in land management practices. Carbon sequestration can be carried out in all the pools, but it requires high cost and complex technology in case of oceanic and geological pool whereas, carbon sequestration in soil and biotic component is a cost-effective means along with ancillary benefits. The carbon sequestration involving these two pools occurs through direct transformation of atmospheric CO_2 into SIC compounds like carbonates thus fixing it and indirect fixation through incorporation of atmospheric CO_2 into plant tissue through photosynthesis and subsequently decomposition of the plant part to be sequestered as SOC (Burras et al., 2001). Among various LUs, the degraded and agricultural soils have a higher potential for SOC sequestration. This is because these soils are depleted of a significant amount of the antecedent organic carbon pool and adoption of restorative management practices such as afforestation or reforestation, allied cropping, agroforestry, and permanent vegetation can significantly boost carbon sequestration (Albrecht & Kandji, 2003; Lal, 2004a, b). Lately carbon sequestration strategy involving combination of the agriculture and forestry sectors has stimulated a lot of attention being a promising source for additional income, food security and as mitigation strategy against CC. Such agricultural system has emerged as a sustainable resilient strategy because along with carbon sequestration they facilitate numerous ecosystem services (ES) from provisional to cultural including regulatory and protectoral services.

8.4 AGROFORESTRY: CONCEPT AND TIMELINE

Agroforestry is not a new practice rather it is a primitive practice which merges trees, herbs, shrubs, and animals along with annual crop production ensuring smooth and continuous supply of food and/or income throughout the year from a same unit of land management. It provides additional benefits of prevention of degradation and maintenance of soil fertility; diversification of income sources; efficient utilization of natural resources like soil nutrients, water, solar radiation and provision of regular employment (Figure 8.4).

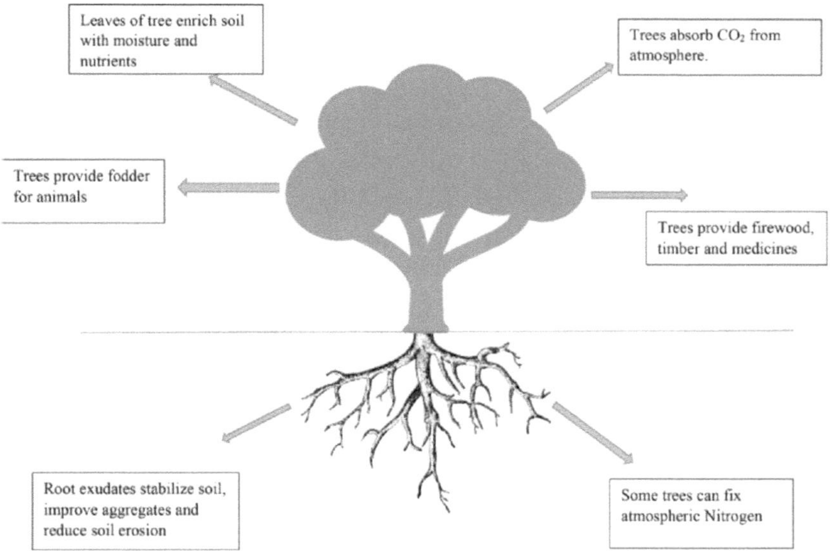

FIGURE 8.4 Benefits of agroforestry systems.

The World Agroforestry Center (www.icraf.cgiar.org) defines it as 'a dynamic, ecologically based, natural resources management system that through the integration of trees on farms and in the agricultural landscape diversifies and sustains production for increased social, economic, and environmental benefits for land users at all levels.' Since many centuries, this practice of growing crops along with trees has been prevailing in many countries particularly under subsistence farming conditions. For instance, home gardening which is a common agroforestry system is one of the oldest forms of agriculture in Southeast Asia, which was practiced by the fishing communities residing in the moist tropical region around 13000–9000 BC (Wiersum, 2004). In Europe it began around 4000 BC when domestic animals were let into forests for feeding. Similarly in Spain the Dehesa system wherein animals are grazed under trees is about 4,500 years old. Even in case of India the ancient scriptures and historical records indicate that trees and forest were always part of Indian culture. The Rishis (or ascetics) in the Aryan civilization often dwelled in forests and their ashrams (cottage) consisted of harmonized agriculture along with pasture with trees, animals, and birds. Towards the end of 19th-century the British empire introduced the "Taungya" method of teak plantation in the colonial countries like South Africa, Burma, Chittagong, and eastern part of India in Bengal where forest was established with help of unemployed and landless laborers who were allowed to grow crop in between the trees.

In spite of being an age-old LU practice, agroforestry gained importance as a contemporary and scientific LU system during 1970s during the period of green revolution (Nair et al., 2008). The green revolution though was responsible for increased productivity, but it was also responsible for associated environmental degradation. It eventually led to faulty land-management practices which resulted in rampant tropical deforestation, fuel wood scarcity, soil degradation, and biodiversity reduction. Additionally, its benefits remained inaccessible to the poor farmers and in low-productive agro-ecological regions. In order to deal with these problems, the traditional method of agroforestry came into rescue. Recognizing the versatile benefits of agroforestry an international organization named International Center for Research in Agroforestry (ICRAF) was established in 1977. The aim of this organization is to support, plan, research, and coordinate globally on the aspects of land-management systems combining agriculture and forestry. Subsequently, during the 1980s and 1990s agroforestry started to be acknowledged and included into national agricultural and forestry research schemes in numerable developing countries (Nair, 1989, 1993). In India scientific and systematic research on agroforestry systems gained momentum and impulsion with establishment of the All India Coordinated Research Project (AICRP) on Agroforestry in 1983 by Indian Council of Agricultural Research, New Delhi. In today's date agroforestry is recognized as a potential and sustainable LU system in both developing and industrialized nations and is a part of course curriculum of forestry and agriculture-courses in the universities.

8.5 TYPES OF AGROFORESTRY SYSTEMS

Agroforestry has global presence but is more widespread in tropical regions due to favorable temperature and moisture conditions along with conducive socioeconomic factors such as dense human population, more labor availability, lesser land-holding size, complicated land tenure, and more proximity to markets (Nair et al., 2008). Some of the well-known and widely practiced agroforestry systems in tropical areas are given in Table 8.2 and the details of some traditional agroforestry systems practiced is given in Table 8.3.

8.6 AGROFORESTRY AND CARBON SEQUESTRATION

Agroforestry sequester carbon as they consume the atmospheric CO_2 during photosynthesis and move the fixed carbon into vegetation, detritus, and eventually into soil pools for "secure" storage. Carbon sequestration in agroforestry occurs in two different segments which are underground and aboveground. The underground component consists living biomass such as roots and other belowground plant parts, soil organisms, stable forms of organic and inorganic carbon stored in various horizon whereas, aboveground encompasses specific plant parts (stem and leaves of trees and herbaceous components). A number of factors like climatic factors, soil type, the type of system (the nature of components and age of perennials such as trees), site quality, and previous land-use greatly influence the carbon sequestration capacity of both the segments. However, on an average, roughly 60% of the total carbon stored in an agroforestry system belongs to soil (belowground) and 30% of it belongs to aboveground segment. Thus, based on this concept it can be said that inclusion of tree along with crop or pastures facilitates greater carbon sequestration due to the combined effect of both long-lived and larger underground and aboveground biomass (AGB) as compared to only crops or pastures. For instance, a homegarden system consisting of multi-layer complex systems has a benefit of a greater number of components which can sequester more carbon as compared to simple row crops and fallow system. In tropical regions consisting of humid climate, perennial crops like coffee or cacao are usually planted with trees such as *Erythrina*

TABLE 8.2 Agroforestry Systems in Tropical Areas

Sl. No.	Agroforestry Systems	Brief Description
1.	Alley cropping (hedgerow intercropping)	Fast-growing woody species, specially leguminous are grown in crop fields as a hedgerow with close (0.5 m) in-row spacing and wide (4 m or more) between-row spacing; the woody species pruned periodically at low height (<1.0 m) to reduce shading of crops; the pruned materials are applied as mulch into the alleys as a source of organic matter and nutrients, or used as animal fodder.
2.	Homegardens	Intimate multi-story combinations of a several trees especially fruit- and nut-producing species and crops in homesteads; livestock may or may not be present; the size of the garden is small (<1 ha), and the garden is managed intensively usually by family labor.
3.	Improved fallow	Fast-growing, preferably leguminous, woody species planted and left to grow during short fallow phases not exceeding 3 years between cropping years; woody species contributes to site improvement and may yield economic products.
4.	Multipurpose trees (MPTs) on farms and rangelands	Fruit trees and other multipurpose trees scattered or planted in some systematic arrangements in crop- or animal-production fields; trees provide products such as fruits, fuelwood, fodder, and timber.
5.	Silvopasture: Grazing systems. Cut and carry system (protein banks)	Integrating trees in animal production systems: • Cattle grazing on pasture under widely spaced or scattered trees. • Stall-feeding of animals with fodder from trees grown in blocks on farms.
6.	Shaded perennial crop systems	Growing shade-tolerant species such as cacao (*Theobroma cacao* L.) and coffee (*Coffea spp.*) under or in between over story shade-, timber-, or other commercial tree crops.
7.	Shelterbelts and windbreaks	Use of trees to protect fields from wind damage, encroachment, floods, etc.
8.	Taungya	Growing agricultural crops during the early stages of establishment of forestry (timber) plantations; cropping is done for 2 or 3 years depending on the rate of tree-canopy development (causing shading of crops) and soil fertility.

spp., *Inga* spp. and *Cordia* spp. giving rise to a complex system with high plant diversity. These complex systems are usually prevalent in countries like Latin America, Southeast Asia and Equatorial Africa.) When carbon stocks were calculated for two tree combinations viz. cacao (*Theobroma cacao*) + laurel (*Cordia alliadora*) and cacao + poro (*Erythrina poeppigiana*) in Costa Rica by studying detailed breakdown of biomass production over 10 years, the Carbon stock stored over 10 years in the system was found to be

TABLE 8.3 Major Traditional Agroforestry Systems in the World

Country	Traditional Agroforestry Systems
India	Scattered trees in crop field (*Acacia nilotica* 'Babul' system of Chhattisgarh, *Prosopis cineraria 'Khejri'* systems of Rajasthan, *Ficus* based systems of Karnataka, *Acacia leucocephala* silvopasture system of Tamil Nadu, Foliage forest '*Soppina bettas'* of Malnad–Karnataka, Coffee under native shade trees in Coorg, Karnataka); Homegardens (Kerala, Northeast India); Bamboo Agroforestry (Konkan region, Ziro valley Arunachal Pradesh); Taungya System (Bengal and North-eastern region during colonial period 1890s).
Sri Lanka	Kandyan forest gardens (ground layer of herbaceous crop and upper canopy of fast-growing multipurpose trees: *Artocarpus heterophyllus* (Jackfruit) and *Cocos nucifera* (coconut).
Indonesia	*Pekarangan* (homegarden intercropping system) and *Kebun-talun* (rotation system of mixed garden and tree planting sequentially combining agricultural crops with tree crops).
Philippines	Multi-story system (coconut-coffee-pineapple-banana mix of Cavite, *Albizia*-coffee/cacao mix of Mindanao, *Gliricidia*-coffee mix in many dispersed areas) and homegardens), Rice terraces: forest agroforestry system (Northern region by Ifugao tribe), Hanunoo Farming (Hanno communities).
Japan	Satoyama: traditional agricultural landscape comprising of several habitats including paddy fields, secondary grasslands, ponds, and streams in sequence from top to bottom of hill.
Spain, Portugal	*Dehesa/Montado*: Manmade ecosystem with two-layer structure showing presence of Savannah-like open tree layer mainly Mediterranean evergreen oaks or to some extent by deciduous oak (*Quercus spp.*) and herbaceous layer consisting of either cultivated cereals or native vegetations for grazing purpose.
England, France, Germany	Pre-verger and Streuobest: Growing of fruit trees such as apples, pears, etc., in arable lands or pasture lands.
North-western Germany	Hauberg: Form of low forest consisting mainly of oak and birch which are burned to release nutrients followed by growing of cereals then fallowed for pastoral use till regeneration of forest.
Greece	*Chania* Systems: intercropping of cereals with fruit trees such as *Olea europaea*, *Castanea sativa*, *Juglans regia*, etc.
Tanzania	Kihamba system or Chagga homegardens (highlands of Mt. Kilimanjaro), Matengo Ngoro pit system (southern Tanzania), *Ngitili* Agroforestry systems (Western Tanzania).
Sahelian zone of West Africa	Parkland System: predominant in semi-arid region dominated by karité (*Vitellaria paradoxa*) and néré (*Parkia biglobosa*) which serve food, fodder, and soil fertility.
Brazil	Swidden fallow agroforestry by Bora Indians, market-orientated agroforestry by Riberenos (*Tamshiyacu*), orchard fallows by Caucasian tribes (Santa Rosa), agroforestry system in the Varzea village of Yanallpa.

TABLE 8.3 *(Continued)*

Country	Traditional Agroforestry Systems
Brazil, Guyana, Peru, Colombia, Venezuela	Amazonian systems: Agro-forestry system consisting of fruit trees, palms, cacao, timber trees, Brazil nut (*Bertholesia excelsa*) and agricultural crops like yams, beans, cassava, etc.
Honduras	The *Quezungual* system which is practiced traditionally by small holder farmers in Lempira area in western region of country, where naturally regenerate trees of different kinds and even shrubs are pollarded to a height of approximately 1.5 m.
South Pacific Islands	Agroforestry systems consisting of tropical fruit and nut trees in upper story, shorter trees like Pacific litchi (*Pometia pinnata*), candle nut tree (*Aleurites moluccana*) and beach hibiscus (*Hibiscus tiliaceus*) in second story and tuber crops like taro, sweet yams in lowest story.

Source: Adapted from Viswanath & Lubina (2017).

11 Mg C ha^{-1} $year^{-1}$ along with 6 Mg C ha^{-1} $year^{-1}$ in the shade trees (Beer et al., 1990). Carbon storage value of as high as 60 Mg ha^{-1} was estimated by Kürsten & Burschel (1993) on similar systems.

In Java, a tropical island of Asia the biomass estimation indicated that about 16 Mg C ha^{-1} could be sequestered in soil if rice fields were replaced by home gardens (Jensen, 1993a, b). Boundary plantings or border hedges such as windbreaks and live fences apart from their primary function of protecting crops from wind damage, significantly contribute to carbon sequestration. A study on capacity of *Calliandra calothyrsus* as a live fence species in Costa Rica revealed that total biomass of about 3–4 Mg dry matter per kilometer of live fence (2 Mg C sequestered per kilometer) is produced in duration of 10 months. Since the width of fences was one meter the total carbon sequestered would value about 20 Mg C ha^{-1} (Baggio & Heuveldop, 1984). Promising carbon sequestration potential has also been observed in case of hedgerow intercropping (HI) where crops are grown between rows of woody species which are coppiced or pruned at regular interval. This can be attributed to the periodic addition of pruning and root turnover into the soil over the years which lead to the accumulation of SOM and nutrient stocks in the soil. A HI trail on a Nigerian Alfisol, consisting of *Gliricidia sepium* and *Leucaena leucocephala*, enhanced surface SOC by 15% (2.38 Mg C ha^{-1}) in contrast to sole crops in a period of 12 years (Kang et al., 1999). In the Philippines, there exists a century old indigenous agroforestry system known as the *Naalad* system that was originated in central Philippines. A study to determine C sequestration potential of *Leucaena leucocephala* during the fallow phase of the *Naalad* system was carried out by Lasco & Suson (1999) using a

6-year-old model fallow. It was observed that above-ground *Leucaena* biomass rose from 4 Mg ha^{-1} in the initial year to 64 Mg ha^{-1} in the final year (end of the fallow). Soils and woody debris in the understory comprised of carbon which was estimated at 25% of the above-ground carbon. Overall, the study revealed that average carbon storage of 16 Mg ha^{-1} took place over the 6-year period. It has been recognized by IPCC and mentioned by Swaminathan (2012); and Dagar et al. (2016) that as compared to rest of land management systems agroforestry systems possess the highest capacity to sequester carbon followed by grazing management, forest management and crop plant management. In general, the potential of agroforestry systems to store carbon was estimated at 12–228 Mg ha^{-1} with a mean of 95 Mg ha^{-1} (Soto-Pinto et al., 2010). Globally about 630 × 10^6 ha of land was reported appropriate for agroforestry (Jose, 2009; Nair et al., 2009). Carbon sequestration potential of agroforestry system in humid tropics estimated was over 70 Mg C ha^{-1}, of which 25 Mg C ha^{-1} was sequestered by top 20 cm soil layer (Mutuo et al., 2005). In India, total potential for soil carbon sequestration was estimated at about 39–49 Mg C $year^{-1}$, of which about 21 billion tons of organic carbon was sequestered in first 30 cm soil depth and 63 billion tons in first one and half meter of soil depth (Lal, 2004b). Though agroforestry systems have tremendous potential sink of carbon, a word of caution is needed here. It is not always true that agroforestry systems will always act as sink; it may sometimes act as significant source of carbon during initial stage and management, especially in the systems which need intensive cultivation and practices like burning, excessive fertilization, regular disturbance and deep tillage.

Carbon dynamics of agroforestry system can be categorized into three phases, the initial phase of establishment where many systems may act as the sources of GHGs (emission of carbon and nitrogen from vegetation and soil). This is followed by the phase of maturation when quick accumulation of tons of carbon takes place in the boles, stems, roots of trees and in the soil. The third phase occurs only if the agroforestry consists of crop rotation, in this phase at the end of the rotation period part of the sequestered carbon will be returned back to the atmosphere, when the trees are harvested, and the land returned to cropping. Thus, the crucial phase in which carbon sequestration occurs is the maturation phase or it can be permanent if the perennial trees are planted for a long term. During the growth and maturation period, the trees accumulate carbon both in their own biomass and the soil by providing plant litters or residues. Plant residues and other organic materials when incorporated into the soil, decompose, and act as a source of carbon and nutrients. A large part of this carbon is lost into atmosphere through soil respiration and rest is sequestered by the living biomass. About one-third part of this soil organic matter (SOM) is recalcitrant in nature and break down very slowly persisting in soil for a long time. Such SOM represent crucial carbon storage and remain in soil for long duration as part of soil aggregates. This SOM which is "secured" from rapid decomposition constitute the carbon sequestered by soil. SOM can be prevented from rapid decomposition through below given processes (Christensen, 1996; Six et al., 2000; von Luetzow et al., 2008):

1. Biochemical Recalcitrance: The resistance to rapid decomposition due to presence of resistant chemical substances likes aromatic polymers and other structures. For instance, lignin which is important constituent of woody plants.
2. Physical Protection: It is the process in which SOM bind to soil aggregates, thus remaining shielded from microbial population and avert decomposition. This aids SOM to remain long time in soil without being lost to atmosphere.
3. Chemical Stabilization: This is also termed as organo-mineral stabilization. It is conversion and binding of SOM with minerals resulting information of organo-mineral complexes. Such complexes stay in soil for long periods and in association with physical protection and biochemical recalcitrance leading to formation of stable SOC. This is the most important factor contributing to recalcitrant SOM. It was reported that at least 86% of SOM was organo-mineral stabilized in forest sub-soil (Mikutta et al., 2006). But the process of complex formation is time taking, this time is provided by occurrence of physical protection and biochemical recalcitrance which facilitate the SOM to stay in soil for extended time.

Agroforestry systems consisting of woody perennials, deep rooted trees aid biological recalcitrance by release of organic materials which are more resistant. The organic exudates released from tree roots and the hyphae of fungi under such vegetation help in formation of aggregates. Formation of aggregates improves soil structure which prevents soil erosion. This in result affects carbon storage by preventing degradation of soil aggregates.

In practical term carbon sequestration through agroforestry will only be effective when there is complete tree growth, and the woody component forms a crucial part of the total biomass as in case of perennial agroforestry systems (perennial-crop combinations, agroforests, windbreaks). One edge that these systems have over others is that sequestration does not end at wood harvest. If the wood is processed into any form of long-lasting products such as furniture, etc., the carbon storage can continue for way beyond. Even if the wood is used as fuel or firewood rather than being in any long-term use, the wood from arable and grazing land can help in mitigating CO_2 emissions. That is because it will – (i) preserve the existing forests and other natural landscapes; (ii) conserve and enhance soil productivity; and (iii) reduce fossil energy consumption due to use of wood as energy sources. Due to the above stated factors agroforestry proves to be an ecologically sustainable alternative for mitigating with land-use problems, such as deforestation, degradation, and desertification, which result in extensive GHG emissions. It is estimated that that in tropical nations the products derived from five hectares of deforestation can be substituted or provided by establishment of one hectare of agroforestry. This implies that up to 10×10^6 ha of forest can be potentially conserved if 2×10^6 ha of new agroforestry systems are established annually. Global deforestation in area of 15×10^6 ha annually, emitting about 1.6 Pg C to the atmosphere (Dixon et al., 1994). Though it is not likely to eliminate deforestation completely, some portion of the 1.6 Pg C deforestation emissions could be minimized by adoption of sustainable agroforestry systems. Thus, under agroforestry system the processes of carbon sequestration are enhanced as compared to mono-cropping.

8.7 ANCILLARY BENEFITS OF AGROFORESTRY

As discussed till now the effective role of agroforestry in carbon sequestration has already proved it to be a potential strategy against CC. However, apart from the gigantic responsibility of carbon sequestration it provides many ancillary benefits against CC. It is seen as a resilient cropping system which can withstand the effect of CC as following:

1. **Resource Utilization:** CC along with expanding population has caused dread of resources, but the perennial tree component of agroforestry system minimizes the risks involved in mono-cropping by efficiently utilizing the limited resources available in CC scenarios.
2. **Tide Over the Water Scarcity and Drought:** The declining rainfall due to CC can lead to low water availability thus bring drought like condition. However, an agroforestry system can buffer the agricultural crops against low water availability by modifying the microclimate by altering radiation flux, air temperature, wind speed,

saturation deficit of under story crops thus eventually changing the rate and duration of photosynthesis, transpiration, and subsequent plant growth (Rao et al., 2007).

3. **Regulate Temperature:** During hot and dry conditions the perennial tress can reduce heat stress through shading, especially in case of heat-sensitive crops like coffee, cocoa, ginger, and cardamom (Lin, 2010).
4. **Protection Against Windstorms:** Trees can reduce evaporation and physical damage to the crops as they act as shelterbelts which slow down the wind speed. Rao et al. (2007) reported that wind speeds can be reduced to 80–95% by windbreaks planted in citrus groves there by lowering the wind damage up to two times the distance of windbreak height.
5. **Prevent Erosion:** The deep and dense rooting of the trees and the improved soil structure prevent soil loss due to erosion. Moreover, in sloping land, tree belt reduces the flow speed of water thus preventing erosion.
6. **Improvised Soil Health:** Agroforestry system can restore and maintain the health of degraded soil through regular incorporation of organic matter and efficient nutrient cycling which improves the physical, chemical, and biological properties of soil subsequently maintaining long-term soil productivity (Nair et al., 2009).

8.8 ROLE OF KYOTO PROTOCOL IN ADOPTION OF AGROFORESTRY

As the concerns over global warming due to greenhouse gases (GHGs) deepened and consequences of CC became more evident, in 1988 an Intergovernmental Panel on Climate Change (IPCC) was established under the patronage of United Nations Environmental Program (UNEP) and the World Meteorological Organization (WMO). The prime motive of this panel was to understand the risk of human-induced CC by assessing relevant technical, scientific, and socio-economic information. Many scientists across the globe contribute to the work of IPCC on a voluntary basis. The two initial assessment report of IPCC indicated the need of establishment of United Nations Framework Convention on Climate Change (UNFCCC) during the Earth Summit held in Rio de Janeiro, Brazil 1992. UNFCCC is an international treaty which aims to combat global warming by reducing atmospheric concentration of GHGs by minimizing the anthropogenic interference. The framework forms foundation of international political effort for the purpose having 197 parties in agreement who commit to reduced GHGs emissions. These nations meet annually in conferences of the parties (COP) to monitor the progress in dealing the CC. The COP and IPCC's second assessment in 1996 resulted in some crucial inputs to the negotiations which paved the way for adoption of Kyoto Protocol.

Kyoto protocol was adopted 11th December 1997 at Kyoto, Japan but entered into force on 16 February 2005. Currently 192 countries are signatories of this international agreement co-operating with each other to reduce the emission of GHGs. The protocol applies to six GHGs which are CO_2, methane (CH_4), nitrous oxide (N_2O), hydrofluorocarbons (HFCs), perfluorocarbons (PFCs), and sulfur hexafluoride (SF_6). It takes into consideration the fact that different countries have different capabilities to deal with CC depending on their economic development. Thus, the obligation to reduce emission of GHGs is moderated based on the amount of greenhouse gases contributed historically into the present atmosphere by respective countries. An approach which was well appreciated in the context of Kyoto Protocol was Land Use, Land Use Change, and Forestry (LULUCF) which acknowledges afforestation and reforestation as GHGs offset activities. Further forest-, crop-, and grazing-land management along with revegetation were included to the list of LULUCF thus making agroforestry a well-recognized strategy of Kyoto protocol.

8.9 ROLE OF CLEAN DEVELOPMENT MECHANISM (CDM) IN AGROFORESTRY

The Article 12 of Kyoto Protocol deals with clean development mechanism (CDM) which facilitates selling of the carbon sequestered via agroforestry systems owned by subsistence farmers in developing countries. This mechanism provides an economic opportunity along with sustainable development for smallholder production systems. CDM provides mutual benefits to both the developed nation through significant reduction in emission thus aiding them to achieve their Kyoto emission goals and to developing nation through contribution to sustainable development. Seling of carbon sequestered by the host nations can benefit the farmers involved in two ways, first one is that they could get profit for the carbon they sequestered by selling it in credit market depending on the amount sequestered and its price. Second is the additional productivity gains linked to sequestering practices. The CDM projects provides farmer the "carbon money" which furthers augments their income through initiating new enterprises such as bee keeping, goat rearing and poultry farming. Thus, CDM provides multiple benefits like income generation, poverty reduction, sustainability, and environmental preservation. Currently CDM has 7,806 projects registered under it and has issued a total of 1.958 billion certified emission reductions (CERs) till date (UNEP DTU Partnership; Hultman et al., 2020).

8.10 MAJOR LOOPHOLES IN CDM

In spite of CDM being a lucrative strategy involving the agroforestry system, it has many loopholes. Its profitability is highly depended on price of carbon in international market, cost associated to monitoring and additional benefits from allied enterprises (Nair et al., 2009). In addition, there are other major limitations like high transaction cost and building of effective institutional capacity for adoption of carbon sequestration projects. It was found that in poor countries like Nepal and Bangladesh the poverty-stricken families were bereft of the benefits from these projects because of large upfront investment costs, follow-up maintenance and operation costs (Boukerche et al., 2013). Least Developed Counties like those in sub-Saharan Africa are least benefitted from CDM and have only 1% of the total CDM portfolio since the share of CERs was issued from registered projects (Spalding-Fecher et al., 2012). As per Kyoto Protocol, each developing nation requires establishing a Designated National Authority (DNA) as a facilitator for connecting link between international investors and local service providers, but many countries lack such institutional capacity to promote viable carbon projects. In addition to absence of supporting policy and legal frameworks some countries also lack a general awareness about carbon payment processes. These countries especially in Africa and Asia lack a system for reasonable, necessary, and transparent assessment of carbon projects and mechanisms to produce abundant revenue through this assessment to finance itself (Jindal et al., 2008).

In some counties CDM projects can in fact have negative impact when sequestration activities are taken up in forest areas by big plantation companies. For example, the commercial plantation projects in Uganda have restricted the locals from harvesting timbers and other forest products leading to loss of their livelihood or in some cases local communities do not receive benefit share of carbon revenue from the projects. At times invasive growth of exotic species in agroforestry systems like *Eucalyptus* interferes with water available to downstream area thus harming local people. CDM was in brink of collapse during 2012 due to the poor price of carbon and the inability of governments to guarantee its prevalence in the future. Since 2012 various refinements have been done in CDM to make it more effective and fairer. A recent review by Hultman et al. (2020) discussed the additional benefits received by the CDM at the local or community level but concluded that the co-benefits are variable and even after recent procedural improvements failed to consistently deliver the co-benefits to local communities. However, in recent years many of the CDM procedures have been refined leading to its better functioning but still further improvement can be done for its fair operation (Jindal et al., 2008). Of late the observance of and lessons learned from CDM has paved its way for the introduction of Article 6 of Paris Agreement which catalyzes blend of carbon finance and sustainable development goals for future. The chapter enables the possibility of nationally determined contributions (NDCs) which could be acquired with market approaches, thus focusing on entire country rather than project or program as in case of CDM.

8.11 AGROFORESTRY AND REDD

REDD which stands for 'reducing emissions from deforestation and degradation' is a United Nations-based program that is supported by the World Bank and was created during the COP in Bali 2007. It aims to popularize any method which will lead to reduction of GHGs emissions by providing financial incentives. REDD encompasses a strategy wherein which countries are remunerated for protecting forests and minimizing emission from forests against an agreed baseline. Any action plan which reduces deforestation or promotes afforestation qualifies for inclusion in this program. As agroforestry systems are one of the promising sustainable land-use systems (LUS), which promote plantation of more trees and reduce the pressure on forest tree, they form an eligible strategy for the REDD program. As per the REDD program developed countries depending on the amount of carbon emitted by them should pay developing countries which are mainly tropical countries, for sequestering carbon and avoiding deforestation. This initiative provides an economic opportunity for poor farmers in the tropical countries by giving them the benefit from carbon payment. The effectiveness of REDD projects depends on how well mitigation results can be measured, reported, and verified (MRV). This means that emission reductions through REDD need to be measured against a long-term baseline (reference scenario) of emission levels without REDD project interventions. The growth of carbon market in tropics impacts the adoption of LU systems which support afforestation. Antle et al. (2007) created a model to analyze how the adoption of agroforestry in the tropical highlands of Peru are influenced by carbon contracts. The results of this analysis suggested that adoption of terraces and other agroforestry practices would increase by participation in carbon contracts depending on amount of carbon sequestered and other key factors impacting land productivity, such as slope of the land. Agroforestry also qualify for being part of REDD + through two ways, i.e., first one being qualified as forest as per criteria set by UNFCCC (minimum area of land of 0.05–1.0 hectares with tree crown cover (or equivalent stocking level) of more than 10–30% with trees with the potential to reach a minimum height of 2–5 meters at maturity *in situ* and second one being the strategy which has the capability for diminishing forest depletion by providing timber and fuelwood which would or else had been procured from adjacent or distant forests thus lessening the extent of deforestation and pressure on natural forests (NFs).

A study conducted in the Guinean rainforest of West and Central Africa (Cote D'Ivoire, Ghana, Nigeria, and Cameroon) showed that about 21,000 km^2 of forests could have been spared and emissions of nearly 1.4 billion t CO_2 could have been reduced if intensification of cacao (*Theobroma cacao* L.) agroforestry systems is carried out through seed-fertilizer technologies and the integration of timber species, had been adopted in the late 1960s. This is in comparison to a baseline of extensive expansion of cacao, cassava, and oil palm into forest areas by about 68,000 km^2 over the same period (Gockowski & Sonwa, 2011). Kimaro et al. (2011) carried out a study in semi-arid Morogoro in Eastern Tanzania, in order to prove the notable contributions of rotational woodlot systems to decrease degradation of forest and offset CO_2 emissions via on-farm wood supply. They used the refence of native vegetation fallows forests), and found that after a 5-year rotation, wood yield (23–51 Mg C ha^{-1}) was enough to suffice household demand for fuelwood. It was also noted that highly productive Acacia fallows (*Acacia crassicarpa* A. Cunn. Ex Benth., *Acacia leptocarpa* A. Cunn. Ex Benth., and *Acacia mangium* Wild) would take four to nine years to replenish carbon lost due to deforestation of Miombo forest or agricultural expansion compared to the long duration of two decades required for re-growing Miombo Woodlands.

The potential of agroforestry being included as REDD program in Africa is plagued by various technical, policy, and economic challenges. Technical challenges include lack of premium quality planting material for desired species, insufficient knowledge and skill of optimal shade management in sustainably intensive and diversified agroforestry systems and lack of skills for processing of products (Jama & Zelia, 2005; Pretty et al., 2011). Economic and policy challenges comprise of vague knowledge about rights to land, trees, and carbon, absence of proper market infrastructure, extended waiting periods for recovery of investments (sometimes up to three years) and inadequate labor supply (Mukadasi et al., 2007). In Mexico where livestock or agriculture using slash and burn practice replaced forest, about 35% of carbon loss was observed (Masera et al., 1997). Also switching from traditional fallow to traditional maize led to a loss of about 94% and agroforestry prototypes were found to maintain carbon in living biomass and increase carbon stock by 22 times living biomass in comparison to the baseline of traditional maize in a period of approximately seven years. Moreover, re-shifting pastures without-trees to silvopastoral (pasture +trees) systems enriched carbon in living biomass on the order of 20 times. Study conducted by Soto-Pinto et al. (2010) in Mexico showed that coffee plantations which were densely populated with growing trees resembling the forest structure, could accumulate high amount of carbon in above ground biomass and in soil and therefore were accepted as a suitable practice for reduction of emissions by deforestation and degradation (REDD).

In Sri Lanka homegarden agroforestry systems form an integral part of the nation's tree population and serve as crucial carbon sinks. These systems consist of rich and diverse mixture of huge and tall coconut trees are mixed with numerous small trees. Coconut (194.4 t C/ha) and homegarden (107.9 t C/ha) sequester high volume of carbon stocks. These agroforestry systems not only add carbon inputs, such as root biomass, litter, and pruning's to the soil, but also control soil erosion which is a major impacting process in the soil carbon dynamics. In addition to this, majority of agroforestry systems are source of non-timber products such as fruits, oils, spices, vegetables, and resins which are harvested without any prominent impact on the carbon stock of a system in comparison to the harvest of wood products (Roshetko et al., 2007). As a consequence of this the load on forest reduces leading to prevention of deforestation. Thus, changes in carbon stocks within home gardens makes them to be considered of immense significance for a national REDD+ Program in Sri Lanka (UN-REDD Program, 2012). Minang et al. (2011) pointed out that improving market infrastructure, strengthening policies on tree rights and ownership would be essential for agroforestry to be fruitful for achieving the goals of REDD+. REDD+ has proven its worth in CC mitigation and ancillary benefits but its consideration for agroforestry still remains incipient at national and sub-national level (Minang et al., 2011). In spite of being an ideal CC mitigation initiative, REDD or REDD+ has following major challenges for its implementation:

- Lack of accurate and standardized methods to estimate carbon sequestration potential of agroforestry systems.
- Poor governance, corruption, low transparency in developing countries act as a hindrance for implementation of REDD in social, economic, and political realms of such countries.
- Uncertainty about reaching of the benefits from carbon payments to forest communities.

For REDD or REDD+ to be successful it needs to be internationally accepted and stringent REDD standards and national and international policies on CC are needed to be implemented. In addition to that there is a dire need for creation of more awareness of REDD mechanisms, specially by the policy-makers along with consistent support from scientists, NGOs as well as stakeholders who are responsible for implementation of the CDM.

8.12 CONCLUSION

The concern of CC is intensifying with each passing year attributing to rise in levels of GHGs in atmosphere. As a consequence, there is rise in sea levels, erratic rainfalls, rising temperature, coral degradation, etc., which is a threat foe existence of life. Under such situation carbon sequestration turns out to be a viable option which can reduce the pace of global warming and thus CC. Agroforestry systems have an edge over other strategies due to its ancillary benefits along with huge potential of carbon sequestration. The associated benefits of agroforestry systems along with carbon sequestration like resource utilization, protection against windstorms, prevention from erosion, improvised soil health, year-round employment and economic benefits make it an attractive and practical solution for the damage caused due to intensified cultivation and reduction of CO_2 concentration from atmosphere thus averting the CC. Different types of agroforestry systems have different potential of carbon sequestration. Apart from directly sequestering carbon, agroforestry indirectly contributes to reduction of emission of CO_2 due to reduced deforestation, as they provide the products that would otherwise have been derived from forests. However, the major challenge in adoption of agroforestry is the slow initial profit from the woody perennials till their establishment. But this limitation can be overcome by selecting fast growing and early maturing trees or plantation of high value seasonal crops by utilizing the profuse sunlight till maturity of woody components. Policies like Kyoto protocol, CDM, and REDD+ at the international level act as backbone for promotion, implementation, and adoption of agroforestry systems. However, these policies are plagued with many loopholes due to which shareholders remain devoid of benefits of carbon sequestration by agroforestry. These hindrances could be dismissed if internationally accepted and stringent REDD+ standards and national and international policies on CC are implemented and more awareness is created among stakeholders about the mechanisms. With proper improvisation in the policies and effective utilization of agroforestry systems a significant amount of carbon can be sequestered which can save the planet and lives in it.

KEYWORDS

- **agroforestry**
- **carbon sequestration**
- **clean development mechanism**
- **climate change**
- **global carbon cycle**
- **soil organic carbon**

REFERENCES

Albrecht, A., & Kandji, S. T., (2003). Carbon sequestration in tropical agroforestry systems. *Agric. Ecosyst. Environ., 99*, 15–27. https://doi.org/10.1016/S0167-8809(03)00138-5.

Antle, J. M., Stoorvogel, J. J., & Valdivia, R. O., (2007). Assessing the economic impacts of agricultural carbon sequestration: Terraces and agroforestry in the Peruvian Andes. *Agric. Ecosyst. Environ., 122*(4), 435–445.

Baggio, A., & Heuveldop, J., (1984). *Initial Performance of Calliandra Calothyrsus Meissm. in Live Fences for the Production of Biomass* (pp. 19–29). Martinus Nijhoff, The Hague, the Netherlands.

Beer, J., Bonnemann, A., Chavez, W., Fassbender, H. W., Imbach, A. C., & Martel, I., (1990). Modeling agroforestry systems of cacao (*Theobroma cacao*) with laurel (*Cordia alliadora*) or poro (*Erythrina poeppigiana*) in Costa Rica. V. Productivity indices, organic material models and sustainability over ten years. *Agrofor. Syst., 12*, 229–249.

Boukerche, S. B., Dulal, H., Brodnig, G., & Quesne, B., (2013). *CDCF Making an Impact: Carbon Finance Delivers Benefits for the Poor*. Working Report 77741, Washington DC, World Bank.

Buckland, S. M., Thompson, K., Hodgson, J. C., & Grime, J. P., (2001). Grassland invasions: Effects of manipulations of climate management. *J. Appl. Ecol., 38*, 301–309.

Burras, C. L., Kimble, J., Lal, R., et al., (2001). Carbon sequestration: Position of the soil science society of America. *Soil Sci. Soc. Am. J., 65*, 1–4.

Cheddadi, K., Guito, J., Jolly, D., (2001). The Mediterranean vegetation: What if the atmospheric CO_2 increased? *Landsc. Ecol., 16*, 667–675.

Christensen, B. T., (1996). Carbon in primary and secondary organomineral complexes. In: Carter, M. R., & Stewart, B. A., (eds.), *Structure and Organic Matter Storage in Agricultural Soils* (pp. 97–165). CRC Press, Boca Raton.

Dagar, J. C., Sharma, P. C., Sharma, D. K., & Singh, A. K., (2016). *Innovative Saline Agriculture* (p. 519). Springer, Dordrecht.

Dixon, R. K., Brown, S., Houghton, R. A., Solomon, A. M., Trexler, M. C., & Wisniewski, J., (1994). Carbon pools and flux of global forest ecosystems. *Science, 263*, 185–190.

Gockowski, J., & Sonwa, D., (2011). Cocoa intensification scenarios and their predicted impacts on CO_2 emissions, biodiversity conservation and rural livelihoods in the guinea rainforest of West Africa. *Environ. Manage., 48*, 307–321.

Hultman, N., Lou, J., & Hutton, S., (2020). A review of community co-benefits of the clean development mechanism (CDM). *Environmental Research Letters, 15*(5), 053002.

IPCC, (2001). *Climate Change 2001: The Scientific Basis* (p. 867). Inter-Government Panel on Climate Change, Cambridge Univ. Press, UK.

IPCC, (2007). *Climate Change: Fourth Assessment Report.* Inter-Government Panel on Climate Change, WMO, Geneva, Switzerland.

Jama, B., & Zeila, A., (2005). *Agroforestry in the Dry Lands of Eastern Africa: A Call to Action.* ICRAF Working Paper Number 1., World Agroforestry Center Nairobi, Kenya.

Jensen, M., (1993a). Soil conditions, vegetation structure and biomass of a Javanese homegarden. *Agrofor. Syst., 24*, 171–186.

Jensen, M., (1993b). Productivity and nutrient cycling of a Javanese homegarden. *Agrofor. Syst., 24*, 187–200.

Jindal, R., Swallow, B., & Kerr, J., (2008). Forestry-based carbon sequestration projects in Africa: Potential benefits and challenges. *Nat. Resour. Forum, 32*, 116–130.

Jose, S., (2009). Agroforestry for ecosystem services and environmental benefits: An overview. *Agrofor. Syst., 76*, 1–10.

Kang, B. T., Caveness, F. E., Tian, G., & Kolawole, G. O., (1999). Long-term alley cropping with four species on an Alfisol in southwest Nigeria- effect on crop performance, soil chemical properties and nematode population. *Nutr. Cycl. Agroecosyst., 54*, 145–155.

Kimaro, A. A., Isaac, M. E., Chamshama, S. A. O., Kumar, B. M., & Nair, P. K. R., (2011). *Carbon Sequestration Potential of Agroforestry Systems; Opportunities and Challenges, Advances in Agroforestry* (Vol. 8). Springer, Dodrecht, Netherlands.

Kürsten, E., & Burschel, P., (1993). CO_2-mitigation by agroforestry. *Water Air Soil Pollut., 70*, 533–544.

Lal, R., (2004a). Soil carbon sequestration in India. *Climatic Change, 65*, 277–296.

Lal, R., (2004b). Soil carbon sequestration in natural and managed tropical forest ecosystems. *J. Sustain For., 21*, 1–30. https://doi.org/10.1300/J091v21n01_01.

Lasco, R. D., & Suson, P. D., (1999). A *Leucaena leucocephala*—Based indigenous fallow system in central Philippines: The naalad system. *Inter. Tree Crop J., 10*, 161–174.

Lin, B. B., (2010). The role of agroforestry in reducing water loss through soil evaporation and crop transpiration in coffee agroecosystems. *Agric. Forest Meteorol., 150*, 510–518.

Masera, O., Ordonẽz, J., & Dirzo, R., (1997). Carbon emissions from Mexican forests: The current situation and long-term scenarios. *Clim. Change, 35*, 265–295.

Mikutta, R., Kleber, M., Torn, M. S., & Jahn, R., (2006). Stabilization of soil organic matter: Association with minerals or chemical recalcitrance? *Biogeochemistry, 77*, 25–56.

Minang, P. A., Bernard, F., Van, N. M., & Kahurani, E., (2011). *Agroforestry in REDD+: Opportunities and Challenges*. ASB Policy Brief No. 26, ASB Partnership for the Tropical Forest Margins, Nairobi, Kenya. Available at: http://www.asb.cgiar.org/PDFwebdocs/ASB_PB26.pdf (accessed on 28 December 2023).

Mukadasi, B., Kaboggoza, J. R., & Nabalegwa, M., (2007). Agroforestry practices in the buffer zone area of Mt Elgon National Park, eastern Uganda. *Afr. J. Ecol., 45*, 48–53.

Mutuo, P. K., Cadisch, G., Albrecht, A., Palm, C. A., & Verchot, L., (2005). Potential of agroforestry for carbon sequestration and mitigation of greenhouse gas emissions from soils in the tropics. *Nutr. Cycl. Agroecosyst., 71*, 43–54.

Nair P. K. R., (1989). *Agroforestry Systems in the Tropics*. Kluwer, Dordrecht, the Netherlands.

Nair P. K. R., (1993). *An Introduction to Agroforestry*. Kluwer Academic Publishers, Netherlands.

Nair P. K. R., Gordon, A. M., & Mosquera-Losada, M. R., (2008). Agroforestry. In: Jorgensen, S. E., & Faith, B. D., (eds.), *Encyclopedia of Ecology* (Vol. 1, pp. 101–110). Elsevier, Oxford, UK.

Nair P. K. R., Kumar, B. M., & Nair, V. D., (2009). Agroforestry as strategy for carbon sequestration. *J. Plant Nutr. Soil Sci., 172*, 10–23.

NOAA, (2020). *Trends in Atmospheric Carbon Dioxide*. Global Monitoring Laboratory. https://gml.noaa.gov/ccgg/trends/ (accessed on 28 December 2023).

Pretty, J., Toulmin, C., & Williams, S., (2011). Sustainable intensification in African agriculture. *Int. J. Agric. Sustain., 9*, 5–24.

Rao, K. P. C., Verchot, L. V., & Laraman, J., (2007). Adaptation to climate change through sustainable management and development of agroforestry systems. *SAT e-Journal, 4*(1), 1–30.

Roshetko, J. M., Lasco, R. D., & Angeles, M. S. D., (2007). Smallholder agroforestry systems for carbon storage. *Mitigation and Adaptation Strategies for Global Change, 12*, 219–242.

Six, J., Elliott, E. T., & Paustian, K., (2000). Soil macro-aggregate turnover and micro-aggregate formation: A mechanism for C sequestration under no-tillage agriculture. *Soil Biol. Biochem., 32*, 2099–2103.

Soto-Pinto, L., Anzueto, M., Mendoza, J., Ferrer, G. J., & De Jong, B., (2010). Carbon sequestration through agroforestry in indigenous communities of Chiapas, Mexico. *Agrofor. Syst., 78*(1), 39.

Spalding-Fecher, R., Achanta, A. N., Erickson, P., Haites, E., Lazarus, M., Pahuja, N., Pandey, N., et al., (2012). *Assessing the Impact of the Clean Development Mechanism, Final Report*. CDM Policy Dialogue.

Swaminathan, M. S., (2012). Agroforestry for evergreen revolution. In: Nair, P. K. R., & Garrity, D., (eds.), *Agroforestry-the Future of Global Land Use, Advances in Agroforestry* (Vol. 9, pp. 7–10). Springer, Dordrecht.

UN-REDD Program, (2012). *Sri Lanka REDD+ Readiness.* Preparation Proposal, UN-REDD Program Eighth Policy Board Meeting. http://www.unredd.net/index.php?option=com_docman&task=doc (accessed on 28 December 2023).

Viswanath, S., & Lubina, P. A., (2017). Traditional agroforestry systems. In: *Agroforestry* (pp. 91–119). Springer, Singapore.

Von, L. M., Kogel-Knabner, I., Ludwig, B., Matzner, E., Flessa, H., Ekschmitt, K., Guggenberger, G., et al., (2008). Stabilization mechanisms of organic matter in four temperate soils: Development and application of a conceptual model. *J. Plant Nutr. Soil Sci., 171*, 111–124.

Wiersum, K. F., (2004). Forest gardens as 'intermediate' land-use system in the nature-culture continuum: Characteristics and future potential. *Agroforest. Syst., 61*, 123–134.

CHAPTER 9

Traditional Agroforestry Systems of North-Eastern India: An Overview

P. L. BHUTIA,[1,2] K. G. BHUTIA,[3] P. PANWAR,[1] N. R. SINGH,[4] S. PAL,[1] and O. P. S. KHOLA[1]

[1]*ICAR–Indian Institute of Soil and Water Conservation, Sector 27A, Madhya Marg, Chandigarh, India*

[2]*ICAR–Research Complex for NEH Region, Nagaland Center, Medziphema, Nagaland, India*

[3]*Rain Forest Research Institute, Jorhat, Assam, India*

[4]*ICAR–Research Complex for NEH Region, Umiam, Meghalaya, Indial*

ABSTRACT

The Northeastern Region (NER) of India constitutes approximately 7.98% of the geographical area of the country holds about 3.77% of India's total population. About 80% of the total population in this region are dependent on agricultural sectors as a source of income and livelihood security. The cropping pattern in the entire NER is food grain-oriented out of which paddy has the major share. Many traditional agroforestry systems (AFS) exist in this region which has been developed over the years based on indigenous knowledge (IK) and skills of local communities. This chapter reviews various traditional agroforestry systems of NER and action to be taken concerning scientific research and administrative reforms to make these traditional agroforestry systems more sustainable and productive.

Sustainable Management and Conservation of Environmental Resources in India.
Sumit Chakravarty, Vineeta, Jahangeer A. Bhat, Munesh Kumar, Gopal Shukla (Eds.)

9.1 INTRODUCTION

The Northeastern Region (NER) of India spreads over 26.22 M ha, from 21.57° and 29.5°N latitude and 89.4° and 97.5°E longitude. This NER covers eight states, such as Assam, Arunachal Pradesh, Manipur, Meghalaya, Mizoram, Nagaland, Sikkim, and Tripura and constitutes approximately 7.98% and 3.77% of the country's total area and populations (2011 census), respectively. The topography of the region is highly variable with a mixture of plains, valleys, plateaus, and mountains, with altitudes from near sea level to over 5,000 m asl. The river Brahmaputra passes through the center of the region from the north-east side to south-west side. This makes a basal line from where the valley develops into plateaus and hills on either side.

The NER experience monsoon type of climate and the majority of the rainfall occurs due to advancing South-West Monsoon (SWM) with heavy rains from late April to October. Here the rainy season coincides with the summer like the rest of India. The winter starts in November and lasts until March and the spring season lasts for short period from mid-March to mid-April. Mean annual rainfall varies within the region with over 11,000 mm in the plateau of Shillong (wettest place on Earth), around 2,000 mm in Brahmaputra valley with local variations. The hills of the region receive on an average 2,000–3,000 mm of rain. Periodic dry spell generally occurs during winter from November to March. There is relatively high temperature and humidity in the plain areas of this region, which decrease with increasing elevation. The average temperature in January is around 16°C and summer temperatures lie between 30 and 33°C in plains. The hills have a cool summer with an average summer temperature of about 20–21°C temperature and a maximum of 25°C and a minimum of 15°C (Dikshit & Dikshit, 2014). The temperature in the higher alpine regions is much lower with the average annual temperature remains under 7°C and mean January temperature under – 1°C. As per the Koeppen classification, the region is categorized into three climatic zones (Dikshit & Dikshit, 2014), which are described in Table 9.1.

Soils of the NER vary depending upon the physiography, climate, parent materials, vegetations, etc. The major soil order of different states of NER of India is given in Table 9.2. Generally, the soil of the region is acidic in reaction, high in organic matter content, insufficient in nitrogen, poor in available phosphorus and moderate to high in potassium content.

TABLE 9.1 The Climatic Zones of North-Eastern Region

Sl. No.	Type	Category	Characteristics
1.	'A' type	Tropical wet-dry climate (Aw).	Tropical rainy climate; dry winters. Location: Part of Tripura, Cachar plain, Mizoram, Manipur located below 25°N latitude.
		a.' Awa'	Tripura and Cachar plains (Barak Valley) of Assam where average temperature in hottest month is above 22°C.
		b. Awb	Mizoram and southern Manipur where average temperature of hottest month is below 22°C.
2.	'C' type	Warm temperate (mesothermal) climate (cf).	North of 25° latitude, where the temperature in the coldest month is less than 18C but above –3°C.
		a. 'Cfa'	Humid subtropical. Location covers whole Brahmaputra valley from Dhubri to Pasighat.
		b. 'Cfb'	Humid mesothermal climate with warm summers. These eastern ranges of Nagaland, Manipur, Sikkim, and Arunachal above 500 m and below 1,500 m ASL comes under this region.
		c. 'Cfc'	Humid mesothermal climate. This covers northern Arunachal Pradesh, Meghalaya, Sikkim, and Nagaland (above 1,500 m ASL).
3.	D' type	–	This covers the alpine region of Arunachal Pradesh, Sikkim where the temperature in the winter months falls below 0°C–3°C.

TABLE 9.2 Percentage Geographical Area Under Major Soil Types of North-East Region

North-East State	Soil Order					
	Alfisol	Entisol	Inseptisol	Mollisol	Ultisol	Miscellaneous
Arunachal Pradesh	0.30	35.60	37.30	–	14.20	12.60
Assam	12.30	32.30	49.30	–	6.10	–
Manipur	0.20	23.10	38.40	–	36.40	1.90
Meghalaya	3.60	10.70	45.70	–	40.00	–
Mizoram	2.60	21.50	37.30	38.60	–	–
Nagaland	4.80	4.00	76.00	–	17.20	–
Sikkim	–	43.00	33.40	23.60	–	–
Tripura	5.00	8.00	80.00	–	7.00	–

Source: Velayutham & Bhattacharya (2000).

The land utilization pattern of different northeastern states is given in Table 9.3.

TABLE 9.3 Pattern of Land Utilization in NER (000 ha) in 2012–2013

North-Eastern States	Geographical Area	Reporting Area for Land Utilization Statistics	Forests	Net Area Sown	Total Cropped Area
Arunachal Pradesh	8,374	7,255	6,748	216	285
Assam	7,844	7,844	1,855	2,809	4,076
Manipur	2,233	2,072	1,728	309	309
Meghalaya	2,243	2,241	946	285	340
Mizoram	2,108	2,094	1,585	116	116
Nagaland	1,658	1,652	863	380	489
Sikkim	710	443	336	77	144
Tripura	1,049	1,049	629	255	474
North-east India	26,219	24,650	5,879	4,448	6,232
All India	3,28,726	3,07,491	71,573	1,39,936	1,94,246

Sources: Agriculture–Statistical Yearbook, Ministry of Statistics and Program Implementation, GOI: Pattern of Land Utilization. http://www.mospi.gov.in/statistical-year-book-india/2018/177.

Majority of the total geographical area (TGA), i.e., 63.26% is under different types of forest varying from tropical moist evergreen to sub-alpine and alpine type forest. Although agriculture is the main source of livelihood, it occupies only 19.24% (net sown area+ current fallow land) of the TGA of NER (varying from 3.51% in Arunachal Pradesh to 36.93% in Assam) as against 50.48% in India. The less area under agriculture and its low productivity are due to high vulnerability from natural calamities like flood, erosion, and other natural calamities, faulty cultivation practices, low input use/infrastructure facilities, lack of market opportunities, land tenure systems, undulating topography and inaccessible terrain. NER is recognized as one of the global biodiversity 'hotspots.' About 40% of the total flora of India is represented in this region. It is a center of origin of many crops like rice, citrus, and their wild relatives.

9.2 LAND TENURE SYSTEM

The NER has a unique and different land tenure system than the rest of the Country. Besides the government holding lands, lands are broadly under two types of land tenure systems as in subsections.

9.2.1 LAND OWNED BY THE VILLAGERS, COMMUNITIES, CLANS, AND CHIEFS

This system of ownership of land is common in the tribal district where the land belongs to different communities or clans are under the custody of the head of the community or clan. Assortment and allotment of land for any type of cultivation to a family within the clan or community is made by the chief in consultation with other residents of that land. The inhabitants of the village need not to pay any fee, but an outsider has to pay a fee to the community for using a plot of land. The individual farmer has no legal rights over the community lands.

9.2.2 LAND OWNED BY INDIVIDUAL FAMILIES

In this system of tenureship, individual especially the eldest member of the family has full legal rights over the land they owned. Generally, in NER small peasants (tenants) take the agriculture land from this landlord on lease for cultivation of crops. This is the main reason for non-adoption of long-term development measures such as land development activities, afforestation, and aquaculture and horticultural plantations by the farmers of NER.

9.3 MAJOR CROPS AND CROPPING PATTERN

Variety of crops and cropping systems exists in this region which is influenced by environmental factors (topography, soil, and climate), the system of farming and food preferences. The major field crops grown in the region are rice, wheat, maize, pulses, oilseeds, jute, and potato. The horticultural crops grown are fruit crops, plantation crops, tuber crops, spices, and vegetables. Overall cereals are most widely cultivated in this region. Among cereals, rice is the staple food, occupying approximately 3310.75 ha, i.e., 91.26% of the cultivated area of cereals. Other important cereals crops are maize, millet, wheat, and their areas vary among different Northeastern states (Table 9.4).

Groundnut, sesamum, rapeseed, mustard, linseed, castor seed are oilseeds grown in this area covering 472.14 ha area. Local beans, pigeon pea and pea are major pulse crops in this region. The other important crops are ginger, sugarcane, potato, and chilies. Horticultural crops encompass almost 18.91% of the cultivated area in NER (Agricultural Statistics at a glance, GOI, 2012). The major fruits grown in NER are

TABLE 9.4 Area Under Principle Crops (000 ha) in 2012–2013

Major Crops	Arunachal Pradesh	Assam	Manipur	Meghalaya	Mizoram	Nagaland	Sikkim	Tripura	Total
Rice	126.085	2488.23	122.69	109.15	14.6	183.33	11.92	254.74	3310.745
Maize	47.486	23.7	19.44	17.31	6.08	68.67	39.97	3.63	226.286
Small millet	22.675	4.43	–	2.28	–	8.58	6.54	–	44.505
Wheat	2.937	33.89	2.4	0.382	–	3.27	0.52	0.65	44.049
Total pulses	9.464	141.24	30.3	3.61	3.09	39.82	6.37	8.454	242.348
Total food grains	208.6	2691.5	174.8	132.7	23.8	305.3	65.9	267.5	3870.1
Total oilseeds	32.262	306.22	44.1	9.952	2.05	64.51	8.24	4.81	472.144
Jute	–	65.1	–	3.9	–	3	–	0.6	72.6
Banana	6.132	51.509	6.58	6.959	10.54	7	1.615	13.58	103.915
Potato	NA	99.772	NA	18.139	0.25	3.8	10.055	8.32	140.336
Black pepper	–	3.72	–	0.89	0.07	0.23	–	0.24	5.15
Chili	2.5	20.59	6.5	1.85	9.02	0.8	–	2.35	43.61
Coconut	–	22.154	–	–	0.02	1.3	–	6.47	29.944

Sources: Agriculture–Statistical Yearbook, Ministry of Statistics and Program Implementation, GOI: Area under Principal Crops. State wise 2012–2013. http://www.mospi.gov.in/statistical-year-book-india/2018/177.

pineapple, banana, guava, mango, peach, plum, pears, coconut, and many minor fruits are also grown in the hills. Plantation crops like tea occupy a greater area and contribute to a good economy in the state of Assam. Apart from that, rubber and coffee cultivation are grown on a large scale in this region. Overall, in the NER of India, almost 90% of the cultivated area is mono-cropped and rainfed and only 10% of the cultivated area is irrigated. Shifting cultivation is the most popular practice followed particularly in the hilly areas, except in Sikkim. The terraced cultivation is also common in some areas of Manipur, Nagaland, and the entire Sikkim. Multitier cropping systems like home gardens are common features throughout NER. Mono cropping of the field crops is the most dominant cropping pattern in the flood plains and valley whereas, the mixed farming system with horticulture and plantation crops are mostly grown in uplands and hills. The productivity of important crops in different states of NER of India is given in Table 9.5.

TABLE 9.5 Productivity of Principle Crops (t/ha) in 2012–2013

Major Crops	Arunachal Pradesh	Assam	Manipur	Meghalaya	Mizoram	Nagaland	Sikkim	Tripura
Rice	2.09	2.06	NA	2.13	2.09	NA	1.79	2.80
Wheat	1.50	1.30	2.50	1.81	NA	1.80	1.06	2.00
Maize	1.44	0.90	NA	1.55	1.31	1.96	1.70	1.29
Small millet	1.03	0.49	NA	0.77	NA	1.12	0.97	NA
Pulses	1.12	0.60	0.94	1.02	1.06	1.10	0.92	0.71
Oil seeds	0.91	0.61	0.73	0.69	1.08	1.05	0.86	0.51

NA Data not available.

Sources: Agriculture–Statistical Yearbook, Ministry of Statistics and Program Implementation; GOI: Production of Principal Crops. State wise 2012–2013. http://www.mospi.gov.in/statistical-year-book-india/2018/177.

9.4 AGRO-CLIMATIC ZONES

The NER have been broadly divided into six broad agroclimatic zones based on agroclimatic factors such as temperature, rainfall, soil types, and water resources (Report of the ICAR Research Review Committee for North-eastern Hill Region, National Agricultural Research Project, Indian Council of Agricultural Research, New Delhi (1984). The agro-climatic zones of NER are described in Table 9.6.

TABLE 9.6 Agro-Climatic Zones of NER

Sl. No.	Zone	Characteristics
1.	Mild tropical plain zone	• This zone covers the plains of Assam, Arunachal Pradesh, Tripura, Nagaland up to an elevation of 200 m asl. • Summer is hot and humid here. This zone encompasses flat, or valley lands interspersed with small hillocks. • Paddy cultivation is dominant. Pineapple, arecanut, betel leaf, other plantation crops are also widely grown. Shifting cultivation is seldom practiced in this zone.
2.	Mild tropical hill zone	• This zone covers southern part of Jaintia hill, northern part of Khasi hills and RiBhoi district, major part of Garo hills, Jiribam area of Manipur, Dimapur, and Ghaspani area of Nagaland and parts of North Cachar hills and Karbi Anglong area of Assam, Jampui hills of Tripura and lower zone of Sikkim within the elevational range of 200–800 m. • Summer is hot and humid. Soils are acidic with high content of iron and sandy loam to clayey loam in texture. Bamboo and canes are common in this zone. • Tropical evergreen, semi-evergreen, and deciduous forest occurs. In this region rice is the most important field crop.
3.	Sub-tropical plain zone	• This zone covers the Imphal valley of Manipur, Bhagti valley of Nagaland and Umkiang area of Meghalaya within the elevational range of 400–1,000 m. • Some of the area under this zone is well facilitated with irrigation systems (Imphal valley). Paddy is the most dominant crop grown here.
4.	Sub-tropical hill zone	• This zone is well distributed among all the northeastern states except Tripura within the elevational range of 1,000–1,500 m. • The soils are loamy silt. • Major crops are rice, maize, wheat, mustard, castor, soybean, pulses, ginger, turmeric, beans, vegetables, etc. Shifting cultivation and terrace cultivation are common in this zone.
5.	Temperate and sub-alpine zone	• This zone covers Tawang, Bomdila, and Dibang valley of Arunachal Pradesh, Sikkim, Shillong plateau of Meghalaya, Mao, Maram, and Ukhrul region of Manipur, Blue Mountain and adjoining area of Mizoram, Phek, Tuensang, Zunheboto, and the higher areas of Mokokchung and Kohima area of Nagaland within the elevational range of 1,500–3,500 m. This zone has been subcategorized into 2 categories.
5.1.	Sub-temperate zone	• This zone encompasses lower elevation range and does not receive any snowfall during winter. • Forest genetic resources are fairly high. Field crops grown are paddy in this zone is potato, maize, and millet.
5.2.	Temperate zone	• The temperate zones cover upper area of Sikkim and Arunachal Pradesh. Snowfall occurs during winter. • Pines, firs, rhododendrons, orchids are very common plants in this region. Millets, maize, barley, wheat, potato, etc., are commonly grown in this zone. • Common fruits are apple, plum, peaches, pears, walnut, etc., livestock are important component of the farm and main source of livelihood.
6.	Alpine zone	• This zone covers part of Sikkim and Arunachal Pradesh (Eastern Himalayas) at an elevation above 3,500 m asl. This zone experiences average minimum temperature of 0–2°C. • The soil is acidic. The forest types vary from temperate evergreen in the lower range to alpine scrub in the higher range. • However, the zone has pasture lands and significant for pastoral nomads.

9.5 TRADITIONAL AGROFORESTRY SYSTEMS

The tradition of raising woody perennials, herbaceous crops and livestock together on the farm has existed since time immemorial but the term 'Agroforestry' given to it is a new one. This practice of mixed farming developed over centuries for meeting different requirements of a family like food grains, fodder, construction materials, firewood, milk, and meat and also for some ecological benefits like soil and moisture conservation (Mishra & Sharma, 2001). Agroforestry is a term refers to a sustainable land-use systems (LUS) where woody perennials like trees, bamboos, shrub, and palms are deliberately grown along with herbaceous crops and or livestock on the piece of land simultaneously or sequentially and managed as per the social as well as cultural compatibility and feasibility of the locals. Agroforestry practices vary considerably from place to place, region to region and country to country. The NER of India is endowed with rich indigenous agroforestry practices, which has evolved over time and space due to wide topographical, ecaphic, climatic variations. In addition to these variations, the evolution of diversity in agroforestry practices has also been greatly influenced by the diverse socioeconomic conditions of the farming communities. Some important traditional AFS of NER of India are discussed in subsections.

9.5.1 TRADITIONAL HOMEGARDENS

Homegardens is one of the most ecologically feasible food producing subsistence agroforestry systems based on the traditional knowledge gained over the years by the locals. This indigenous agroforestry system is present throughout the NER of India. It is a common practice in the region to grow different types of trees, shrubs, vines, bamboos, canes, herbaceous crops with or without animal components in dense and random arrangements in/nearby homestead (Figure 9.1). The main purpose of this traditional system is to meet the household requirements like food, timber, fodder, and fuelwood all year round. But recently with the advent of market for certain agriculture products, it is also inclined towards the production of cash crops for income generation.

Structurally, homegardens is a very complex system composed of three-five vertical canopy strata, each layer having different forms of plants occupying different niches. In general, all the homegardens have an herbaceous species on the lower layer which is usually partitioned into the lowermost layer mainly comprised of vegetables, foodgrains, herbal plants of less

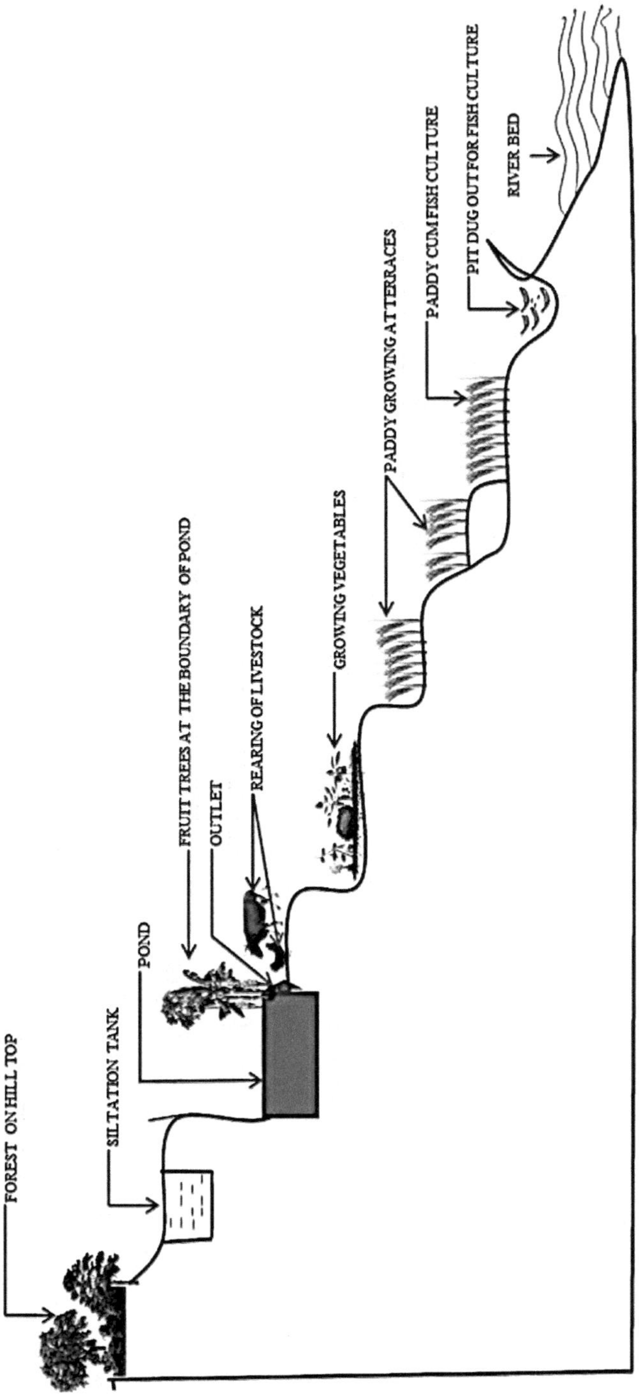

FIGURE 9.1 Schematic diagram of *Zabo* farming system.

than one-meter height and the successive layer of plants like papaya, yam, banana, and cassava usually of one to three meters in height. The intermittent layer of 5–10 m height consists of various fruit species. The upper layer is further stratified into topmost layers consisting of mature timber and fruits trees species of over 25 m height and the second layer comprising of trees of 10–20 m height (Fernandes & Nair, 1986).

Apart from random stratified structure, some large homegardens have zonal characteristics (Das & Das, 2005; Tynsong & Tiwari, 2010) where gardens are divided into different micro-zones or management zones based on the composition, function, and location as – (i) bamboo groves; (ii) betel groves; (iii) banana groves; (iv) vegetable garden; and (v) dense or forest zone. The bamboo, betel, and banana groves are usually composed of pure stands of bamboo, betel, and banana plants. The forest groves are mainly planted with different multipurpose tree species to provide fodder, timber, fruits, and fuelwood. The larger area of the homegardens is allotted to the betel zones pertinent to its high commercial value. The betel, banana, and vegetable zones are placed near the home. Religious zone is also a very important part of homegardens in Barak valley where sacred plants like *Ficus spp*, *Aegle marmelos, Terminalia arjuna, Ocimum sanctum* are planted and raised for religious purposes. Fruits trees are also an important component usually planted in the boundary or grown mixed in betel or banana zones.

The area and species diversity of homegardens were found to be mainly determined by the socio-economic conditions of the families who maintained them. The dominant plant species (excluding herbaceous component) present in different homegardens of NER of India are given in Table 9.7.

The area under homegardens in different parts of NER of India varied from 0.02 to 1.20 ha (Das & Das, 2005; Deb et al., 2009; Sahoo, 2009; Tynsong & Tiwari, 2010; Tangjang & Nair, 2016). Generally, poor families have smaller homegardens and subsequently less diversity and vice versa. The species grown are mainly perennial. Species compositions of the homegardens are governed to a large extent by two major factors like environmental factors that decide the appropriateness of the habitat for the species and socio-economic factors like culture, tradition, household preferences, requirements, food habits and market demands. Apart from the herbaceous component which is more or less similar in most of the homegardens, there is a wide difference in the species composition of homegardens of Nyshi tribes of Nirjuli and Doimukh area of Arunachal Pradesh and homegardens of Kalita communities of Barak valley due to their cultural differences. While Nishi farmers favor growing more of woody species and bamboo for construction/housing

TABLE 9.7 Dominant Species, Major Use Category, and Major Cash Crops of Traditional Homegardens in NER

Place	Inhabitants	Dominant Homegarden Plants	Major Use Category	Major Cash Crops	References
Barak Valley, Assam	Bengali, tea tribes, tea garden laborers.	*Areca catechu*, *Musa* sps., *Artocarpus heterophyllus*, *Mangifera indica*, *Toona ciliate*, *Psidium guajava*, *Citrus maxima*, *Carica papaya*, *Hibiscus rosa*, *Nyctanthes arboritritis*, *bamboo;* Herbaceous crops: vegetables, tubers, rhizomes, ornamentals, medicinal crops; Rare species: *Aquilaria malaccensis*, *Vatica lanceaefloia*	Fruit yielding crops followed by timber and miscellaneous.	Arecanut, Banana, Jack fruit, bamboo.	Das & Das (2005)
Selesih, Sairang, and Thingsulthliah of Mizoram	Mizo	*Albizia procera*, *Artocarpus heterophyllus*, *Citrus reticulate*, *Mangifera indica*, *Parkia timoriana*, *Tamarindus indica*, *Cajanus cajan*, *Citrus sinensis*, *Musa paradisiac*, *Osidium guajava*, *Carica papaya*, *Citrus anamensis*, *Citrus limon*, *Clerodendron colebrokianum*, *Thysanolaena maxima*; herbaceous crops: Cereal crops, vegetables, tubers, rhizomes, etc.	Vegetable, fruit yielding species, medicinal plants, multipurpose tree species, miscellaneous use species, timber, and spice.	Vegetables, seasonal fruits, rhizomatous crops, arecanut, broom grass, tree bean.	Sahoo (2009)
Ziro valley	Apatani	*Pinus wallichiana*, *Phyllostachys bambusoides*	Construction materials.	Pine and bamboo.	Tangjang & Nair (2016)
Nirjuli and Doimukh	Nyishi	*Livistona jenkinsiana*, *Grevilia robusta*, *Artocarpus heterophyllus*, *Mangifera indica*, *Bambusa spp.*, *Carica papaya*, *Psidium guajava*, *Citrus spp.*, *Ananus comosus*; Herbaceous species: cereals, vegetables, tubers, rhizomes, medicinal plants, etc.	Vegetables, medicinal plants, Fruit, and timber yielding plant.	Local vegetables, bamboo sps, timber producing species.	Deb et al. (2009)
South Meghalaya	Khasi	*Areca catechu*, *Artocarpus heterophyllus*, *Artocarpus lakoocha Mangifera indica Citrus lemon*, *Citrus maxima*, *Duabanga grandiflora*, *Cinnamomum tamala*, *Bambusa sps.*, *Psidium guajava*, *Baccaurea sapida*, *Clerodendron colebrookianum*, *Averrhoa carambola*, *Hibiscus rosa-sinensis*, *Hibiscus sabdariffa*, *Manihot esculenta*, *Piper betle*, *Passiflora edulis*, *Elaeagnus latifolia*, *Houttuynia*	Fruit yielding crops, medicinal plants, timber.	*Piper betle*, *Mangifera indica*, *Litchi chinensis*, and *Areca catechu.*	Tynsong & Tiwari (2010)

purposes, plantation crops like arecanut and fruits yielding species are major components in the larger homegardens in Harmutty, Assam (Tangjang & Arunachalam, 2009). Apart from being a cash crop, arecanut plants play a very important role in the rituals and festivals of the locals. Homegardens in the Barak valley, Assam like most of the plains/valley area of the NER also maintain ponds for rearing fish in their home yards. Multipurpose trees (MPTs), palms, bananas, and papaya are planted surrounding the pond and many tubers and vegetables are inter-planted forming an eco-friendly production system. Livestock components like cattle, pig, goat, sheep, and birds like chicken and duck are also very significant components of the homegardens of NER and are used for nutritional requirements of the family (meat, eggs, milk), soil enrichment (manure) and for income generation.

One of the unique homegardens of NER is 'Bamboo-Pine Homegardens' which is indigenous to Apatani tribe of Ziro Valley (Tangjang & Nair, 2016). This indigenous system is an excellent example of natural resource management practice and has been developed over the years based on the traditional knowledge of its bio-cultural interaction. Here, the monopodial bamboo (*Phyllostachys bambusoides*) is planted along with the pine (*Pinus wallichiana*) on the homestead area which is generally located on the uphill sides. The mixture is based on the traditional knowledge that no trees other than pines could survive and grow in bamboo groves. Likewise, no plant other than bamboo could establish and grow with the pine trees. This indigenous homegardens is capable of securing livelihood of the farmers of Ziro valley by providing several ecosystem services (ES) as mentioned below:

1. **Provisioning Services:**
   - Providing materials for household needs like construction, fencing, roofing, handicrafts, medicine, and food.
   - Income generation by sale of products.
   - Increased employment.

2. **Regulating and Supporting Services:**
   - Soil enrichment, efficient soil nutrient cycling, and soil and water conservation, improve water storage, biodiversity conservation and carbon sequestration.

3. **Cultural Services:**

- Enhancing recreational value, culture preservation, provision of social securities.

Homegarden is also considered an important site for *in situ* conservation of plant genetic diversity (Perera & Rajapakse, 1991; Gajaseni & Gajaseni, 1999) and also can serve as a storehouse of the gene pool for many indigenous species. Apart from native traditional domesticated and wild plant species, rare species like *Aquilaria malaccensis* and *Vatica lanceaefloia* were also conserved in the homegardens (Das & Das, 2005). Konyak tribes successfully domesticated *Aquilaria agallocha* in homegardens of Konya tribes of Nagaland (Ramakrishnan et al., 1996). Other rare species like *Caryota urens* and *Licuala peltate* were also successfully conserved. Homegardens are also efficient in preserving underutilized edible species like *Baccaurea sapida, Garcinia sp., Flacourtia jangomus, Spondia spinnata* and *Meyna spinosa.* The traditional homegardens of NER is sustainable LUS that are not only important in providing daily household needs like food, fruits, medicine, and income of the farmers but also play a very important role in sustaining the entire ecosystem by means of soil and water conservation, biodiversity conservation, carbon sequestration and water regulation. Thus, this traditional farming system needs to be conserved and some research on proper design and management can be conducted to make the system more productive.

9.5.2 SHIFTING CULTIVATION

Shifting cultivation, also called jhum, is an indigenous and the most primitive form of land use (LU) system in NER of India (Figure 9.2).

a) b)

FIGURE 9.2 Shifting cultivation area in Nagaland, India – (a) maize with beans grown in Jhum land; and (b) deforestation caused by shifting cultivation.

This system has been in practice for thousands of years across the hilly terrain of NER (Chauhan, 2001) except Sikkim. The total area under shifting cultivation in NER of India is around 676.54 ha (Table 9.8) which includes both current jhum (300.10 ha) and abandoned jhum (376.44 ha).

TABLE 9.8 Area (000 ha) Under Shifting Cultivation in North-Eastern States (2015–2016)

State	Current Shifting Cultivation	Abandoned Shifting Cultivation	Total Area Under Shifting Cultivation
Arunachal Pradesh	50.91	120.42	171.33
Assam	5.26	8.21	13.47
Manipur	50.00	29.86	79.86
Meghalaya	23.79	42.27	66.06
Mizoram	69.15	101.14	170.29
Nagaland	97.93	68.12	166.05
Tripura	3.06	6.42	9.48
Total	300.10	376.44	676.54

Source: Wasteland Atlas of India (2010; https://dolr.gov.in/documents/wasteland-atlas-of-india).

Among different NE states, Arunachal Pradesh has the maximum area under shifting cultivation followed by Mizoram, Nagaland, Manipur, Meghalaya, Assam, and Tripura. However, Nagaland is having the highest percentage of TGA (10.01%) under jhum cultivation followed by Mizoram (8.08%). The Jhum is a sequential agrisilviculture system wherein at a particular period of time either agriculture crops or woody perennials occupy the space and perform the function. Jhum is mostly practiced in the hilly areas endowed with moist deciduous, evergreen, or bamboo forests in their secondary successional stages. The land for shifting cultivation is normally selected as per the ease of vegetation clearing and high soil fertility (Siddiqui & Chohan, 2015). The village chief or council holds the power to allot the land to the cultivators and the allotment is done as per the size of the family. The allotted area is clear-felled during winter months like November-December and debris is allowed to dry and slashes are burnt before the onset of monsoon. All the members of the community take part in clearing the selected/allotted jhum plots in most of the places, while in some places; individual family clears their allotted plot. Generally, different crops are cultivated to obtain food throughout the year. The Jhum system is characterized by a mixed cropping system that involves a growing of 25–35 species consisting of food grain, oilseeds, leaf, and fruit vegetables, tubers, and rhizomes species (Maikhuri & Ramakrishna, 1991) and the genetic diversity of crops is maintained through the process of seed exchange and conservation measures. The land is used for the cultivation of crops for a short period of one to three years and then allowed to rest for a longer period than the cropping duration. Thus, during the fallow period, the land regains its original cover and fertility status. The fallow period varies among the regions.

Jhuming is a complex system that differs across the regions as per variations in the ecology and cultures of different tribes. This practice is strongly embedded with the cultural identity of the ethnic people of the region. The knowledge of jhuming has been evolved through numerous observations and trials carried out by the tribal community and the same have been transmitted from generation to generation. The jhum lands are mostly owned by the community so the jhumias has to follow the customary rules related to the use and management of resources, laid down by the community (Tripathi & Barik, 2003). This system of farming is also considered very imperative in community life across the NER of India. In the hills of NER, jhuming is still a dominant economic activity where jhumias either depend very less or not at all on outside activity for fulfilling their livelihood needs (Gupta, 1994). The illiterate and older farmers have a greater affinity to practice shifting cultivation as they are more connected to their tradition compared to educated and young farmers (Rahman et al., 2011). The dependency of farmers on shifting cultivation has increased significantly which was mainly attributed to an increase in population and lack of opportunities for earnings from off-farm activities. The shifting cultivation is being practiced by the farmers because it gives more or less stable income every year. Shifting cultivation provides a cultural identity tied with rituals and festivals which revolve around the Jhum besides giving economic security (Punitha et al., 2018).

The Jhum cultivation was a sustainable LU practice when the jhum cycle was more than 20 years. During the long fallow period, the vegetation gets ample time to emerge and establish, and the soils get sufficient time to rejuvenate and restore their health and productive capacity. But the jhum cycle has been reduced to 2–4 years at present due to an increase in population pressure on land resources. This short fallow period or reduced jhum cycle is too inadequate to provide recuperation to the site and restore the ecological damage. As a result, there is severe land degradation caused by soil erosion. Research conducted in Meghalaya indicates that jhuming resulted to a great loss of soil from the cultivated plots where magnitude of the soil erosion was influenced by the degree of slope, soil texture and permeability. Soil erosion on hill slopes of 50–60% slope under first-year jhum, second year jhum, abandoned jhum and bamboo forest has been estimated as 146.6, 170.2, 30.2, and 8.2 tons/ha, respectively (Singh & Singh, 1981). Shifting cultivation caused large scale deforestation and denudation of hills slopes. The reduction in shifting cultivation cycles has created a sense of livelihood insecurity among the ethnic communities due to poor crop yield. The shortening of shifting cultivation cycles adversely affected the entire ecosystem. Thus, the reduction in jhum cycles is a major challenge for the farming communities, policymakers, researchers, and the farmers of the region.

9.5.3 ALDER-BASED FARMING SYSTEM

Alder based farming system (Figure 9.3) is a unique agroforestry system developed by the indigenous communities namely Angami, Chakhesang, Yimchunger, Konyak, and Chang of Nagaland (Gokhale et al., 1985; Singh, 1992; Das et al., 2009).

a) b)

c) d)

FIGURE 9.3 Alder based farming system – (a) burnt pollarded alder field ready for cultivation; (b) crop sown in alder land; (c) Alder fallow land; and (d) firewood from pollard coppice shoots of Alder.

This system was developed by these communities through numerous years of trials and errors to ameliorate the jhum land, with the idea that planting of Alnus tree in the jhum land can increase the soil fertility and control soil erosion. *Alnus nepalensis* (Alder) is a fast-growing pioneer species and can establish and grow in open degraded habitats with low soil fertility. The trees can fix atmospheric nitrogen in the soil and are efficient in providing stability to slopes that tend to slip and erode by virtue of its extensive lateral root system (Rathore et al., 2010; Das et al., 2012). In this system, crops like cereals, spices, vegetables, tuberous, and rhizomatous crops are grown along with alder tree in typical four-year cycle, where two years of cropping between the alder trees followed by two years of fallow which allows the soil to regain its fertility and the coppices are encouraged to grow (Das et al., 2012). Alnus seedlings whether naturally grown or intentionally planted become established during the first cropping year of the jhum cycle. In the first two years, primary crops like rice and secondary crops like Colocasia, chili, tapioca, Amaranthus, and potato are grown intermixed with the trees. Subsequently, the field is left fallow for a few years to let the alder tree grow.

The alder trees grow rapidly reaching 70–80 cm in diameter and 10 meters in height and develop rough fissured bark during the first eight to nine years. This is the optimum age for first pollarding. The age of first pollarding is also determined by the jhum cycle. In case of jhum cycle of four years, alder tree reaches 4 years by the time plot is reopened for cultivation. Since the trees will be too young for pollarding only trimming of side branches is done to avoid shading to crops. In the next cropping period, the tree will be eight years old and almost the proper age for pollarding and if the tree is still small then pollarding can be delayed until the next cropping period. After the first pollarding, subsequent pollarding is done once every 4–6 years. Trees are pollarded at a height of seven to eight feet from the ground from late January to early February. The tree is pollarded with utmost care to avoid splitting of stump head and the head is covered with mud or straw to avoid excessive drying and cracking. A stone is placed on top of the stump to protect the wound from sunlight, wind, and frost and to promote the uniform sprouting of new shoots around the stump. Entire sprouted shoots in the pollarded tree are allowed to grow for a year thereafter only 5–6 shoots are left, and the rest are slashed. The bigger branches are used as firewood. The remaining slash is raked into piles at safe distances from the alder trunks and burned in late February and ashes are scattered around, and then hoeing of soil is done trimming many fine rootlets and nodules. After this, the roots and residual slash are raked into piles for burning and then crops are raised and the same process is repeated in the following cropping phases.

Alder trees along with crops form a very productive and lucrative agroforestry system in the region. The alder-based farming system has a huge role in stabilizing jhum land and rehabilitation of the degraded jhum land (Cairns, 2007). The alder tree by virtue of its nitrogen-fixing ability, spreading lateral roots and ability to produce a good amount of litter is a good agroforestry tree species that help in the enrichment of the soil fertility and stabilization of the slopes against erosion. Alder based agroforestry, if exploited with proper research strategies, can become a potential farming model globally.

9.5.4 *ZABO FARMING*

Zabo is an integrated farming system practiced by the Chakesang tribe of Kikruma village in the Phek district of Nagaland which is located at an altitude of 1,270 m above mean sea level. About 958 ha of area are under this farming system. In the Chakhesang dialect, the word "*Zabo*" means impounding runoff water (Singh et al., 2015a). The people of Kikruma village were facing the problem of water scarcity despite having sufficient rainfall due to heavy surface runoff. To solve this problem, they were forced to come out with an idea or technique for water harvesting/conservation which eventually gave rise to *Zabo*. This farming system involves forest, agriculture, horticulture, livestock, and fisheries practices with well-founded soil and water conservation base (Figure 9.4). The important inherent aspects of this system are water resource development, water management, and protection of the environment (Sharma & Singh, 1994; Sharma et al., 1994; Das et al., 2009). This system is a viable practice of resource management and maintenance of ecological balance (Pulamte, 2008).

a) b)

FIGURE 9.4 Traditional homegardens of Assam – (a) major components of homegarden (cattle, arecanut, and some local woody and herbaceous species); and (b) fishpond with ducks in homegarden.

In this farming system, the forest land occupies the hilltop, water harvesting ponds in the middle, and cattle yard and paddy fields on the lower side of the hill. The forest lands, on hilltops, act as catchments and water is carried through inlet channels from the catchment area and then collected in a pond with seepage control. Silt retention tanks are built at several locations above the ponds to retain silts from entering the ponds along with the runoff water and water is allowed to remain in the silt retention tanks for two or three days before releasing it to the main ponds and the silt retention tanks are cleaned every year. Vegetables and fruits like squash, Colocasia, cucurbits, banana, papaya, oranges, and citrus are cultivated on the periphery of the pond. Livestock such as cattle, goat, sheep, pig, and poultry are commonly reared beside the pond. The cattle yards are fenced with bamboo and wooden branches. The water from the ponds is released either by open channels or by bamboo pipes. Water is allowed to pass through the cattle-yard before reaching it to the fields for irrigation thereby carries the dung and urine of the animals along with it, thus helps in maintaining soil fertility (Singh, 2007).

In the *Zabo* farming system, the paddy fields are generally located at the lower elevations. To avoid percolation and seepage of water the field embankments are thoroughly rammed by beating with wooden sticks and paddy husk is also used with the mud to reduce seepage losses from the bunds. Commonly "*Tanyekemuga*" – a local variety of paddy which matures in about 180 days is cultivated in paddy fields. The paddy is transplanted in June and July. Usually, two supplementary irrigations are given from the ponds, in case of poor rains. The water level is maintained in the fields at about 10 cm height and the productivity of paddy ranges between 3 and 4 t/ha (Sharma & Sharma, 2003). Paddy cum fish culture is a common practice which is one of the components of *Zabo* farming systems where farmers raise the fishes in their paddy field terraces. The fingerlings are released during July and small pits are dug out in the center of the field to provide shelter for fish when the field dries up. By the end of October when the paddy is mature and ready for harvesting, the paddy field becomes dry. In case if there is excess water in the field at the time of harvesting of paddy then the water is drained out of the field before harvesting. The fishes move into the pit dug in the middle of the field as the field dries up and remains in the pit from where they are harvested. Farmers obtain about 60–75 kg of fish per hectare from paddy cum fish culture from the paddy cum fish culture. The celebration starts with thanksgiving and offering prayers to Mother Nature for bestowing this gift on them. In the tradition of the Chakesang tribe, fish signifies money and wealth and is also regarded as holy and clean.

9.5.5 LARGE CARDAMOM-BASED AGROFORESTRY SYSTEM

A large cardamom-based agroforestry system is a traditional agroforestry system (AFS) prevalent among the people living in the hills and mountains of eastern Himalayas (Figure 9.5). The crop is usually grown under the shade of nitrogen-fixing Himalayan alder (*Alnus nepalensis*) and other tree species as the growth of the crop is greatly affected by direct exposure to the sun (Partap *et al.*, 2014). The alder tree is grown as a shade tree in large cardamom based AFS that provides greater ecological sustainability as the alder tree easily comes up well on areas damaged by landslides and within the similar agro-climatic zone as cardamom (Sharma *et al.*, 2000). The large cardamom (*Amomum subulatum*) is a shade loving, perennial, and important cash crop usually planted under the tree on mountain slopes. It is usually grown in the region having mean annual rainfall from 1,500–3,500 mm. The crop is largely cultivated between an elevation of 600–2,000 m above msl from subtropical area to the temperate zones where the alder is also occurring naturally (Sharma *et al.*, 2007). The native 'Lepcha' indigenous community of Sikkim is believed to initiate the cultivation of large cardamom which was later adopted by other communities like Bhutias and Nepalis of Sikkim and later on the neighboring Darjeeling district of India, parts of Bhutan and eastern Nepal also started its cultivation (Sharma *et al.*, 2000). The cultivation of large cardamom has later expanded to other NER and also to Uttarakhand (Srinivasa, 2006; Tangjang & Sharma, 2018).

FIGURE 9.5 Cardamom-based agroforestry system.

The alder is a nitrogen-fixing tree species and is commonly used as shade trees wherein about 70% of the cardamom-based AFS has alder as shade tree while 30% has mixed-tree species. Besides alder tree, some other commonly used shade trees in the large cardamom based AFS are *Terminalia myriocarpa, Schima wallichii, Eurya acuminata, Leucosceptrum canum, Engelhardtia acerifolia, Bucklandia spp., Edgeworthia gardneri, Viburnum erubescens, Maesa cheria, Symplocus theifolia, Albizia lebbeck, Erythrina indica, Castanopsis indica, Cryptomeria japonica, Betula alnoides, Litsea polyantha, Macaranga denticulata, Ficus nemoralis, Nyssa sessiliflora, and Osbeckia paniculata* (Sharma et al., 2007, 2009). Trees are regularly pruned at a height of 4–5 m to allow the spread of its new vegetative growth, thereby providing a uniform shade (Jaiswal, 2014).

Sikkim alone produces about 45% of the world's large cardamom and it is one of the important livelihood sources for the mountain people (Das *et al.*, 2012). *Alnus nepalensis* being nitrogen-fixing in nature helps in maintaining the soil fertility which in turn enhances the productivity of large cardamom through a faster rate of nutrient cycling. Cardamom based AFS is sustainable land management and production system of sloping land on the mountainous region which helps in conserving soil and water, biodiversity, increasing soil fertility and productivity, reducing soil erosion and serves as a carbon sink at a higher rate compared to any other LUS in the region and augment the living standards of the people by increasing the farm incomes and also provides esthetic values for the mountain communities (Buckingham, 2004). Based on various data regarding rainfall events revealed that cardamom based AFS had significantly low soil loss, overland flow, and nutrient loss compared to other crops-based systems (Table 9.9).

TABLE 9.9 Comparison of Soil and Nutrient Loss Under Cardamom-based and Other Crops-based System

Rainfall Event	Cardamom-based Agroforestry System	Other Crops-based System
Soil loss (kg/ha)	30.00	477
Overland flow (% of precipitation).	2.17	9.55
Nutrient Loss (kg/ha)		
Nitrogen	0.41	2.35
Organic carbon	1.86	11.86
Total phosphorus	0.02	1.49

Source: Sharma *et al.* (2000).

In India, Sikkim is the largest producer of large cardamom and the second-largest producer in the world after Nepal. It contributes US $500–1,700 to annual household cash income in Sikkim, based on the landholding size of the farmers and effective farm management activities (Sharma *et al.*, 2000; Sharma, 2013). Sharma *et al.* (2007) reported that the total input costs of 123 US $per hectare were involved and an output of 1,619 USD per hectare was achieved in the cardamom based AFS. The output/input ratio in the cardamom based AFS was 13.16 while, it was only 1.85 in rainfed agriculture. Cardamom based AFS was 5.7 times more beneficial to farmers compared to rainfed agriculture based on cost-benefit analysis (Sharma *et al.*, 2002). According to the study, it was reported that the 5–40-year-old alder trees were capable of adding 58–155 kg/ha of nitrogen to the soil through nitrogen fixation where 10–15-year-old trees were observed to be highly potential in adding the highest amount of nitrogen to the soil (Sharma *et al.*, 2009). The cardamom based AFS sequest 3.5 times higher carbon as compared to rainfed agriculture indicating possible climate change (CC) mitigation potential (Mukherjee, 2012). This system is not only a vital source of revenue for the mountain communities but also performs various ES and also provides employment opportunities through ecotourism (Sharma *et al.*, 2009). Hence, this system is ecologically as well as economically viable land-use system in mountain slopes of Northeast India. However, the area, as well as the production of the large cardamom, has drastically reduced over the last decade as a result of CC (Sharma et al., 2002, 2016; Negi *et al.,* 2018). Development of scientific agro-techniques, disease-free planting materials, and maintenance of soil fertility are the important aspects that need to be addressed for sustaining and enhancing the production capacity of the large cardamom in the region (Singh et al., 2015b; Sharma et al., 2016).

9.6 WAY FORWARD

The traditional agroforestry systems of NER of India have been developed and practiced by the indigenous inhabitants since time immemorial. These farming systems came into existence to cater various household's needs of the local people and were shaped by their socioeconomic conditions and the available natural resources. However, these farming systems can be modified or improvised through some administrative reforms and scientific research to make the system more sustainable and productive for which the following strategies can be adopted:

1. Development of proper land use policy and land tenure systems for better management of land use and also to reduce the adoption of unsustainable farming systems like shifting cultivation.
2. Provision of financial assistance/subsidy making terraces and construction of rainwater harvesting structure in the hilly area to promote sustainable sedentary farming systems in hills.
3. Integrated Watershed Management Program involving local inhabitants should be promoted.
4. The highly sustainable and remunerative farming systems like Zabo farming, Alder based farming systems; Cardamom based agroforestry systems should be replicated to other areas of the region having similar agroclimatic conditions.
5. The modification or improvisation of the traditional farming systems should be done taking into consideration the significance of traditional knowledge and sentiments of the local people for easy acceptance, e.g., promotion of Alder based farming systems in shifting cultivated areas.
6. Need for scientific research on CC adaptation of these systems, impacts of trees-based systems on soil fertility to understand the flux and dynamics of soil nutrients, development screening of early maturity and high yielding local varieties of crops, identification of superior genotypes of trees for reclamations of fallow lands in shifting cultivated areas and mass multiplication for large scale supply to farmers.
7. Development of rural infrastructures and market for agriculture products.
8. Provision of good planting material through state forest department.
9. Promotion of capacity building program for local youths in agriculture or non-agricultural related fields.

9.7 CONCLUSION

NER has a wide range of climatic, edaphic, topographical, biological, and cultural diversity. Thus, various traditional AFS are prevalent in this part of the country. These systems have been evolved over many decades based on the knowledge of the ethnic communities to meet their livelihood needs and are an integral part of the socioeconomic and cultural identity of the local communities. Most of the traditional agroforestry systems of

NER are diversified and highly sustainable comprising of traditional water harvesting systems, soil fertility management and erosion control. Yet, these systems need to be further improved with suitable scientific interventions involving community participation in on-farm research activities exploring and recognizing their traditional knowledge and establishing Institutions that can develop strong coherent linkage between the stakeholders and the Institute. There is also a need for administrative reforms and pro-people holistic developmental policies for these traditional LUS. The knowledge on highly sustainable and remunerative traditional agroforestry systems needs to be disseminated into new areas of similar agro-ecological conditions to meet the increasing household demands and improving the socio-economic conditions of local communities.

KEYWORDS

- **agro-climatic zones**
- **agroforestry**
- **alder-based farming system**
- **land use**
- **north-east India**
- **resource conservation**
- **traditional agroforestry**
- **Zabo farming**

REFERENCES

Agricultural Statistics at a Glance, Government of India, (2012). Directorate of Economics and Statistics, Department of Agriculture and family Welfare, Ministry of Agriculture and family Welfare, Government of India, New Delhi.

Buckingham, S., (2004). *Synthesis Report on Cardamom Cultivation* (p. 76). Fauna and Flora International Community-based Conservation in the Hoang Lien Mountains, Darwin Initiative of the UK Government in Vietnam.

Cairns, M., (2007). *The Alder Managers: The Cultural Ecology of a Village in Nagaland, N.E. India*. Ph. D. Dissertation, Australian National University, Canberra.

Chauhan, B. S., (2001). *Shifting Cultivation in Perspective*. Nagaland University.

Das, A., Munda, G. C., Ghosh, P. K., Patel, D. P., Ngachan, S. V., & Tripathi, A. K., (2009). Natural resource conservation through indigenous farming systems in North-East India. In: *Abstract: 4th World Congress on Conservation Agriculture* (pp. 355, 356). New Delhi. NASC complex.

Das, A., Ramkrushna, G. I., Choudhary, B. U., Munda, G. C., Patel, D. P., Ngachan, S. V., Ghosh, P. K., et al., (2012). Natural resource conservation through indigenous farming systems: Wisdom alive in Northeast India. *Indian Journal of Traditional Knowledge, 11*(3), 505–513.

Das, T., & Das, A. K., (2005). Inventorying plant biodiversity in homegardens: A case study in Barak Valley, Assam, Northeast India. *Current Science, 89*(1), 155–163.

Deb, S., Arunachalam, A., & Das, A. K., (2009). Indigenous knowledge of Nyishi tribes on traditional agroforestry systems. *Indian Journal of Traditional Knowledge, 8*(1), 41–46.

Dikshit, K. R., & Dikshit, J. K., (2014). *North-East India: Land, People and Economy* (p. 800). Advances 149 in Asian Human-Environmental Research, Springer Netherlands. doi: 10.1007/978-94-007-7055-3_6.

Fernandes, E. C. M., & Nair, P. K. R., (1986). An evaluation of the structure and function of tropical home gardens. *Agricultural Systems, 21*, 279–310.

Gajaseni, J., & Gajaseni, N., (1999). Ecological rationalities of the traditional homegarden system in the Chao Phraya Basin, Thailand. *Agroforestry Systems, 46*, 3–23.

Gokhale, A. M., Zeliang, D. K., Kevichusa, R., Angami, T., & Bendangnungsang, S., (1985). *The Use of Alder Tree*. State Council of Educational Research & Training, Government of Nagaland, Kohima.

Gupta, M. D., (1994). Continuity and change in tribal societies: A study of the economic aspects of tribal Jhumia society in Tripura. In: Sangma, M. B., (ed.), *Essays on Northeast India* (pp. 147–164). New Delhi, Indus publishing company.

Jaiswal, A., (2014). Large cardamom cultivation in India. *MANJARI: A Monthly Newsletter of NTFP, 1*(2), 1–16.

Maikhuri, R. K., & Ramakrishnan, P. S., (1991). Comparative analysis of the village ecosystem function of different tribes living in the same area in Arunachal Pradesh living in North-Eastern India. *Agricultural Systems, 9*, 57–72.

Mishra, A. K., & Sharma, U. C., (2001). Traditional wisdom in range management for resource and environment conservation in North-eastern region of India. *Envis Bulletin: Himalayan Ecology and Development, 9*(1), 31–36.

Mukherjee, D., (2012). Resource conservation through indigenous farming system in hills of West Bengal. *Journal of Crop and Weed, 8*(1), 160–164.

Negi, B. K., Joshi, R. K., & Pandey, A., (2018). Status of large cardamom (*Amomum subulatum* Roxb.) farming systems in the changing scenario of modern economics of Sikkim, Himalaya. *Global Journal of Bio-science and Biotechnology, 7*(1), 189–199.

Partap, U., Sharma, G., Gurung, M. B., Chettri, N., & Sharma, E., (2014). *Large Cardamom Farming in Changing Climatic and Socioeconomic Conditions in the Sikkim Himalayas*. International Center for Integrated Mountain Development, Kathmandu, Nepal.

Perera, A. H., & Rajapakse, R. M. N., (1991). A baseline study of Kandyan forest gardens of Sri Lanka: Structure, composition, and utilization. *Forest Ecology and Management, 45*, 269–280.

Pulamte, L. (2008). Indigenous agricultural systems of Northeast India. *Report of India Science and Technology. National Institute of Science and Technology Development Studies (NISTADS), CSIR, 346.* https://scholar.google.com/scholar?&q=Pulamte%20

L%20%282008%29%20Indigenous%20agricultural%20systems%20of%20Northeast%20India.%20Report%20of%20India%20Science%20and%20Technology.%20National%20Institute%20of%20Science%20and%20Technology%20Development%20Studies%20%28NISTADS%29%2C%20CSIR%2C%20p%20346 (accessed on 28 December 2023).

Punitha, P., Ansari, M. A., Pandey, D. K., Ram, D., Datt, S., Sharma, P. K., Aheibam, M., et al., (2018). Shifting cultivation in Northeast India: Social dimension, cross cultural reflection, and strategies for improvement. *Indian Journal of Agricultural Sciences, 88*(6), 811–819.

Rahman, S. A., Rahman, F., & Sunderland, T., (2011). Causes and consequences of shifting cultivation and its alternatives in the hill tracts of eastern Bangladesh. *Agroforestry System*. doi 10.1007/10457-011-9422-3.

Ramakrishnan, P. S., Das, A. K., & Saxena, K. G., (1996). *Conserving Biodiversity for Sustainable Development*. Indian National Science Academy, New Delhi.

Rathore, S. S., Karunakaran, K., & Prakash, B., (2010). Alder based farming system a traditional farming practices in Nagaland for amelioration of jhum land. *Indian Journal of Traditional Knowledge, 4*, 677–680.

Sahoo, U. K., (2009). Traditional homegardens and livelihood security in North-East India. *Journal of Food, Agriculture & Environment, 7*(2), 665–670.

Sharma, E., Sharma, R., Singh, K. K., & Sharma, G., (2000). A boon for mountain populations: Large cardamom farming in the Sikkim Himalaya. *Mountain Research and Development, 20*(2), 108–111.

Sharma, G., (2013). *Opportunities and Challenges of Large Cardamom Farming, Beekeeping, and Pollination System in Sikkim*. Report submitted to ICIMOD.

Sharma, G., Sharma, R., & Sharma, E., (2009). Traditional knowledge systems in large cardamom farming: Biophysical and management diversity in Indian mountainous regions. *Indian Journal of Traditional Knowledge, 8*(1), 17–22.

Sharma, G., Uma, P., Dahal, D. R., Sharma, D. P., & Sharma, E., (2016). Declining large-cardamom production systems in the Sikkim Himalayas: Climate change impacts, agro economic potential, and revival strategies. *Mountain Research and Development, 36*(3), 286–298.

Sharma, R., Sharma, G., & Sharma, E., (2002). Energy efficiency of large cardamom grown under Himalayan alder and natural forest. *Agroforestry Systems, 56*(2), 233–239.

Sharma, R., Xu, J., & Sharma, G., (2007). Traditional agroforestry in the eastern Himalayan region: Land management system supporting ecosystem services. *Tropical Ecology, 48*(2), 189–200.

Sharma, U. C., & Sharma, V., (2003). The "Zabo" soil and water management and conservation system in northeast India: Tribal beliefs in development of water resources and their impact on society- a historical account of a success story. The Basis of Civilization- Water Science? In: *Proceeding: UNESCO/IAHS/IWHA Symposium Rome*.

Sharma, U. C., & Singh, K. A., (1994). Agriculture with alder- a potential indigenous farming system of Nagaland. In: *Proceeding National Seminar on Development in Soil Science*. New Delhi.

Sharma, U. C., Prasad, R. N., & Sonowal, (1994). An indigenous technique of soil and water, conservation in northeastern region- the Zabo system of farming. Soil and water conservation challenges and opportunities. In: Bhushan, L. S., Abrol, I. P., & Rama, M. R. M. S., (eds.), *Proceeding: 8th, ISCO Conference*. Oxford and IBH, publication Co. Pvt. Ltd., New Delhi.

Siddiqui, S., & Chohan, S., (2015). *Legal Response to Institutionalizing Participatory Land Use Planning in Nagaland: Developing Strategies for Mainstreaming Sustainable Jhum Practices into Existing Policy and Legal Framework of Nagaland.* UNDP Report, the Printing Wala.

Singh, A. K. (2007). Indigenous water management system by the farmers of north eastern hill region. *LEISA INDIA, 9*(1), 15. https://scholar.google.com/scholar?hl=en&as_sdt=0%2C5&q=+Indigenous+water+management+system+by+the+farmers+of+north+eastern+hill+region&btnG= (accessed on 28 December 2023).

Singh, A., & Singh, M. D., (1981). Effect of various stages of shifting cultivation on soil erosion from steep hill slopes, *Indian Forester, 106*(2), 115–121.

Singh, M., Gupta, B., Sarvade, S., & Avasthe, R., (2015b). Biomass and carbon sequestration potential in different agroforestry systems in Giri catchment of Northwestern Indian Himalaya. *Indian Journal of Agroforestry, 17*(2), 42–48.

Singh, N. P., (1992). Natural farming in Nagaland- an outline, In: Somani, L. L., Totawat, K. L., & Baser, B. L., (eds.), *Proceedings National Seminar on Natural Farming.* Udaipur.

Singh, R. K., Singh, V., Rajkhowa, C., & Deka, B. C., (2015a). Zabo: A traditional way of integrated farming. In: Deka, B. C., Patra, M. K., Trigugnanavel, A., Chatterjee, D., Borha, T. R., & Ngachan, S. V., (eds.), *Resilient Shifting Cultivation: Challenges and Opportunities.* ICAR, Nagaland.

Srinivasa, H. S., (2006). *Large Cardamom Cultivation in India.* Report of the Spices Board, Regional Office, Gangtok, Sikkim, India.

Tangjang, A., & Sharma, A., (2018). Problem faced by the large cardamom growers during production and marketing: A case study of Tirap District of Arunachal Pradesh, India. *International Journal of Current Microbiology and Applied Sciences, 7*(5), 2561–2573.

Tangjang, S., & Arunachalam, A., (2009). Role of traditional home garden systems in Northeast India. *Indian Journal of Traditional Knowledge, 8*(1), 47–50.

Tangjang, S., & Nair, P. K. R., (2016). Integrated bamboo + pine homegardens: A unique agroforestry system in Ziro Valley of Arunachal Pradesh, India. *International Journal of Environmental and Agriculture Research, 2*(2), 25–34.

Tripathi, R. S., & Barik, S. K., (2003). Shifting cultivation in Northeast India. In: Bhat, B. P., Bujarbaruah, K. M., Sharma, Y. P., & Patiram, (eds.), *Approaches for Increasing Agricultural Productivity in Hill and Mountain Ecosystem.* ICAR.

Tynsong, H., & Tiwari, B. K., (2010). Plant diversity in the homegardens and their significance in the livelihoods of War Khasi Community of Meghalaya, North-east India. *Journal of Biodiversity, 1*(1), 1–11.

Velayutham, M., & Bhattacharya, T., (2000). Soil resource management. In: Yadava, J. S. P., & Singh, G. B., (eds.), *Natural Resource Management for Agriculture Production in India* (pp. 1–135). New Delhi, Alfa Printers.

CHAPTER 10

Evidence-Based Agroforestry Systems as Climate-Resilient Farming Practices for Improving Livelihood and Sustainability in India

AJAY KUMAR MISHRA,[1] RAJESWARI DAS,[2] DEBABRATA NATH,[2] VANDANA KUMARI,[2] SARITA MISHRA,[3] BISWORANJITA BISWAL,[4] and RAKHI TYAGI[5]

[1] *Terrestrial Ecosystem Management Laboratory, Graduate School of Global Environmental Studies, Kyoto University, Kyoto, Japan*

[2] *Department of Soil Science, Dr. Rajendra Prasad Central Agricultural University, Pusa, Samastipur, Bihar, India*

[3] *Department of Chemistry, Udai Pratap College, Varanasi, Uttar Pradesh, India*

[4] *Department of Agronomy, National Dairy Research Institute, Karnal, Haryana, India*

[5] *Department of Environmental Sciences, Sri Guru Govind Singh Tricentenary University, Gurgaon, Haryana, India*

ABSTRACT

Agroforestry systems play a substantial role in rejuvenating degraded and low fertility soils by modifying the microclimate and linkage between soil plants and microbes. In addition, agroforestry systems offer various economic and ecological services that boost climate resilience and sustainability in different production landscapes. This chapter highlights ecosystem services (ES),

Sustainable Management and Conservation of Environmental Resources in India.
Sumit Chakravarty, Vineeta, Jahangeer A. Bhat, Munesh Kumar, Gopal Shukla (Eds.)

esthetic value, and climate change (CC) mitigation, agroforestry systems adapted and suited to Indo-Gangetic Plains (IGP). Furthermore, agro-ecology-specific rotation years, temporal variations and synergy of the crops are crucial for securing higher returns in agroforestry models. Vertical stratification of soil carbon stock under agroforestry is one of the most options for long-term carbon sequestration and mitigating GHG emissions. In Trans-IGP, poplar and *Eucalyptus*-based agroforestry systems also enhanced soil carbon sequestration, increased soil organic carbon (SOC) and buffer soils from the emergence of salt in salt-affected lands. Future research should identify-specific tree species based on local needs and commercial values to diversify farm income. The development of a site-specific agroforestry success model will incline other farmers of the region to adopt agroforestry systems.

10.1 INTRODUCTION

Anthropogenic exploitation of natural resources and deforestation in South Asia fueled climate change (CC) and food insecurity pose major constraints in achieving the goal of sustainable development and affecting the livelihood of millions of people. Concentration of carbon dioxide (CO_2) has increased globally from 280 ppm in 1,750 to 412.89 ppm in late 2020, causing global warming. It is expected to reach 700 ppm in 2085, primarily due to human activities. The increase in the global atmospheric concentration of carbon represents an imbalance among the carbon sources and sinks in the biosphere (Arya et al., 2018), mainly due to urbanization and deforestation. Furthermore, the decline in soil fertility and organic matter content caused soil degradation in South Asia, which is the cradle of agroforestry in its long history of tree-based farming systems under multiple agro-ecological conditions (Kumar et al., 2021).

The Indo-Gangetic Plains (IGP), which make up approximately 13% of the country's total land area and are among the world's largest fluvial plains, supply roughly half of the food grains needed to sustain the country's 40% population. The productivity of the rice-wheat cropping system in IGP has stagnated in the last few decades. Reasons are mainly three: first, excessive loss of soil organic carbon (SOC) due to cultivation (oxidation of SOC due to deep tillage, lack of cover to soil during hot, dry summer) and unscientific crop management (burning of straw and crop residue, and lack of cereal-legume rotation); second, excessive flood irrigation from the canal which causes a rise in the water table and thus soil salinity (upward movement of salts during hot, dry summer season) and third, excessive use of fertilizers particularly urea, which promotes oxidation of organic matter and exacerbates the emission of CO_2 from soil to the atmosphere (Kumar et al., 2016). Soil sodicity is another factor that disperses carbon from clay minerals and makes carbon prone to oxidation (Srivastava et al., 2016). To overcome the-mentioned challenges; site-specific agroforestry systems based on landscape, ecology, climate, and species compatibility need to be designed and implemented to develop synergies and trade-offs between profitability and environmental safeguard. Adoption of agroforestry systems has been advocated as a means for mitigating increased CO_2 in the atmosphere and offerings been advocated for mitigating increased CO_2 in the atmosphere and offering diverse ES that govern micro-climate of the region (Kumar et al., 2019).

Furthermore, agroforestry systems serve as a tool to combat CC and strengthen farmers' resilience to natural disasters through biodiversity conservation, carbon sequestration, water resource protection, soil erosion, and pollution (Chavan et al., 2015; Chaudhari et al., 2015). The key players for the success story of agroforestry systems in western-IGP are market availability and wood-based industrial development. This study attempts to build evidence-based agroforestry systems as climate-resilient farming practices for improving livelihood and sustainability.

10.2 CLIMATE CHANGE'S (CCS) IMPACT ON AGRICULTURE: OPENING THE DOOR TO AGROFORESTRY

Agriculture is one of the most vulnerable sectors to global CC, leaving food security at stake. This is a major concern for humanity, and it has prompted developed global power countries to develop strategies to mitigate and adapt to CC. Agriculture is the predominant pay in developing countries that account majorly for rural communities having limited access to alternative financial and biophysical resources (Reppin et al., 2020). Dependency of agriculture on climate concerns the economic development of developing countries to adapt to unpredictable, less hospitable weather phenomena. CC has a greater negative impact in the tropics, with changing precipitation and temperature patterns resulting in lower crop yields, increased pest incidence, and decreased pasture quality. Agricultural contribution of GHG is equivalent to 14% of total anthropogenic emissions (IPCC, 2013) which is estimated to be 5–7 $GtCO_2$/year (Scholes et al., 2014). However, agriculture can help mitigate CC through an array of practices that enhance carbon sequestration in soil and biomass (Scholes et al., 2014).

Agroforestry is a low-cost solution to the world's CC and food security challenges. One of the six global research projects of the World Agroforestry Center is Agroforestry Systems for Climate Change Adaptation and mitigation intending to improve the livelihood and resilience of small and marginal farmers' farming system climate variability and CC (Table 10.1). Trees act as a buffer against several biotic and abiotic stress and deep-rooted perennials with huge reserves and less susceptible to climate variability than annual crops (Kumar et al., 2018). They can also withstand extreme events like droughts and floods, lowering production risks for small farmers. Agroforestry systems are considered one of the climate-smart agriculture (CSA) practices that is interweaved for improving soil health, carbon sequestration, diversity, water, and soil conservation, resource distribution, and conservation (Kumar et al., 2014; Newaj et al., 2015). A judicious and planned combination of tree species and agricultural crops that interact and are interdependent is a major contributor to global carbon (GC) sequestration both in biomass and in soil (Kumar et al., 2020a, b). The tropical agroforestry systems have tremendous potential for sequestrating 12–228 Mg/ha (Pandey, 2007). India has 25 million ha (8.2% of total geographical area (TGA)) under agroforestry, including trees outside forests and other scattered trees (Dhyani, 2014). The carbon sink potential of agroforestry depends on species, age, geography, local climate and management (Mishra et al., 2013; Chaudhari et al., 2014).

TABLE 10.1 Basics of Mitigation and Adaptation Strategies of Agroforestry System

Mitigation **(Activities that reduce GHG emission into atmosphere and enhance GHG storage in ecosystem)**	**Adaptation** **(Actions to eliminate negative effects and benefit from positive effects of climate change)**
Reduce GHG emission: • Reduce fossil fuel burning. • Reduce methane emission by improving forage quality. • Increase GHG sink. • Reduce nitrous oxide emission by improved nutrient recycling and reduced nitrogen fertilizer use in agroforestry system.	Increase adaptive capacity. Focuses on climate variability. Amelioration of crop microclimate to survive extreme weather events. Enhance biodiversity of native fauna and flora. Helps in crop diversification. Protect natural resources by structural and functional diversity.
Accumulates carbon in woody biomass.	Enhances risk bearing ability and stability under variable climate.

10.3 REDESIGNING CLIMATE RESILIENT AGROFORESTRY SYSTEMS

Redesigning agroforestry systems is a new avenue that offers climate resilience and diversification of farm income for achieving food security and livelihood. If transformed into innovative, existing monotonous cropping systems will reap multifaceted benefits (Giri et al., 2019). An example of such transformation can be a riparian buffer that is a natural barrier or buffer to pollutants and other anthropogenic activities from contaminating nearby streams. These buffers mitigate GHG emissions, enhance methane uptake and sequester carbon can be redesigned to provide CC services. Widening the existing zones with suitable species of trees will harvest more biomass, protect stream under increasing strange rainfall events and provide additional economic goods. The carbon-rich woody biomass converted to biochar and reused as soil amendment for improving soil quality. They too benefit beneficial insects, moderate local temperature, provide wildlife corridors and adapt to CC (Schoeneberger et al., 2012).

Another example is water breaks to address flood intensity and increasing frequency in the face of CC by strategically planting them in flood plains to mitigate flood events (Wallace et al., 2000). These agroforestry systems trap sediments from getting into crop fields; assure additional income from food, fodder, and timber. Therefore, redesigning agroforestry in a new planned fashion keeping in mind the existing situation and upcoming climate-related variability and long-term change can serve multiple objectives of food, feed, fiber, energy, and other ES presented in Table 10.2 and Figure 10.1.

10.4 CRITICAL CONSEQUENCES OF CLIMATE CHANGE (CC) ON LIVELIHOOD AND SUSTAINABILITY

Climate change (CC) has not only had a severe impact on people's quality of life by interfering with natural resources, economic activities, food security, and health, but it has also put a strain on the planet's ecosystem's long-term sustainability for future generations (Rockström et al., 2009). These ill effects are found to be more specific on rural livelihoods as compared to urban ones as the former are more dependable on natural resources (Podder & Samanta, 2019). Livelihood refers to the means of making a person's or family's living and livelihood is sustainable when it can maintain assets and resources for the present and the future and enable it to cope with, and recover from, external shocks such as CC impacts and other natural hazards (Scoones, 2009). As we know, livelihood plays a pertinent role in maintaining social and economic balances; whoever is the victim of CC, be

TABLE 10.2 Different Types of Agroforestry in India Under Different Situations

Types	Regions	References
	Traditional Agroforestry	
Alder-cardamom, soil conservation hedges, silvipasture, plantation-based.	North-east hill and highlands.	Handa et al. (2016)
Homestead, shelter belts, wind breaks, coconut based, bamboo-based.	Coastal belt	Nair (1989)
Khejri based, wind breaks, shelter belts, MPTS for fuel, fodder.	Arid and semi-arid.	Newaj et al. (2015)
Gmelina and *acacia*-based, homestead, plantations, multi-story garden, alley, and intercropping.	Humid and sub-humid.	Handa et al. (2016)
Poplar-based.	Indo-Gangetic plains.	Chaudhari et al. (2015)
	Commercial Agroforestry	
Pulpwood-based.	Punjab, Haryana, Uttar Pradesh, Andhra Pradesh, Gujarat, Tamil Nadu.	Handa et al. (2016)
Timber-based	Kerala, Maharashtra, Tamil Nadu, Madhya Pradesh.	Handa et al. (2016)
Willow-based	J&K, HP, Uttarakhand, Punjab.	Handa et al. (2016)
Mulberry and muga (*Antheraea assamensis*) based.	North-east hills.	Newaj et al. (2015)
Lac and gum-based.	Semi-arid and arid (Madhya Pradesh).	Newaj et al. (2015)

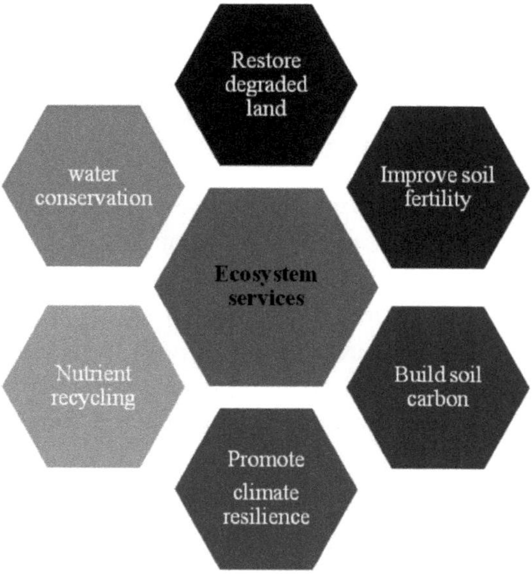

FIGURE 10.1 Various ecosystem services offered by agroforestry systems.

it rural or urban, but the social and economic balance will be affected as a whole, which will further deteriorate sustainability. Therefore, it's high time to discuss the cause, impact, and mitigation of CC effects through thorough surveys and analysis on this crucial topic. Before preparing for the mitigation, knowledge about the root cause and the Scenario of its ill effects across boundaries is inevitable.

10.4.1 WORLD SCENARIO

CC has not only had a severe impact on quality of life by interfering with natural resources, economic activities, food security, and health, but it has also put a strain on the planet's ecosystem's long-term sustainability for future generations. The Intergovernmental Panel on Climate Change (IPCC), which includes over 1,300 scientists from the United States and other countries, predicts a 2.5–10°F temperature rise over the next century (USGCRP, 2014, 2016). The IPCC, which consist of over 1,300 scientists from the United States and other countries, predicted a temperature rise of 2.5–10°F over the next century. As global temperatures increase, net annual costs will increase over time (IPCC, 2013). Taking into account the predictions, the situation will be worse if precautions and solutions do not find their way out to CC.

Changing forest cover has been driving the severe impacts of climate globally as deforestation and forestation are studied to have serious influences on rates of evapotranspiration (ET) and surface albedo (Prevedello et al., 2019). The thought is alarming that the present generation is making the global forest change fact unseen, and the consequence is a deadly impact on local CC (Duveiller et al., 2018). The strong influence of forests on local land surface temperature (LST) may cause them to lower surface albedo and higher ET (Li et al., 2014; Bright et al., 2017).

Few case studies on the arctic, methane emission, tropical forests, wildfires, North American forests and ocean carbon absorption by NASA (Rasmussen & Ramsayer, 2015) have concluded the impact of present land-use policies on the local climate of few countries. One of the most striking examples is in Brazil, where the impact of deforestation on the local environment was found to be greater than that of forestation (0.38°C versus 0.18°C, respectively), due to unrestrained deforestation versus forestation (Prevedello et al., 2019). A drastic change in the local climates is the reason for the migration of people from place to place over the globe in search of a livelihood to pursue, which is under threat and needs utmost attention (Table 10.3).

TABLE 10.3 Impact of Climate Change on Livelihoods of Several Continents Over the Globe and Future Opportunities

Country	Effect of Climate Change	Impact on Livelihoods	Opportunities	References
Africa	Challenges to marine diversity in Mozambique.	Sustainability of artisanal fisheries.	Understanding of fluctuations in marine fish stocks, the sustainable use of marine resources, and fisheries management.	Boon (2009); United Nations Human Development Program (2007)
Asia	Shrinkage of glaciers in Central Asia and downstream river runoff. Impact on future precipitation.	Deterioration of economy due to interference in agriculture.	Better water management strategies, i.e., water saving technologies in irrigation, building reservoirs, increasing water productivity, changing cropping systems, and water reuse.	Schlüter et al. (2010); Nakaegawa et al. (2013); Hijioka et al. (2014)
Australia	Widespread draft in southeast Australia (1997–2009). Floods in eastern Queensland in early 2011. Heat waves in 2014–2015.	Resulted in substantial economic loss. Floods destroyed hundreds of lives and more than 5 billion dollars of revenue. Caused damage to infrastructure, health, and agriculture.	Enhancing energy opportunities, great use of renewable energy, eco-friendly industrial approaches and enhanced carbon uptake in landscape could be promising.	Reisinger et al. (2014)
Antarctica	Western peninsula of Antarctica is being affected by fossils burning thousands of miles away. Sea ice appears latter but disappears faster. So much is changing so fast in Antarctica that it is almost impossible the scientists to predict.	No human settlement is prominent in western peninsula due to this faster climate change impacts. Even animals living there have been suffering and struggling for their existence. Adelie penguin population have almost shrieked by 90% which is a matter of concern.	As western peninsula is Antarctica's hotspot. It needs attention for at least its future and sustainability to be predicted.	Bromwich (2012)
Europe	Prolonged drought in central Europe where 34% of total land is under agricultural practices from 2018 to 2019 due to inexorable rise in GHG emission.	60% of the crops were affected over the region. Hence, affected the health as well as livelihood of the people.	Reduction in emission of GHGs could lower the risk of damaging dry periods. Planting more trees and agroforestry can be a mitigating strategy to be adopted.	Barredo (2010)

10.4.2 INDIAN SCENARIO

As per World Bank data of 2020 majority of the Indian population (65%) is living in rural areas and are also getting affected due to CC. Since CC has been seen to have its vicious circle of influence over the world. But it has more impact in a country like India, which is primarily a country with an agrarian economy and climate being the sole factor influencing agriculture in our country. The agriculture sector in India accounts for 18% of the country's GDP and employs 50% of the workforce. Along with agriculture, a reduction in forest productivity and a shift in the forest type boundaries along altitudinal and rainfall gradients have been found. On the other hand, marine and lake productivity are also being hampered by sea ice, rapid warming, and higher organic inputs while combined impacts of wildfire and insect outbreaks decrease forest productivity (Kumar et al., 2020a).

"CC is likely to mean monsoon systems affect larger areas over longer timescales, and rainfall during monsoon season is likely to intensify, becoming less predictable," according to several news articles on the effect of CC on the monsoon in India. The most significant effect, which is already being observed, is a hike in the year-to-year variability of monsoon strength and associated rainfall extremes (Nandi, 2019). Krishna Achuta Rao, an Indian climate scientist, said "The IPCC and other scientific literature has already highlighted that global monsoon systems will be impacted, monsoon rain will increase, so will monsoon variability"(Nandi, 2019). All the emerging uncertainties due to CC have catalyzed the problems of future food security within the country (Kumar et al., 2020a). The matter is of great concern that millions of people worldwide face the serious impact of CC over their livelihoods and income. Additionally, CC mostly targets agriculture and drives the predictability of undernourishment of the world population. Emphasis should be given to adapting and mitigating CC, but understanding and predictability regarding the impact and opportunities of CC are even more critical. Summarizing, a greater sense of awareness about CC should be unavoidable to address the significant challenges.

10.5 CONTEMPORARY AGRICULTURE: BOTH DRIVER AND SURVIVOR OF CLIMATE CHANGE (CC)

Climate is an important factor in agricultural growth. The primary role of agriculture in human well-being is concerned with the potential effects of

CC on agricultural production. It is essential to understand that the climate scenarios used in the study remain somewhat unpredictable. The designs are important for predicting potential opportunities but do not reflect current circumstances. Agriculture is a dynamic field with multiple driving criteria (environmental, economic, and social). The fact that the production of crops is exceptionally vulnerable to CC (McCarthy et al., 2001) has now been well understood with different regional consequences. The IPCC (Third Assessment Report) predicts a general decline in likely crop yields and decreased agricultural water supply and population water resources. Biophysical impacts are the primary drivers of agricultural reactions to CC. Crop development is biophysically influenced by changes in meteorological factors, including increased temperature and evolving precipitation regimes. One of the key problems is to grasp the effects from climate scenarios used as inputs. Uncertainties over the scale of planned improvements in all regions contribute to delays in the agricultural assessments. For example, in certain countries, rainfall forecasts may be positive or negative, which is the primary variable for agriculture, depending on the climate scenario adopted. The long-term average of regional temperature fluctuates naturally. For example, rainfall variability occurs in terms of timing and quantity, affecting agriculture each year. It is clear that changes have taken place in the past and will continue to occur, and CC is changing these patterns of variability, resulting in more droughts and floods, for instance. Nonetheless, there are several uncertainties, especially regarding rainfall scenarios for the future.

CO_2 is a plant photosynthesis factor and hence affects the development of biomass. The opening of plant stomata's is also regulated and thus affects the transpiration of plants. As a result, plants growing in theoretical terms would yield more biomass and use less water under increased CO_2 conditions. Greenhouse studies confirm that the plant's activity is nevertheless less apparent than theoretical due to the various associations of the physiological processes. Changes are much lower in field conditions. The majority of crop models used to measure CC include the impact of CO_2 increase on crop yield and water use, with the option to simulate (Rosenzweig & Iglesias, 1998). The crop model's outcome is difficult to verify, since there are only a few such studies around the world that create doubt regarding the simulated results. Elevated CO_2 could lead to enhance agricultural productivity too.

Atmospheric CO_2 levels can increase some crop's growth, rate of transpiration and water use efficiency (WUE). Depending on their photosynthetic pathways, plants are classified as C3, C4, or CAM. Extra CO_2 is likely to benefit C3 plants such as rice, potato, soybean, wheat, and vegetables, as well as most trees several studies' findings have revealed that increased CO_2 levels often benefit the majority of crops. The availability of plant nutrients, the type of crop, temperature, precipitation, and other environmental conditions are all known to have an impact on the response (IPCC, 1996). The majority of C4 plants are of tropical origin, and they include grasses and important crops like maize, millet, sorghum, and sugarcane. Less advantage from increasing CO_2 is anticipated for C4 plants. CAM plants are a subset of C4 plants, and it is unlikely that they will be harmed. Increased temperature can also produce beneficial effects. The expansion of the available growing season for plants and a decrease of the growing cycle for certain plants for maturation will be a significant consequence of increasing temperatures, particularly where low average temperature slow average temperatures restrict agricultural production currently limiting agricultural production. This favors high-altitude agriculture with rises in yields and the diversity of crops produced and high latitude regions where the pole ward change of agriculture's thermal limits increases productive potential. Soils and other influences will, however, not achieve any of this ability. In certain areas, increased rainfall may also encourage increased output and provide more irrigation water.

10.6 EMERGING AGROFORESTRY SYSTEMS AS CLIMATE-RESILIENT PRACTICE

Climate is highly variable, particularly in arid regions of continents; many observers have blamed recent land degradation on CC (Wessels et al., 2007). In fact, it is almost probable that predicted future CC would have a detrimental effect on agricultural resources across many regions of the continent (Beddington et al., 2012; DeSouza et al., 2012; Gonzalez et al., 2012; IPCC, 2012; Vermeulen et al., 2012). Agroforestry has gained popularity as a land-use strategy that can help address global CC while also providing other environmental, economic, and social benefits (Garrity, 2004; Jose, 2009; Jose & Bardhan, 2012; Jose et al., 2012; Smith et al., 2012; Waldron et al., 2017). The incorporation of trees on farms can improve farming systems while mitigating the effects of CC (Brown et al., 2018). Various studies indicate that agroforestry can help to prevent the degradation of the environment, enhance agricultural productivity, improve carbon sequestration, produce high-quality water, and support healthy soil and ecosystems, all while generating stable incomes and other human welfare benefits (Jose, 2009; Garrity et al., 2010; Kim et al., 2016; Miller et al., 2020).

Different agroforestry systems have emerged in different agro-ecological zones, depending on the climatic, environmental, economic, and socio-cultural niches where they occupy, and multiple systems with range of agroforestry systems have emerged in different agroecological zones. Many systems of various compositions can perform essentially the same functions for both livelihoods and landscapes. Multiple characteristics must be included in order to comprehend all of the primary distinguishing characteristics of agroforestry systems (Nair, 1993; Sinclair, 1999; Torquebiau, 2000; Chaudhari et al., 2014). In the majority of successful case studies of agroforestry establishment, tree-based systems outperform treeless alternatives in terms of productivity, sustainability, and responsivity to people's cultural or material demands. Agroforestry is not, however, being used everywhere. More knowledge is required about the productive and ecological performance of agroforestry systems, the political, social, and cultural conditions that must be realized for them to be established, and the trade-offs farmers must make when deciding which land-use practices to implement.

10.7 AGROFORESTRY: ECOSYSTEM SERVICES (ES) AND STRATEGIES FOR GHG EMISSION MITIGATION

In a multifunctional working landscape, agroforestry can be a practical land-use alternative that, in addition to reducing poverty, provides several ecosystem services (ES) (carbon sequestration, biodiversity conservation, soil enrichment, air, and water quality, and agroforestry esthetic values). Though evidence for many of these claims was lacking in the agroforestry literature until recently, it is now regarded as providing ES, environmental benefits, and economic commodities as part of a multifunctional working landscape (Giri et al., 2019). Schroth et al. (2004) compiled the first comprehensive synthesis of the significance of agroforestry systems in biodiversity conservation in tropical landscapes, using examples from a variety of countries. Schroth & Sinclair (2003) discussed how agroforestry practices can improve soil fertility. Montagnini (2006) used case studies from around the world to examine the carbon sequestration potential of agroforestry systems. Increasing net greenhouse gas (GHG) emissions that lead to global CC threaten ecosystems and people's livelihoods (IPCC, 2013). Atmospheric concentrations of the GHGs particularly CO_2, CH_4, and N_2O have increased since 1750, and by 2014, the concentrations of these GHGs had reached 398 ppm, 1,833 ppb, and 327 ppb, exceeding their pre-industrial levels by about

43%, 154%, and 21%, respectively (Kim et al., 2016; Table 10.4). Between 1880 and 2012, the global average temperature, including land and ocean surface temperatures, increased by about 0.85C. This temperature increase has led to changes in many extreme weather events observed since the 1950s (IPCC, 2013). Many natural systems are being impacted by CC, particularly temperature increases (IPCC, 2001), and people's livelihoods are being impacted by diminishing water resources, decreased crop productivity, and an increase in the frequency of flood and storm damage (IPCC, 2001).

TABLE 10.4 Observed Net Methane (CH_4) and Nitrous Oxide (N_2O) Emissions Under Various Agroforestry Systems

Types of Agroforestry Systems	CH_4*	N_2O**	$CH_4 + N_2O$***
Improved fallow	–3.2 ± 1.0 (4)	10.1 ± 5.9 (10)	2,901 ± 1,792
Intercropping	–	3.2 ± 0.4 (4)	–
Shaded perennial-crop systems	–1.3 ± 1.0 (3)	9.7 ± 7.8 (5)	2,847 ± 2,358
Shifting cultivation	–2.6	1.3	299
Tree plantations on arable land	–0.14	1.4	412
Riparian buffer	–0.28	9.1	2,702
Average	–1.6 ± 1.0	7.7 ± 3.3	2,240 ± 1,017

*kg CH_4 ha^{-1} y^{-1}; **kg N_2O ha^{-1} y^{-1}; ***kg CO_2 eq ha^{-1} y^{-1}.

Source: Modified from: Kim et al. (2016).

Kim et al. (2016) have compiled and analyzed various observations and have shown that carbon sequestration in biomass and the soil with changes in net emissions of CH_4 and N_2O emissions estimated that a shift to agroforestry could mitigate 27.2 ± 13.5 t CO_2 eq ha^{-1} y^{-1}, at least over the early growth stage (up to a maximum age of 14 years). There is a paucity of data on carbon sequestration rate and GHG emissions in various agroforestry systems, so useful research can be conducted using appropriate low-cost technologies such as biomass estimation over successive samplings. Instead of using destructive sampling, biomass carbon stocks can be estimated with globally or regionally available allometric equations (Chave et al., 2005; Návar, 2009). N_2O and CH_4 fluxes can be measured with manual chambers to identify control factors and approximately quantify emissions such as automatic chambers connected to a gas analyzer (e.g., gas chromatograph, photoacoustic spectroscopy, tunable) diode laser gas analyzer) and isotope methods.

10.8 AGRICULTURAL PRODUCTIVITY AND PROFITABILITY UNDER AGROFORESTRY SYSTEMS

Agroforestry interventions encourage farmers to take up several complementary practices to achieve different social-ecological objectives (Waldron et al., 2017). For example, the planting of tree species can be promoted simultaneously with crops and shrubs that provide benefits in the future. The first necessary step is successful farmer mobilization and engagement. The second step represents a given intervention, such as farmer capacity development or facilitating access to appropriate tree germplasm for significant and proper adoption of the promoted agroforestry practices and tree germplasm. Following such adoption, several intermediate outcomes are expected, such as improvement in soil health and other ES, including water filtration, increased crop productivity, or reduced production costs, and thus increased returns. Some intervention participants may discover that increased availability and usage of tree/shrub fodder resulted in increased milk and other livestock production and returns. Other agroforestry products like timber, firewood, and fruits, are expected to boost and aid in the diversification of food and income sources (Mbow et al., 2014a; Sharma et al., 2016; Waldron et al., 2017; Van Noordwijk et al., 2018). Together, such kind of intermediate outcomes is expected to interact to bolster household resilience to shocks and overall household income, food and nutritional security. These positive benefits and the broader context in which this stylized theory of change is embedded will affect further household investment in agroforestry (Table 10.5). For instance, it helps enhance livelihoods in arid areas of India where local knowledge has been acquired over a long period of time. Arid regions are especially notable for their use of ethnoforestry techniques and their indigenous understanding of how to grow farm-town trees. According to Tewari et al. (2001), agroforestry, a much-improved Jujube-based pasture system, can be a significant source of livelihood in many arid regions. This approach meets the communities demands for production in terms of food for people, feed for livestock, and energy for cooking.

10.9 EVIDENCE-BASED REMUNERATIVE AGROFORESTRY SYSTEMS IN IGP OF INDIA

Covering more than 15% India's total land area, IGP are fertile lands covering most of the parts of northern and eastern India. It was formed due to the carried away sedimentations of Himalaya by the Ganga and Indus rivers and

TABLE 10.5 Major Ecosystem Services and Environmental Benefits Associated with Different Types of Agroforestry Systems

Ecosystem Services	Agroforestry Systems	Improvements	References
Carbon sequestration	Fodder bank agroforestry; alley cropping system.	Increase in global potential to sequester 1.1–2.2 Pg of carbon (vegetation and soil) over 50 years; 3,91,000 Mg C y^{-1} by 2010 and 5,86,000 Mg C y^{-1} by 2040.	Dixon (1995); IPCC (2000); Nair et al. (2009)
Agroforestry for soil enrichment.	Alley-cropping system; N-fixing red alder (*Alnus rubra*) in a maize; alley cropping systems involving pecan and cotton; shade coffee agroforestry systems.	Higher soil organic matter and microbial biomass; improved soil aggregates stability, soil carbon, soil nitrogen, and soil enzyme activity; increased soil pH, CEC, Ca, and Mg.	Seiter et al. (1995); Lee & Jose (2003); Schroth & Sinclair (2003); Udawatta et al. (2008)
Biodiversity conservation	Multistrata Cacao (*Theobroma cacao*) agroforestry systems.	Habitat for species, preserve germplasm of sensitive species, reduce rates of conversion of natural habitat, connectivity by creating corridors between habitat remnants, conserve biological diversity by providing other ecosystem services such as erosion control and water recharge.	Kumar & Nair (2004); McNeely (2004); Schroth et al. (2004); Harvey et al. (2006); Kabir & Webb (2009)
Improved air and water quality.	Switch grass (*Panicum virgatum*) and woody stem (*Populus deltoides* X *nigra*). Silvopastoral systems.	limiting wind erosion and particulate matter in the air, reducing noise pollution, and mitigating odor from concentrated livestock operations; reduction in nitrate-N by 72% at a depth of 0.9 m.	Udawatta et al. (2002); Lee et al. (2003); Allen et al. (2004); Tyndall & Colletti (2007); Anderson et al. (2009)

their tributaries and IGP provides nearly 50% of total food grains to more than a quarter of the human population and provides shelter nearly a quarter livestock population (Pathak et al., 2014). The whole area has been classified into four meso-level regions viz., upper Gangetic plain (UGP), Middle Gangetic plain (MGP), lower Gangetic plain (LGP) and Trans Gangetic plain (TGP). LGP covers about 80% of West Bengal (WB) and Jharkhand, MGP covers eastern districts of Uttar Pradesh (UP) and northern parts of Bihar, UGP covers the central and western half of UP and TGP covers the states of Punjab, Haryana, Chandigarh, Delhi, and Ganganagar district of Rajasthan (Pathak et al., 2000, 2014). Due to its fertility status, the soils of IGP have been overexploited in terms of their natural diversity, including soil, further pre and post green revolution use of high yielding varieties (HYVs), fertilizers, pesticides, and other agrochemicals have reported to set their hazardous footprints on both soil and atmosphere in terms of emission of GHGs, global warming and declined soil health (Mishra et al., 2014, 2016). On the same context, the natural resources of IGP are under constant threat due to heavy biotic and demographic pressure (Pathak & Dagar, 2001). The agroforestry system could be considered one of the potential solutions to address the ever-increasing problems of the area. Various traditional and modern agroforestry systems have already been practiced in the region. Scattered trees on-farm (the practice of growing crops under scattered tress mostly in Punjab), home gardens (north-east India, West Bengal and Sikkim), trees on farm boundaries (mainly in Punjab and Haryana), Silvopastoral system (trees grown on field bunds and branches have mostly eloped for nutritious fodder purpose in Kandi areas of Punjab) are few examples of traditional agroforestry systems practiced in IGP (Pathak et al., 2006, 2014; Jeet et al., 2011; Kumar et al., 2019). Various agroforestry systems according to different regions of IGP have been planned (Table 10.6).

In the 1970s organized and systematic agroforestry research was started in IGP mainly through industrial plantations of *Populus deltoides*. But the momentum in the study was gained after the formation of a coordinating center for the All India Coordinated Research Project (AICRP) on Agroforestry (AICRP AF) at BCKV (Jhargram), NDUAT (Faizabad), RAU (Pusa), GBPUAT (Pantnagar), CSUAT (Kanpur), CCSAU (Hissar) and PAU (Ludhiana) in 1983. Initially, there were three core projects under AICRP AF viz., Diagnostic survey and appraisal of existing agroforestry practices, collection, and evaluation of promising multipurpose trees (MPTs) and management of agroforestry system (Pathak et al., 2014). During its early stages, AICRP AF diagnostic and design survey was organized by different

TABLE 10.6 Region Wise Allocations of Agroforestry Systems with Preferred Species for Plantation

Region	Agroforestry System	Preferred Species
Punjab lower hill zone.	Agri-silvi	• **Tree:** *Acacia nilotica, Dalbergia sissoo, Acacia catechu, Butea monosperma.* • **Crops:** Pearl millet, Sorghum, Sugarcane, Wheat.
	Farm boundary	• **Tree:** *Populus deltoides, Eucalyptus* spp. • **Crops**: Rice, wheat, cotton, potato.
Haryana	Agri-silvi	• **Tree:** *Populus deltoides*, Litchi, Mango. • **Crops:** Rice, wheat, potato, sugarcane.
	Farm boundary	• **Trees:** *Eucalyptus hybrid, Populus deltoids.*
	Agri-horti	• **Trees:** *Zizyphus* spp.
	Scattered trees	• **Trees**: *Prosopis cineraria, A. nilotica, Dalbergia sissoo.* • **Crops:** Pearl, millet, sunflower.
Western UP	Agri-silvi	• **Trees:** *Dalbergia sissoo, Acacia nilotica,* Mango, Neem. • **Crops:** Rice, wheat.
	Horti-silvi	• **Trees:** *Populus Deltoids, Dalbergia Sissoo, Acacia Nilotica, Morus Alba, Mangifera Indica, Psidium Guajava.* • **Crops:** Rice, wheat.
	Block plantation	• **Trees:** *Populus deltoides*, *Eucalyptus* spp.
	Farm boundary	• **Trees:** Poplars and *Eucalyptus* Spp. • **Crops:** Rice, wheat.
Central UP	Agri-silvi	• **Trees:** *Azadirachta indica, Dalbergia sissoo, Mangifera indica, Psidium guajava* • **Crops:** Rice, wheat.
	Farm boundary	• **Trees:** *Acacia nilotica, Prosopis juliflora*, *Eucalyptus* spp., *Derris indica, Acacia catechu, Bamboo* spp.

TABLE 10.6 *(Continued)*

Region	Agroforestry System	Preferred Species
Eastern UP	Farm boundary	• **Trees:** *Dalbergia sissoo, Eucalyptus* sp., mango. • **Crops:** Rice, wheat.
	Agri-horti-silvi	• **Trees:** *Dalbergia sissoo,* mango, guava. • **Crops:** Vegetables, wheat
	Orchards homestead	• **Trees:** *Madhuca latifolia, D. sissoo*, neem, bamboo.
Bihar (north-west districts)	Shelter belt	• **Trees:** *D. sissoo, Dendrocalamus strictus.* • **Crops:** Wheat, winter maize, kharif paddy
	Farm boundary	• **Trees:** *D. sissoo, Wendanlandia exerta.* • **Crops:** Wheat, gram, rajmah, castor.
	Agri-silvi	• **Trees:** *D. sissoo, Bombax ceiba, Tectona grandis, Zizyphus jujube.* • **Crops:** Fodder maize, local grass.
	Silvi-pastoral	• **Trees:** *D. sissoo, A. nilotica, M. indica.*
	Agri-horti-silvi	• **Trees:** *Morus alba, Terminalia arjuna, Litchi chinensis, Emblica officinalis.* • **Crops:** Ginger, turmeric.
	Homestead	• **Trees:** *Azadirachta indica, D. sissoo,* mango, jackfruit, litchi, guava.
	Homestead	• **Trees:** *Azadirachta indica, D. sissoo, Leucaena leucocephala,* mango, jackfruit, guava, bael.
West Bengal	Farm boundary	• **Trees:** *A. nilotica, Azadirachta indica, Terminalia arjuna, Butea monosperma, Zizyphus mauratiana.* • **Crops:** Rice, red gram, mustard, maize, sunhemp.

Source: Pathak et al. (2000, 2014); Planning Commission (2001).

centers, as mentioned earlier in this section. By generating important information, the survey identified suitable agroforestry practices for various regions in IGP. Like in LGP region *Eucalyptus hybrid >Acacia auriculiformis >Gmelina arborea >Acacia nilotica >Azadirachta indica*, in MGP region *Populus deltoides >Anthocephalus cadamba >Eucalyptus hybrid > Dalbergia sissoo >Acacia nilotica*, in UGP region *Populus deltoides >Eucalyptus hybrid >Dalbergia sissoo >Anthocephalus cadamba >Leucaena leucocephala* and in TGP region *Populus deltoides >Eucalyptus hybrid >Dalbergia sissoo>Melia azedarach>Leucaena leucocephala* MPTs are grown based on priority.

Due to its compatible nature with various crops, poplar (*Populus deltoides*) has been grown in the states of Punjab, Haryana, Uttarakhand, and UP in over 0.03-million-hectare land (Gandhi & Dhiman, 2010). Eucalyptus-based agroforestry system is mostly seen in the former four states for plywood and timber catering to industries mainly grown on the farm bunds. In IGPs agroforestry system has played an important role in – (a) rehabilitation of degraded lands due to salinization, wind, and water erosions (Dagar, 2012); (b) rehabilitation of salt-affected and water logged soils (Indian Farming Forestry Development Cooperative Limited, UP); (c) in biofuel and bioenergy generation from tree borne oil seed (TBOs) (Planning Commission, 2004); and (d) in carbon sequestration and mitigating CC. Agroforestry s been seen to have a significant effect on the income and employment generation of the farmers of IGP by generating more employment opportunities over the years. Few remunerative agroforestry systems have been recommended for IGP in terms of cost is to benefit ratio (B:C ratio) of various agroforestry systems in different durations ranging from 7–20 years of continuous use practice under various regions as shown in Table 10.7.

10.10 AGROFORESTRY SYSTEMS FOR ADAPTATION AND MITIGATION

Agroforestry systems include a collection of soil management practices, including home gardens, crop rotation systems for land conservation, crop diversification, border plantings, perennial crops, live structures, improved fallows, or combined strata agroforestry (Neupane & Thapa, 2001; Stigter et al., 2002; Albrecht & Kandji, 2003; Thangata & Alavalapati, 2003; Oke & Odebiyi, 2007; Rice, 2008). The two sides of the coin for agroforestry systems are presented in Figure 10.2.

TABLE 10.7 Remunerative Agroforestry Systems Recommended for Indo-Gangetic Plains

Zones	Type of Land	Agroforestry System		Duration of Practice (Years)	Benefit: Cost Ratio at 15% Discount Factor
		Tree	Crop		
Lower Gangetic plain	Rainfed	*Acacia auriculiformis*	Jute (*Corchorus olitorius*)	12–15	1.2
Middle Gangetic plain	Rainfed	*Dalbergia sissoo*	Sesamum (*Sesamum indicum*)	15–20	1.2
Middle Gangetic plain	Non-arable land	Bamboo (*Bambusa vulgaris*)	Marvel grass (*Dichanthium annulatum*)	8–10	1.8
Upper Gangetic plain	Irrigated	*Populus deltoides*	Wheat (*Triticum aestivum*)	7–10	3.3
Trans Gangetic plain	Rainfed	*Acacia nilotica*	Wheat (*T. aestivum*, guar (*Cyamopsis tetragonoloba*)	15–20	2.6

Source: Planning Commission (2001); Pathak et al. (2014).

	MITIGATION POSITIVE	MITIGATION NEGATIVE
ADAPTATION POSITIVE	Soil carbon sequestration, improved water holding capacities, use of manure instead, mixed agroforestry for commercial products, income diversification with trees, reduced nitrogen fertilizer, fire management	Dependence on biomass energy, overuse of ecosystem services, Increased use of mineral fertilizers Poor management of nitrogen and manure, over extraction of non-timber products, timber extraction
ADAPTATION NEGATIVE	Integral protection of forest reserves, limited rights to agroforestry trees, ForestPlantation excluding harvest	Use of forest fires for pastoral and land management, tree exclusion in farming lands,

FIGURE 10.2 Positive and negative implications of agroforestry practices for adoption or mitigation of climate change.

Source: Adapted from: Mbow et al. (2014b).

According to Verchot et al. (2006), agroforestry systems have the following specific benefits for mitigating and adapting to CC threats in the coming years: Creating microclimatic conditions which enhance extensive flora and fauna.

- Their deep root systems can explore a larger soil volume for nutrition and water, which will aid during droughts.
- Increased soil total porosity, reduced run-off, and increased land cover lead to increased water entry into soil and retention in the soil profile, reducing moisture scarcity during low rainfall years.
- Agroforestry products can contribute to the economic development of a large number of poor farmers by enhancing food security and eliminating poverty.

10.10.1 ECOSYSTEM RESILIENCE THROUGH ADOPTION OF AGROFORESTRY SYSTEMS

Through microclimate buffering and water flow regulation, well-managed agroforestry can play a significant role in enhancing resilience to uncertain climates (Nguyen et al., 2013). Management choices in agroforestry include tree pruning and steps to lessen belowground competition, especially for water (Jackson et al., 2000); for example, trees tap into deep groundwater instead of high soil moisture, which annual plants rely on. Attention is paid to the effect of agroforestry on management and ecosystem functions that are favorable. Agroforestry will help preserve and protect natural resources by, for instance, mitigating non-point source (NPS) pollution, reducing soil erosion, preserve, and protecting natural resources by mitigating NPS pollution, reducing soil erosion, and creating wildlife habitat (Molua, 2005). It eases answers to changes in conditions while, at precisely the exact same time restoring or keeping water and soil sources (Powell et al., 1996; Du-Toit et al., 2004; Molua, 2005). As trees can buffer conditions that influence crop development, advancement through agroforestry has a significant effect. Specifically, the effects of trees may buffer atmospheric and temperature equilibriums, deficit-reducing vulnerability to temperatures below which developmental and physiological processes and yield become vulnerable (Lott et al., 2009). Scattered trees in agroforestry farms can improve the growth by lowering air incident solar radiation and soil temperature while enhancing WUE and, water status, gas exchange (Bayala et al., 2008). These claims derive from a couple of examples and have to be substantiated in study. The contribution of agroforestry to ecosystem functions in water recycling by rainfall-enhanced utilization in contrast to cropping systems. Lott et al. (2003) reported that about 25% of the water generated by trees could be utilized throughout the rainy season, suggesting they are in a position to use off-season rain (containing 15–20% of the overall yearly rainfall) and residual soil water following the cropping period, with the remainder being lost by evaporation (40%) or heavy drainage (33–40%). This complementarity between crops and trees extends the chances of soil moisture uptake making land resource use more efficient than in monoculture that is pristine (Lehmann et al., 1998; Jackson et al., 2000).

10.10.2 CARBON SEQUESTRATION THROUGH ADOPTION OF AGROFORESTRY SYSTEMS

Agroforestry has a specific part to play in the reduction of the atmospheric build-up of GHGs (IPCC, 2000). Agroforestry had the greatest potential for sequestration of carbon of any land use (LU) evaluated by the IPCC's Land-Use Change and Forestry report. Agroforestry has a possibility, since there is, although not as it's the land-use practices with the carbon density. Improved agroforestry systems that decrease the vulnerability of small farmers and assist them to accommodate that decrease the vulnerability of small farmers and assist them in accommodating changing conditions frequently meet the requirements for a qualified afforestation/reforestation action from clean development mechanism (CDM). These programs may be encouraged to fulfill the needs of environmental and social advantages of environmental and social advantages produced by CDM projects and to make synergies between adaptation and mitigation. Carbon stored in soil and biomass and sequestered by trees reduces GHG concentrations in the air. Estimates of agroforestry's carbon sequestration potential range from 100 MT CO_2 per year by 2030 to over 2000 MT CO_2 per year over 30 years. Irrespective of the amount, methods tend to sequester amounts of carbon with no trees compared to agroforestry systems. The carbon stock differed with the region and the agroforestry system within the region. The carbon stocks were 17–36 Mg C ha^{-1} in home gardens of Kerala (Kumar & Nair, 2011). Other studies have proved that agroforestry system mitigate GHG emissions (Sathaye & Ravindranath, 1998; Verchot et al., 2006; Arora et al., 2013; Goswami et al., 2013; Kanime et al., 2013; Verma et al., 2014). Recent research projected the carbon sequestration potential of tropical agroforestry systems is projected between 12 and 228 Mg ha^{-1} using an average value of 95 Mg ha^{-1} (Pandey, 2007). Hence, based on international estimates of this region acceptable for agroforestry (585–1,215 × 10^6 ha), 1.1–1.2 Pg C can be saved in the terrestrial ecosystems for the next 50 years (Albrecht & Kandji, 2003).

In India, evidence is emerging that agroforestry systems are potential land-use systems (LUS) for increasing and conserving aboveground and soil carbon stocks in order to mitigate CC. Agroforestry's average potential has been estimated to be 25 t C ha^{-1} over 96 M ha (Sathaye & Ravindranath, 1998). As a result, India's total carbon-storage potential is approximately 2,400 million tons. Singh & Pandey (2011) characterized the agroforestry for carbon sequestration as appealing as: (i) it sequesters carbon in plants and soils based upon the pre-production soil carbon; (ii) the intensive utilization of the land for agricultural production lowers the demand to get slash and burn or shifting cultivation; (iii) the timber products generated under agroforestry function as a replacement for comparable goods unsustainably harvested in the natural woods; (iv) to the extent which agroforestry raises earnings of the farmers, it decreases the incentive for more extraction in the natural forest (NF) for earnings improvement and finally; and (v) agroforestry practices could have double reduction benefits as fodder species using high nutritive value can help intensify diets of methane-producing ruminants while they are also able to sequester CO_2. Many studies have been conducted in India to determine the total carbon storage capacity of various agroforestry systems (Table 10.8).

TABLE 10.8 Total Carbon Storage Under Agroforestry Systems in Different Regions of the Country

Regions	Agroforestry Systems	C Storage (t C ha–1)
Semi-arid region	Silvipastoral system (age five years).	9.5–17.0
	Acacia nilotica + natural pasture.	
	A. nilotica + established pasture.	19.7
	Dalbergia sissoo + natural pasture.	12.4
	D. sissoo + established pasture.	17.2
	Hardwickia binata + natural pasture.	16.2
	H. binata + established pasture.	17.0
North-western India	Silvipastoral system (age six years).	6.8–18.55
	Acacia/Dalbergia/Prosopis + *Desmostacya* *Acacia/Dalbergia/Prosopis* +*Sporobolus*.	1.5–12.32
Central India	Block plantations (age six years).	24.12–31.12
	Gmelina arborea.	
Arid region (Rajasthan)	Agrisilvicultural system (age eight years) *Emblica officinalis* + *Vigna radiata*.	12.7–13.0
	Hardwickia binata + *Vigna radiata*.	8.6–8.8
	Colophospermum mopane + *Vigna radiata*.	4.7–5.3
Semi-arid region	Agrisilvicultural system (age 11 years).	26.02
	Dalbergia sissoo + crop.	
	Agrisilvicultural system (age five years).	34.77
	Albizia procera + black gram-mustard.	
	Albizia procera + green gram-wheat.	35.13
North-Western Himalaya	Silvipastoral system.	2.17
	Agrihortipastoral.	1.15
	Hortipastoral.	1.08

Source: Adapted from: Newaj et al. (2014).

10.11 STATUS OF AGROFORESTRY AFTER NATIONAL AGROFORESTRY POLICY IMPLEMENTATION (2014)

"Agroforestry" is defined by Lundgren & Raintree (1982) as "an integration of LUS and technologies in which woody perennials (trees, shrubs, palms, bamboos, etc.) are deliberately used on the same piece of land management units as crops and animals, and it has been practiced since ancient times in some form of spatial arrangement or temporal sequences, and there are both ecological and economic interactions between the different components." In India, it has been practiced for thousands of years (Puri & Nair, 2004) for the benefit of different interdependent components, i.e., trees, crops, food, fodder, fertilizer, and fiber (FSI, 2013). Agroforestry has been depicted by a number of studies in various parts of the country to be more profitable to farmers than agriculture or forestry for a specific area of land. A number of studies in various parts of the country found that farming was more profitable for farmers than agriculture or forestry for a specific land area. Agroforestry has numerous advantages, and it is well known that the IPCC lists agroforestry as one of the tools for combating CC. Agroforestry systems have the potential to be an effective tool in the mitigation and adaptation to CC (Chavan et al., 2014).

Today, the increasing population of India and increasing demand for food, feed, and fodder put pressure on the farming system, leading to the degradation of natural resources and CC (Dhyani et al., 2013). To address the problem of increasing population, a sustainable management system that is capable of food production from marginal agricultural lands while also maintaining and increasing the quality of the producing environment must be devised (Dobriyal, 2014). Agroforestry is an economical and ecologically viable option because it works simultaneously by offering large-scale diversification in agriculture. It provides fuel food, fodder, and fibers on one hand and stabilizes the ecosystems by increasing tree cover and reducing pressure on the forests on the other hand (Bijalwan, 2014). Agroforestry systems in India encompass both traditional and modern LU systems and have great potential and use in many aspects like providing employment to rural and urban populations through production, processing employment to rural and urban populations through production, processing, and value addition. Agroforestry systems have the potential to fulfill the 65% of the country's timber demand by producing 100 million cubic meters of timber/pulpwood for domestic and industrial use (Chavan et al., 2015), 70–80% of plywood demand and 60% of the raw material demand for paper and pulp.

Despite Agroforestry's massive potential in India and its multiple benefits, the adoption rate is still low; the majority of farmers face challenges and have been hesitant to adopt these systems on large, primarily because of certain apprehensions about the tree component such as long rotation, lack of market infrastructure, reduction in gross area and complicated legal procedures involved in tree farming trade and market fluctuations. In 2014, the Government of India removed these challenges by adopting the Agroforestry policy of the country, which is expected to increase farm productivity. It improves the livelihood of farmers by adopting National Agroforestry Policy (2014). India is the first nation in the world to adopt a comprehensive policy for agroforestry, paving the way for agroforestry to be used to transform the lives of rural farming populations, protect ecosystems, and ensure food security through sustainable means. Adoption of the new policy shows a significant shift in opinion towards agroforestry because earlier, agroforestry was considered a part of forestry, but now the policy approach views it as part of agriculture. National Agroforestry Policy (2014) eliminates most of the challenges of agroforestry, i.e., shortage of good planting material and seeds, insufficient research on agroforestry models, lack of marketing infrastructure and cumbersome legislation process (Planning Commission, 2001; National Agroforestry Policy, 2014) but the primary challenge is to transform the policy from paper to ground level (Figure 10.3). Instead of having a significant impact on the Indian agroforestry system, there are still

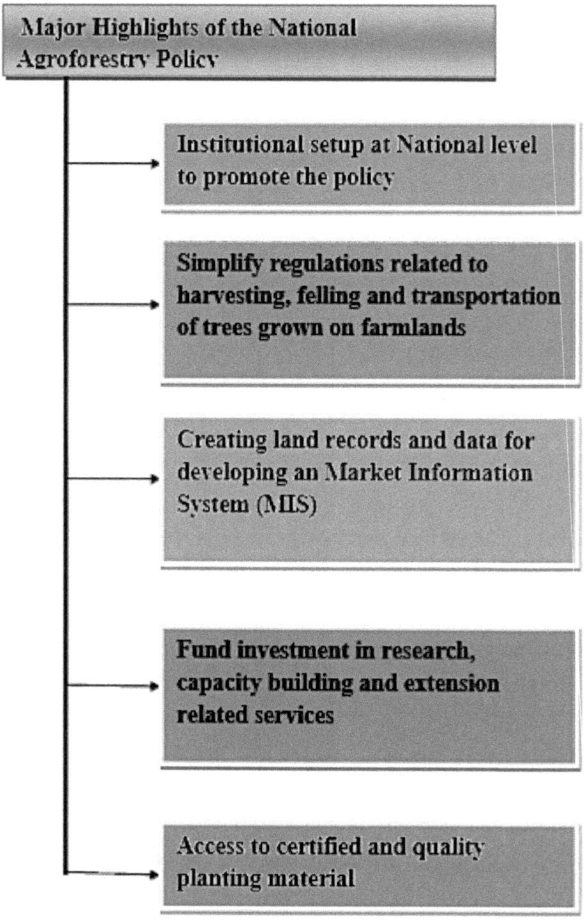

FIGURE 10.3 Major highlights of National Agroforestry Policy (2014).

some gaps, such as a lack of data on the selected species of trees (absence of agroforestry tree manual for farming community); no proper mechanism for the certification process of nurseries and propagating material; no proposal for blending of biofuel in diesel and the establishment of a dedicated agroforestry extension system; and less emphasis given to unique and high-tech agroforestry systems, such as aqua forestry (Kumar et al., 2017).

10.12 CONCLUSION AND FUTURE THRUST

Agroforestry systems have multifaceted benefits in diversifying the system, enhancing climate resilience and increasing the 'farmers' income. Agro-ecology-specific rotation years, temporal variations and synergy of the crops are crucial for securing higher returns in agroforestry models. Vertical stratification of soil carbon stock under agroforestry is one of the most viable options for long-term carbon sequestration and mitigating GHG emissions. Extensive studies recommended poplar-wheat-vegetables, poplar-maize, poplar-fruit crops and poplar-turmeric cropping systems in the salt-affected ecologies. These are becoming popular and are extensively being adopted by the farmers of Trans-IGP.

Similarly, it is also established that combining Eucalyptus-based agroforestry system with sugarcane, wheat, sorghum, and shade-tolerant crops like turmeric helps farmers to build climate resilience and a sustainable way to secure their livelihood during extreme weathers. In the Trans-Indo-Gangetic Plains, the integration of Poplar and Eucalyptus in agroforestry systems has been observed to significantly improve soil carbon sequestration. This enhancement in soil organic carbon (SOC) also serves as a protective measure against the rise of salt in salt-affected lands. For future research, it is vital to pinpoint tree species that are specifically suited to local conditions, factoring in both the needs of the community and potential commercial benefits. This approach would aid in diversifying sources of income for farmers, leading to more sustainable and economically viable agricultural practices. Policymakers and government agencies need to encourage the provision of incentives for creating esthetic values and ES by adopting agroforestry systems. Periodic evaluation and refinement of the National Agroforestry Policy (2014) needs to be adapted to better understand the impacts. The development of a site-specific agroforestry success model will encourage other farmers in the area to adopt agroforestry systems.

KEYWORDS

- **agroforestry**
- **agroforestry models**
- **carbon sequestration**
- **climate change**
- **Indo-Gangetic plains**
- **policy makers**
- **soil fertility**

REFERENCES

Albrecht, A., & Kandji, S. T., (2003). Carbon sequestration in tropical agroforestry systems. *Agriculture, Ecosystems & Environment, 99*, 15–27.

Allen, S., Jose, S., Nair, P. K. R., Brecke, B. J., Nkedi-Kizza, P., & Ramsey, C. L., (2004). Safety net role of tree roots: Experimental evidence from an alley cropping system. *For. Ecol. Manage., 192*, 395–407.

Anderson, S. H., Udawatta, R. P., Seobi, T., & Garrett, H. E., (2009). Soil water content and infiltration in agroforestry buffer strips. *Agrofor. Syst., 75*, 5–16.

Arora, G., Chaturvedi, S., Kaushal, R., Nain, A., Tewari, S., Alam, N. M., & Chaturvedi, O. P., (2013). Growth, biomass, carbon stocks and sequestration in age series *Populus deltoides* plantations in Tarai region of central Himalaya. *Turk. J. Agric. For., 38*, 550–560.

Arya, R., Mishra, A. K., & Chaudhry, S., (2018). Variation in soil properties and carbon stocks under roadside plantation and rice-wheat cropping system in North Western Haryana, India. *Int. J. Curr. Microbiol. App. Sci., 7*(4), 1939–1949. https://doi.org/10.20546/ijcmas.2018.704.222.

Barredo, J. I., (2010). No upward trend in normalized windstorm losses in Europe: 1970–2008. *Natural Hazards and Earth System Sciences, 10*, 97–104.

Bayala, J., Heng, L. K., Noordwijk, M. V., & Ouedraogo, S. J., (2008). Hydraulic redistribution study in two native tree species of agroforestry parklands of West African dry savannahs. *Acta Oecolo.* doi: 10.1016/j.actao.06.2008010.

Beddington, J. R., Asaduzzaman, M., Clark, M. E., Bremauntz, A. F., Guillou, M. D., Howlett, D. J. B., Jahn, M. M., Lin, E., Mamo, T., Negra, C., et al., (2012). What next for agriculture after durban? *Science, 335*, 289–290.

Bijalwan, A., (2014). Vegetation status of agroforestry systems in their District of Garhwal Himalaya, India. *Asian Journal of Science and Technology, 4*(12), 11–14.

Boon, E. K., (2009). *Area Studies (Regional Sustainable Development Review): Africa* (Vol. I, p. 335). EOLSS Publication, Oxford, UK.

Bright, R. M., Davin, E., O'Halloran, T., Pongratz, J., Zhao, K., & Cescatti, A., (2017). Local temperature response to land cover and management change driven by non-radiative processes. *Nat. Clim. Change, 7*, 296–302.

Bromwich, D. H., et al., (2012). Tropospheric clouds in Antarctica. *Rev. Geophys., 50*, RG1004. doi: 10.1029/2011RG000363.

Brown, S. E., Miller, D. C., Ordonez, P. J., & Baylis, K., (2018). Evidence for the impacts of agroforestry on agricultural productivity, ecosystem services, and human well-being in high-income countries: A systematic map protocol. *Environmental Evidence, 7*(1), 24.

Chaudhari, S. K., Kumar, P., Mishra, A. K., Singh, K., Rai, P., Singh, R., & Sharma, D. K., (2015). Labile carbon fractions buildup and dynamics under vertical stratification of *Populus deltoides* and *Eucalyptus tereticornis* based agroforestry systems in trans-Gangetic plains of India. *Ann Agric Res., 36*(1), 1–9.

Chaudhari, S. K., Kumar, P., Singh, K., Mishra, A. K., Rai, P., Singh, R., & Sharma, D. K., (2014). Aggregate fractions and organic carbon dynamics in partially reclaimed sodic soils growing *Eucalyptus tereticornis*. *Journal of Soil Salinity and Water Quality, 6*(2), 96–100.

Chavan, S. B., Keerthika, A., Dhyani, S. K., Handa, A. K., Newaj, R., & Rajarajan, K., (2015). National agroforestry policy in India: A low hanging fruit. *Current Science, 108*(10), 1826–1834.

Chavan, S., Newaj, R., Keerthika, A., Ram, A., Jha, A., & Kumar, A., (2014). Agroforestry for adaptation and mitigation of climate change. *Popular Kheti, 2*(3), 214–220.

Chave, J., Andalo, C., Brown, S., Cairns, M. A., Chambers, J. Q., Eamus, D., Fölster, H., et al., (2005). Tree allometry and improved estimation of carbon stocks and balance in tropical forests. *Oecologia, 145*, 87–99.

Dagar, J. C., (2012). Utilization of degraded lands/habitats and poor-quality water for livelihood security and mitigating climate change. *Indian J. Agrofor., 14*(1), 1–16.

DeSouza, H. N., DeGoede, R. G. M., Brussaard, L., Cardoso, I. M., Duarte, E. M. G., Fernandes, R. B. A., Gomes, L. C., & Pulleman, M. M., (2012). Protective shade, tree diversity and soil properties in coffee agroforestry systems in the Atlantic Rainforest biome. *Agric. Ecosyst. Environ., 146*, 179–196.

Dhyani, S. K., (2014). Agroforestry research in India: Experience of hundred years with focus on organized national program. In: *Proceedings World Agroforestry Congress*. New Delhi.

Dhyani, S. K., Handa, A. K., & Uma, (2013). Area under agroforestry in India: An assessment for present status and future perspective. *Indian J. Agroforestry, 15*(1), 1–11.

Dixon, R. K., (1995). Agroforestry system: Sources or sinks of greenhouse gases? *Agrofor. Syst., 31*, 99–116.

Dobriyal, M. J. R., (2014). Agroforestry practices for non-wood forest products and rural development. In: Raj, A. J., & Lal, S. B., (eds.), *Agroforestry: Theory and Practices* (p. 540). Scientific Publishers, India.

Du-Toit, J. T., Walker, B. H., & Campbell, B. M., (2004). Conserving tropical nature: Current challenges for ecologists. *Trends Ecol. Evol., 19*, 12–17.

Duveiller, G., Forzieri, G., Robertson, E., Li, W., Georgievski, G., Lawrence, P., Wiltshire, A., et al., (2018). Biophysics and vegetation cover change: A process-based evaluation framework for confronting land surface models with satellite observations. *Earth Syst. Sci. Data, 10*, 1265–1279.

FSI, (2013). *India State of Forest Report*. Forest Survey of India, Ministry of Environment and Forests, Dehra Dun, India.

Gandhi, J. N., & Dhiman, R. C., (2010). Performance of wheat intercropped under different tree spacings of poplar (*Populus deltoides* Marsh) plantations. *Indian J. Agrofor., 12*(2), 28–31.

Garrity, D. P., (2004). Agroforestry and the achievement of the millennium development goals. *Agrofor. Syst., 61*, 5–17.

Garrity, D. P., Akinnifesi, F. K., Ajayi, O. C., Weldesemayat, S. G., Mowo, J. G., Kalinganire, A., et al., (2010). Evergreen Agriculture: A robust approach to sustainable food security in Africa. *Food Security, 2*, 197–214.

Giri, A., Kumar, G., Arya, R., Mishra, S., & Mishra, A. K., (2019). Carbon sequestration in *Populus deltoides* based agroforestry system in Northern India. *International Journal of Chemical Studies, 7*(1), 2184–2188.

Gonzalez, P., Tucker, C. J., & Sy, H., (2012). Tree density and species decline in the African Sahel attributable to climate. *J. Arid Environ., 78*, 55–64.

Goswami, S., Verma, K. S., & Kaushal, R., (2013). Biomass and carbon sequestration in different agroforestry systems of a Western Himalayan watershed. *Biol. Agric. Hortic., 30*(2), 88–96.

Handa, A. K., Toky, O. P., Dhyani, S. K., Chavan, S. B., & Toky, I. D. O., (2016). Innovative agroforestry for livelihood security in India. *World Agriculture, 7*, 7–16.

Harvey, C. A., Gonzales, J. G., & Somarriba, E., (2006). Dung beetle and terrestrial mammal diversity in forest, indigenous agroforestry systems and plantain monocultures in Talamanca, Costa Rica. *Biodiver. Conserv., 15*, 555–585.

Hijioka, Y., Lin, E., Pereira, J. J., Corlett, R. T., Cui, X., Insarov, G. E., Lasco, R. D., et al., (2014). Asia. In: Barros, V. R., Field, C. B., Dokken, D. J., Mastrandrea, M. D., Mach, K. J., Bilir, T. E., Chatterjee, M., et al., (eds.), *Climate Change 2014: Impacts, Adaptation, and Vulnerability* (pp. 1327–1370). Part B: Regional aspects. Contribution of working group II to the fifth assessment report of the intergovernmental panel on climate change, Cambridge University Press, Cambridge, United Kingdom and New York, NY, USA.

IPCC, (1996). *Climate Change 1995: Impacts, Adaptations, and Mitigation of Climate Change: Scientific-Technical Analyses*. Contribution of Working Group II to the Second Assessment Report of the Intergovernmental Panel on Climate Change.

IPCC, (2000). *Land Use, Land-Use Change, and Forestry* (p. 375). A special report of the IPCC, Cambridge University Press, Cambridge.

IPCC, (2001). *Climate Change 2001: Impacts, Adaptation and Vulnerability*. Contribution of Working Group II to the Third Assessment Report of the Intergovernmental Panel on Climate Change., Cambridge University Press, Cambridge.

IPCC, (2012). Summary for policy makers: Special report on managing the risks of extreme events and disasters to advance climate change adaptation. In: Barros, C. B., Stocker, T. F., Qin, D., Dokken, D. J., Ebi, K. L., Mastrandrea, M. D., Mach, K. J., Plattner, G. K., Allen, S. K., et al., (eds.), *A Special Report of Working Groups I and II of the Intergovernmental Panel on Climate Change*. Cambridge, IPCC.

IPCC, (2013). Summary for policymakers. In: Stocker, T. F., Qin, D., Plattner, G. K., Tignor, M., Allen, S. K., Boschung, J., Nauels, A., et al., (eds.), *Climate Change 2013: The Physical Science Basis; Contribution of Working Group I to the Fifth Assessment Report of the Intergovernmental Panel on Climate Change*. Cambridge University Press, Cambridge, United Kingdom and New York, NY, USA.

Jackson, N. A., Wallace, J. S., & Ong, C. K., (2000). Tree pruning as a means of controlling water use in an agroforestry system in Kenya. *Forest Ecol. Manage., 126*, 133–148.

Jeet, R., Dagar, J. C., Lal, K., Singh, G., Toky, O. P., Tanwar, R. S., Dar, S. R., & Kumar, M., (2011). Bio-drainage to combat water logging, increase farm productivity and sequester carbon in canal command area of North-West India. *Curr. Sci., 100*(11), 1673–1680.

Jose, S., & Bardhan, S., (2012). Agroforestry for biomass production and carbon sequestration: An overview. *Agrofor. Syst., 86:* 105–111.

Jose, S., (2009). Agroforestry for ecosystem services and environmental benefits: An overview. *Agroforestry Systems, 76*, 1–10.

Jose, S., Gold, M. A., & Garrett, H. E., (2012). The future of temperate agroforestry in the United States. In: Nair, P. K. R., & Garrity, D., (eds.), *Agroforestry-the Future of Global Land Use* (pp. 217–245). Dordrecht, Springer Netherlands.

Kabir, E. M., & Webb, E. L., (2009). Can homegardens conserve biodiversity in Bangladesh? *Biotropica, 40*, 95–103.

Kanime, N., Kaushal, R., Tewari, S. K., Raverkar, K. P., Chaturvedi, S., & Chaturvedi, O. P., (2013). Biomass production and carbon sequestration in different tree-based systems of Central Himalayan Tarai region. *Forests, Trees, and Livelihoods, 22*(1), 38–50.

Kim, D. G., Kirschbaum, M. U., & Beedy, T. L., (2016). Carbon sequestration and net emissions of CH_4 and N_2O under agroforestry: Synthesizing available data and suggestions for future studies. *Agriculture, Ecosystems & Environment, 226*, 65–78.

Kumar, B. M., & Nair, P. K. R., (2004). Tropical homegardens: A time-tested example of sustainable agroforestry. *Advances in Agroforestry* (Vol. 3). Springer, Dordrecht.

Kumar, B. M., & Nair, P. K. R., (2011). Carbon sequestration potential of agroforestry systems: opportunities and challenges. *Advances in Agroforestry* (p. 307). Springer Science, the Netherlands.

Kumar, G., Kumari, R., Kishore, B. S. P. C., Saikia, P., Kumar, A., & Khan, M. L., (2020a). Climate change impacts and implications: An Indian perspective. In: Roy, N., Roychoudhury, S., Nautiyal, S., Agarwal, S., & Baksi, S., (eds.), *Socio-economic and Eco-biological Dimensions in Resource use and Conservation; Environmental Science and Engineering*. Springer, Cham. https://doi.org/10.1007/978-3-030-32463-6_2.

Kumar, P., Chaudhari, S. K., Mishra, A. K., Singh, K., Rai, P., Singh, R., & Sharma, D. K., (2014). Labile carbon dynamics and soil amelioration in six-year-old *Eucalyptus tereticornis* plantation in sodic soils. *Journal of Soil Salinity and Water Quality, 6*(2), 91–95.

Kumar, P., Mishra, A. K., Chaudhari, S. K., Basak, N., Rai, P., Singh, K., Singh, R., et al., (2018). Carbon pools and nutrient dynamics under eucalyptus-based agroforestry system in semi-arid region of North-West India. *Journal of the Indian Society of Soil Science, 66*(2), 188–199.

Kumar, P., Mishra, A. K., Chaudhari, S. K., Sharma, D. K., Rai, A. K., Singh, K., Rai, P., & Singh, R., (2020b). Carbon sequestration and soil carbon build-up under eucalyptus plantation in semi-arid regions of North-West India. *Journal of Sustainable Forestry*. https://doi.org/10.1080/10549811.2020.1749856.

Kumar, P., Mishra, A. K., Chaudhari, S. K., Singh, R., Singh, K., Rai, P., Pandey, C. B., & Sharma, D. K., (2016). Biomass estimation and carbon sequestration in *Populus deltoides* plantations in India. *Journal of Soil Salinity and Water Quality, 8*(1), 25–29.

Kumar, P., Mishra, A. K., Kumar, M., Chaudhari, S. K., Singh, R., Singh, K., Rai, P., & Sharma, D. K., (2019). Biomass production and carbon sequestration of *Eucalyptus tereticornis* plantation in reclaimed sodic soils of North-West India. *Indian Journal of Agricultural Sciences, 89*(7), 1091–1095.

Kumar, Y., Thakur, T. K., & Thakur, A., (2017). Socio-cultural paradigm of agroforestry in India. *Int. J. Curr. Microbiol. App. Sci., 6*(6), 1371–1377.

Lee, K. H., & Jose, S., (2003). Soil respiration and microbial biomass in a pecan–cotton alley cropping system in southern USA. *Agrofor. Syst., 58*, 45–54.

Lee, K. H., Isenhart, T. M., & Schultz, R. C., (2003). Sediment and nutrient removal in an established multi-species riparian buffer. *J. Soil Water Conserv., 58*, 1–8.

Lehmann, J., Peter, I., Steglich, C., Gebauer, G., Huwe, B., & Zech, W., (1998). Below-ground interactions in dry land agroforestry. *Forest Ecol. Manage., 111*, 157–169.

Li, Y., Zhao, M., Motesharrei, S., Mu, Q., Kalnay, E., & Li, S., (2014). Local cooling and warming effects of forests based on satellite observations. *Nat. Commun., 6*, 6603. pmid: 25824529.

Lott, J. E., Khan, A. A. H., Black, C. R., & Ong, C. K., (2003). Water use in a *Grevillea robusta* maize overstorey agroforestry system in semi-arid Kenya. *Forest Ecol. Manage., 180*, 45–59.

Lott, J. E., Ong, C. K., & Black, C. R., (2009). Understorey microclimate and crop performance in a *Grevillea robusta* based agroforestry system in semi-arid Kenya. *Agric. Forest Meteorol., 149*, 1140–1151.

Lundgren, B. O., & Raintree, J. B., (1982). Sustainable agroforestry. In: Nestle, B., (ed.), *Agricultural Research for Development- Potential and Challenges in Asia*. ISNAR.

Mbow, C., Smith, P., Skole, D., Duguma, L., & Bustamante, M., (2014b). Achieving mitigation and adaptation to climate change through sustainable agroforestry practices in Africa. *Current Opinion in Environmental Sustainability, 6*, 8–14.

Mbow, C., Van, N. M., Luedeling, E., Neufeldt, H., Minang, P. A., & Kowero, G., (2014a). Agroforestry solutions to address food security and climate change challenges in Africa. *Current Opinion in Environmental Sustainability, 6*, 61–67.

McCarthy, J., Canziani, O., Leary, N., Dokken, D., & White, K., (2001). *Climate Change 2001: Impacts, Adaptation, and Vulnerability.* New York, Cambridge University Press.

McNeely, J. A., (2004). Nature vs nurture: Managing relationships between forests, agroforestry, and wild biodiversity. *Agrofor. Syst., 61*, 155–165.

Miller, D. C., Ordoñez, P. J., Brown, S. E., Forrest, S., Nava, N. J., Hughes, K., & Baylis, K., (2020). The impacts of agroforestry on agricultural productivity, ecosystem services, and human well-being in low- and middle-income countries: An evidence and gap map. *Campbell Systematic Reviews, 16*(1), e1066.

Mishra, A. K., Chaudhari, S. K., Kumar, P., Singh, K., Rai, P., & Sharma, D. K., (2014). Consequences of straw burning on different carbon fractions and nutrient dynamics. *Indian Farming, 64*, 11, 12.

Mishra, A. K., Shinjo, H., Jat, H. S., Jat, M. L., Jat, R. K., & Funakawa, S., (2016). Farmers perspective on adaptation and up-scaling of conservation agriculture-based management practices in Indo-Gangetic plains of India. In: *Int. Conf Cons Agri Sustain Land Use*.

Mishra, A. K., Singh, J., Kumar, V., Srivastava, R., & Srivastava, S., (2013). Standing carbon stock estimation in different tree species grown in dry tropical forests of Vindhyan highland, Mirzapur, India. *Ecology, Environ Conservation, 19*(2), 401–407.

Molua, E. L., (2005). The economics of tropical agroforestry systems: The case of agroforestry farms in Cameroon. *Forest Policy Econ., 7*, 199–211.

Montagnini, F., (2006). *Environmental Services of Agroforestry Systems* (p. 126). Food Products Press, USA.

Nair, P. K. R., (1989). *Agroforestry Systems in Major Ecological Zones of the Tropics and Sub-tropics.* In: International Workshop on the Applications of Meteorology to Agroforestry Systems Planning and Management, Nairobi, Kenya, ICRAF.

Nair, P. K. R., (1993). *An Introduction to Agroforestry.* University of Florida Gainesville, Kluwer Academic Publishers.

Nair, P. K. R., Kumar, B. M., & Nair, V. D., (2009). Agroforestry as a strategy for carbon sequestration. *J. Plant Nutr. Soil Sci., 172*, 10–23.

Nakaegawa, T., Kitoh, A., & Hosaka, M., (2013). Discharge of major global rivers in the late 21st century climate projected with the high horizontal resolution MRIAGCMs. *Hydrological Processes, 27*(23), 3301–3318.

Nandi, J., (2019). *Climate Change to Impact Monsoon in India: Report.* Hindustan Times, New Delhi.

National Agroforestry Policy, (2014). Department of Agriculture and Cooperation, Ministry of Agriculture, Government of India.

Návar, J., (2009). Biomass component equations for Latin American species and groups of species. *Ann. For. Sci., 66*, 1–21.

Neupane, R. P., & Thapa, G. B., (2001). Impact of agroforestry intervention on soil fertility and farm income under the subsistence farming system of the middle hills, Nepal. *Agric. Ecosyst. Environ., 84*, 157–167.

Newaj, R., Chavan, S. B., & Prasad, R., (2015). Climate-smart agriculture with special reference to agroforestry. *Indian Journal of Agroforestry, 17*(1), 96–108.

Newaj, R., Dhyani, S. K., Chavan, S. B., Rizvi, R. H., Rajendra, P., Ajit, B. A., & Handa, A. K., (2014). *Methodologies for Assessing Biomass, Carbon Stock and Carbon Sequestration in Agroforestry Systems.* National Research Center for Agroforestry, Jhansi.

Nguyen, Q., Hoang, M. H., O'born, I., & Noordwijk, M. V., (2013). Multipurpose agroforestry as a climate change resiliency option for farmers: An example of local adaptation in Vietnam. *Climate Change, 117*, 241–257.

Oke, D. O., & Odebiyi, K. A., (2007). Traditional cocoa-based agroforestry and forest species conservation in Ondo State, Nigeria. *Agric. Ecosyst. Environ., 122*, 305–311.

Pandey, D. N., (2007). Multifunctional agroforestry systems in India. *Current Science, 92*(4), 455–463.

Pathak, P. S., & Dagar, J. C., (2001). *Traditional Agroforestry Systems: An Over View.* Landmarks in Botany in India CAS in Botany, BHU, Varanasi.

Pathak, P. S., Dagar, J. C., Kaushal, R., & Chaturvedi, O. P., (2014). Agroforestry inroads from the traditional two crop systems in heartlands of the Indo-Gangetic plains. In: Dagar, J. C., Singh, A. K., & Arunachalam, A., (eds.), *Agroforestry Systems in India: Livelihood Security and Ecosystem Services, Advances in Agroforestry* (Vol. 10, pp. 87–116).

Pathak, P. S., Pateria, H. M., & Solanki, K. R., (2000). *Agroforestry Systems in India: A Diagnosis and Design Approach* (p. 166). NRCAF and ICAR, Jhansi.

Pathak, P. S., Solanki, K. R., Rai, P., Handa, A. K., & Pateria, H. M., (2006). Agroforestry In: *Handbook of Agriculture.* Indian Council of Agriculture Research, New Delhi.

Planning Commission, (2001). *Report of Task Force on Greening India for Livelihood Security and Sustainable Development.* Govt. of India, New Delhi.

Planning Commission, (2004). *Report of the Committee on Development of Biofuels.* Planning Commission, New Delhi, Government of India.

Podder, S. K., & Samanta, D., (2019). Impact of climate changes on rural livelihoods and re-orienting the situation through human resources towards a sustainable society. In:

Proceedings International Conference on Sustainable Computing in Science, Technology & Management (SUSCOM-2019) (pp. 952–957). Elsevier SSRN.

Powell, J. M., Rivera, F. S., Hiernaux, P., & Turner, M. D., (1996). Nutrient cycling in integrated rangeland/cropland systems of the Sahel. *Agric. Syst., 52*, 143–170.

Prevedello, J. A., Winck, G. R., Weber, M. M., Nichols, E., & Sinervo, B., (2019). Impacts of forestation and deforestation on local temperature across the globe. *PLoS One, 14*(3), e0213368. https://doi.org/10.1371/journal.pone.0213368.

Puri, S., & Nair, P. K. R., (2004). Agroforestry research for development in India: 25 years of experiences of a national program. *Agroforestry Systems, 61*, 437–452.

Rasmussen, C., & Ramsayer, K., (2015). *Seven Case Studies in Carbon and Climate*. Global Climate Change Newsletter, NASA.

Reisinger, A., Kitching, R. F., Chiew, F., Hughes, L., Newton, P. C. D., Schuster, S. S., Tait, A., & Whetton, P., (2014). Australasia. In: Barros, V. R., Field, C. B., Dokken, D. J., Mastrandrea, M. D., Mach, K. J., Bilir, T. E., & Chatterjee, M., (eds.), *Climate Change 2014: Impacts, Adaptation, and Vulnerability* (pp. 1371–1438). Part B: Regional aspects. Contribution of working group II to the fifth assessment report of the intergovernmental panel on climate change. Cambridge University Press, Cambridge, United Kingdom and New York, NY, USA.

Reppin, S., Kuyah, S., De Neergaard, A., Oelofse, M., & Rosenstock, T. S., (2020). Contribution of agroforestry to climate change mitigation and livelihoods in Western Kenya. *Agroforestry Systems, 94*(1), 203–220.

Rice, R. A., (2008). Agricultural intensification within agroforestry: The case of coffee and wood products. *Agric. Ecosyst. Environ., 128*, 212–218.

Rockström, J., Steffen, W., Noone, K., Persson, A., Stuart, C. F., Lambin, E. F., Lenton, T. M., et al., (2009). A safe operating space for humanity. *Nature, 461*(7263), 472–475.

Rosenzweig, C., & Iglesias, A., (1998). The use of crop models for international climate change impact assessment. In: Tsuji, G. Y., Hoogenboom, G., & Thornton, P. K., (eds.), *Understanding Options for Agricultural Production* (pp. 267–292). Kluwer Academic Publishers, Dordrecht, The Netherlands.

Sathaye, J. A., & Ravindranath, N. H., (1998). Climate change mitigation in the energy and forestry sectors of developing countries. *Annu. Rev. Energy Environ., 23*, 387–437.

Schlüter, M., Hirsch, D., & Pahl-Wostl, C., (2010). Coping with change: Responses of the Uzbek water management regime to socio-economic transition and global change. *Environmental Science & Policy, 13*(7), 620–636.

Schoeneberger, M., Bentrup, G., De Gooijer, H., Soolanayakanahally, R., Sauer, T., Brandle, J., et al., (2012). Branching out: Agroforestry as a climate change mitigation and adaptation tool for agriculture. *Journal of Soil and Water Conservation, 67*(5), 128A–136A.

Scholes, R. J., Palm, C. A., & Hickman, J. E., (2014). *Agriculture and Climate Change Mitigation in the Developing World*. CCAFS Working Paper No. 61. CGIAR research program on climate change, Agriculture and Food Security (CCAFS)., Copenhagen, Denmark. Available at: https://ccafs.cgiar.org/ (accessed on 28 December 2023).

Schroth, G., & Sinclair, F., (2003). *Trees Crops and Soil Fertility: Concepts and Research Methods* (p. 464). CABI, Wallingford, UK.

Schroth, G., Da Fonseca, G. A. B., Harvey, C. A., Gascon, C., Vasconcelos, H., & Izac, A. N., (2004). *Agroforestry and Biodiversity Conservation in Tropical Landscapes*. Island Press.

Scoones, I., (2009). Livelihoods perspectives and rural development. *J. Peasant Stud., 36*(1), 171–196. https://doi.org/10.1080/03066150902820503.

Seiter, S., Ingham, E. R., William, R. D., & Hibbs, D. E., (1995). Increase in soil microbial biomass and transfer of nitrogen from alder to sweet corn in an alley cropping system. In: Ehrenreich, J. H., Ehrenreich, D. L., & Lee, H. W., (eds.), *Growing a Sustainable Future* (pp. 56–158). University of Idaho, Boise, ID.

Sharma, N., Bohra, B., Pragya, N., Ciannella, R., Dobie, P., & Lehmann, S., (2016). Bioenergy from agroforestry can lead to improved food security, climate change, soil quality, and rural development. *Food and Energy Security, 5*, 165–183.

Sinclair, F. L., (1999). A general classification of agroforestry practice. *Agroforest Systems, 46*(2), 161–180.

Singh, V. S., & Pandey, D. N., (2011). *Multifunctional Agroforestry Systems in India: Science Based Policy Options*. RSPCB Occasional Paper No. 4/2011, (2011). Rajasthan State Pollution Control Board.

Smith, J., Pearce, B. D., & Wolfe, M. S., (2012). Reconciling productivity with protection of the environment: Is temperate agroforestry the answer? *Renew. Agric. Food Syst., 28*, 80–92.

Srivastava, K., Jat, H., Meena, M., Choudhary, M., Mishra, A. K., & Chaudhari, S. K., (2016). Long term impact of different cropping systems on soil quality under silty loam soils of Indo-Gangetic plains of India. *Journal of Applied and Natural Science, 8*(2), 584–587. https://doi.org/10.31018/jans.v8i2.841.

Stigter, C. J., Mohammed, A. E., Al-amin, N. K. N., Onyewotu, L. O. Z., Oteng'i, S. B. B., & Kainkwa, R. M. R., (2002). Agroforestry solutions to some African wind problems. *J. Wind Eng. Ind. Aerodyn., 90*, 1101–1114.

Tewari, J. C., Tripathi, D., & Narain, P., (2001). Jujube: A multipurpose tree crop for arid land farming systems. *The Botanica, 51*, 121–126.

Thangata, P. H., & Alavalapati, J. R. R., (2003). Agroforestry adoption in southern Malawi: The case of mixed intercropping of *Gliricidia sepium* and maize. *Agric. Syst., 78*, 57–71.

Torquebiau, E. F., (2000). A renewed perspective on agroforestry concepts and classification. *Comptes Remdus Academie des Sciences Paris, Sciences de la Vie/Life Sciences, 323*, 1009–1017.

Tyndall, J., & Colletti, J., (2007). Mitigating swine Odor with strategically designed shelterbelt systems: A review. *Agrofor. Syst., 69*, 45–65.

Udawatta, R. P., Kremer, R. J., Adamson, B. W., & Anderson, S. H., (2008). Variations in soil aggregate stability and enzyme activities in a temperate agroforestry practice. *Appl. Soil Ecol., 39*, 153–160.

Udawatta, R. P., Krstansky, J. J., Henderson, G. S., & Garrett, H. E., (2002). Agroforestry practices, runoff, and nutrient loss: A paired watershed comparison. *J. Environ. Qual., 31*, 1214–1225.

United Nations Development Program, (2007). *Human Development Report*. Palgrave Macmillan, New York.

USGCRP, (2014). In: Melillo, J. M., Terese, R., & Gary, W. Y., (eds.), *Climate Change Impacts in the United States: The Third National Climate Assessment*. United States Global Change Research Program.

USGCRP, (2016). In: Crimmins, A., Balbus, J., Gamble, J. L., Beard, C. B., Bell, J. E., Dodgen, D., Eisen, R. J., et al., (eds.), *Impacts of Climate Change on Human Health in the United States: A Scientific Assessment* (p. 312). U.S. Global Change Research Program, Washington, DC.

Van, N. M., Duguma, L. A., Dewi, S., Leimona, B., Catacutan, D. C., & Lusiana, B., (2018). SDG synergy between agriculture and forestry in the food, energy, water, and income nexus: Reinventing agroforestry? *Current Opinion in Environmental Sustainability, 34*, 33–42.

Verchot, L. V., Noordwijk, M. V., Serigne, K., Tomich, T., Chin, O., Albrecht, A., Jens, M., et al., (2006). Climate change: Linking adaptation and mitigation through agroforestry. *Mitig. Adapt. Strat. Glob. Change, 12*, 901–918.

Verma, A., Kaushal, R., Alam, N. M., Mehta, H., Chaturvedi, O. P., Mandal, D., Tomar, J. M. S., et al., (2014). Predictive models for biomass and carbon stocks estimation in *Grewia optiva* on degraded lands in Western Himalaya. *Agroforestry Systems, 88*(5), 895–905.

Vermeulen, S. J., Campbell, B. M., & Ingram, J. S. I., (2012). Climate change and food systems. *Annu. Rev. Environ. Resour., 37*, 195–222.

Waldron, A., Garrity, D., Malhi, Y., Girardin, C., Miller, D. C., & Seddon, N., (2017). Agroforestry can enhance food security while meeting other sustainable development goals. *Trop. Conserv. Sci., 10*, 1–6.

Wallace, Douglas, C., Geyer, Wayne A., & Dwyer, John, P., (2000). "Waterbreaks: Managed Trees for the Floodplain." *Agroforestry Notes (USDA-NAC).* 19. https://digitalcommons.unl.edu/agroforestnotes/19 (accessed on 28 December 2023).

Wessels, J. K., Prince, D. S., Malherbe, J., Small, J., Frost, E. P., et al., (2007). Can human-induced land degradation be distinguished from the effects of rainfall variability? A case study in South Africa. *J. Arid Environ., 68*, 271–297.

CHAPTER 11

Sericulture-Based Agroforestry Systems for Livelihood and Nutritional Security in India: An Overview

DEBNIRMALYA GANGOPADHYAY

Department of Sericulture, Raiganj University, Raiganj, North Dinajpur, West Bengal, India

ABSTRACT

Sustainable management of natural resources becomes most effective strategy to address the growing concern on livelihood and nutritional security of developing countries like India where around a fifth of the global population resides in the country. It is projected that India may become the world's most overpopulated country by 2024. Agriculture sector alone cannot sustain the diversified food demand of such an overpopulated country due to scarcity of land, poor soil health and declining agricultural productivity. The continuous degradation of forest also reduces the livelihood opportunities and intensifies the risk of malnutrition among the poor people who are solely dependent on forest. Agroforestry promotes an efficient utilization of natural resources bridging the gap between agriculture and forestry by developing an integral system to address various social, economical, and environmental objectives. India is home to diverse species of flora and fauna having economic importance. Development of an agroforestry model integrating agro-based rural industries like sericulture, especially, may be very useful to improve socioeconomically a more sustainable and better landscape. For instance, practicing sericulture cultivating trees like *Terminalia arjuna*, *Terminalia tomentosa*, *Shorea robusta*, *Persea bombycina*, *Litsea monopetela*, *Ricinus communis* and *Manihot esculanta* that are hosts for diverse silkworm fauna

Sustainable Management and Conservation of Environmental Resources in India.
Sumit Chakravarty, Vineeta, Jahangeer A. Bhat, Munesh Kumar, Gopal Shukla (Eds.)

in fields or homestead gardens can serve both the purpose of bio-resource conservation and non-farm output debilitating the impacts of climate change (CC) in agriculture. In this chapter, the status of sericulture industry in India and its trend has been reviewed. Besides, some critical issues like scope of the sector in creating livelihood opportunity and nutritional security as viable agroforestry model have been identified. The chapter would be helpful in recognizing potential of the sericulture-based agroforestry systems to formulate policies for mitigation of CC that are compatible with safeguarding both food security and livelihood in rural India.

11.1 INTRODUCTION

Agroforestry is the scientific management of land where trees or other woody perennial plants are cultivated in combination with crops or animals on the same land (Nair, 2011). It is an ecologically-based sustainable land management practice that provides an affordable alternative to the expensive conventional conservation measures for greater agricultural productivity (Pandey, 2007; Kumar et al., 2012). The main aim of agroforestry is to exploit the complimentary relationship among plants, crops, and animals or livestock to such an extent so that the system provides productivity, profitability, and sustainability (Chaturvedi et al., 2016). It not only promotes efficient utilization of natural resources but also bridges the gap between agriculture and forestry by developing an integral system to address various social, economical, and environmental objectives (Brown et al., 2018). Agroforestry provides a number of ecological and environmental benefits. For example, it supports biodiversity, mitigates climate change (CC) via carbon sequestration and improves soil health and quality of water (Dollinger & Jose, 2018; Jose & Bardhan, 2012; Mbow et al., 2014). Besides, it offers more livelihood opportunities particularly to the rural people who practice agroforestry on their farms (Pandey, 2007; Handa et al., 2016). As per the estimates of World Bank, there are approximately 1.2 billion rural people across the world engaged in agroforestry related activities for their livelihood (World Bank, 2004).

In India, tree-based farming system combining trees into agriculturally productive landscapes is most common traditional practices (Dhyani & Handa, 2013). Such cultivation practice of economically important plant species serves dual purpose of bio-resource conservation and non-farm output debilitating the impacts of CC in agriculture. The most traditional practices of agroforestry involve planting of trees in rows separately in crop fields or along the bunds. These trees not merely provide construction materials, timber, fuel, food, fodder, and raw materials for the forest-based cottage enterprises but also help in maintaining a well-balanced ecological niche (Ghosh et al., 2011; Parihaar et al., 2014; Handa et al., 2016). Many times, these trees enrich the soil nutrient protecting soil erosion (Moench, 1991). However, agroforestry management is more complex in nature since multiple species having varied agronomical and physiological requirements are involved (Kumar et al., 2012). The success of agroforestry therefore depends on selection of suitable plant species and agricultural crops combination. Generally, the plant species that possess multipurpose benefits like nitrogen fixing, short gestation period, easy to grow, adaptable to harsh climatic condition and economically important are preferred (Jhariya et al., 2015; Chaturvedi et al., 2016).

Intensification of agriculture through selection and breeding of high yielding crop varieties has resulted in a change from poly-cultural system to more temporal or spatial monoculture. The uniformity of genetic materials in monoculture reduces biodiversity with a concurrent hazardous incidence of pest problems (Shanker & Solanki, 2000). India is endowed with a diverse and abundant sericultural bioresources, marked by a variety of silkworm species and the cultivation of several host plants. This bioresource wealth is not only economically significant but also culturally and environmentally important. The integration of sericulture into agroforestry systems holds immense potential for sustainable and diversified agricultural practices. In this sericulture-based agroforestry approach, silkworms are cultivated alongside specific tree species, fostering a mutually beneficial relationship that positively impacts both the sericulture industry and the broader agroecosystem. This chapter presents a comprehensive review of the present state and developments in India's sericulture industry. Additionally, it explores key aspects, including the industry's potential to generate livelihood opportunities and enhance nutritional security through a viable agroforestry model. The insights provided in this chapter can be useful in developing suitable policies addressing CC, with a focus on aligning these policies with the dual objectives of ensuring food security and supporting livelihoods in rural India.

11.2 STATUS OF SERICULTURE INDUSTRY IN INDIA AND ITS TREND

Sericulture is an old tradition in India. It is a farm-based industrial activity that involves cultivation of host plants, rearing of silkworms and collection of silk thread for the textile industry. In India, sericulture research work was initiated during the late 1989s by the Central Silk Board; Bangalore through World Bank supported National Sericulture Project. The scientific approach to sericulture has been realized recently. During the past few decades, the Indian silk industry has witnessed a remarkable progress. The low-yielding polyvoltine hybrids have been largely replaced by superior crossbred and bivoltine hybrids (Gangopadhyay, 2008). Raw silk productivity has witnessed a quantum jump from mere 18,475 MT (2006–2007) to 35,820 MT (2019–2020) (CSB, 2019–2020). The new technology, in addition to maximizing silk yield has also resulted in qualitative improvement in cocoon production with considerable increase in average productivity rate. The silk statistics of India is given in Table 11.1.

TABLE 11.1 Silk Statistics of India

Particulars	**Unit**	**Year**					
		2014–2015	**2015–2016**	**2016–2017**	**2017–2018**	**2018–2019**	**2019–2020**
Production of silk (Total)	MT	28,708	28,523	30,348	31,906	35,468	35,820
Mulberry	MT	21,390	20,478	21,273	22,066	25,344	25,239
Tasar	MT	2,434	2,819	3,268	2,988	2,981	3,136
Eri	MT	4,726	5,060	5,637	6,661	6,910	7,204
Muga	MT	158	166	170	192	233	241
Mulberry acreage	Ha.	2,19,819	2,08,947	2,16,810	2,23,926	2,35,001	2,39,967
Employment	Lakh person	80.3	82.5	85.1	86.04	91.78	94.3

Source: Central Silk Board (2019–2020).

India stands second position in the world next only to China in terms of silk production. The Indian silk market is driven by both export and strong domestic demand. At present, India imports around 6,000–8,000 tons of raw silk mainly from China to meet the growing domestic demand, besides exporting silk accounting only 4–5% market share in the global silk trade. Generally, the silk products like fabrics, readymade garment, carpet, bed-cover, and cushion-cover are exported from India to USA, UK, Germany, France, Italy, and Spain. The trend of silk export over the years is presented in Figure 11.1.

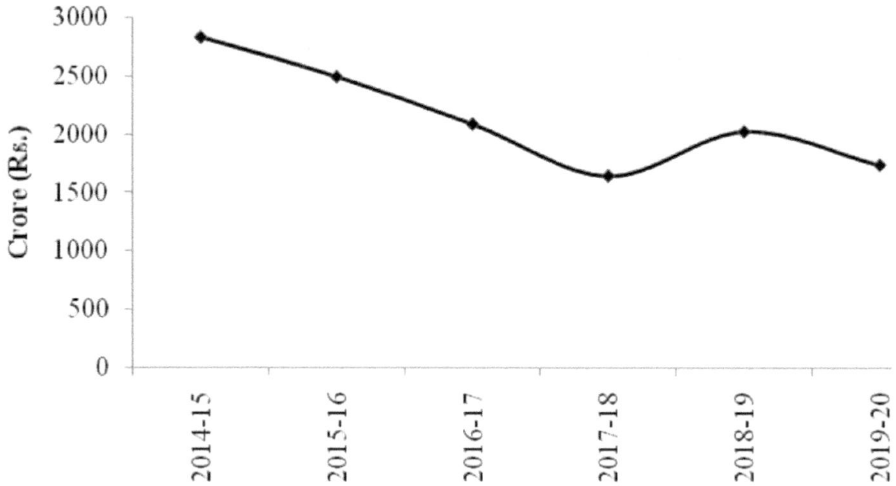

FIGURE 11.1 Year-wise silk export earnings.

Source: DGCIS (2019–2020).

11.3 SCOPE OF SERICULTURE SECTOR AS POTENTIAL AGROFORESTRY

The population of India has been growing at a much faster rate. For instance, the population had doubled since 1975 to more than 1.2 billion as per record of the 2011 Census of India (Chandramouli & General, 2011). It is projected that India may become the world's most overpopulated by 2024 (Gladstone, 2015). In India more than 70% of the poor people are residing in the countryside below the poverty line and mainly dependent on agriculture. Forest supports around 200 million[1] rural poor who are living in 1.73 lakh[2] villages for their livelihood (Bhatia & Yousuf, 2013). Recently, potentiality of the agriculture sector becomes limited to create new job opportunities in rural areas because of scarce capital and lack of economic stimulus, inadequate farm infrastructure and volatile prices of agricultural commodities. Sustainability of agricultural production system is further hampered by rapid increase in human population growth and scarcity of land (Handa et al., 2016). Continuous degradation of forest also reduces the livelihood opportunities and intensifies the risk of malnutrition among the poor people who are solely dependent on forest. As per the report of the National Forest Commission of India more

[1] There is no census data available for the forest-dependent population in India. Estimates collected from various sources indicate the data from 200 to 400 million (MoEF, 2009; ICFRE, 2012).

[2] According to the census report, there are around 6.41 lakh total villages in India.

than 41% forest cover of the country appeared to be deteriorated and lost its productivity (MoEF, 2006). It is therefore necessary to build an integrated system that can address multiple aspects of productivity, environment, and socio-economic objectives. Promotion of sericulture-based agroforestry can be an ideal land use (LU) option to deal with the issues of poverty alleviation and malnutrition among the rural poor in India. The sericulture-based agroforestry system can be adopted in diverse agro-climatic zones varying from tropical to sub-tropical and temperate zones integrating various agroforestry plant species (*viz.*, perennial woody plants, shrubs, fruits, vegetables, and food crops, etc.) with sericulture crops and livestock production system for higher income. The scope of sericulture-based agroforestry for livelihood opportunity and nutritional security is discussed in subsections.

11.3.1 SERICULTURE-BASED AGROFORESTRY FOR LIVELIHOOD OPPORTUNITY

In India, sericulture has a close link with the socio-economic life of the people. Being an agro-based cottage industry, sericulture proves to be a profitable economic avenue to the rural entrepreneurs, where farmers are resource-constrained by small land holding size (Gangopadhyay, 2016). This is largely because of some significant economic activities associated with sericulture practices. For example, sericulture farm and non-farm activities such as propagation of food plants, silkworm rearing and processing of silk thread generate enormous employment opportunities to the rural artisans. This is suitable for both the small and large landholders. It is recorded that sericulture system alone has the potential to generate employment @ 11 man-days for the production of one kg of raw silk throughout the year (Gangopadhyay, 2008). Various economic activities associated with sericulture are presented in Table 11.2. Secondly, sericulture is a profitable venture since it requires minimum capital investment potentially with higher returns. The huge demand of silk (approximately 85% of silk commodities are consumed in the domestic market) also ensures stable market (CSB, 2019–2020). Thirdly, sericulture being the tradition of the country is not new to most parts of the country and thus, easy to adopt and preferred by the farmers.

TABLE 11.2 Components of Economic Activities Associated with Sericulture

Economic Activity	Skill Required	Beneficiary Type
Production of industrial silkworm seed.	High	Government agencies/LSPs
Production of basic silkworm seed.	High	Seed rearers/BSFs
Raising of nursery for mulberry sericulture.	Low	Farmers
Propagation of mulberry plants.	Low	Farmers
Agro-processed food products form mulberry fruits.	High	SMEs
Vermicompost preparation recycling farm waste.	Moderate	Farmers
Rearing of silkworm silk production.	Moderate	Farmers
Preparation of disinfectants for silkworm rearing.	High	SMEs
Reeling and post reeling of silk yarn.	High	Reelers
Manufacturing of silk garments.	High	Weavers, printers, etc.

Abbreviations: LSPs: Licensed seed producers; BSFs: basic seed farms; and SMEs: small and medium enterprises.

Source: Adapted from: Gangopadhyay (2016).

India is home to many agroforestry plant species that are hosts to silkworms (Table 11.3). There are four types of commercially important silk crops namely, mulberry, tasar, muga, and Eri. Mulberry based agroforestry systems have been found economically viable in many parts of the country. Farmers of Uttaranchal often integrate a variety of woody perennial plants with mulberry crops and livestock production system to enhance their income level (Rama et al., 2004). Cultivation of mulberry (*Morus* sp.) trees intercropped with soybeans and green gram is preferred by the sericulture farmers for higher profitability in Karnataka (Yadav & Kumar, 1998). Mulberry is a strong woody perennial plant belonging to the genus *Morus* of the family Moraceae. However, size of the tree is usually kept shorter by pruning for easy cultivation. It plays an important role to safeguard the environment like improves the air quality, water resources and soil health remedying heavy metals from soil and atmosphere and conserving beneficial soil microflora and fauna (Lu et al., 2004; Yao et al., 2004; Ghosh et al., 2017).

TABLE 11.3 List of Agroforestry Trees That are Hosts for Silkworms

CN, SN, F	Silk Production	Silkworm Species	Other Uses
Mulberry *Morus alba*; Moraceae	Mulberry	*Bombyx mori*	Fodder, fuel-wood, timber for construction and sports items.
Arjun *Terminalia arjuna*; Combretaceae Asan *Terminalia tomentosa*; Combretaceae Bahada *Terminalia bellerica*; Combretaceae Sidha *Lagerstroemia parviflora*; Combretaceae Sal *Shorea robusta*; Dipterocarpaceae Ber *Ziziphus jujuba*; Rhamnaceae Mahua *Madhuca latifolia*; Sapotaceae	Tasar	*Antheraea mylitta*	Fodder, fuel, timber, gum, medicine, dye, fodder, resin, dry leaves for production of leaf plates, turning, making agricultural implements, high quality charcoal; liquor from flowers and berries consumed, seed-oil for soap industry.
Som *Persea bombycina*; Lauraceae Soalu *Litsea monopetela*; Lauraceae Digloti *Litsaea salicifolia*; Lauraceae Castor *Ricinus communis*; Euphorbiaceae Tapioca *Manihot esculanta*; Euphorbiaceae Borkesseru *Ailanthus excelsa*; Simaroubaceae	Muga	*Antheraea assamensis*	Stem is used for construction work like cabinet making, fenching, leaves as fodder; wood for agricultural implements, construction work, furniture, seeds for industrial use and oil extraction, leaves for medicine, synthetic resins, plastics, fibers, paints, varnishes, manioc is used in laundry products, making matchwood boxes and match splints.

Abbreviations: CN: Common name; SN: scientific name; and F: family.

Shifting cultivation through diverging forest lands for agriculture is predominantly practiced by the tribal communities in hilly areas of north-east India due to scarcity of agricultural land and inadequate livelihood opportunities. However, shifting cultivation contributes significantly to forest loss and land degradation (Kumar et al., 2012; Rahman et al., 2012). To address both the environmental and societal objectives, a mulberry and muga sericulture-based agroforestry systems has been developed for possible adoption in the hilly zones of north-east India (Dhyani et al., 1996). Sericulture with fruit plants and fodder grass model demonstrated higher preference among the farmers followed by sericulture with uplands field crops (Figure 11.2).

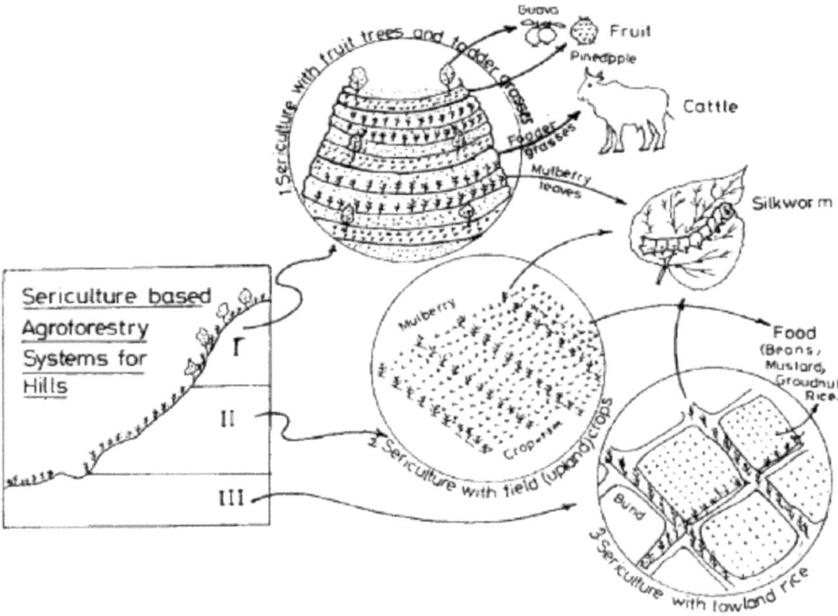

FIGURE 11.2 Sericulture based agroforestry systems for hilly areas of north-east India.

Source: Reprinted with permission from: Dhyani et al. (1996).

Tropical tasar culture is practiced by the tribal communities in the dense humid tropical forests sprawling over the central and southern plateau of India. The tasar silkworm, *Antheraea mylitta* is reared mainly on primary food plants, Arjun (*Terminalia arjuna*), Asan (*Termanalia tomentosa*), and Sal (*Shorea robusta*). These agroforestry plant species are found abundantly in the forest areas and intrinsic sources of food, fuel, timber, non-timber, and plant-derived medicines. A cost-benefit analysis has revealed that tasar food plants propagated in dense plantation (2,800 plants/ha with 2 m × 2 m spacing) can significantly bolster the rural economy (Dhyani, 2018). However, rapid industrialization and deforestation pose a grave threat to the conservation of these important plant species. In this context, sericulture-based agroforestry can play an important role. In India, around 114 million hectares area is under degraded and wasteland and unfit for cultivation (Maji et al., 2010). If relatively less productive lands are brought under agroforestry plantation of silkworms' host plants and intercropped with vegetables and pulses may open up more income generating avenues to the farmers all-round the year rejuvenating soil health.

11.3.2 SERICULTURE-BASED AGROFORESTRY FOR FOOD AND NUTRITIONAL SECURITY

In India, the increasing trend of population becomes a serious concern over food security. It demands for enhancing food production supply substantially despite of several limitations of land scarcity, poor soil health, declining agricultural productivity and CC. The food demand may increase more than 70% of the current food requirement to feed the growing population of the country's 2 billion people by 2050 (Hamshere et al., 2014). Though the production of rice and wheat in India has increased remarkably during the last decade, however, production of coarse grains and pulses witnessed only a marginal change. Increase in production of protein rich food, particularly pulses are essential to meet the increasing need of protein among the poor and middle classes (Handa et al., 2016). Sericulture based agroforestry systems can provide desired diversification to increase the food production in the country. Crops like cereals, pulses, oil seeds, fruits, vegetables, and fodder grasses can be intercropped successfully with agroforestry trees those are hosts to the silkworms even in rainfed and hilly regions (Dhyani et al., 1996; Yadav & Kumar, 1998; Singhvi & Katiyar, 2009; Pandey & Dhar, 2013; Liu et al., 2015). These crops in addition to the regular sericulture crops can provide nutritional security and higher return than the common field crops to the farmers (Dhyani, 2018). Besides, silkworm pupae which constitute major by-products of sericulture crops and are consumed as food supplements can address the critical issues of malnutrition and increasing demand for protein among the poor and middle classes (Mishra et al., 2003; Ray & Gangopadhyay, 2019).

11.4 CONCLUSION

Practicing sericulture cultivating trees that are hosts for silkworms in fields or homestead gardens is historically a source of livelihood particularly in rural India. Sericulture based agroforestry when developed properly has the potential to improve socio-economically a more sustainable and better landscape. The trees can play significant role towards adapting to varied climate as important carbon sequester to mitigate environmental pollution. Besides, it provides livelihood and employment opportunities to the underprivileged farmers who are largely dependent on natural resources for food, fodder, and fuel wood particularly in rainfed areas. The growing demand of diversified food requirements through appropriate tree crop or legume combination may be more sustainable to such economic enterprises of silk production, thereby higher economic profitability of farmers protecting the environment for future generation. However, success of such endeavors depends on the governmental and non-governmental support system to educate the farmers about alternative agricultural practices simultaneously creating better market avenues for the produce.

KEYWORDS

- **agricultural practices**
- **agroforestry**
- **environmental pollution**
- **legume**
- **livelihood**
- **nutritional security**
- **sericulture**

REFERENCES

Bhatia, N. K., & Yousuf, M., (2013). Forest insect industry in collaborative forest management: An overview. *Int. J. Indust. Entomol., 27*(1), 166–179.

Brown, S. E., Miller, D. C., Ordonez, P. J., & Baylis, K., (2018). Evidence for the impacts of agroforestry on agricultural productivity, ecosystem services, and human well-being in high-income countries: A systematic map protocol. *Environ. Evid., 7*(1), 1–16.

Chandramouli, C., & General, R., (2011). *Census of India 2011 Provisional Population Totals*. Gov. of India, New Delhi.

Chaturvedi, O. P., Mishra, P. K., & Kaushal, R., (2016). Agroforestry for natural resource conservation and sustainable production. In: Tewari, et al., (eds.), *Holistic Development of Agroforestry* (pp. 15–41). Jaya Publishing House: Delhi.

CSB: Annual Report, (2019–2020). Ministry of Textiles, Government of India.

DGCIS: Annual Report, (2019–2002). Ministry of Commerce, Government of India.

Dhyani, S. K., & Handa, A. K., (2013). Area under agroforestry in India: An assessment for present status and future perspective. *Ind. J. Agrofor., 15*(1), 1–11.

Dhyani, S. K., (2018). Agroforestry in Indian perspective. In: Rao, R., et al., (eds.), *Agroforestry Opportunities for Enhancing Resilience to Climate Change in Rainfed Areas* (pp. 12–28). ICAR – Central Research Institute for Dry land Agriculture, Hyderabad, India.

Dhyani, S. K., Chauhan, D. S., Kumar, D., Kushwaha, V., & Lepcha, S. T., (1996). Sericulture-based agroforestry systems for hilly Areas of North-east India. *Agrofor. Syst., 34*(3), 247–258.

Dollinger, J., & Jose, S., (2018). Agroforestry for soil health. *Agrofor. Syst., 92*(2), 213–219.

Gangopadhyay, D., (2008). Sericulture industry in India: A review. In: Banerjee, P., (ed.), *India Science & Technology* (pp. 334–339). National Institute of Science Technology and Development Studies, CSIR, New Delhi.

Gangopadhyay, D., (2016). Information and communication technology (ICT) for sericulture knowledge management: Case of Indian sericulture. *Int. J. Inf. Sc. Comput., 3*(2), 59–66.

Ghosh, A., Gangopadhyay, D., & Chowdhury, T., (2017). Economical and environmental importance of mulberry: A review. *Int. J. Plant Environ., 3*(2), 51–58.

Ghosh, S. R., Wadud, M. A., Mondol, M. A., & Rahman, G. M. M., (2011). Optimization of plant density of Akashmoni (*Acacia auriculiformis*) for production of fuel wood in the bunds of crop land. *J. Agrofor. Environ., 5*(2), 1–6.

Gladstone, R., (2015). *India Will be Most Populous Country Sooner Than Thought, UN Says.* The New York Times.

Hamshere, P., Sheng, Y., Moir, B., Gunning-Trant, C., & Mobsby, D., (2014). What India wants: Analysis of India's food demand to 2050. *Aust. Bur. Agric. Resour. Econ. Sci.* Research Report No. 14.15.

Handa, A. K., Toky, O. P., Dhyani, S. K., Chavan, S. B., & Toky, I. D. O., (2016). Innovative agroforestry for livelihood security in India. *World Agric., 7*, 7–16.

Jhariya, M. K., Bargali, S. S., & Raj, A., (2015). Possibilities and perspectives of agroforestry in Chhattisgarh. In: Zlatic, M., (ed.), Intech *Precious Forests-Precious Earth* (pp. 237–257). Open.

Jose, S., & Bardhan, S., (2012). Agroforestry for biomass production and carbon sequestration: An overview. *Agrofor. Syst., 86*(2), 105–111.

Kumar, B. M., Singh, A. K., & Dhyani, S. K., (2012). South Asian agroforestry: Traditions, transformations, and prospects. In: Nair, P. K. R., & Garrity, D., (eds.), *Agroforestry-The Future of Global Land Use* (pp. 359–389). Springer, Dordrecht.

Liu, Y. J., Ni, J. P., Zhang, Y., & Zhou, C., (2015). Effects of different crop-mulberry intercropping systems on nutrients in arid purple soils in the three gorges reservoir area. *Acta Prataculturae Sinica, 6*(2), 178–185.

Lu, M., Wang, R. Q., & Qi, X. S., (2004). Reaction of planting tree species on chlorine pollution in the atmosphere. *J. Shandong Univ., 39*, 98–101.

Maji, A. K., Reddy, G. O., & Sarkar, D., (2010). *Degraded and Wastelands of India: Status and Spatial Distribution.* ICAR, New Delhi.

Mbow, C., Van, N. M., Luedeling, E., Neufeldt, H., Minang, P. A., & Kowero, G., (2014). Agroforestry solutions to address food security and climate change challenges in Africa. *Curr. Opin. Environ. Sustain., 6*, 61–67.

Mishra, N., Hazarika, N. C., Narain, K., & Mahanta, J., (2003). Nutritive value of non-mulberry and mulberry silkworm pupae and consumption pattern in Assam, India. *Nutri. Res., 23*(10), 1303–1311.

MoEF, (2006). *Report of the National Forest Commission, Ministry of Environment and Forests, Government of India, New Delhi.*

Moench, M., (1991). Soil erosion under a successional agroforestry sequence: A case study from Idukki District, Kerala, India. *Agrofor. Syst., 15*(1), 31–50.

Nair, P. R., (2011). Agroforestry systems and environmental quality: Introduction. *J. Environ. Qual., 40*(3), 784–790.

Pandey, D. N., (2007). Multifunctional agroforestry systems in India. *Current Science, 92*(4), 455–463.

Pandey, R. K., & Dhar, A., (2013). Mulberry intercropping for sustainable livelihood in Kandi area of Rural Sub-Himalayan, India. *Res. J. Agric. Sci., 4*(4), 484–487.

Parihaar, R. S., Bargali, K., & Bargali, S. S., (2014). Diversity and uses of ethno-medicinal plants associated with traditional agroforestry systems in Kumaun Himalaya. *Ind. J. Agric. Sci., 84*(12), 1470–1476.

Rahman, S. A., Rahman, M. F., & Sunderland, T., (2012). Causes and consequences of shifting cultivation and its alternative in the hill tracts of Eastern Bangladesh. *Agrofor. Syst., 84*(2), 141–155.

Rama, K., Chakrabarty, S., Dhyani, S. K., Mourya, G. S., & Upadhayay, R. K., (2004). Mulberry (*Morus alba* L.) as an agroforestry plant in Uttaranchal. *Ind. For., 130*(8), 939–944.

Ray, M., & Gangopadhyay, D., (2019). Preliminary report on nutrient composition and antioxidant activity in the Pupae of Eri Silkworm, *Samia ricini. Sericologia, 59*(3, 4), 143–148.

Shanker, C., & Solanki, K. R., (2000). Agroforestry: An ecofriendly land-use system for insect management. *Outlook Agric., 29*(2), 91–96.

Singhvi, N. R., & Katiyar, R. L., (2009). Intercropping of mulberry with garlic, onion, and carrot in Maharashtra. *Plant Arch., 9*(1), 265–266.

World Bank, (2004). *Sustaining Forests: A Development Strategy*. The World Bank, Washington, DC, USA.

Yadav, B. R. D., & Kumar, T. D. N., (1998). Effect of row arrangement on yield and monetary benefits in mulberry (*Morus indica*) plus soybean (*Glycine max*) and mulberry plus green gram (*Phaseolus radiatus*) intercropping. *Ind. J. Agric. Sci., 68*(3), 149–151.

Yao, F., Ni, W., & Yang, X., (2004). Genotype resources and ecological adaptation of mulberry plants (*Morus indica* L.) and their application foreground. *Bull. Sc. Tech., 20*(4), 289–292.

CHAPTER 12

Potential of Different Soil Organic Amendments in Agroforestry Practices to Enhance Carbon Sequestration and Reduce Greenhouse Gas Emissions

KHUSHBOO GUPTA, RAUSHAN KUMAR, and NIRMALI BORDOLOI

Department of Environmental Sciences, Central University of Jharkhand, Brambe, Ranchi, Jharkhand, India

ABSTRACT

Agroforestry is production of trees, cattle, and crop on the same piece of land. In this chapter, we are reviewing and updating knowledge about increasing carbon sequestration potential and reducing greenhouse gas (GHG) emission through application of organic amendment (biochar, manure, crop residue and compost) in land ecosystem particularly the agroforestry land use (LU) management. The main area of attention were: (i) disturbance in soil must be reduced and increase the tree biomass below and above ground so that is can fix more amount of soil organic carbon (SOC) which is one of the objectives of agroforestry system; and (ii) the durable storage of SOC is finite and it rely on the accessibility of binding sites that is the soil mineral, composition, and depth. Introducing biochar obtained from various feedstock increase organic carbon (OM) content and CH_4 uptake and decrease CO_2 and N_2O (84%) emissions. On the other hand, compost used in agroforestry emits less CO_2 and N_2O gasses in comparison to pellet manure. In summary, amplifying plants and organic amendment application increased carbon sequestration and reduce GHG emissions is better option for increasing tree and crop biomass and production in agroforestry. However, there is insufficient data to support these practices. Furthermore, studies are also needed to quantify

Sustainable Management and Conservation of Environmental Resources in India.
Sumit Chakravarty, Vineeta, Jahangeer A. Bhat, Munesh Kumar, Gopal Shukla (Eds.)

and improve the quality of organic material used in agroforestry and thereby decreased the dependency of mineral fertilizers for both enhancing the carbon sequestration and mitigating greenhouse gas emission.

12.1 INTRODUCTION

Agroforestry practices involve trees along with crops and cattle used in agricultural system (Shrestha et al., 2018). Interaction among tress, crop, and domestic animal provides lots of benefit and services, for instance utmost crop yield, boosting environmental services, increasing food security and agro-ecosystem flexibility (Ajayi et al., 2011; Lorenz & Lal, 2014). Agroforestry systems develop as an aggregate land use (LU) practice that increases carbon sequestration and decrease greenhouse gasses (GHGs) emission from land ecosystem, assist in mitigation of global CC. Agriculture, subsequent to energy sector is the second largest contributor of GHGs (30%) globally (McIntire, 2009; Shrestha et al., 2018). Agricultural GHGs such as carbon dioxide (CO_2), methane (CH_4) and nitrous oxide (N_2O) participate in global warming (Thangarajan et al., 2013). Methane generated mainly by methanogenic archaebacteria (methanogenic bacteria are the members of the domain archaea and are strict anaerobes) while, degrading organic matter in the absence of oxygen, application of crop residue may lead to rise in CH_4 during paddy farming (Figure 12.1).

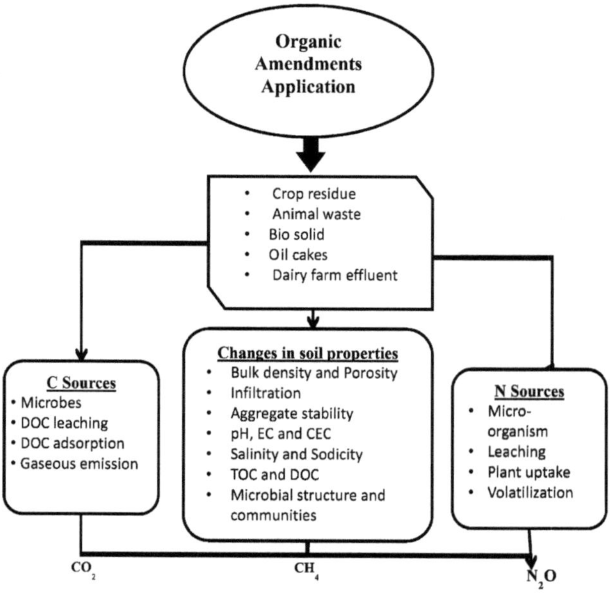

FIGURE 12.1 Greenhouse gas released by organic amendment.

Source: Modified from Lee et al. (2020).

Emission of CH_4 depends upon carbon availability (Lee et al., 2020), and is influenced by the stability between CH_4 oxidation and production processes. CO_2 mainly emitted from biogenic sources such as rhizomicrobial respiration, basal respiration, root respiration, during degradation of plant remaining by microorganism. Nitrous oxide (N_2O) emission takes place by different process like nitrification and denitrification (Figure 12.1). Organic amendments provide nutrition to the microorganism which cause increase in their biomass and deforestation rates due to decrease in redox potential of the soil. Agroforestry practices identifies as a sustainable approach, that integrates both forestry and agriculture LUs to gain maximum profit by increasing resource efficiency and CC mitigation and adaptation (Schoeneberger et al., 2012; Cubbage et al., 2013). Many studies mentioned about agroforestry practices occurred mainly in the tropical regions due to supporting climatic condition and different socio-economic factors. The potential of agroforestry regarding CC adaptation and mitigation is growing in temperate and boreal forest system (Nair et al., 2008; Schoeneberger et al., 2012). Rising the biomass of agroforestry for biofuels and biopower which replaces the use of fossil fuel in order to decrease the atmospheric CO_2 (Jose & Bardhan, 2012). There were several global programs which comprises agroforestry such as Reducing Emission from Deforestation and Forest Degradation incorporates the function of sustainable management of forest, conservation, and enrichment of carbon stock in forest (REDD+) associated with adaptation and mitigation of CC (Nair & Garrity, 2012).

According to IPCC report, carbon sequestration was supported by the reforestation and afforestation (Nair et al., 2010). Soil carbon sequestration plays crucial role in renovating soil quality and food security (Paustian et al., 2016; Smith, 2016; Fuss et al., 2018; Sykes et al., 2020) which lead towards the achievement of sustainable development goals (Keesstra et al., 2016; Chabbi et al., 2017). Carbon sequestration is similar across different terrestrial ecosystem (Nair et al., 2010). Several crop rotation practices and application of different organic amendments boost carbon input in the soil. Incorporating biochar under maize-rice cropping system enhanced soil organic carbon (SOC) sequestration. This type of amendment significantly enhanced SOC by 22 and 19% in rice-rice system and by 30 and 31% under maize-rice system. Addition of biochar shows more (9%) concentration of SOC under maize-rice in contrast to rice-rice cropping system (Mehmood et al., 2020). Changes in the LU pattern and crop rotation could vary soil stability and arrangements of SOC under various agricultural practices, finally changes in the potential of SOC sequestration (Wang et al., 2014; Zhang et al., 2017).

Mixture of biochar and straw increases rice yield when added into the soil. Incorporating biochar increases crop production and environmental sustainability under maize-rice cropping system. Fang et al. (2020) reported that application of organic amendments (sorghum stubble or sugarcane bagasse), gypsum and nutrients improved the soil performance and crop yield in sodic subsoil (below 20 cm). In comparison to topsoil (0–10 cm), subsoil with higher concentration of clay have more capacity to store carbon. Wang et al. (2015a–c) documented that the rate of SOC mineralization decreases with the addition of organic amendments.

Several studies were devoted on biochar applications as organic amendment for agriculture purposes. It imparts various benefits such as it promotes nutrient retention capacity and humification process when added with compost (Guo et al., 2020). Collective application of compost and biochar to soil give sound agronomic effects, increasing microbial biomass, restraining emission of GHG and ammonium, improving immobilization process in organic pollutants and heavy metals, reduce salinity stress and increasing plant growth and properties (physical, chemical, and biological) of soil. Biochar having distinctive physical and chemical attributes attracted attention of many researchers. Biochar amendment at the rate of 30 ton/hectare increased SOC sequestration in the topsoil horizon (15 cm depth) by 16% in maize-based cropping system over two years (Yang et al., 2020). Biochar integrated with corn residue reduced GHG (N_2O and CO_2) production. Application of slow-release fertilizer (SRF) was reported to improve nitrogen use efficiency (NUE) in the soil (Gupta et al., 2020). Coating urea particles with starch polymers in fertilizers slows down the nutrient release rate to provide them during the period of crop requirement. Moreover, SRF have the potential to diminish GHG mainly N_2O production by bringing modification in the nitrification and denitrification process in soil. In the following section different variety of organic amendments used in the agroforestry will be discussed, which support in the restraining GHG production and further help in climate mitigation and adaptation and incorporating organic amendment into the soil determinately increases soil carbon stock.

12.2 SOIL ORGANIC AMENDMENTS

It is the configurations of organic moieties originate from living biomass. It generally involves green manure, biochar, wood chip, husk, compost, animal manure, sewage manure, straw, and geo-textile. These materials are particularly rich in organic matter and micro- and macro-elements that enhance soil fertility by improving microclimate condition and supply substrate for microbial growth (Maiti & Ahirwal, 2019). Application of organic amendments in agriculture makes soil more effective in nature. It is helpful in recovering soil quality, fertility, and maximizes crop productivity. Various amendments are used in organic agriculture that aids materials with natural origin, i.e., organic or minerals (Al-Busaidi et al., 2020). Different studies were conducted on the characteristics of organic amendments, when applied to the soil, cause alteration in the soil carbon stock, manufacture, carbon quantity, pH ratio and C/N ratio (Chenu et al., 2019). Many studies had discussed about the impact of residue quality on SOC stock. The recalcitrant materials help in degrading compound very easily and contribute further to soil organic matter (SOM) on longer periodic basis for example clayey soils which are rich in lignin (Cotrufo et al., 2013). Long-term SOM basically resulting from microbial substances are easily degradable (Miltner et al., 2012; Kallenbach et al., 2015). Moreover, soil amendment influences different soil properties for instance Al-Busaidi et al. (2020) reported that compost or manure increases the values of electrical conductivity (EC) when incorporated into soil. Tekwa et al. (2013) observed maximum EC in cropped soil in contrast to un-cropped soil. Soil amended with organic crop residue lead to increase in SOM (Hannachi et al., 2015). Soil physical parameter like water holding capacity (WHC) was found to be elevated in the soil having no-tillage (Bello et al., 2010).

12.3 TYPES OF ORGANIC AMENDMENTS

There are numerous sources and categories of organic amendments that make improvement in the soil quality and its property. Organic amendments applied in agriculture are municipal solid waste, yard waste, humic substances, bone meal and blood, animal manure and crop residue. A huge variety of organic amendments value has grown into the market that eventually increases organic farming system.

12.3.1 BIOCHAR

Biochar released from pyrolysis treatment made from biomass and amended with soil for improving health and productivity. According to Trupiano et al. (2017) soil supplemented with biochar has altered soil chemical properties like pH and EC whereas, some of the studies also showed that soil amended with biochar shows Na^+ adsorbing properties and lower soil EC (Thomas et al., 2013;

Akhtar et al., 2015; Lashari et al., 2015; Ali et al., 2017). Numerous studies are done on biochar application revealed that improve soil water storage capacity, organic matter and total organic carbon (TOC). It also increases plant growth traits and yield attributes for example leaf number, plant height, final mass, crop yield, crop growth and productivity and chlorophyll content (Hua et al., 2012; Agegnehu et al., 2015). Thus, biochar effect on soil is not universal, it mainly depends upon biochar type, kind of soil, crop variety and experiment condition (Liu et al., 2013). Biochar produced from various source contain porous carbons, and composed of micropores less than 2 nm, meso-pores 2–50 nm, and macro-pores greater than 50 nm. Pore shaving size 2–20 nm provides large surface area and better adsorption potential of biochar. These properties, makes biochar as a tool for eradicating organic pollutants from polluted soil and water resources (Singh et al., 2016) while, also have many agronomic benefits when applied as soil amendments (Cao et al., 2018). It is used to boost the soil properties, soil fertility, plant growth attributes and carbon sequestration (Lehmann et al., 2011); improve crop yield parameters and reduce plant stress mainly due to salinity and drought (Kanwal et al., 2018); immobilize heavy metals (Puga et al., 2015) and remediate soil contaminated by organic pollutant (Chen et al., 2017; Rao et al., 2017).

12.3.2 COMPOST

For improving soil health and productivity one of the traditional organic amendments, compost is broadly used in agriculture. In many studies composts demonstrated the desirable results regarding plant health and its productivity by increasing pH and EC. The organic material present in the soil have potential to reduce the pH of the soil by mineralization processes in which organic acid is released which finally decreases the soil pH. Wang et al. (2018) reported that composting is the finest way to use the organic matter again, which present within the organic wastes. Much research paper showed that addition of biochar with compost reduces GHG emissions and generate valuable slow-release organic fertilizer based on biochar (Zafar et al., 2020). Pandit et al. (2019) mentioned that co-composite (composting material mixed with organic amendment during composting process) mixed with biochar increases the physicochemical properties and nutrient availability of the soil (Vandecasteele et al., 2016). Co-composting process involves organic coating on the biochar particles, which diminish hydrophobicity of biochar particles and support in retaining nutrients that improves the agronomic performance of the soil (Kammann et al., 2015; Hagemann et al., 2017; Joseph et al.,

2018). The organic cover on the biochar particles may influence the soil redox potential (Hagemann et al., 2017). Redox potential was found to be higher in the compost amended with biochar during composting stage in contrast to non-biochar amended compost.

12.3.3 CROP RESIDUE

Crop residue plays a key role in improving physical, chemical, and biological characteristic of soil (Turmel et al., 2015). Several studies were conducted on the crop residues in developing countries which showed positive result on soil quality, carbon sequestration, SOM, bulk density, enhancing nutrient cycling, moisture retaining capacity and reduce soil losses and environmental risk. Soil amended with crop residue produced higher grain yield and biomass leaving greater crop residue after harvest (Govaerts et al., 2005). Furthermore, crop yield is positively correlated with the crop residue, hence more quantity of crop residue is available for mixing in soil, animal feed or other purposes (Figure 12.2).

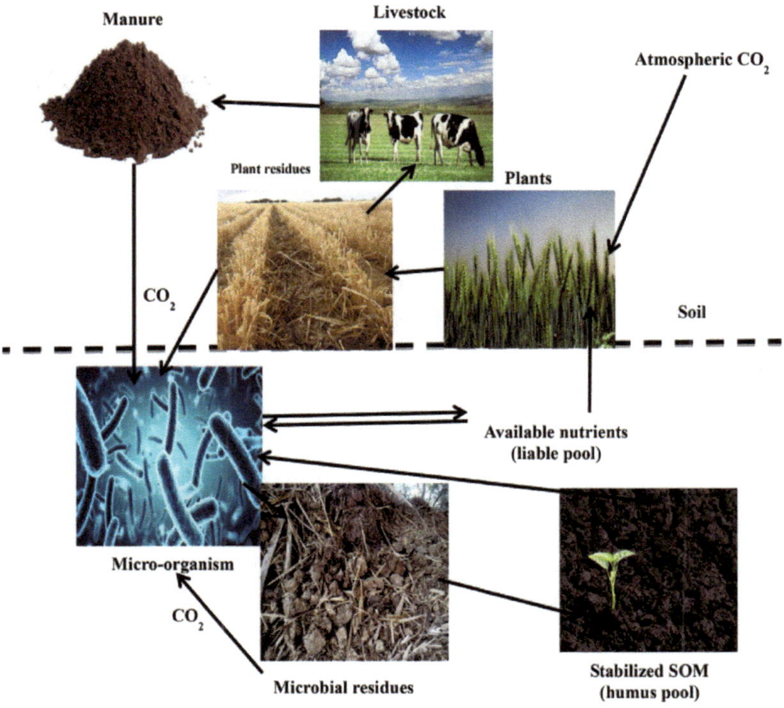

FIGURE 12.2 Plant residue transformations by soil microbes.

Source: Modified from Turmel et al. (2015).

12.3.4 ANIMAL MANURE

Maillard & Anders (2013) reported that soil applied with animal manure was found to be significantly higher in build-up of SOC. Several studies have reported that supply of carbon increases in the soil by mixing crop residues and animal manure. They are the direct source of carbon input for the soil. Parham et al. (2002) reported that use of application of animal manure was reported to increase soil pH from 4.7 to 5.7 which is in contrast to chemical fertilizer application making soil more acidic. Application of manure also regulates phosphorous cycling, microbiological activities and provide adequate nutrients for plant development and production (Parham et al., 2002). Therefore, long term function of cattle manure increases the phosphorus cycle up to 0–30 cm soil depth and promote microbial performance in the soil.

Poultry manure (PM) is a component of poultry litter, is an outcome of poultry farming. It also comprises of feather, spilled feed, sawdust, and straw that is used in the breeding. Poultry litter contains high moisture around 40–45% with high ash properties such as high chlorine, sodium, potassium, and phosphorous. It is regarded as a fuel that causes intense fouling and slagging of heat transfer surfaces (El-Haggar, 2007). Boateng et al. (2009) reported positive effect of PM on maize crop growth. PM is rich in calcium content that increases pH level from 4.3 to 4.6. It also has high EC and low ammonium content. Soil treated with PM (8t PM/ha) resulted in high moisture and organic matter content in comparison to control (soil without treatment). Several studies promoted the use of PM and regarded as precious fertilizer. Integrating synthetic fertilizer with the PM is one of the best ways to minimize the use of synthetic fertilizer and also to gain maximum yield.

12.3.5 BIOFERTILIZER

Biofertilizer are eco-friendly source of plant nutrient and cost effective in nature which improves plant nutrients and soil quality (Singh et al., 2016). Dębsk et al. (2016) reported that application of biofertilizer in the arable soils help to improve soil fertility and enhance stability of organic matter. Filamentous blue green algae (BGA) (algal biofertilizer) area natural biofertilizer based on renewable energy resource and protect environment from pollution. Application of biofertilizer amplifies CO_2 sequestration in rice cultivation. Moreover, it increases plant height, length of leaf and grain yields (Wang et al., 2015c). Among all biofertilizers, phosphate solubilizing bacteria (PSB) is prominent (Rao, 2014). Biofertilizers are mostly nitrogen fixer, phosphate-solubilizing, and cellulolytic microbes. Nitrogen fixer biofertilizers fix atmospheric nitrogen into soil in available forms. These include Rhizobium, Azospirillum, Azotobacter, Azolla, and BGA. Rhizobiums develop symbiotic relationship with the root nodules of legumes for nitrogen fixing. PSB absorb several organic acids which increase phosphorous uptake by the plant by dissolving tricalcium phosphate (TCP) and rock phosphate in soil (Dotaniya et al., 2015). Rhizobium biofertilizer improve yield traits in pulse crop in temperate climate. Use of Azospirillum in agroforestry significantly improves the harvest index, yield traits and leaf area index (Singh et al., 2016). *Azotobacter chroococcum* combined with phosphate fertilizer increased soil enzymes performance and alkaline phosphatase function in the soil (Singh et al., 2016). In a greenhouse experiment, combined application of *Glomus fasciculatum* + *A. chroococcum* + 50% of recommended phosphorous increased the onion bulb size (Singh et al., 2016).

12.4 MINFLUENCE OF ORGANIC AMENDMENTS

12.4.1 EFFECT ON SOIL PHYSICAL PROPERTIES

For determining the soil resistance to erosion and sustainable crop production, soil structure is considered as an important component in agricultural system (Panachuki et al., 2011). Physical stabilization of SOM and soil structure may be influenced by many factors like use of machinery, residue amendment, rainfall, and tillage practices (Blanco-Canqui et al., 2006). Addition of crop residue can recover soil structure by – (i) enhancing soil aggregation by adding organic matter to the topmost soil; (ii) shielding soil aggregate from the raindrop force; and (iii) preserve soil from compaction caused by the rainfall (Verhulst et al., 2010).

Soil stability and its aggregate may affect the soil porosity, root progression and motion of gas, water, and nutrients within the soil system. Soil aggregate act as a cover for the organic matter and form complex organo-mineral which is hardly bounded which even microbes cannot degrade (Edwards & Bremner, 1967). Retention of crop residue on the surface protect from soil loss against water, reduce surface runoff (Panachuki et al., 2011), lessen evaporation which maintain the water content in the soil making it drought resilient (Verhulst et al., 2011). Adding crop residues are helpful in reducing the daytime surface temperature of the soil (Verhulst et al., 2010).

This cooling effect of soil hot tropical climate is beneficial for optimum plant growth and development.

12.4.2 EFFECT ON SOIL CHEMICAL PROPERTIES

SOC is regarded as an important parameter of soil quality and performing sustainable agriculture as it increases the water retention ability, soil aggregation and nutrient reservoir for plant (Liu et al., 2006). Crop residue is the key to increase and retain the SOC level (Govaerts et al., 2009). Decomposition rates of accumulated SOC may be affected by many climatic factors either it is present at the surface or incorporated inside the soil. Soil pH is also affected by the composition of crop residues and soil characteristics (Butterly et al., 2011). Residues of legumes crops, and ash alkalinity have greater effect on soil pH in comparison to wheat residue, because wheat residue has lower ash alkalinity and nitrogen concentration. Addition of the legumes root decrease the soil pH as root contain mainly total nitrogen content in the plant system (Xu & Coventry, 2013). Cation exchange capacity (CEC) is an indicator of soil fertility. Crop residue increases the SOM and therefore increases the CEC, which depends on pH (Turmel et al., 2015). Crop retention affects the nutrient content in the soil. Addition of legumes plant residue with low carbon nitrogen ratio increases the mineralization rate of nitrogen while, cereals residues having high carbon nitrogen ratio increases nitrogen immobilization during its decomposition (Govaerts et al., 2006).

12.4.3 EFFECT ON BIOLOGICAL PROPERTIES

Soil biodiversity plays a crucial role in agroecosystems by affecting quality of crop, soil borne pest presence, nutrient cycling and water movement (Brussaard et al., 2007). Soil biodiversity comprises of soil fauna and microflora (fungi and bacteria). Soil fauna classified according to size like microfauna (nematodes and protozoa), mesofauna (acarids and enchytraea) and macrofauna (earthworms, termites, and large arthropods). These organisms degrade and decompose organic matter (Roger-Estrade et al., 2010). Breaking down of SOM by the microbes affects the water and nutrient availability in agroforestry. Microbial biomass in the soil referred as a live portion of SOM (Salinas-Garcia et al., 2002). Addition of crop residue help in stimulates the soil microbial biomass as well as microbial activity.

12.5 INCREASING SOC SEQUESTRATION IN AGROFORESTRY

Carbon sequestration rate and magnitude can be changed by livestock, tress, and crop production management system (Nair et al., 2010). For instance, net rise in SOC pool depends on the choice of soil management practices and agroforestry system (Table 21.1) that affects the quality and amount of carbon input particularly beneath the tree canopy.

TABLE 12.1 Rate of SOC Sequestration in Agroforestry Systems

Agroforestry System	Soil Depth (cm)	Age (Years)	Sequestration Rate (MgC/ha/yr)	References
Shelterbelt agroforestry system.	0–20	–	0.92	Hubner et al. (2021)
	20–40	–	0.72	
	40–60	–	0.52	
Alley cropping system.	0–40	13	1.25	Lorenz & Lal (2014)
Shaded cacao systems.	0–100	30	302	Gama-Rodrigues et al. (2010)
Shaded coffee, *Coffea robusta* L. Linden þ *Albizia* spp.	0–40	13	97.3	Dossa et al. (2008)
Alley cropping system: *Erythrina poeppigiana* (Walp.) O.F. Cook + maize and bean (*Phaseolus vulgaris L.*).	0–40	19	2.34	Oelbermann et al. (2006)
–	0–20	–	1.79	–
Faidherbia albida (Delile) A. Chev. parkland.	0–100	35	33.3	Takimoto et al. (2008)
Intercropping system: hybrid poplar + barley.	0–20	19	1.04	Peichl et al. (2006)
Multistrata agroforest: cacao + canopy trees.	0–15	15	0.06	Isaac et al. (2005)

Stand density of trees may affect the SOC sequestration for example, advanced stocking levels of trees was reported to increase the vegetation carbon pool (Nair et al., 2010). In contrast, thinning and trimming of trees reduce litter production and thus SOC sequestration. Higher plant biomass results in more organic carbon pool in the soil. Stabilization and lower decomposition rate of SOM amplify in SOC pool (Sollins et al., 2007). In contrast to monocultures, agroforestry systems are efficient in utilizing the available resource present in the surrounding for biomass growth and boosting soil carbon build up. Some agroforestry practices accelerate the direct carbon input (Lorenz & Lal, 2014). These involve returning of trimmed biomass back into the soil for decomposition. Grazing should be allowed so as to add animal manure (cow) dung in the soil. Integration of trees litter and animal manure increases crop production.

12.6 STABILIZATION OF ORGANIC CARBON IN SOIL

Sequestration of SOC mainly depends upon the carbon input and soil stabilization processes. Among all plant root, mainly legumes root gives larger input to SOC sequestration. Addition of SOC is particularly the outcome of incomplete degradation, microbial products and fire residues. Carbon residues present at the soil surface may be mixed into mineral soil by physical mixing and transport, solubilization, and adsorption (Lorenz & Lal, 2005). One of the primary vectors is plant root (litter and rhizodeposition) that allows carbon in the SOC pool (Rass et al., 2005). The organic residues reside inside and outside of the soil aggregates. The comparative significance of root (litter and rhizodeposition) against other amending processes for SOC circulation and motion rely upon soil, climate, and types of vegetation. The turnover of small and fine roots transmits large fraction of net primary productivity (NPP) in the soil (Guo et al., 2008). Rhizodeposition explain about how organic carbon compound released from roots (Jones et al., 2009). Many isotopes labeling was used to estimate the quantity of fixed carbon by plant through photosynthesis separation below the ground have targeted on young plants at initial stage this partitioning is highly affected by age of plant.

Adsorption of organic compound that can be dissolved in the soil profile is a direct carbon input. Dissolve organic carbon (DOC) is finally returned as CO_2 in the atmosphere (Bolan et al., 2011). There are some sources of litter depositor include humus, crop residue and uses of organic amendment like biosolids and manure are significant DOC resource used in the agroforestry system. DOC resides in the subsoil and is linked with insufficient presence of iron and aluminum (hydro) oxides along with high surface area. Therefore, translocation of DOC helps in the development of mineral bonding in the subsoil (Schrumpf et al., 2013). Some DOC leach in the nearby water bodies (Bolan et al., 2011). This carbon loss by leaching is vital for carbon equilibrium in agroforestry system (Kindler et al., 2011).

Stability of SOC in an ecosystem is mainly depends on the biotic and abiotic components of environment (Schmidt et al., 2011). Residence time of SOC is determined by organic matter and molecular structure of input plant. Materials derive from microbes play vital role in SOC stabilization (Kleber et al., 2011). Turnover of SOC is the function of microbial ecology and availability of resource within limited physical soil environment. Hence, major centennial scale stabilization mechanism is a slowdown process for SOC (Sanderman et al., 2010). The significant processes which lead to long term stabilization of SOC are: (i) practices which use for physical protection; provide organic matter spatially inaccessible to decomposers; and (ii) interactions of organic matter with metal ions, minerals, and organic substances leads to the formation of organomineral complexes. Physical protection and biochemical recalcitrance allow SOC to persist in soil foe longer period for organomineral complexes to form (Six et al., 2000). Hence, SOC stabilization is a combination of short- and long-term procedure in agroforestry system.

12.7 EFFECTS OF ORGANIC AMENDMENTS ON SOIL CARBON DYNAMICS IN AGROFORESTRY

Intensive agriculture with refining crop productivity can enhance soil C sequestration due to better yields of root biomass, crop residues as well as root exudates to the soil (Mandal et al., 2020). Agro-ecosystems have been considerably sequestrated C into soil that can increased C input to the soil or reduction of C emissions, or both (Nieder & Benbi, 2008). Benbi & Brar (2009) reported intensive agriculture patterns enhanced the SOC by 38% with progressive increase in crop yield. Availability of SOC in the soil depends upon the decomposition rate of the SOC along with quantity as well as composition of crop residues, plant roots and organic material (Follett, 2001). Above ground crop residues are significantly enhancing the SOC content as compared to below ground crop residues (Bolinder et al., 2020). C contents in agroecosystems can be accomplished in a number of ways such as selection of suitable crops varieties which produce high biomass, recycling of residue, reduced tillage intensity, improved nutrient and water management practices, application of organic materials (e.g., animal manure, compost, sludge, and green manure), reducing summer or winter fallow, changing from monoculture to rotation cropping, and switching from annual crops to perennial vegetation (Benbi, 2013).

Organic amendments *v.i.z* animal manure, plant manure, farmyard manure, etc., in soil enhanced the crop yield, since it is a source of organic matter which leads to greater crop-mediated C input to the soil (Diacono

& Montemurro, 2011). An experiment conducted by Pathak et al. (2011) showed that application of organic amendments along with NPK fertilizer on an average sequester C at the rate of 0.17 t C ha^{-1} yr^{-1}. Moreover, organic amendments in the soil also improved the formation of macro-aggregates leading to C storage inside the aggregates, which is protected from decomposition (Sodhi et al., 2009). Soil microorganism plays an important role in the formation of soil C pools and labile form of C has been largely influenced by the activity and population of microorganisms. When organic material is added into the soil, the decomposition process will take place and at same time there is more carbon being consumed by soil microorganisms than nutrients released in the soil. Nevertheless, the ability of these microorganisms to release plant nutrients, such as nitrogen is influenced by the quality and quantity of organic matter inputs added into the soil. Therefore, the application of crop residue or organic materials capture and store more carbon than nutrients like nitrogen, then soil microbes consume all the nutrients contained within OM for their own growth. At the same time, if the added crop residues or organic material comprise of more nutrients than required by soil microbes, the superfluous will be released into the soil which are available for plants. A higher percentage of OM in the soil can increase the capacity of soil to hold soil nitrogen which supports different microbial diversity and also enhances the growth of arbuscular mycorrhizal fungi (AMF) (Pimentel et al., 2005).

12.8 ROLE OF AGROFORESTRY IN MITIGATION OF GHGS

Agroforestry plays a dynamic function for both in the possible land-management systems and mitigation of GHG emissions (Nath et al., 2020). Likewise, agroforestry practices contribute towards agricultural GHG mitigation activities such as sequestration of C in terrestrial land biomass, decrease GHG emissions and avoiding emissions through reduced fossil fuel and energy usage (Schoeneberger et al., 2017). The mitigation of GHG emissions has been benefited for the ecological services and managed the agroforestry operations that can helps to increase the C capture and storage and fixed the functioning of nutrient cycle in agroforestry as equated with conventional practices (Olson et al., 2000). Several agroforestry practices influenced the GHG emissions through how the trees, crops, livestock components or a combination of these three showed advanced mitigation of GHG. Somehow, agroforestry systems have unique properties to mitigate GHG beyond the

C sequestration described before involves knowledge of its impacts on the additional major GHGs in agriculture such as N_2O and CH_4. However, many researchers have been related the impacts of agroforestry on these two GHGs with limited resources. But the proper functioning of nutrient cycle with respect to the greater spatial and temporal diversity in agroforestry systems (Olson et al., 2000) that would be very supportive towards the foundation of agroforestry with neutral benefit effects of reducing of these two GHGs emissions. In addition, the application of fertilizers and grazing in agroforestry will influence the trends and magnitude of mitigation potential with other management activities.

12.8.1 *MITIGATION OF NITROUS OXIDE (N_2O)*

Nitrous oxide (N_2O) is the main GHG released from different agricultural practices and also considered the major anthropogenic sources and is measured to account for around 69% of the calculated annual atmospheric N_2O (IPCC, 2013). In conventional practices, the excess use of fertilizers can lead to an increase in the amounts of nitrate in soils that may be further unutilized by crops with frequently released off-gas as N_2O (Hoben et al., 2011). Therefore, agroforestry practices have been considered as good approaches for the mitigation of N_2O. It was reported that tree-based intercrops (agroforestry) can limit N_2O emissions than herb-lands (Baah-Acheamfour et al., 2017). The reason of that, use of nitrogen-based fertilizer increased N_2O flux from herbaceous lands. Thus, agroforestry practices gained attention to recuperate nitrates (Thevathasan et al., 2004; Dougherty et al., 2009). In agroforestry practices, residual nitrate left from nitrogen fertilizer applications can be assimilated by roots of tree and reduce the emission of N_2O. Evers et al. (2010) reported decreased N_2O emissions from agroforestry was around 1.2 kg N_2O ha^{-1} yr^{-1}. Similar results also found by Amadi et al. (2016) reported about four times decreased N_2O from shelterbelts agroforestry. This is because trees of shelterbelts agroforestry have a large potential to elimination of N fertilization and also great soil aeration capacity which all make favorable condition for reduce N_2O emission. Several other literatures have been also documented reduce N_2O emission from different agroforestry, e.g., afforestation plantings (Allen et al., 2009), windbreaks (Ryszkowski & Kedziora, 2007), and riparian forest buffers (Kim et al., 2009). These studies stated that tree-based intercropping (TBI) eliminate the N-fertilizer inputs and ultimately decrease N_2O emissions. It was also reported that compact nutrient cycling in agroforestry practices should also play a vital role in dipping N_2O emissions (Olson et al., 2000). Other studies also confirmed lower N_2O emission from agroforestry when compared with those of treeless cropping and grazing systems (Allen et al., 2004; Nair et al., 2007; Bambo et al., 2009; López-Díaz et al., 2011).

12.8.2 *MITIGATION OF METHANE*

CH_4 is one of the major GHG emitted from different agricultural practices, livestock, manure management and decomposition of organic material. Agroforestry has the higher potential to affect the CH_4 oxidation and reduction by influencing microbial assisted soil activities (Schoeneberger et al., 2017). However, higher CH_4 uptake has been seen in the forested areas compared with that of cropped fields (Mosier et al., 2006; Gacengo et al., 2009). Lower CH_4 emission is due to lessen soil disturbance, superior soil macro-porosity and lessen nitrogen N fertilizer applications in agroforestry practices compared with cropped fields. Agroforestry system such as silvopasture has also capacity to influence CH_4 emissions. The discussions strongly suggested that agroforestry systems are clever to offset CH_4 emissions whereas conventional cropping practices may aggravate CH_4 emissions (Schoeneberger et al., 2017). Several studies have reported that observations of CH_4 uptake in agroforestry lower than that of forests but greater than for cropping land (Tsuruta et al., 2000). Study showed that average methane consumption through soil is the highest in native and logged forests ranging from 28 to 38 g C m^{-2} h^{-1}, followed by agroforestry systems averaging about 22 g C m^{-2} h^{-1} and least in a cassava/Imperata cropping system at about 15 g C m^{-2} h^{-1} in Sumatra (Tsuruta et al., 2000). In the humid tropics, agroforestry practices mitigated N_2O and CO_2 emissions from soils and enhanced the CH_4 uptake, compared with sole cropping systems (Mutuo et al., 2005). Emission of N_2O in agroforestry system was found as low as 3%, CO_2 emissions 70% and CH_4 uptake found almost twice of the cropping system (Wang et al., 2015a).

12.9 CONCLUSION

Agroforestry with different soil organic amendments has appeared as a holistic LU practice creating a favorable setting for increasing C storage and pay important role in GHGs mitigation in agricultural landscapes. From several literature surveys, we found that agroforestry practices have a large potential to increased vegetative C sequestration, SOC stocks and also decreased N_2O and CH_4 emissions compared to cropland. Different organic amendments such as compost, animal manure, plant residue, etc., in agroforestry have considered as good strategies to increase carbon sequestration with reducing GHG emissions from soil. Studies showed that the process of C sequestration and GHG emissions in agroforestry practices are very complex and depend on various biophysical factors such as climatic conditions, soil properties, water regime, vegetation characteristics, and site-specific management practices. However, extensive research is needed to develop good strategies as well as examining different types of agroforestry systems that have the capacity to sequester C and reduce GHG emissions from agricultural land. Furthermore, studies are also needed to quantify and improve the quality of organic material used in agroforestry and thereby decreased the dependency of mineral fertilizers for both enhancing the C sequestration and mitigating GHG emissions.

KEYWORDS

- **carbon sequestration**
- **climate change**
- **fertilizers**
- **greenhouse gas**
- **organic amendment**
- **organic carbon**

REFERENCES

Agegnehu, G., Bass, A. M., Nelson, P. N., Muirhead, B., Wright, G., & Bird, M. I., (2015). Biochar and biochar-compost as soil amendments: Effects on peanut yield, soil properties and greenhouse gas emissions in tropical North Queensland, Australia. *Agric. Ecosyst. Environ., 213*, 72–85.

Ajayi, O. C., Place, F., Akinnifesi, F. K., & Sileshi, G. W., (2011). Agricultural success from Africa: The case of fertilizer tree systems in southern Africa (Malawi, Tanzania, Mozambique, Zambia, and Zimbabwe). *Int. J. Agric. Sustain., 9*(1), 129–136.

Akhtar, S. S., Andersen, M. N., & Liu, F., (2015). Biochar mitigates salinity stress in potato. *J. Agron. Crop Sci., 201*(5), 368–378.

Al-busaidi, W., Khan, M., Janke, R., Al-Sheriqi, M., & Al-Yahyai, R., (2020). Efficacy of soil amendments. In: Organic farming systems. *Agrociencia., 54*, 1–31.

Ali, S., Rizwan, M., Qayyum, M. F., Ok, Y. S., Ibrahim, M., Riaz, M., Arif, M. S., et al., (2017). Biochar soil amendment on alleviation of drought and salt stress in plants: A critical review. *Environ. Sci. Pollut. Res., 24*(14), 12700–12712.

Allen, D. E., Mendham, D. S., Cowie, A., Wang, W., Dalal, R. C., & Raison, R. J., (2009). Nitrous oxide and methane emissions from soil are reduced following afforestation of pasture lands in three contrasting climatic zones. *Soil Res., 47*(5), 443–458.

Allen, S. C., Jose, S., Nair, P. K. R., Brecke, B. J., Nkedi-Kizza, P., & Ramsey, C. L., (2004). Safety-net role of tree roots: Evidence from a pecan (*Carya illinoensis* K. Koch)–cotton (*Gossypium hirsutum* L.) alley cropping system in the southern United States. *For. Ecol. Manag., 192*(2, 3), 395–407.

Amadi, C. C., Van, R. K. C., & Farrell, R. E., (2016). Soil-atmosphere exchange of carbon dioxide, methane and nitrous oxide in shelterbelts compared with adjacent cropped fields. *Agric. Ecosyst. Environ., 223*, 123–134.

Baah-Acheamfour, M., Chang, S. X., Bork, E. W., & Carlyle, C. N., (2017). The potential of agroforestry to reduce atmospheric greenhouse gases in Canada: Insight from pair wise comparisons with traditional agriculture, data gaps and future research. *For. Chron., 93*(2), 180–189.

Bambo, S. K., Nowak, J., Blount, A. R., Long, A. J., & Osiecka, A., (2009). Soil nitrate leaching in silvopastures compared with open pasture and pine plantation. *J. Environ. Qual., 38*(5), 1870–1877.

Bello, O. S., Solomon, M. G., & Iyapo, O. A., (2010). The effects of over cultivation on some soil properties, nutrients response and yields of major crops grown on acid sand soils of Calabar South-Southern part of Nigeria. In: *19th World Congress of Soil Science, Soil Solutions for a Changing World.*

Benbi, D. K., & Brar, J. S., (2009). A 25-year record of carbon sequestration and soil properties in intensive agriculture. *Agron. Sustain. Dev., 29*(2), 257–265.

Benbi, D. K., (2013). Greenhouse gas emissions from agricultural soils: Sources and mitigation potential. *J. Crop Improv., 27*(6), 752–772.

Blanco-Canqui, H., Lal, R., Post, W. M., Izaurralde, R. C., & Owens, L. B., (2006). Rapid changes in soil carbon and structural properties due to stover removal from no-till corn plots. *Soil Sci., 171*(6), 468–482.

Boateng, S. A., Zickermann, J., & Kornahrens, M., (2014). Poultry manure effect on growth and yield of maize. *West African J. of App. Eco., 9*, 45682.

Bolan, N. S., Adriano, D. C., Kunhikrishnan, A., James, T., McDowell, R., & Senesi, N., (2011). Dissolved organic matter: Biogeochemistry, dynamics, and environmental significance in soils. *Adv. Agron., 110*, 1–75. Academic Press.

Bolinder, M. A., Crotty, F., Elsen, A., Frac, M., Kismányoky, T., Lipiec, J., & Kätterer, T., (2020). The effect of crop residues, cover crops, manures, and nitrogen fertilization on soil organic carbon changes in agroecosystems: A synthesis of reviews. *Mitigation and Adaptation Strategies for Global Change, 25*(6), 929–952.

Brussaard, L., De Ruiter, P. C., & Brown, G. G., (2007). Soil biodiversity for agricultural sustainability. *Agric. Ecosyst. Environ., 121*(3), 233–244.

Butterly, C. R., Kaudal, B. B., Baldock, J. A., & Tang, C., (2011). Contribution of soluble and insoluble fractions of agricultural residues to short-term pH changes. *Eur. J. Soil Sci., 62*, 718–727.

Cao, Y., Gao, Y., Qi, Y., & Li, J., (2018). Biochar-enhanced composts reduce the potential leaching of nutrients and heavy metals and suppress plant-parasitic nematodes in excessively fertilized cucumber soils. *Environ. Sci. Pollut. Res., 25*(8), 7589–7599.

Chabbi, A., Lehmann, J., Ciais, P., Loescher, H. W., Cotrufo, M. F., Don, A., & Rumpel, C., (2017). Aligning agriculture and climate policy. *Nat. Clim. Chang., 7*(5), 307–309.

Cheng, J., Lee, X., Gao, W., Chen, Y., Pan, W., & Tang, Y., (2017). Effect of biochar on the bioavailability of difenoconazole and microbial community composition in a pesticide-contaminated soil. *Appl. Soil Ecol., 121*, 185–192.

Chenu, C., Angers, D. A., Barré, P., Derrien, D., Arrouays, D., & Balesdent, J., (2019). Increasing organic stocks in agricultural soils: Knowledge gaps and potential innovations. *Soil Till. Res., 188*, 41–52.

Cotrufo, M. F., Wallenstein, M. D., Boot, C. M., Denef, K., & Paul, E., (2013). The microbial efficiency matrix stabilization (MEMS) framework integrates plant litter decomposition with soil organic matter stabilization: Do labile plant inputs form stable soil organic matter? *Glob. Change Biol., 19*(4), 988–995.

Cubbage, F., Balmelli, G., Bussoni, A., Noellemeyer, E., Pachas, A. N., Fassola, H., & De Silva, M. L., (2012). Comparing silvopastoral systems and prospects in eight regions of the world. *Agrofor. Syst., 86*(3), 303–314.

Dębska, B., Długosz, J., Piotrowska-Długosz, A., & Banach-Szott, M., (2016). The impact of a bio-fertilizer on the soil organic matter status and carbon sequestration- results from a field-scale study. *J. Soils Sediments, 16*(10), 2335–2343.

Diacono, M., & Montemurro, F., (2011). Long-term effects of organic amendments on soil fertility. *Sustainable Agriculture, 2*, 761–786. Springer, Dordrecht.

Dotaniya, M. L., Datta, S. C., Biswas, D. R., & Meena, H. M., (2015). *Phosphorus Dynamics Mediated by Bagasse, Press Mud and Rice Straw in Inceptisol of North India, 59*(4), 358–369.

Dougherty, M. C., Thevathasan, N. V., Gordon, A. M., Lee, H., & Kort, J., (2009). Nitrate and *Escherichia coli* NAR analysis in tile drain effluent from a mixed tree intercrop and monocrop system. *Agric. Ecosyst. Environ., 131*(1, 2), 77–84.

Edwards, A. P., & Bremner, J. M., (1967). Microaggregates in soils. *J. Soil Sci., 18*(1), 64–73.

El-Haggar, S. M., (2007). Sustainability of agricultural and rural waste management. *Sustainable Industrial Design and Waste Management*, 223–260.

Evers, K. A., Bambrick, A., Lacombe, S., Dougherty, M. C., Peichl, M., Gordon, A. M., & Bradley, R. L., (2010). Potential greenhouse gas mitigation through temperate tree-based intercropping systems. *Open Agric., 4*(1).

Fang, Y., Singh, B. P., Farrell, M., Van, Z. L., Armstrong, R., Chen, C., & Tavakkoli, E., (2020). Balanced nutrient stoichiometry of organic amendments enhances carbon priming in a poorly structured sodic subsoil. *Soil Biol. Biochem.*, 107800.

Follett, R. F., (2001). Soil management concepts and carbon sequestration in cropland soils. *Soil Till Res., 61*(1, 2), 77–92.

Fuss, S., Lamb, W. F., Callaghan, M. W., Hilaire, J., Creutzig, F., Amann, T., & Luderer, G., (2018). Negative emissions- Part 2: Costs, potentials, and side effects. *Environ. Res. Lett., 13*(6), 063002.

Gacengo, C. N., Wood, C. W., Shaw, J. N., Raper, R. L., & Balkcom, K. S., (2009). Agroecosystem management effects on greenhouse gas emissions across a coastal plain catena. *Soil Sci., 174*(4), 229–237.

Govaerts, B., Sayre, K. D., & Deckers, J., (2005). Stable high yields with zero tillage and permanent bed planting? *Field Crops Res., 94*(1), 33–42.

Govaerts, B., Sayre, K. D., Ceballos-Ramirez, J. M., Luna-Guido, M. L., Limon-Ortega, A., Deckers, J., & Dendooven, L., (2006). Conventionally tilled and permanent raised beds with different crop residue management: Effects on soil C and N dynamics. *Plant Soil, 280*(1, 2), 143–155.

Govaerts, B., Verhulst, N., Castellanos-Navarrete, A., Sayre, K. D., Dixon, J., & Dendooven, L., (2009). Conservation agriculture and soil carbon sequestration: Between myth and farmer reality. *CRC Crit. Rev. Plant Sci., 28*(3), 97–122.

Guo, D., Li, H., Mitchell, R. J., Han, W., Hendricks, J. J., Fahey, T. J., & Hendrick, R. L., (2008). Fine root heterogeneity by branch order: Exploring the discrepancy in root turnover estimates between minirhizotron and carbon isotopic methods. *New Phytol., 177*(2), 443–456.

Guo, X. X., Liu, H. T., & Zhang, J., (2020). The role of biochar in organic waste composting and soil improvement: A review. *Waste Manage., 102*, 884–899.

Gupta, K., Bordoloi, N., Baruah, K. K., & Gorh, D., (2020). Role of starch polymer-coated urea in the mitigation of greenhouse gas emissions from rice and wheat ecosystems. In: *Ecological and Practical Applications for Sustainable Agriculture* (pp. 131–148). Springer, Singapore.

Hagemann, N., Joseph, S., Conte, P., Albu, M., Obst, M., Borch, T., & Kappler, A., (2017). *Composting-Derived Organic Coating on Biochar Enhances its Affinity to Nitrate* (p. 10775). EGUGA.

Hannachi, N., Cocco, S., Fornasier, F., Agnelli, A., Brecciaroli, G., Massaccesi, L., & Corti, G., (2015). Effects of cultivation on chemical and biochemical properties of dryland soils from southern Tunisia. *Agric. Ecosyst. Environ., 199*, 249–260.

Hoben, J. P., Gehl, R. J., Millar, N., Grace, P. R., & Robertson, G. P., (2011). Nonlinear nitrous oxide (N_2O) response to nitrogen fertilizer in on farm corn crops of the US Midwest. *Glob. Change Biol., 17*(2), 1140–1152.

Hua, L., Chen, Y., & Wu, W., (2012). Impacts upon soil quality and plant growth of bamboo charcoal addition to composted sludge. *Environ. Technol., 33*(1), 61–68.

IPCC, (2013). *Climate Change- The Physical Science Basis- Findings and Lessons Learned.* IPCC.

Jones, D. L., Nguyen, C., & Finlay, R. D., (2009). Carbon flow in the rhizosphere: Carbon trading at the soil–root interface. *Plant Soil., 321*(1, 2), 5–33.

Jose, S., & Bardhan, S., (2012). Agroforestry for biomass production and carbon sequestration: An overview. *Agrofor. Syst., 86*(2), 105–111.

Joseph, S., Kammann, C. I., Shepherd, J. G., Conte, P., Schmidt, H. P., Hagemann, N., & Mitchell, D. R., (2018). Microstructural and associated chemical changes during the composting of a high temperature biochar: Mechanisms for nitrate, phosphate and other nutrient retention and release. *Sci. Total Environ., 618*, 1210–1223.

Kallenbach, C. M., Grandy, A. S., Frey, S. D., & Diefendorf, A. F., (2015). Microbial physiology and necromass regulate agricultural soil carbon accumulation. *Soil Biol. Biochem., 91*, 279–290.

Kammann, C. I., Schmidt, H. P., Messerschmidt, N., Linsel, S., Steffens, D., Müller, C., & Joseph, S., (2015). Plant growth improvement mediated by nitrate capture in co-composted biochar. *Sci. Rep., 5,* 11080.

Kanwal, S., Ilyas, N., Shabir, S., Saeed, M., Gul, R., Zahoor, M., & Mazhar, R., (2018). Application of biochar in mitigation of negative effects of salinity stress in wheat (*Triticum aestivum L.*). *J. Plant Nutr., 41*(4), 526–538.

Keesstra, S. D., Bouma, J., Wallinga, J., Tittonell, P., Smith, P., Cerdà, A., & Bardgett, R. D., (2016). The significance of soils and soil science towards realization of the United Nations sustainable development goals. *Soil, 2*(2), 111–128.

Kim, D. G., Isenhart, T. M., Parkin, T. B., Schultz, R. C., & Loynachan, T. E., (2009). Nitrate and dissolved nitrous oxide in groundwater within cropped fields and riparian buffers. *Biogeosci. Discuss., 6*(1).

Kindler, R., Siemens, J. A. N., Kaiser, K., Walmsley, D. C., Bernhofer, C., Buchmann, N., & Heim, A., (2011). Dissolved carbon leaching from soil is a crucial component of the net ecosystem carbon balance. *Glob. Change Biol., 17*(2), 1167–1185.

Kleber, M., Nico, P. S., Plante, A., Filley, T., Kramer, M., Swanston, C., & Sollins, P., (2011). Old and stable soil organic matter is not necessarily chemically recalcitrant: Implications for modeling concepts and temperature sensitivity. *Glob. Change Biol., 17*(2), 1097–1107.

Lashari, M. S., Ye, Y., Ji, H., Li, L., Kibue, G. W., Lu, H., Zheng, J., & Pan, G., (2015). Biochar-manure compost in conjunction with pyroligneous solution alleviated salt stress and improved leaf bioactivity of maize in a saline soil from central China: A 2-year field experiment. *J. Sci. Food Agric., 95*(6), 1321–1327.

Lee, J. H., Lee, J. G., Jeong, S. T., Gwon, H. S., Kim, P. J., & Kim, G. W., (2020). Straw recycling in rice paddy: Trade-off between greenhouse gas emission and soil carbon stock increase. *Soil Tillage Res., 199,* 104598.

Lehmann, J., Rillig, M. C., Thies, J., Masiello, C. A., Hockaday, W. C., & Crowley, D., (2011). Biochar effects on soil biota: A review. *Soil Biol. Biochem., 43*(9), 1812–1836.

Liu, X., Herbert, S. J., Hashemi, A. M., Zhang, X., & Ding, G., (2006). Effects of agricultural management on soil organic matter and carbon transformation-a review. *Plant Soil Environ., 52*(12), 531.

Liu, X., Zhang, A., Ji, C., Joseph, S., Bian, R., Li, L., Pan, G., & Paz-Ferreiro, J., (2013). Biochar's effect on crop productivity and the dependence on experimental conditions- a meta-analysis of literature data. *Plant Soil, 373*(1, 2), 583–594.

López-Díaz, M. L., Rolo, V., & Moreno, G., (2011). Tree's role in nitrogen leaching after organic, mineral fertilization: A greenhouse experiment. *J. Environ. Qual., 40*(3), 853–859.

Lorenz, K., & Lal, R., (2005). The depth distribution of soil organic carbon in relation to land use and management and the potential of carbon sequestration in subsoil horizons. **Adv. Agron.,** *88,* 35–66.

Lorenz, K., & Lal, R., (2014). Soil organic carbon sequestration in agroforestry systems. A review. *Agro. Sustain. Develop., 34*(2), 443–454.

Maillard, É., & Angers, D. A., (2013). Animal manure application and soil organic carbon stocks: A meta-analysis. *Glob. Change Biol., 20*(2), 666–679.

Maiti, S. K., & Ahirwal, J., (2019). Ecological restoration of coal mine degraded lands: Topsoil management, pedogenesis, carbon sequestration, and mine pit limnology. In: *Phytomanagement of Polluted Sites* (pp. 83–111).

Mandal, A., Majumder, A., Dhaliwal, S. S., Toor, A. S., Mani, P. K., Naresh, R. K., & Mitran, T., (2020). Impact of agricultural management practices on soil carbon sequestration and its monitoring through simulation models and remote sensing techniques: A review. *Crit. Rev. Environ. Sci. Technol., 1,* 49.

McIntyre, B. D., (2009). *International Assessment of Agricultural Knowledge, Science, and Technology for Development (IAASTD): Synthesis Report with Executive Summary: A Synthesis of the Global and Sub-Global IAASTD Reports No. E14-197.*

Mehmood, I., Qiao, L., Chen, H., Tang, Q., Woolf, D., & Fan, M., (2020). Biochar addition leads to more soil organic carbon sequestration under a maize-rice cropping system than continuous flooded rice. *Agric. Ecosyst. Environ., 298,* 106965.

Miltner, A., Bombach, P., Schmidt-Brücken, B., & Kästner, M., (2012). SOM genesis: Microbial biomass as a significant source. *Biogeochem., 111*(1–3), 41–55.

Mosier, A. R., Halvorson, A. D., Reule, C. A., & Liu, X. J., (2006). Net global warming potential and greenhouse gas intensity in irrigated cropping systems in north-eastern Colorado. *J. Environ. Qual., 35*(4), 1584–1598.

Mutuo, P. K., Cadisch, G., Albrecht, A., Palm, C. A., & Verchot, L., (2005). Potential of agroforestry for carbon sequestration and mitigation of greenhouse gas emissions from soils in the tropics. *Nutr. Cycl. Agroecosyst., 71*(1), 43–54.

Nair, P. K. R., & Garrity, D., (2012). Agroforestry research and development: The way forward. *Agroforestry-The Future of Global Land Use* (pp. 515–531). Springer.

Nair, P. K. R., Gordon, A. M., & Mosquera-Losada, M. R., (2008). Agroforestry. In: Jorgensen, S. E., & Fath, B. D., (eds.), *Ecological Engineering, Encyclopedia of Ecology* (Vol. 1, pp. 101–110). Elsevier, Oxford.

Nair, P. R., Nair, V. D., Kumar, B. M., & Showalter, J. M., (2010). Carbon sequestration in agroforestry systems. In: *Advances in Agronomy* (pp. 237–307).

Nair, V. D., Haile, S. G., Michel, G. A., & Nair, P. K., (2007). Environmental quality improvement of agricultural lands through silvopasture in south-eastern United States. *Scientia. Agricola, 64*(5), 513–519.

Nath, A. J., Sileshi, G. W., Laskar, S. Y., Pathak, K., Reang, D., Nath, A., & Das, A. K., (2020). Quantifying carbon stocks and sequestration potential in agroforestry systems under divergent management scenarios relevant to India's nationally determined contribution. *J. Clean. Product.,* 124831.

Nieder, R., & Benbi, D. K., (2008). *Carbon and Nitrogen in the Terrestrial Environment.* Springer Science & Business Media.

Olson, R., Schoeneberger, M., & Aschmann, S., (2000). An ecological foundation for temperate agroforestry. In: Garrett, H. E., Rietveld, W. J., & Fisher, R., (eds.), *North American Agroforestry: An Integrated Science and Practice* (pp. 31–61). ASA Special Publication. Madison, WI. *American Society of Agronomy.*

Panachuki, E., Bertol, I., Alves, S. T., Oliveira, P. T. S. D., & Rodrigues, D. B. B., (2011). Soil and water loss and water infiltration in red latosol under different management systems. *Rev. Bras. Ciênc. Solo., 35*(5), 1777–1786.

Pandit, N. R., Schmidt, H. P., Mulder, J., Hale, S. E., Husson, O., & Cornelissen, G., (2019). Nutrient effect of various composting methods with and without biochar on soil fertility and maize growth. *Arch. Agron. Soil Sci.*

Parham, J. A. S. P., Deng, S., Raun, W., & Johnson, G., (2002). Long-term cattle manure application in soil. *Biol. Fertil. Soils, 35*(5), 328–337.

Pathak, H., Byjesh, K., Chakrabarti, B., & Aggarwal, P. K., (2011). Potential and cost of carbon sequestration in Indian agriculture: Estimates from long-term field experiments. *Field Crops Res., 120*(1), 102–111.

Paustian, K., Lehmann, J., Ogle, S., Reay, D., Robertson, G. P., & Smith, P., (2016). Climate-smart soils. *Nature, 532*(7597), 49–57.

Pimentel, D., Hepperly, P., Hanson, J., Douds, D., & Seidel, R., (2005). Environmental, energetic, and economic comparisons of organic and conventional farming systems. *Bioscience, 55*(7), 573–582.

Puga, A. P., Abreu, C. A., Melo, L. C. A., & Beesley, L., (2015). Biochar application to a contaminated soil reduces the availability and plant uptake of zinc, lead, and cadmium. *J. Environ. Manage., 159*, 86–93.

Rao, D. L. N., (2014). Recent advances in biological nitrogen fixation in agricultural systems. *Proc. Natl. Acad. Sci., 80*(2), 359–378.

Rao, M. A., Simeone, G. D. R., Scelza, R., & Conte, P., (2017). Biochar based remediation of water and soil contaminated by phenanthrene and pentachlorophenol. *Chemosphere, 186*, 193–201.

Rass, D. P., Rumpel, C., & Dignac, M., (2005). Is soil carbon mostly root carbon? Mechanisms for a specific stabilization. *Plant Soil, 269*, 341–356.

Roger-Estrade, J., Anger, C., Bertrand, M., & Richard, G., (2010). Tillage and soil ecology: Partners for sustainable agriculture. *Soil Till. Res., 111*(1), 33–40.

Ryszkowski, L., & Kędziora, A., (2007). Modification of water flows and nitrogen fluxes by shelterbelts. *Ecol. Engg., 29*(4), 388–400.

Salinas-Garcıa, J. R., De Velázquez-García, J. J., Gallardo-Valdez, M., Dıaz-Mederos, P., Caballero-Hernández, F., Tapia-Vargas, L. M., & Rosales-Robles, E., (2002). Tillage effects on microbial biomass and nutrient distribution in soils under rain-fed corn production in central-western Mexico. *Soil Till. Res., 66*(2), 143–152.

Sanderman, J., Farquharson, R., & Baldock, J., (2010). *Soil Carbon Sequestration Potential: A Review for Australian Agriculture.* CSIRO Sustainable Agriculture National Research Flagship.

Schmidt, M. W., Torn, M. S., Abiven, S., Dittmar, T., Guggenberger, G., Janssens, I. A., et al., (2011). Persistence of soil organic matter as an ecosystem property. *Nature, 478*(7367), 49–56.

Schoeneberger, M. M., Bentrup, G., & Patel-Weynand, T., (2017). *Agroforestry: Enhancing Resiliency in US Agricultural Landscapes Under Changing Conditions*. Gen. Tech. Report WO-96. US Forest Service Department of Agriculture, Washington, DC.

Schoeneberger, M., Bentrup, G., De Gooijer, H., Soolanayakanahally, R., Sauer, T., Brandle, J., Zhou, X., & Current, D., (2012). Branching out: Agroforestry as a climate change mitigation and adaptation tool for agriculture. *J. Soil Water Conserv., 67*(5), 128A–136A.

Schrumpf, M., Kaiser, K., Guggenberger, G., Persson, T., Kögel-Knabner, I., & Schulze, E. D., (2013). Storage and stability of organic carbon in soils as related to depth, occlusion within aggregates, and attachment to minerals. *Biogeosciences, 10*, 1675–1691.

Shrestha, B. M., Chang, S. X., Bork, E. W., & Carlyle, C. N., (2018). Enrichment planting and soil amendments enhance carbon sequestration and reduce greenhouse gas emissions in agroforestry systems: A review. *Forests, 9*(6), 369.

Singh, M., Dotaniya, M. L., Mishra, A., Dotaniya, C. K., Regar, K. L., & Lata, M., (2016). Role of biofertilizers in conservation agriculture. *Conservation Agriculture, 4*, 113–134.

Six, J. A. E. T., Elliott, E. T., & Paustian, K., (2000). Soil macro-aggregate turnover and micro-aggregate formation: A mechanism for C sequestration under no-tillage agriculture. *Soil Biol. Biochem., 32*(14), 2099–2103.

Smith, P., (2016). Soil carbon sequestration and biochar as negative emission technologies. *Glob. Change Biol., 22*(3), 1315–1324.

Sodhi, G. P. S., Beri, V., & Benbi, D. K., (2009). Soil aggregation and distribution of carbon and nitrogen in different fractions under long-term application of compost in rice–wheat system. *Soil Till. Res., 103*(2), 412–418.

Sollins, P., Swanston, C., & Kramer, M., (2007). Stabilization and destabilization of soil organic matter—A new focus. *Biogeochem., 85*(1), 1–7.

Sykes, A. J., Macleod, M., Eory, V., Rees, R. M., Payen, F., Myrgiotis, V., et al., (2020). Characterizing the biophysical, economic, and social impacts of soil carbon sequestration as a greenhouse gas removal technology. *Glob. Change Biol., 26*(3), 1085–1108.

Tekwa, J., Garila, Y., & Bashir, A., (2013). An assessment of the physico-chemical properties and fertility potentials of fallow and cultivated vertisols in Numan Area, Adamawa State-Nigeria. *Agric. Sci. Res. J., 3*(3), 104–109.

Thangarajan, R., Bolan, N. S., Tian, G., Naidu, R., & Kunhikrishnan, A., (2013). Role of organic amendment application on greenhouse gas emission from soil. *Sci. Total Environ., 465*, 72–96.

Thevathasan, N. V., Gordon, A. M., Simpson, J. A., Reynolds, P. E., Price, G., & Zhang, P., (2004). Biophysical and ecological interactions in a temperate tree-based intercropping system. *J. Crop Improv., 12*(1, 2), 339–363.

Thomas, S. C., Frye, S., Gale, N., Garmon, M., Launchbury, R., Machado, N., et al., (2013). Biochar mitigates negative effects of salt additions on two herbaceous plant species. *J. Environ. Manage., 129*, 62–68.

Trupiano, D., Cocozza, C., Baronti, S., Amendola, C., Vaccari, F. P., Lustrato, G., Di Lonardo, S., Fantasma, F., Tognetti, R., et al., (2017). The effects of biochar and its combination with compost on lettuce (*Lactuca sativa* L.) growth, soil properties, and soil microbial activity and abundance. *Int. J. Agron.*

Tsuruta, H., Ishizuka, S., Ueda, S., & Murdiyarso, D., (2000). Seasonal and spatial variations of CO_2, CH_4, and N_2O fluxes from the surface soils in different forms of land-use/cover in Jambi, Sumatra. *The Impacts of Land-Use/Cover Change on Greenhouse Gas Emissions in Tropical Asia* (pp. 7–30). Global change impacts center for Southeast Asia and National Institute of Agro-Environmental Sciences.

Turmel, M. S., Speratti, A., Baudron, F., Verhulst, N., & Govaerts, B., (2015). Crop residue management and soil health: A systems analysis. *Agric. Syst., 134*, 6–16.

Vandecasteele, B., Sinicco, T., D'Hose, T., Nest, T. V., & Mondini, C., (2016). Biochar amendment before or after composting affects compost quality and N losses, but not P plant uptake. *J. Environ. Manag., 168*, 200–209.

Verhulst, N., Carrillo-García, A., Moeller, C., Trethowan, R., Sayre, K. D., & Govaerts, B., (2011). Conservation agriculture for wheat-based cropping systems under gravity irrigation: Increasing resilience through improved soil quality. *Plant Soil, 340*(1, 2), 467–479.

Verhulst, N., Govaerts, B., Verachtert, E., Castellanos-Navarrete, A., Mezzalama, M., Wall, P., et al., (2010). Conservation agriculture, improving soil quality for sustainable production systems. *Advances in Soil Science: Food Security and Soil Quality.* CRC Press, Boca Raton, FL, USA.

Wang, G., Welham, C., Feng, C., Chen, L., & Cao, F., (2015a). Enhanced soil carbon storage under agroforestry and afforestation in subtropical China. *Forests, 6*(7), 2307–2323.

Wang, H., Boutton, T. W., Xu, W., Hu, G., Jiang, P., & Bai, E., (2015b). Quality of fresh organic matter affects priming of soil organic matter and substrate utilization patterns of microbes. *Sci. Rep.*, *5*(1), 1–13.

Wang, H., Guan, D., Zhang, R., Chen, Y., Hu, Y., & Xiao, L., (2014). Soil aggregates and organic carbon affected by the land use change from rice paddy to vegetable field. *Ecol. Engg.*, *70*, 206–211.

Wang, Q., Awasthi, M. K., Ren, X., Zhao, J., Li, R., Wang, Z., et al., (2018). Combining biochar, zeolite, and wood vinegar for composting of pig manure: The effect on greenhouse gas emission and nitrogen conservation. *Waste Manag.*, *74*, 221–230.

Wang, R., Peng, B., & Huang, K., (2015c). The research progress of CO_2 sequestration by algal bio-fertilizer in China. *J. CO_2 Util.*, *11*, 67–70.

Xu, R. K., & Coventry, D. R., (2013). Soil pH changes associated with lupin and wheat plant materials incorporated in a red-brown earth soil. *Plant Soil, 250*(1), 113–119.

Yang, W., Feng, G., Miles, D., Gao, L., Jia, Y., Li, C., & Qu, Z., (2020). Impact of biochar on greenhouse gas emissions and soil carbon sequestration in corn grown under drip irrigation with mulching. *Sci. Total Environ.*, 138752.

Zafar-ul-Hye, M., Tahzeeb-ul-Hassan, M., Abid, M., Fahad, S., Brtnicky, M., Dokulilova, T., et al., (2020). Potential role of compost mixed biochar with rhizobacteria in mitigating lead toxicity in spinach. *Sci. Rep., 10*(1), 1–12.

Zhang, T., Chen, A., Liu, J., Liu, H., Lei, B., Zhai, L., & Wang, H., (2017). Cropping systems affect paddy soil organic carbon and total nitrogen stocks (in rice-garlic and rice-fava systems) in temperate region of southern China. *Sci. Total Environ., 609*, 1640–1649.

CHAPTER 13

Restoration of Saline and Sodic Soil Through Using Halophytes as Agroforestry Components

MANISH MATHUR[1] and PREET MATHUR[2]

[1]*ICAR–Central Arid Zone Research Institute, Jodhpur, Rajasthan, India*

[2]*Jodhpur Institute of Engineering and Technology, Jodhpur, Rajasthan, India*

ABSTRACT

The current chapter is on the global problem of salty sand sodic soils, which has an impact on the numerous ecosystem services (ES) of land and soil. The restoration of such lands is the most difficult task for scientific communities, and it can not only improve soil quality but also increase the area under cultivation for food and fodder crops. The following 10 primary headings are associated with the concept of saline and sodic soils; salinity and sodicity influenced soils: Land degradation in terms of soil salinity and sodicity, restoration, and its dynamics, halophytes: characteristics, distribution, classification, and halo-remediation; adaptation mechanism of halophytes under saline conditions; ecological significance of halophytes under saline conditions; potential use of halophytes under saline conditions The role of agroforestry in soil reclamation, as well as experience from the Indian subcontinent, are highlighted. Agroforestry and its role in soil reclamation are further elaborated with scientific information on tree species selection; halophytes for agroforestry; an assortment of multifunctional species for salt-affected areas; and multifunctional agroforestry systems such as silvipastoral, silvi-agric, horti-agricultural, and sequential agroforestry systems. In this

Sustainable Management and Conservation of Environmental Resources in India.
Sumit Chakravarty, Vineeta, Jahangeer A. Bhat, Munesh Kumar, Gopal Shukla (Eds.)

effort, the scientific explorations for saline and sodic soil remediation from the Indian subcontinent were explained under the following sub-headings: salt-tolerant plants/halophytes, suitable planting techniques, agroforestry systems on saline areas, grassland/pastoral systems, afforestation of coastal saline areas, knowledge on afforestation/agroforestry on alkali and or sodic soils, planting techniques. This is the first endeavor from India's arid region to elaborate on the restorative dynamics of saline and salt-damaged soils, namely reclamation, remediation, and rehabilitation. Furthermore, thorough and comparative classifications, as well as species from other halophytic categories, were discussed. Finally, research gaps in the usage of several halophytes for restoration of such lands were identified. Our understanding of the physiological and biochemical mechanisms underlying halophytes under varying salinities is particularly limited. Thus, physiological and molecular research into the underlying mechanisms of these processes is critical. It would be important to find the genes that are linked to the high biomass phenotype and then make transgenic plants with better features for cleaning up pollution.

13.1 CONCEPT OF SALINE AND SODIC SOILS

Excess concentrations of NaCl, Na_2CO_3, Na_2SO_4, or magnesium salts are commonly associated with soil salinity issues. As electrical conductivity (EC), salinity quantifies the concentration of all soluble salts in soil or water (EC). A soil is classified as saline when its EC at 25°C equals or exceeds 4 dS m^{-1}. Sodicity, on the other hand, monitors sodium ions (in soil or water relative to calcium and magnesium ions). It is either stated as a sodium adsorption ratio (SAR) or as a percentage of exchangeable sodium (ESP). If SAR is greater than or equal to 13 (mmoles $L^{-1})^{0.5}$ or ESP is greater than or equal to 15, the soil is considered sodic. The entire land surface area on the planet 13.2×10^9 of which only 7×10^9 ha is arable and only 1.5×10^9 ha is cultivated (Massoud, 1981). About 0.34×10^9 ha (23%) of the cultivated fields are saline, whereas the remaining 0.56×10^9 (37%) are sodic (Shahid, 2013). There are two types of soil salinity: "dryland salinity," which occurs on unirrigated land, and "irrigated-land salinity." Many types of saline soil classification are available in scientific literature, including the US Salinity Laboratory Staff Classification (Richards, 1954), the FAO-UNESCO Classification (FAO-UNESCO, 1974), the Australian Salinity Classification (Isbell, 1998), and the Russian and Chinese Saline Soil

Classification (Shahid, 2013). The USDA technique was used to determine slight, moderate, and strong categories based on EC (4.0–8.0, 8.1–30.0, and > 30 dS m^{-1}.), pH (8.5–9.0, 9.1–9.8, and > 9.8), and ESP (<15, 15–40, > 40) values (Table 13.1) (Richards, 1954). The codes A through H were used to describe physiographic units such as "A" for alluvial, "B" for aeofluvial/aeolian/arid, "C" for deltaic, "D" for coastal, "F" for peninsular, "G" for mudflats/mangrove swamps, and "H" for others. Thus, 15 kinds of soil were identified (Table 13.2). It had three saline soil types (sl, s2, and s3), three sodic soil categories (nl, n2, and n3), and nine saline-sodic soil classifications (slnl, sln2, sln3, s2nl, s2n2, s2n3, s3nl, s3nl, s3n2, s3n3, and s3n3).

TABLE 13.1 MDegree of Salinity and Sodicity

Degree	Sodicity		Salinity
	pHs	ESP	ECe (dSm–1)
Slight	8.5–9.0	<15	4.0–8.0
Moderate	9.1–9.8	15–40	8.1–30.0
Strong	> 9.8	> 40	> 30

TABLE 13.2 Description of Saline and Sodic Soils

Code and Categories of Salt Affected Soil	Sodicity Class	ESP Range	pH Range	Salinity Class	Ec (dS m–1)
sl-slightly saline.	Low	Traces	Low	Slight	4.0–8.0
s2-moderately saline.	Low	Traces	Low	Moderate	8.1–30.0
s3-strongly saline.	Low	Traces	Low	Strong	>30.0
n1-lightly sodic.	Slight	< 15	8.5–9.0	Low	Low
n2-moderately sodic.	Moderate	15–40	9.1–9.8	Low	Low
n3-strongly sodic.	Strong	> 40	> 9.8	Low	Low
s1n1-slightly saline, slightly sodic.	Slight	< 15	8.5–9.0	Slight	4.0–8.0
s1n2-slightly saline, moderately sodic.	Moderate	15–40	9.1–9.8	Slight	4.0–8.0
s1n3-slightly saline, strongly sodic.	Strong	> 40	> 9.8	Slight	4.0–8.0
s2n1-moderately saline, slightly sodic.	Slight	< 15	8.5–9.0	Moderate	8.1–30.0
s2n2-moderately saline, moderately sodic.	Moderate	15–40	9.1–9.8	Moderate	8.1–30.0
s2n3-moderately saline, strongly sodic.	Strong	> 40	> 9.8	Moderate	8.1–30.0
s3n1-strongly saline, slightly sodic.	Slight	< 15	8.5–9.0	Strong	>30.0
s3n2-strongly saline, moderately sodic.	Moderate	15–40	9.1–9.8	Strong	>30.0
s3n3-strongly saline, strongly sodic.	Strong	> 40.0	> 9.8	Strong	>30.0

Soil salinization can be caused by either – (i) primary or natural salinization, which occurs naturally; or (ii) secondary salinization, which is caused by human activity (Carrow & Duncan, 2011). Here are some examples of natural or primary salinization:

- Salt build-up in the soil over time due to weathering of salt parent materials, particularly in dry places where natural leaching is inadequate due to low and sporadic precipitation. This is the cause of the most salt-affected soils, which include many turfgrass areas located on naturally salinized soils.
- Fading ocean or coastal beds that have left salt deposits at the surface or at distinct soil horizons.
- Sea salt deposition transported onto neighboring land regions by wind, rain, or flooding (e.g., storm surges or recurrent high tides).
- Salt movement into the plant root zone caused by a high saline water table or large tidal swells was common in coastal swamps or marshes.

Secondary salinization is mostly caused by human actions such as irrigation and drainage techniques. Understanding the factors linked to secondary salinization is useful since it can help to reduce negative effects. The following are the different types of secondary salinization:

- Saline water irrigation prevents salt accumulation in the plant root zone when there is insufficient leaching or drainage.
- Saline water irrigation in places where surface drainage occurs to some extent, as well as spillovers in groundwater salinization (aquifers).
- Saline water irrigation, specifically in arid or semiarid areas where evaporative need continuously exceeds leaching requirements and irrigation system potential, resulting in upper soil profile salt accumulation and layering.

13.2 SALINITY AND SODICITY AFFECTED SOILS: GLOBAL AND INDIAN SCENARIO

On a global basis, salinity or related pressures damage around 20% of cultivated land and nearly half of all irrigated fields to some extent. Soil salinization is expected to damage 30% of worldwide land within the next 25 years, with the figure rising to 50% by 2050 (Wang et al., 2003). According to the world soil map, the regional estimate of saline and sodic soils is 397 and 434 million hectares, respectively, which accounts for around 15% of the global land area (Mandal et al., 2010). Salinity affects approximately 200 × 10^6 hectares of land in the Southwest United States and Mexico. Salt water intrusion into aquifers is substantial in Spain, Portugal, Greece, and Italy. Amplified groundwater and soil salinity affect 20 × 10^6 hectares of land in the Middle East. In Egypt, 1 × 10^6 ha of cultivable land near the Nile River is damaged by salt. Due to salinity, 25 × 10^6 hectares of land in Iran are unproductive. 80 × 10^6 hectares of Africa are saline, sodic, or saline/sodic. Due to salt, portions of Pakistan (10 × 10^6 ha) and Bangladesh (3 × 10^6) are unproductive. In Thailand, salt affects 3.58 × 10^6 ha, of which 3.0 and 0.58 × 10^6 ha are attributable to inland and coastal saline soils, respectively. The overall land area impacted by salt in China is 26 × 10^6 ha (Inner Mongolia, Yellow River basin, tidal coastal regions), while the extent of saline soils in Australia is 357 × 10^6 ha. In India, salt-affected soils can be found in 14 different states, as well as the Andaman and Nicobar Islands. According to Mandal (2016), salt affects approximately 6.73 million ha of land in India, with the majority of such regions located in the states of Uttar Pradesh, Bihar, Haryana, Punjab, and West Bengal. It also included a large area in the arid and semiarid regions of Gujarat, Rajasthan, Madhya Pradesh, and Maharashtra, as well as the peninsular regions of Maharashtra, Karnataka, Tamil Nadu, Andhra Pradesh, and Orissa. A significant region is spread throughout the east and west coasts of Andhra Pradesh, Orissa, Tamil Nadu, Kerala, Karnataka, and Maharashtra. The saline soils have spread across 12 states, covering 2.95 million ha (34%) of the 6.73 million acres. These are inland and coastal saline soils found in arid/semiarid regions with poor quality ground water and coastal areas with regular seawater inundation. Inland salinity covers approximately 1.75 million ha, with 1.2 million ha scattered along India's east and west coasts. Gujarat, Andhra Pradesh, West Bengal, Orissa, Tamil Nadu, and the Andaman and Nicobar Islands are all highlighted. Kachchh, Banaskantha, Jamnagar, Bhavnagar, Rajkot, and Surendranagar districts of Gujarat have a large area. These are slightly (44%), moderately (26%), and severely (18%) saline in character. These soils can be found on the East Coast in the districts of East and West Godavari, Krishna, and Srikakulam in Andhra Pradesh; Midnapur and 24-Parganas in West Bengal; and Kendrapara, Puri, Bhadrak, Baleswar, Ganjam, and Jagatsinghapur in Orissa. The island areas are heavily salty and are located in the South, Middle, and North Andaman. The proportion of salty soils to the total geographical area (TGA) in several states in the inland plains is as follows: Gujarat (17%), Rajasthan (6.6%), Maharashtra (5.9%), Karnataka (0.06%), Haryana (1.6%), and Bihar (1.6%). These soils are commonly found in Jalor, Bhilwara, Ajmer, Churu,

Barmer, and Pali districts of Rajasthan, Jind, Hisar, Sirsa, Rohtak, and Jhajjar (districts of Haryana), Surendranagar, Ahmadabad, Jamnagar, Rajkot, and Amreli (districts of Gujarat) and These are also found in, primarily in the districts of Pune, Thane, Sangli, Satara, and Rayagad (West Maharashtra), as well as in the Bihar district of Saran (Figure 13.1).

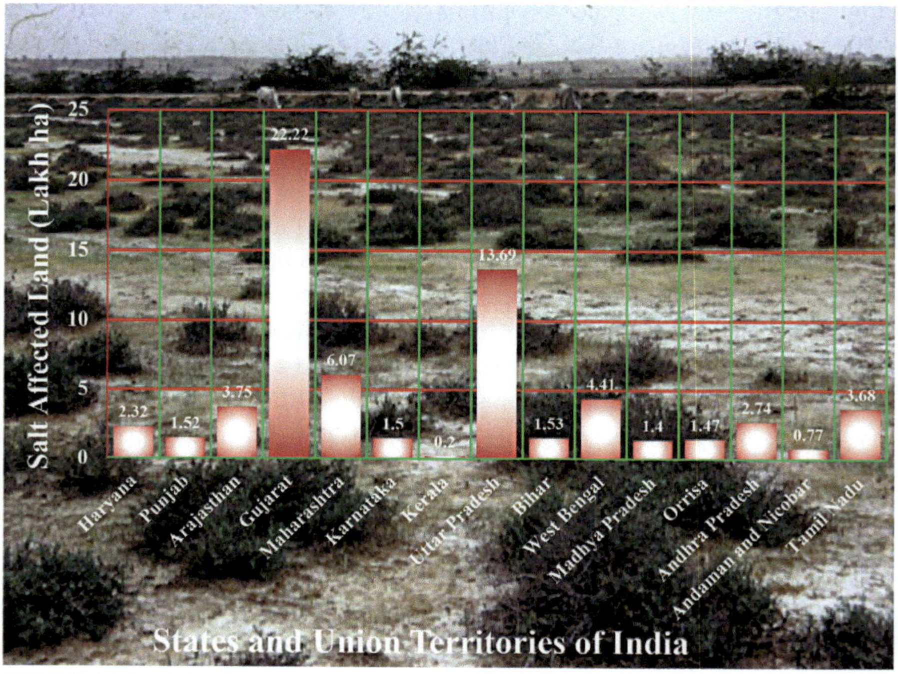

FIGURE 13.1 Salt affected lands in different states of India.

Source: Adapted and redraw from: Sahab et al. (2021).

Sodic soils cover 3.77 million acres, or 66% of the total land area. These soils are most common in the "Indo-Gangetic Plain," as well as arid and semiarid regions of Western and Central India, and the peninsular region of Southern India. Andhra Pradesh, Bihar, Karnataka, Haryana, Gujarat, Maharashtra, Madhya Pradesh, Punjab, Rajasthan, Tamil Nadu, and Uttar Pradesh were found to have a larger proportion of these soils. These soils cover 1,787,759 hectares in the Indo-Gangetic Plain (47% of the total area in the states of Uttar Pradesh, Haryana, Punjab, and Bihar). These are primarily found in the Upper, Middle, and lower Gangetic plains (LGP) of Uttar Pradesh. In Punjab, significant land is located in the Trans-Gangetic Plain districts of Firozpur, Faridkot, Sangrur, Mansa, Amritsar, and Patiala. Sodic soils are common in Haryana's Central Haryana, which includes the districts of Sonipat, Kaithal, Panipat, Karnal, Jind, and Kurukshetra. These can be found in the districts of Ahmadabad, Banas kantha, Junagadh, Mehsana, Surendra Nagar, and others in Gujarat. These are both severely (18%) and mildly (5%) saline in character. In Rajasthan's central and northern districts, mildly (41%) and moderately (6%) sodic soils are found in Chittorgarh, Ganganagar, Hanumangarh, Jaisalmer, and Udaipur. Sodic soils are found in Western Maharashtra and the Vidarbaha region, which includes the districts of Ahmednagar, Jalgaon, Akola, Solapur, Buldana, Amravati, Nasik, Kolhapur, and Aurangabad. Particularly in arid and semi-arid areas of Rajasthan, salt affected areas can be divided into two categories: In the eastern part of the state, salt lakes such as Sambhar, Didwana, and Kuchaman; and (ii) salt basins such as Pachpadara, Thob, Sanwarla, Bap (Bap Malhar), Pokaran, Lunkaransar, Lanela (Kharia, Kanodwala, and Mitha Ranns (Sen et al., 1982). These salt ponds have less than 1% slope and gather local runoff. This water evaporates, and the basins dry up. Dehydration encloses the surface layers in a white crust of salt in the form of patches of clay soil with deposits such as sodium chloride, sodium sulfate, potassium chloride, magnesium sulfate, gypsum, nitrates, and other salts (Sodha et al., 2018).

13.3 LAND DEGRADATION WITH RESPECT TO SOIL SALINITY AND SODICITY

When an ecosystem responds to a disturbance and maintains its condition of dynamic equilibrium, it is said to be stable (Gandolfi et al., 2007). A degraded ecosystem is less resilient to perturbations, resulting in the extinction of critical species and interactions. Human action is critical and urgent when the environment suffers serious damage, such as the extinction of vital species and the acceleration of degradation (diseases, erosion, leaching, and inbreeding; Rudolf & Lafferty, 2011). Soil degradation is a progressive process that results in desertification (Figure 13.2). Other land degradation processes, such as soil erosion, acidification, and structural decline, are also linked to soil salinity (Gupta & Goyal, 2017). According to available information, salinity, and sodicity factors, combined with soil erosion, pose an alarming threat to global agriculture and other ecosystem services (ES) of natural resources (Fitzpatrick, 2002; Pitman & Läuchli, 2002; Hajkowicz & Young, 2005). Saline soils have significant socioeconomic and environmental implications that are linked to a variety of provisional and cultural

ES. These soils decreased the amount of land that could be used for farming; took away nutrients; stopped the succession process by destroying and breaking up habitats; and drove up the cost of reclamation and agricultural inputs (Gupta & Goyal, 2017; Sahab et al., 2021).

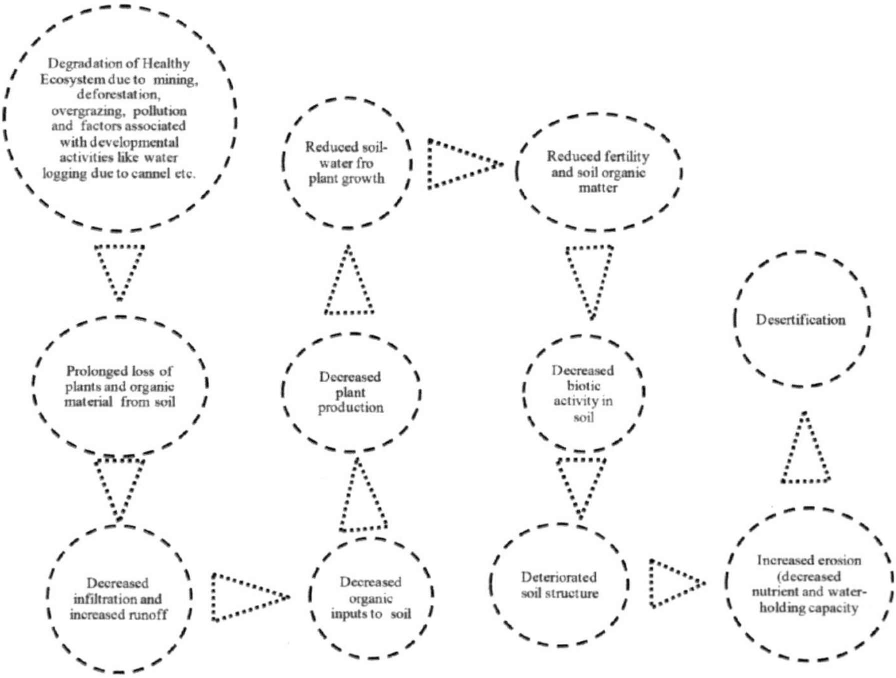

FIGURE 13.2 Soil degradation cycle.

13.4 RESTORATION AND ITS DYNAMICS

Ecological restoration does not necessarily return an ecosystem to its previous state, but rather reinstates and rebuilds an ecosystem to a healthy state with structure and functional components. In other words, every ecosystem has two essential components: structure and function (or a series of functions). When the ecosystem is harmed, both of these are lost. A simple graph with structure and function as the two axes can show this loss and the restoration process (Figure 13.3). An ecosystem's structural unit is made up of both living (biotic) and nonliving (abiotic) components. Populations of organisms (species diversity) and the living resources they utilize are examples of living components. Non-living components include non-living resources (such as space) and non-living physical environment characteristics (e.g., temperature, humidity, habitat complexity). Abiotic factors, producers, consumers, and decomposers are the four functional components of an ecosystem.

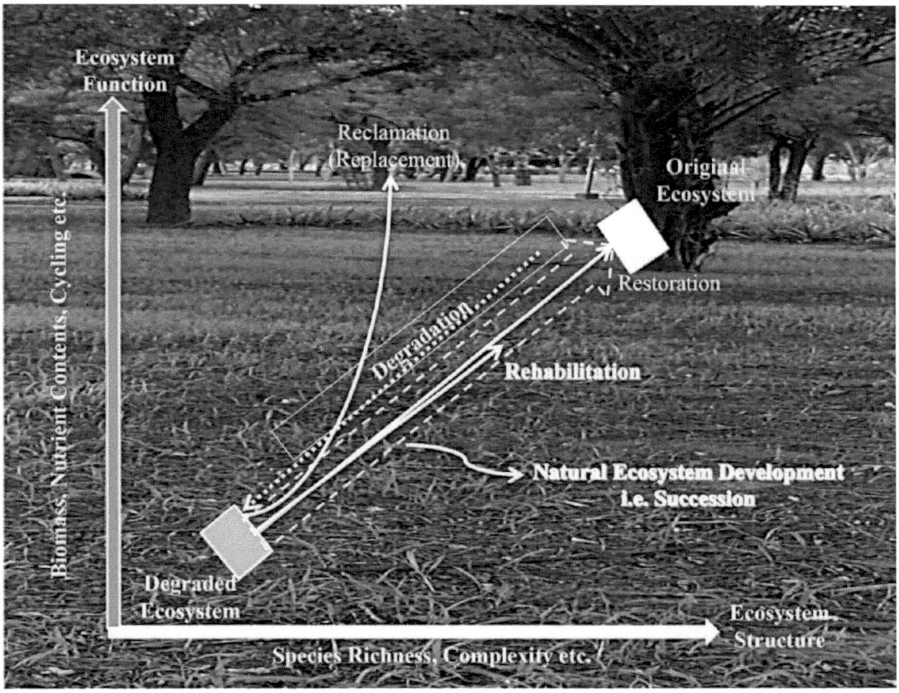

FIGURE 13.3 A diagrammatic representation of the degradation of ecosystems and the different processes of restoration that are possible encompassing of both structure and function.

Source: Adapted from: Cortina et al. (2006) and reproduce with permission of publisher.

Many conceptual models have been established by restoration ecologists to describe how ecosystem structure/composition and functioning are related (Bradshaw, 2002). The Linear Structure vs. Function model (or "LSF") established for the reclamation of destroyed land is one of the most prominent and frequently utilized by restoration ecologists and practitioners. This model assumes a simultaneous shift in structure and function in aggravating and degrading ecosystems, i.e., a linear increase in ecological function as the complexity of their structure increases (Figure 13.3). Structure can be any portrayal of community composition and the way organisms are organized in the context of the LSF model, and function includes ecosystem functioning (standing biomass, nutrient accumulation). This approach defines restoration as a "concurrent increase in structure and function supported by human intervention, and it mimics changes occurring throughout secondary succession." The many approaches to restoring a degraded ecosystem can be defined in terms of two major characteristics: structure and function. When both attributes degrade, they are frequently lowered, though not always equally. When used narrowly, restoration means "restoring the ecosystem to its original or former state in terms of both structure and function." Then there are other options, such as rehabilitation, "in which pre-disturbed conditions are not completely achieved, and replacement of the original by something else." Many people use the term "reclamation" to refer to all of these options. The "four Rs" are commonly used to address a faulty system: reclamation, remediation, rehabilitation, and restoration. The goal of reclamation was to return the place to productive usage. It is primarily concerned with land surface qualities and the capacity to use the site for a certain land use (LU) activity (redevelopment). Remediation aims to eradicate or treat soil, air, and water contamination in order to reduce or eliminate threats to human health and other living forms. The goal of rehabilitation is to improve site conditions so that degraded ES can be restored. Restoration is a broad term that frequently refers to any activity undertaken to restore site conditions. Thus, restoration includes reclamation, remediation, and rehabilitation. Ecological restoration focuses on the entire ecosystem, including flora and fauna as well as general ecosystem functioning, rather than just soil, air, and water quality. In general, the purpose of ecological restoration is to restore the environment to its natural state. The primary goals of remediation and reclamation are to improve site conditions in order to reduce human health risks and facilitate future human activities. Furthermore, rehabilitation is frequently centered on increasing ecological services for human well-being. In other words, they are concerned with people. Ecological restoration, on the other hand, focuses on the ecosystem itself, its functioning, and general health. Table 13.3 shows the comparative descriptions of the four Rs.

13.5 HALOPHYTES: CHARACTERISTICS, CLASSIFICATION, DISTRIBUTION, AND HALO-REMEDIATION

Halophytes (halo = salt + phyte = plant) are plant species that thrive and survive in severely saline environments. Gycophytes (glyco = sweet + phyte = plant) on the other hand, are "sweet-water plants" that thrive on fresh water and via biomass and yield generally decreases with increased salt levels. Oligohalophytes (oligo = few) or "salt-resisting glycophytes" that can

TABLE 13.3 Comparative Descriptions of Four R's Related with Ecosystem Disturbance and Management

Reclamation	Rehabilitation	Remediation	Restoration
The act of restoring a land to its former, better condition. Land reclamation in the context of a wasteland refers to the conversion of wasteland into productive land. Repairing damaged land in order to improve its productive function. For example, salinized cropland that is unsuitable for agriculture can be planted with salt-tolerant species (halophytes), preventing soil erosion and providing wildlife habitat.	This refers to restoring a damaged area to some degree of its previous state through reinstating ecosystem functionality, with the goal of recovering ecosystem services rather than complete ecological restoration. For example, applying organic amendments to a damaged soil region in order to enhance crop planting and subsequent food production.	Removing or treating site contamination (site clean-up) in order to protect both human health and the environment. This has a specific meaning for hazardous waste disposal facilities. Excavation of contaminated soils, pumping, and treating contaminated groundwater, and dirt are some examples.	The activity of restoring and restoring degraded, damaged, or destroyed ecosystems to their original (pre-disturbance) state. A catch-all word for any and all efforts carried out at the site to restore conditions. Restoration includes the subsets of reclamation and remediation.

withstand salty circumstances but do not grow well once exposed to fresh water. Miohalophytes (mio = partially) are "salt-tolerant plants" capable of maintaining productivity up to a certain salt level threshold. Miohalophyte crop application (such as sugar beet, dates, Lucerne, and olives) has slowed salt-induced productive land loss in some marginal locations. There is currently research into the development of "salt-loving plants," i.e., halophilic (philic = loving) halophytes, also known as euhalophytes (eu = true) or "true halophytes" because they have increased productivity with increasing salt levels and actually grow better under salty conditions than under freshwater conditions. Halophytes were identified by Greenway & Munns (1980) as "a type of native flora of salty soils that contain solutions with a Psi of at least 3.3 bar, corresponding to 70 mM monovalent salts." Non-halophytes are plants that cannot thrive in these environments. Halophytes are classified ecologically as – (i) obligatory; (ii) facultative; and (iii) habitat-indifferent halophytes (Walter, 1961; Sengbusch, 2003). Their growth patterns alter under saline conditions (Figure 13.4).

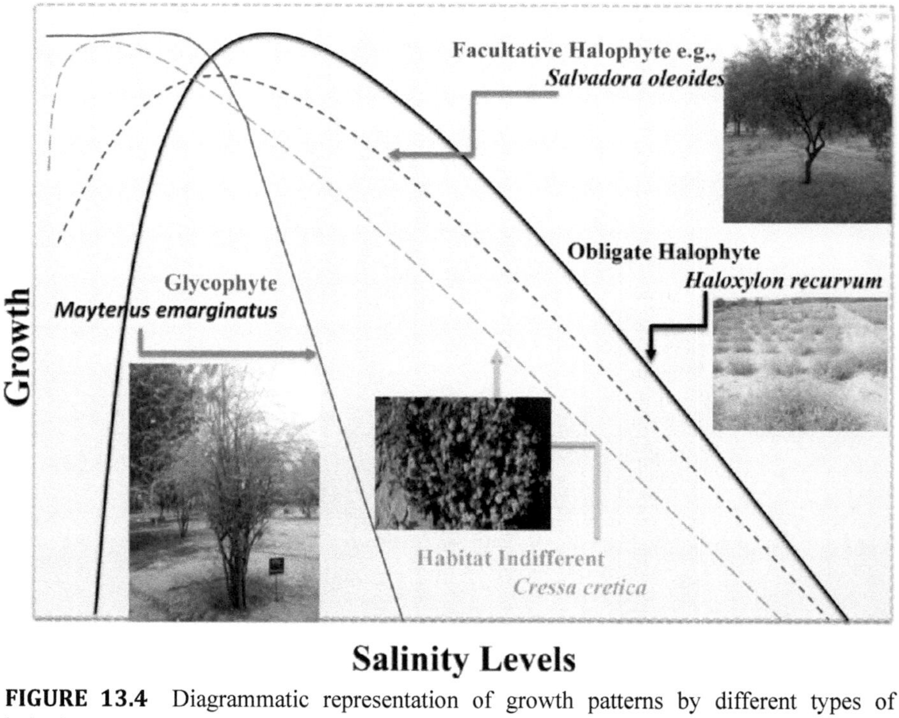

FIGURE 13.4 Diagrammatic representation of growth patterns by different types of halophytes.

Source: Adapted and redraw from: Hasanuzzaman et al. (2014).

Obligate halophytes can only thrive in saline environments. This class includes many Chenopod plant species. However, facultative halophytes can establish themselves on salty soils, although their best growth occurs in salt-free or low-salt circumstances. This group includes most Brassicaceae, Cyperaceae, and Poaceae species, as well as many dicotyledons such as *Aster tripolium, Glaux maritima, and Plantago maritima.* Plants that are resistant to their environment have a proclivity to grow on salty soil, resulting in fierce competition for species that are unable to make such physiological adjustments. *Chenopodium glaucum*, *Myosurus minimus*, and *Potentilla anserina* can thrive in a variety of environments. Many species, including *Festuca rubra, Agrostis stolonifera*, and *Juncus bufonius*, have genetic differences between populations living on saline and salt-free soils. However, when compared to glycophytes, all three types of halophytes do better in terms of development (Figure 13.4). Walter (1961) and Hasanuzzaman et al. (2014) have classified the halophytes into three types (Table 13.4).

TABLE 13.4 Classes or Types of Different Halophytes

Classes/Types	Characteristics	Examples
Salt excluding	In these plants, the root system possesses an ultrafiltration apparatus, and such traits help them for their establishment at high saline areas.	*Rhizophora mucronata*, *Ceriops candolleana*, *Bruguiera gymnorrhiza*, and *Kandelia candel*
Salt excreting	These plants adjust internal salt levels through foliar glands.	*Avicennia officinalis*, *Avicennia alba*, *Avicennia marina*, *Aegiceros corniculatum*, and *Acanthus ilicifolius*, *Excoecaria agallocha*, *Limnitzera racemosa*, *Pentatropis sianshoides*
Salt accumulating	They accrue high concentrations of salt in their cells and tissues and overcome salt toxicity through developing succulence.	*Sonneratia apetala*, *Sonneratia acida*, *Sonneratia alba*, *Salvadora persica*, *Sesuvium portulacastrum*, and *Suaeda nudiflora*

Halophytes can grow in a variety of habitats, including marshes, estuaries, cliffs, and dunes, and make up about 1% of the world's vegetation (Flowers & Colmer, 2008). Halophytes, particularly those in the Chenopodiaceae family, can acquire more salt than other plants cultivated in similar conditions (Flowers et al., 1977). Furthermore, many halophytes adapt to high salinity conditions in their environment by increasing the salt content of their tissue (Abbas, 2005; Shekhawat et al., 2006). The majority of halophyte biomass growth in a saline soil is aided by inorganic ions. Increases in organic matter also occur, but not in proportion to increases in inorganic ions and water (Flowers et al., 1977). This gives halophytes an advantage in bioremediation of salt-affected areas (Shekhawat et al., 2006). According to plant modular analysis, more than 90% of Na^+ in halophytes is detected in the shoot and at least 80% in the leaves. In highly vacuolated halophytes, the vacuole of the leaf cell can account for up to 95% of the cell volume, with salt concentrations in the vacuole ranging from 500 to 700 mM. The most salt-tolerant halophytes may maintain an osmotic pressure two to three times higher than the external medium's osmotic pressure (Glenn et al., 1999; Flowers & Flowers, 2005). Dagar & Singh (2007) describe the diversity of halophytes in the hot desert region, and according to their criteria, around 350 species in western Rajasthan (which accounts for more than half of the hot dry region) are salt tolerant. About 4%, 24%, and 72% of all salt-tolerant plant species are obligatory, facultative, or transitional halophytes, respectively. These salt-tolerant species are quite diverse in terms of taxonomy, plant morphology, and environment. Taxonomically, the Asteraceae, Cyperaceae, Chenopodiaceae, and Poaceae families account for a large proportion of salt-bearing species in India's north-western hot and dry regions. Poaceae is the most common herbaceous plant family. The orders with the most halophytes include Caryophyllales (where the Chenopodiaceae family stands out with about 381 species of halophytes) and Alismatales (Flowers & Colmer, 2008; Flowers et al., 2010).

Plant families that include woody perennial halophytic species include Chenopodiaceae, Tamaricaceae, and Salvadoraceae. Salt-tolerant shrub/tree genera include *Haloxylon, Suaeda, Tamarix, Salvadora*, and *Prosopis. Sporobolus, Sesuvium* (true halophyte); *Cyperus, Heliotropium, Eragrostis, Pulicaria, Portulaca* (facultative halophyte); and *Trianthema, Euphorbia, Indigofera, Aristida, Cenchrus, Corchorus*, and *Tribulus* are the major herbaceous genera with the most taxa (glycophyte). Table 13.5 depicts the key halophytic plant species of Rajasthan and Gujarat's dry and semi-arid regions, classifying them according to their nature, i.e., varieties of halophytes.

To thrive in saline circumstances, halophytes can use a variety of processes. They can be either "salt tolerant" or "salt averse." Some common traits that lead to overall good salt tolerance include: (i) exclusion of Na^+ and chlorine ions; (ii) selective uptake of such ions; (iii) preservation of ions in the leaf sheath; (iv) driving ions into aged leaves; (v) salt glands to expel excess salt; (vi) organic solute synthesis; (vii) succulence; and (viii) ion compartmentalization (Ball, 1988; Gul et al., 2000; Flowers & Flowers,

TABLE 13.5 Plant Species Belongs to Different Types of Halophytes

Types of Halophytes	Under-Shrubs, Shrubs, and Trees	Grasses and Herbs
True halophytes	*Haloxylon recurvum, Salsola baryosma, Suaeda fruticosa, Suaeda nudiflora, Tamarix dioica, T. ericoides, Tamarix indica*	*Aeluropus lagopoides, Sporobolus helvolus, S. ioclados* (Syn. *S. marginatus*), *Urochondra setulosa, Atriplex stocksii, Cressa cretica, Heliotropium curassavicum, Sesuvium portulacastrum, S. sesuvioides*
Facultative halophyte	*Alhagi maurorum, Barleria acanthoides, B. prionitis, Withania somnifera, Aerva javanica* var. *javanica* (syn. *A. persica*), *Aerva javanica* var. *bovei* (syn. *A. pseudotomentosa*), *Capparis decidua, Haloxylon salicornicum, Opuntia dillenii, O. elatior, O. ficus indica, Periploca aphylla, Tamarix aphylla, Ephedra ciliata* (syn. *E. foliata*), *Prosopis chilensis, P. juliflora, Salvadora oleoides, S. persica, Tamarix aphylla* (syn. *T. articulata*)	*Brachiaria ramosa, Chloris barbata, C. virgata, Dactyloctenium aegypticum, Digitaria ciliaris* (syn. *D. adscendens, D. pennata, Echinochloa colona, Eragrostis ciliaris, E. pilosa, E. tenella, E. tremula, Sporobolus coromandelianus, S. diander, S. maderaspatanus, S. tremulus, Cyperus bulbosus, C. compressus, C. conglomeratus, C. iria, C. rotundus, C. tuberosus, Kyllinga bulbosa* (syn. *Cyperus triceps*), *Pycreus flavidus* (syn. *Cyperus flavidus*), *Schoenoplectus grossus* (syn. *Scripus grosus*), *Cassia italica, Chenopodium album, C. ambrosioides, Chrozophora prostrata, Cleome brachycarpa, Convolvulus prostratus, Fagonia indica, Heliotropium bacciferum, H. marifolium, H. supinum, Portulaca oleracea, P. quadrifida, P. pilosa* (syn. *P. tuberosa*), *Pulicaria crispa, P. foliolosa, P. rajputanae, P. wightiana, Sphaeranthus indicus, Zygophyllum simplex*
Glycophyte	*Abutilon bidentatum, A. fruticosum, Taverneira cuneifolia, Abutilon indicum, Acacia jacquemontii, Calligonum polygonoides, Calotropis gigantea, C. procera, Clerodendrum phlomidis, Cordia gharaf, Grewia tenax, Indigofera oblongifolia, Leptadenia pyrotechnica, Lycium barbarum, Maytenus emarginatus, Mimosa hamata, Sarcostemma viminale* (syn. *S. acidum, Sericostoma pauciflorum, Ziziphus nummularia, Cocculus pendulus, Maerua oblongifolia, Pentatropis spiralis, Pergularia daemia, Acacia nilotica, Prosopis cineraria, P. glandulosa, Tecomella undulata*	*Aristida adscensionis, A. funiculata, A. hystricula, Cenchrus biflorus, C. ciliaris, C. prieurii, C. setigerus, Dactyloctenium scindicum, Ochthochloa compressa* (syn. *Eleusine compressa*), *Panicum antidotale, P. turgidum, Saccharum spontaneum, Arnebia hispidissima, Bergia suffruticosa, Commicarpus verticillatus* (syn. *Boerhavia verticillata*), *Citrullus colocynthis, Corchorus depressus, C. tridens, Euphorbia granulata, E. heyneana* (syn. *E. microphylla*), *Farsetia hamiltonii, Indigofera cordifolia, I. sessiliflora, Peganum harmala, Polygala eriocarpa, Rhynchosia capitata, T. portulacastrum* L., *Tribulus pentandrus* (syn. *T. alatus*), *T. terrestris, Zaleya govinda* (syn. *Trianthema govinda*), *Zaleya radimita*

2005; Seaman, 2005; Ye et al., 2005; Masters et al., 2007). Land degradation caused by salinity and/or sodicity can be treated or remedied using a variety of mechanisms or changes such as agronomic methods, the adoption of salt-tolerant crop cultivars, and phytoremediation. According to Ashraf et al. (2005, 2010); Rabhi et al. (2008), phytoremediation has the potential to be a cheap and environmentally friendly way to clean up salt-affected sites.

13.6 ADAPTATION MECHANISM OF HALOPHYTES UNDER SALINE CONDITION

Salts absorbed by halophytes do not directly affect plant growth by modifying turgor, photosynthesis, or the activity of one or a few enzymes. The accumulation of salts in aged leaves hastened their demise. The absence of leaves reduces the delivery of assimilates or hormones to the growing organs, affecting growth (Munns et al., 1995). However, because of their polyphyletic characteristics (taxonomic groups with similar traits generally descend from one or more ancestral lineages), halophytes have evolved the same basic method of osmotic adjustment: accumulation of inorganic salts, primarily sodium chloride, in the vacuole and accumulation of organic solutes within the cytoplasm. The routes by which Na^+ and Cl^- enter halophyte cells are less well understood, but they may include ion channels and pinocytosis in addition to Na^+ and Cl-transporters. To provide the proton motive force, Na^+/H^+ antiporters within the tonoplast, H+ ATPases, and maybe PPIases are required. Tonoplast antiporters are found in halophytes, but they must be activated by NaCl in salt-tolerant glycophytes, and they may be lacking in salt-sensitive glycophytes. Halophyte vacuoles may have changed lipid content to prevent Na+ leakage back to the cytoplasm. It should also be mentioned that halophytes frequently have huge vacuoles. *Suaeda maritime*, a potential halophyte, for example, fills 77% of vacuole mesophyll cells, allowing it to accumulate higher salt concentrations up to 500 mM (Glenn et al., 1999).

13.7 ECOLOGICAL SIGNIFICANCE OF HALOPHYTES

The negative effects of global climate change (CC) can be linked to an increase in salty areas, requiring scientists to produce extremely salt-tolerant crop varieties. Because of their distinct characteristics, halophytes can make substantial contributions in industrial, ecological, and agricultural settings.

In agronomic field trials, they are examined as vegetable, forage, and oilseed crops. *Tamarix aphylla*, the most productive species, yields 10–20 t ha^{-1} of biomass on seawater irrigation, similar to conventional crops. *Salicornia bigelovii*, an oilseed halophyte, produced two tons of seed per hectare with 28% oil and 31% protein, which is comparable to soybean output and seed quality. Many plant species have traditionally been employed as herbs and vegetables, and hence, rediscovery of the potential of several prospective halophytic plant species to be farmed as leafy vegetables has been ongoing for several decades. Some halophytes make good animal feed, so they can be used in areas where salt is a problem.

Over the previous three decades (Dagar, 1995a), halophytes have gained popularity as a source of fodder (Mathur et al., 2007; Singh et al., 2009), oilseeds (Reddy et al., 2008; Ruan et al., 2008), alkaloids (Hussain et al., 2008), biofuel (Qadir et al., 2008), and biomass for renewable energy (Dagar, 1995a). They are also used for phytoremediation of salt-affected soils (Qadir & Oster, 2004), decontamination of heavy metal polluted soils (Lewis & Devereux, 2009), lowering water level (Barnett-Lennand, 2002), restoration of overgrazed pastures and improving soil fertility (Qadir et al., 2008), and carbon sequestration (Geesing et al., 2000). Furthermore, Ventura et al. (2015) examined the potential of halophytes for usage in agrarian systems. Different halophytes, according to them, can be used for various properties such as food and vegetables (*Crithmum maritimum, Portulaca oleracea, Salicornia* spp., and *Aster tripolium, Salsola soda, Crambe maritime, Beta maritime, Chenopodium quinoa, Atriplex hortensis, Tetragonia tetragonioides, Inula crithmoides*); fodder (*Atriplex nummularia, Distichlis spicata, Leptochloa fusca, Cynodon dactylon, Salicornia bigelovii, Salicornia herbacea*) and multiple applications (*Aster tripolium, Mesembryanthemum crystallinum, Atriplex hortensis* and *Crithmum maritimum* for vegetable and ornamental, *Salicornia/Sarcocornia* spp. for biofuel/oilseed crop and bioremediation and *Pennisetum clandestinum* for Ground cover and fodder biomass).

Aside from direct use, halophytes are also being pursued for environmental restoration. Qadir & Oster (2004) demonstrated halophytes as a bioremediation method by comparing the results of 14 studies with gypsum use and vegetative reclamation of sodic soils. Yunusa & Newton (2003) also advised the use of halophytes rather than chemical amendments for the improvement of saline soils. Furthermore, phytoremediation improved soil structure and created macropores, which improved air and water infiltration. Soil under the canopy *of Haloxylon recurvum* in hot, dry areas had a lower pH (8.8), EC (1.01 dS m^{-1}), and better carbon (0.28%) than soil in open areas, which had a pH of 9.1, an EC of 1.21 dS m^{-1}, and 0.16% organic carbon. These findings indicate that the species has the ability to reduce pH and salinity, making it a viable candidate for phytoremediation of salt-affected soils. The establishment of *Haloxylon salicornicum* enhanced silt and clay content; improved water holding capacity (WHC); organic carbon; available nitrogen; available phosphorus; and EC; and decreased pH and bulk density, according to ICAR-CAZRI results. Surface soils (0–5 cm) under its canopies had 69% more organic carbon, 47% more available nitrogen, and 41% more available phosphorus than open regions. Also, Rathore et al. (2015) found that the density and biomass of herbaceous plants in *H. salicornicum* alleys were 1.2 and 1.6 times greater than in the open area.

13.8 POTENTIAL USE OF HALOPHYTES UNDER SALINE CONDITION

Because the restoration of salt-affected soils is not always feasible or cost-effective. As a result, researchers are focusing their efforts on bio-saline agriculture in order to gain a better understanding of how naturally adapted plants (halophytes) deal with salts. The study of halophytes is frequently beneficial from three perspectives (Gleen et al., 1999). First, halophytes adopt tactics to survive and maintain productivity in saline water, and such knowledge is always valuable in developing salt-tolerant novel types. Second, halophytes produced in an agronomic setting frequently fail to assess the general practicality of high-salinity agriculture. Third, halophytes have the potential to become an immediate supply of alternative crops (Hasanuzzaman et al., 2014). When compared to glycophytes, halophytes are frequently better adapted to metal-affected habitats, making them an honest choice for an eco-friendly and sustainable solution to contaminated coastal settings. Haloremediation is concerned with the biological reduction of salts in which aerial plant parts accumulate a large amount of salt. Chaudhury et al. (1964) developed this method to remove salt from soil using *Suaeda fruiticosa*. This species has a minimal irrigation demand, can thrive in very salty soils, and can remove 1,088 kg in 0.4 acres. Under greenhouse circumstances, Rabhi et al. (2009) investigated the potential of *Suaeda fruticosa* and *Tecticornia indica* for soil desalination. Trees, shrubs, and grasses grown in salt-affected areas and/or with high saline water tables will provide two main benefits: (i) control of taking over to or withdrawal from groundwater; and (ii) use of saline drainage water or piped spring water to reduce salty water tables. By intercepting channel seepage, trees placed beside irrigation channels can also help to keep water tables low. Many species may utilize water from water tables (and soils) with an EC of up to 10 dS m^{-1} without a significant decrease in transpiration rate. However, the amount of spring water directly used by trees and shrubs has only been determined in a few studies, mostly using isotopic methods. The knowledge of the equilibrium root-zone salinities predicted to develop under the rainfall and/or irrigation schedules utilized, as well as plant genotypes capable of growing at acceptable rates in salinity/sodicity environments, might aid in reclaiming the lands (Thorburn, 1996).

Singh et al. (1993) screened a large number of tree germplasm and found trees of the *Prosopis* genus, followed by *Acacia* and *Casuarina*, to be extremely promising for salt land rehabilitation. Similarly, *Leptochloa fusca* and *Brichiaria mutica* have been identified as potential grasses for thriving in salt areas (Sandhu et al., 1981; Kumar & Abrol, 1986). Some systematic research is being conducted to identify plants and grasses in salt areas (Tomar & Gupta, 1986; Singh et al., 1989; Singh, 1995a). Singh (1995a) created an agroforestry approach based on *Leptochloa fusca* for the development of salt soils. This study demonstrates that *Prosopis juliflora* and *L. fusca* are suitable plants for saline soils. The study also discovered possibilities for fuel wood and feed production when these two species were planted together in an irrigated agroforestry system. Salt-resistant multifunctional plants can aid in the bio-amelioration of degraded agriculture and wastelands in alkali soils (Singh et al., 2011, 2016). Qadir et al. (1996a) investigated salt-tolerant kallar grass (*Leptochloa fusca*) for remediation of salt-affected land and found that bioremediation performed similarly to gypsum remediation and was superior to the impact of leaching with no cropping. Kallar grass cultivation with 50% gypsum can help to reduce soil salinity. In 1995, DeVilliers et al. discovered the bio-efficacy of *Mesembryanthemum* in removing Na^{++} and Cl^{-} from saline mining soil. This species can absorb more Na^{++}, Cl^{-}, Mn^{++}, and chloride-containing compounds from the contaminated soil and accumulate these chemical compounds in their aerial parts. Qadir et al. (1996b) investigated different salt-tolerant forages (sesbania: *Sesbania aculeate*, kallar grass: *Leptochloa fusca*, millet rice: *Echinochloa colona*, and ragi: *Eleusine coracana*) for saline soil reclamation and reported that biomass production was directly proportional to the reclamation efficiency of salt-affected land and was better than the control (leaching without gypsum or crop). These species' soil amelioration qualities were discovered in the following order: *Sesbania aculeata* gypsum > *Leptochloa fusca* > *Echinochloa colona* > *Elusine coracana*. Thus, *Sesbania aculeata* emerged as the most promising biotic material for growing on salt-affected soils, capable of producing high-quality feed while also reducing soil saliniation and sodication processes (Qadir et al., 1996b).

Phytoremediation using trees and grasses is beneficial since these plants are frequently used as food, wood, and fuel (Hasanuzzaman et al., 2014). Halophytic grasses are also particularly good at revegetating brine-contaminated soil, which is often the outcome of gas and oil mining. *Glycyrrhiza glabra*, also known as licorice, is salt tolerant and has been utilized in phytoremediation of saline soils (Kushiev et al., 2005). Positive effects on soil are being realized as a result of the spread of this species. The salt concentration of the soil on which Glycyrrhiza glabra grows has been studied, and it shows a steep drop. Abandoned soils are frequently brought back to life at a minimal cost and in a short period of time by the use of licorice (Kushiev et al., 2005). *Suaeda maritima, Sesuvium portulacastrum, Clerodendron inerme, Ipomoea pes-caprae, Heliotropium curassavicum*, and one tree species, *Excoecaria agallocha*, were studied by Ravindran et al. (2007). Among the species tested, *S. maritima* and *S. portulacastrum* had higher salt build-up in their tissues, resulting in better reclaimed salt-damaged soil. Rabhi et al. (2009) assessed the desalinization capacity of two native (Tunisia) species, *Arthrocnemum indicum* and *Suaeda fruticosa*, in their rhizospheric zone under non-leaching conditions. They compared the salt removal abilities of these two species to that of an introduced halophyte, *Sesuvium portulacastrum*. Their pot experiment demonstrated that *Sesuvium portulacastrum* was the most suitable species for salt removal in arid and semi-arid locations where precipitation is insufficient to leach salts from the rhizosphere. *Suaeda salsa* was discovered to be suitable for saline soil restoration because the plant absorbs salt from the soil and stores it in plant parts (Panta et al., 2016). Halophytes are frequently used in man-made wetlands (Farzi et al., 2017). Farzi et al. (2017) investigated the use of built wetlands (CWs) for water salinity reduction and evaluated three halophytic plants of the Chenopodiaceae family (*Salicornia europaea, Salsola crassa*, and *Bienertia cycloptera*). Experiments were conducted with these species at three salinity gradients [EC 2, 6, and 10 dS/m]. With a one-week retention time, EC and concentrations of calcium (Ca), magnesium (Mg), sodium (Na), and chlorine (Cl) were determined before and after phytoremediation. Their findings suggest that these species have the capacity to survive and thrive across all salinity gradients. The potential of Portulaca oleracea to remove and accumulate considerable levels of salt and chlorine has also been evaluated in a hydroponic environment (De Lacerda et al., 2015). Lamb's quarters (*Chenopodium album*) species also generates high dry biomass in salty conditions and has been discovered and used as a potential species for saline soil rehabilitation and clean-up. Under greenhouse circumstances, the phytoremediation capability of six halophytic plants was evaluated: *Atriplex lentiformis, Suaeda fruticosa, Suaeda nudiflora, Parkinsonia aculeata, Portulaca oleracea*, and *Xanthium strumarium* (Devi et al., 2015). According to the results of scientific research into their phytoremediation capacity, *Suaeda fruticosa* and *Atriplex lentiformis* were the best salt hyperaccumulator species, whereas *Xanthium strumarium* and Parkinsonia aculeata were the least hyperaccumulator. Rathore et al. (2015) investigated the potential of two natural shrubs from the Indian arid region, *Calligonum polygonoides* and *Haloxylon salicornicum*, for the regeneration of degraded lands. Planting these shrubs increases the percent content of silt + clay (30–39%), increases water holding capacities (24–30%), increases levels of organic carbon (48–69%), available nitrogen (31–47%), available phosphorus (32–41%), and EC (21–33%), and decreases pH (7–12%) and bulk density (5–6%) in the surface layer of soil beneath the canopy. Furthermore, soil property changes were more pronounced in the uppermost layer (0–5 cm) than in the deeper layer (5–20 cm) and were more pronounced in *Haloxylon salicornicum* than in *Calligonum polygonoides*.

13.9 AGROFORESTRY AND ITS ROLE IN SOIL RECLAMATION

Agroforestry is an important approach for dealing with salinization (Jianfeng et al., 2004). Agroforestry practices provide considerable financial and environmental benefits. Agroforestry can improve farm profitability in numerous ways: (i) The total output per unit area of tree/crop/livestock combinations is greater than any single element alone; and (ii) crops and livestock are more productive when they are protected from wind damage. Agroforestry contributes to the conservation and sustainability of natural resources by, for example, preventing soil erosion and establishing wildlife habitats. The persistence of indigenous herbaceous plants and trees ensures a continual foliage and crown cover, preventing soil exposure to direct sun radiation. This lowers evapotranspiration (ET) and topsoil cracking, which would otherwise diminish water retention and damage soil texture, both of which are required for optimal development and soil amelioration. Shrub and tree canopies around fields act as both physical and physiological wind barriers. Furthermore, taming the wind slows the rate of ET from growing crops, which is especially important in arid locations. It has been proven that the more trees there are, the less evaporation occurs, and the less salt accumulates on topsoil. Afforestation has been shown to be beneficial in revegetating saline landscapes, giving valuable foodstuffs to local pastoral people from marginal degraded land, utilizing otherwise low-quality water, unproductive fields, and lowering the raised groundwater level through bio-drainage (Heuperman et al., 2002; Khamzina et al., 2006).

13.9.1 TREE SPECIES SELECTION

Multipurpose trees (MPTs), especially nitrogen-fixing and salt-tolerant species that are well adapted to saline soil conditions, have various advantages over herbaceous sources of organic matter, nitrogen, and feed. They also offer a higher feed value during the dry season and can be planted in between crops or in conjunction with them without replacing them. They can provide microclimatic benefits such as windbreak effects and reduced evaporation if properly selected and positioned in relation to specific crops. Furthermore, while performing their farm service functions, these trees can provide food, fuel wood, construction materials, fodder, and other immediately beneficial by-products. Finally, because they are a sort of standing capital, they can serve as a source of variable savings for emergency needs, which is a viable and appealing motivation for individual farmers to plant trees. The principles of species selection are as follows: increased tolerance to salinity so that it can survive under salt stress; rapid growth to cover the land and reduce evaporation as soon as possible; and multipurpose, which provides farmers with additional economic benefits in addition to the ecological effect. Tables 13.5 and 13.6 display list of tree and shrub species that can handle different levels of saltiness and sodicity (Table 13.7).

TABLE 13.6 Tree and Shrub Species Suitable for Saline Soils

Tolerance (ECe dS m^{-1})	Trees and Shrubs
High (>25)	*Acacia nilotica, Acacia farnesiana, Prosopis juliflora, Parkinsonia aculeate, Salvodora persica, Salvodora oleiodes,* and *Tamarix articulata*
Moderate (10–25)	*Acacia nilotica, Acacia tortilis, Acacia pennatula, Callistemon lanceolatus, Casuarina equisetifolia, Casuarina glauca, Casuarina obesa, Casuarina cunninghamiana, Eucalyptus camaldulensis, Leucaena leucocephala, Feronia limonia* and *Pithecellobium dulce*
Sensitive (<10)	*Acacia auriculiformis, Albizzia caribea, Caesalpinia eriostachys, Crescentia alata, Eucalyptus tereticornis, Guazuma ulmifolia, Haematoxylon brasiletto, Leucaena shannonii, Samanea saman, Senna atomaria, Syzygium cuminii, Tamarindus indica, Terminalia arjuna*

TABLE 13.7 Relative Tolerances of Tree Species for Soil Sodicity

Soil pH	Fuel Wood/Fodder/Timber Species	Fruit Species
8.2–9.0	*Azadirachta indica, Grevillia robusta, Hardiwickia binata, Kigelia pinnata, Leucaena lencocephata, Mangfifera indica, Melia azedarach, Moringa loiefera, Morus alba Populus deltoids, Tectona grandis,* and *Pyrus communis*	*Aegle marmelos, Ficus spp., Grewia subinaequalis, Prunus persica, Manigifera indica, Morus alba, Sapindus laurifolium* and *Vitis vinifera*
9 1–9.5	*Cordia rothii, Albizia lebbeck, Cassia siamea, Pongamia pinnata, Sesbania sesban, Parkinsonia aculeata, Dalbergia sissoo, Kigelia pinnata, Butea monosperma, Syzygium cumini, Tectona grandis, Populus deltoides*	*Achrus japota Feronia limonia, Punica granatum, Phoenix dactylifera, Tamarindus indica, Syzygium cuminii*
9.6–10.0	*Albizia lebbeck, Cassia siamea, Casuarina equisetifolia, Capparis decidua, Eucalyptus tereticornis, Pithecellobium dulce, Prosopis alba, Prosopis cineraria, Salvadora persica, Salvadora oleoides,* and *Terminalis arjuna*	*Carissa congesta, Emblica officinalis, Psidium guajava* and, *Zizyphus mauritiana*
>10	*Acacia nilotica, Casuarina equisetifolia, Prosopis juliflora, Tamarix articulate,* and *Prosopis cineraria*	Not recommended

13.9.2 HALOPHYTES FOR AGROFORESTRY

Halophytes may withstand salinity by altering internal water relations, ionic compartmentation in cell vacuoles, the buildup of companionable organic solutes, succulence, and salt-secreting glands and bladders (Shabala & Mackay, 2011). Halophyte species can remove salts from soils by salt excluding, excreting, or accumulating via morphological, anatomical, and physiological changes in their organelles as well as cellular level (Hasanuzzaman et al., 2014). Halophytic plants may also be a good candidate for heavy metal removal from both normal and hypersaline soils (Manousaki & Kalogerakis, 2009; Nedjimi & Daoud, 2009). Many halophytic plant species, including *Arthrocnemum indicum*, *Atriplex nummularia, Atriplex halimus, Suaeda maritima, Sesuvium portulacastrum, Suaeda fruticosa,* and *Tamarix aphylla*, have been utilized to clear polluted soils because their tissues can engross more salts, lowering salts from saline land (Ravindran et al., 2007). According to Glenn et al. (2013), halophytes may maintain high agricultural land production up to a root-zone salinity of 70 gl^{-1} TDS, which is double the salinity of seawater. This would, however, necessitate regular irrigation to keep the shallow-root zone at field capacity and exact estimations of leaching requirements to minimize salt accumulation. Biosaline agriculture allows species such as *Cynodon dactylon, Inula crithmoides*, and *Ruppia maritima* to collect biomass, according to Ztürk et al. (2014). Similarly, *Salicornia europaea*, a high-quality edible oil-producing species with a high economic value, can be utilized to efficiently reclaim highly salinized soils (Muscolo et al., 2014). Hundreds of terrestrial halophytes are grown on seawater or in hypersaline salt marshes in the Arabian Gulf region, where salinity levels are usually higher than in seawater. *Arthrocnemum macrostachyum, Atriplex farinosa, Halopeplis perfoliata, Halocnemum strobilaceum, Haloxylon persicum, Salicornia europaea, Suaeda maritima, Suaeda vermiculata, Salsola imbricate, Sesuvium verrucosum, Tamarix aphylla,* and *Zygophyllum mandavielli* are examples Böer (2002) identified 76 plant species of halophytes for the UAE based on a review of the literature. Around 14 are seawater-tolerant halophytes, 29 are halophytes, 31 are semi-halophytes, and two are parasitic species that live on the roots of Chenopodiaceae and Zygophyllaceae members. In the same way, Boer & Al Hajiri (2002) found 49 halophytes in Qatar. These included one species of mangrove, 12 seawater-tolerant halophytes (tolerant to salinity), 17 halophytes (tolerant to salinity), 20 semi-halophytes (tolerant to salinity but probably dependent on brackish to fresh water for normal development), and two parasitic plants that live off the roots of true halophytes.

13.9.3 SELECTION OF MULTIFUNCTIONAL TREE SPECIES FOR SALT AFFECTED SOILS

Researchers have challenges in introducing and growing seedlings in salty and sodic settings. The selection of acceptable tree species for high biomass and bioenergy production in salt-affected soils is determined by the species' tolerance to salinity and sodicity, adaptability to the local agro-climate, and plantation aim. Several studies were undertaken in India to assess the performance of a wide range of tree species in salty and sodic settings. Yadav (1980) recommended many afforestation approaches, including the use of *Acacia nilotica, Eucalyptus tereticornis,* and *Prosopis juliflora* on sodic soils. Gupta et al. (1988) tested the halo-remediation properties of *Eucalyptus tereticornis, Acacia nilotica, Araucaria cunninghamii, Casuarina equisetifolia, Dalbergia sissoo*, and *Pongamia pinnata* in pot culture. However, *Acacia nilotica, Eucalyptus tereticornis*, and *Casuarina equisetifolia* grew successfully at 30.6 ESP, whereas *Dalbergia sissoo, Pongamia pinnata*, and *Araucaria cunninghamii* grew successfully at 15.2 ESP. *Casuarina equisetifolia* tolerated salt moderately well. *Acacia nilotica* and *Eucalyptus camaldulensis* growth was reduced by 50% at 5.0 dSm^{-1} salinity in clay soil, although they grew well at ECe 10.0 dSm^{-1} in sandy soil, according to Yadav & Singh (1986). *Acacia auriculiformis*, on the other hand, did not survive beyond ECe 2.5 dSm^{-1}. At high sodicity (pH > 10.0), 30 forest tree species were tested. Only 13 out of 30 species survived after seven years of sowing. Only *Prosopis juliflora, Tamarix articulata*, and *Acacia nilotica* were found to be suited for such soils among the 13 surviving species. *Eucalyptus tereticornis* demonstrated good survival and height development but no significant biomass. However, *Dalbergia sissoo, Kigelia pinnata, Pithecellobium dulce, Terminalia arjuna*, and *Parkinsonia aculeata* survived but did not produce economically viable biomass (Dagar & Tomar, 2002). At the Central Soil Salinity Research Institute (CSSRI) Regional Research Station in Lucknow, the performance of 10 tree species on sodic soils with ESP 89 was studied. After 10 years of field research, only three species, *Prosopis juliflora, Acacia nilotica*, and *Casuarina equisetifolia*, achieved economic biomass survival rates of more than 90%. During the first four years, *Eucalyptus tereticornis* performed well, then its growth rate slowed. *Azadirachta indica, Dalbergia sissoo*, and *Melia azedarach* all performed poorly.

13.9.4 MULTIFUNCTIONAL AGROFORESTRY SYSTEMS

13.9.4.1 SILVIPASTORAL SYSTEM

The silvipastoral system is defined as the planting of MPTs with grasses and legumes in an integrated system and their utilization through cut and carry in early years, followed by in situ grazing. According to CSSRI, Karnal (CSSRI, 2014), Rhodes grass (*Chloris gayana*), kallar grass (*Leptochloa fusca*), and para grass (*Brachiaria mutica*), are high biomass producing grass species. The silvipastoral practices of mesquite (*Prosopis juliflora*) and kallar grass have been shown to be the most promising for firewood and feed production, as well as soil improvement (Singh, 1995b, 2017). Grewal & Abrol (1989) reported that trees such as *Acacia nilotica, Eucalyptus tereticornis*, and *Parkinsonia aculeata* were planted on ridges, and kallar grass was established in trenches between ridges in their Silvipastoral model for rainwater conservation and production of fuel and forage from alkali lands. During the monsoon, this method collects rainwater, which helps the trees and kallar grass grow more.

13.9.4.2 SILVI-AGRICULTURE SYSTEM

In this approach, trees are grown for a reasonable amount of time before agricultural production begins. Singh (1998) planted wheat and oats in pots filled with topsoil (30 cm) obtained from 24-year-old *Albizia lebbeck, Acacia nilotica, Eucalyptus tereticornis, Prosopis juliflora*, and *Terminalia arjuna* plantations on a very alkali soil and a reclaimed sodic soil. The organic carbon content and nutrient status of soil under 24-year-old plantations were significantly higher than those of farm soil recovered using gypsum. Soil improvement was greatest with *Prosopis* and least with *Eucalyptus.*

13.9.4.3 SILVI-HORTI-PASTURE OR HORTI-AGRICULTURAL SYSTEM

The Narendra Dev University of Agriculture and Technology, Faizabad, has investigated horticultural species-based agroforestry models for alkali soils (Dagar & Singh, 2007). In this model, the growth rate of *Psidium guajava + Eucalyptus globulus + Leucaena leucocephala* was faster, and the production of fruit, fodder, and fuel wood was higher. Bottle gourd, tomato, cabbage, and spinach intercrops have been planted effectively alongside Psidium guajava trees. *Carissa carandas* (karonda), *Emblica officinalis* (aonla), *Syzygium cumini* (jamun), *Psidium guajava* (guava), and *Ziziphus mauritiana* are examples of fruit species that can be grown effectively in alkaline soils (ber). If the soil pH is 9.5, we can grow aromatic and medicinal crops like *Anethum graveolens, Matricaria recutita, Psyllium*, and *Ocimum* in between fruit trees.

13.9.4.4 SEQUENTIAL AGROFORESTRY SYSTEM

Instead of producing trees and arable crops concurrently, this technique grows them in a specific order. This approach generally accelerates soil fertility, which has declined due to continuous cropping or where the soil's inherent fertility status is low, as in sandy desert and salt-affected soils. Short-lived, fast-growing, nitrogen-fixing trees such as *Sesbania, Leucaena*, and *Cajanus* that have been cultivated for 4–5 years are harvested for fire wood or feed, and the soil is used for arable farming. Rao & Gill (1990) studied this land-management method on a recovered alkali soil with a surface pH of 8.8 from 1985 to 1990. *Sesbania* was farmed for four years before switching to a rice-wheat farming pattern. Only phosphorus and zinc were applied at the recommended rates to the crops. The initial rice harvest yielded 6.4 t ha^{-1} in *Sesbania* plots without supplemental nitrogen. Wheat yields in *Sesbania* plots averaged 2.2 t ha^{-1} even without nitrogen administration, compared to 1.35 t ha^{-1} in control plots. Mineralization of organic wastes yielded 0.85 Mg ha^{-1} extra grains and 17 kg ha^{-1} additional nitrogen ha^{-1}. Sesbania's residual effect was also observed on the next rice harvest. Crop nitrogen intake was 142 kg ha^{-1} in the control plots and 222 kg ha^{-1} in the Sesbania plots. Even after crops had grown and no nitrogen was added, the available nitrogen status in *Sesbania* plots was comparable to plots fertilized with 120 kg N ha^{-1}. As an alternative, the use of woody vegetation for managing water fluxes to recharge is known as "bio-drainage" and was previously advocated for dryland salinity control (caused by a shallow water table) (Marcar & Khanna, 1997; Stirzaker et al., 1999; Turner & Ward, 2002; INCID, 2003; Kapoor, 2014). Tree plantations have also been found to be effective at controlling seepage from canals and rivers (Kapoor & Denecke, 2001; Manjunatha et al., 2005; Jeet-Ram et al., 2011; Sprenger et al., 2013), resulting in increased agricultural production on farms. In regions where water remains stagnant, such as in Bangladesh, wetlands form. According to Ghosal et al. (2004), these areas can be successfully used for agroforestry. Kushiev et al. (2005) said that leguminous liquorice (Glycyrrhiza glabra) could be grown in abandoned salty areas in Central Asia's Hungry Steppes where high-water tables were caused by poor irrigation management and a lack of drainage infrastructure. Dagar et al. (2015) said that liquorice grown on moderately alkaline soil could be used to make money.

13.10 EXPERIENCES FROM THE INDIAN SUBCONTINENT

Eucalyptus is an extensively planted tree species due to its fast growth rate, good wood characteristics, and carbon sequestration. Field studies in the Indian subcontinent have revealed substantial variance in tree provenances for frost resistance, water logging, and salinity tolerance; wood characteristics; number of lignotubers; regeneration via coppice formation; and growth rate (Roy Chowdhury et al., 2011; Dagar, 2014). Some of the ways that tree planting has been used to fix waterlogged, salty soils are described in subsections.

13.10.1 SALT-TOLERANT PLANTS-HALOPHYTES

Dagar & Singh (2007) reported 1140 salt-tolerant vascular species distributed in India under 541 genera and 131 families, as well as a number of potentially economically important woody and herbaceous halophytic taxa (including coastal) appropriate for agroforestry (Dagar, 2003, 2005a, b, 2014, 2015). The species include *Acacia, Arthrocnemum, Atriplex, Avicennia, Azadirachta, Borassus, Brachiaria, Bruguera, Carissa, Cassia, Casuarina, Ceriops, Chloris, Coccoloba, Cressa, Crithmum, Dichanthium, Distichlis, Eucalyptus, Juncus, Kochia, Leucaena, Leptochloa, Limonium, Lumnitzera, Maireana, Melia, Pongamia, Panicum, Paspalum, Pennisetum, Porterasia, Prosopis, Rhizophora, Salicornia, Salvadora, Seteria, Simmondsia, Sonneratia, Spergularia, Sporobolus, Suaeda, Syzygium, Tamarix, Taxodium, Terminalia, Vetiveria, Xylocarpus, Ziziphus*, and *Zostera*.

13.10.2 SUITABLE PLANTING TECHNIQUES

Planting on high ridges was frequently thought to be good for waterlogged saline soil, where the most salts concentrate at the surface of the soil and decrease with depth into the water table. To take advantage of low salinity and better soil moisture regimes in sub-surface layers, Tomar & Gupta (1984–1994) and Tomar (1997) attempted and compared sub-surface sapling planting (at a depth of 30 cm below the surface) with ridge planting (40 cm high). Significantly more salt was collected along the ridges, resulting in poor seedling survival and growth. The more salt accumulated within the root zone (one meter depth) of ridge planted trees, the greater the surface area of the ridge. Another downside of ridge planting was the difficulty in retaining rainwater on the ridge tops, as well as the presence of salts, which increased soil erosion susceptibility. Under the sub-surface planting method, on the other hand, the roots were exposed to a less salty zone and got most of their water from the phreatic zone.

13.10.3 AGROFORESTRY SYSTEMS ON SALT AFFECTED LANDS

During the 19th-century, salt-affected regions in the Indian subcontinent were planted with trees in an attempt to establish profitable plantations. These species provided numerous benefits, including soil improvement, environmental benefits, and the prevention of water table rise in wet areas. Nonetheless, tree planting did not become an element of agricultural output or concerted restoration initiatives until the second half of the 20th-century. In India and Pakistan, several salt tolerant tree species have been identified (Jain et al., 1985; Sheikh, 1987; Singh et al., 1989; Aslam et al., 1993), including highly salt tolerant species (*Prosopis juliflora, P. chilensis, P. alba,*

Acacia ampliceps, Salvadora persica, Tamarix articulata) and moderately salt tolerant species (*Acacia nilotica, A. tortilis, A. cambagei, Azadirachta indica, Casuarina equisetifolia, Eucalyptus camaldulensis, E. tereticornis,* and *E. microtheca*). Among shrubs, *Atriplex, Suaeda, Haloxylon,* and *Salsola* were shown to have high salt tolerance, while among grasses, *Leptochloa fusca* was found to be the most tolerant to both salinity and water logging (Malik et al., 1986). In waterlogged salt-affected areas, many species of *Paspalum, Coix, Echinochloa, Panicum, Phragmites, Brachiaria, Panicum, Atriplex, Salicornia, Juncus, Typha, Cyperus,* and *Scirpus* show potential.

13.10.4 GRASSLAND OR PASTORAL SYSTEMS

Aeluropus lagopoides, Chloris barbata, Chloris gayana, Chloris virgata, Cynodon dactylon, Dichanthium caricosum, Echinochloa colonum, E. crus-galli Eragrostis spp., *Leptochloa fusca, Sporobolus helvolus, S. marginatus, S. spicatus* and *Leptochloa fusca* grass has been shown to provide greater benefits in terms of both fodder production and soil amelioration. *Azoarcus*, an associative nitrogen-fixing bacteria, lives as an endophyte in the roots of *L. fusca* grass, a pioneer species of alkali soils that produces 9–12 Mg ha^{-1} of dry biomass without the use of fertilizer; *Azoarcus* provides about half of the plant nitrogen (90–120 kg ha^{-1}) (Malik & Zafar, 1984; Malik et al., 1986). *Aeluropus lagopoides, Sporobolus helvolus, Cynodon dactylon, Brachiaria ramosa, B. mutica, Paspalum* spp., *Echinochloa colonum, E. crusgalli, Dichanthium annulatum, Digitaria ciliaris, Vetiveria zizanioides,* and *Eragrostis* spp. are other important grasses that are Camel, sheep, and goats eat *Atriplex, Kochia, Suaeda, Salsola, Haloxylon,* and *Salvadora*, which are significant forage shrub species in salty environments. Most of the region in the Rann of Kutch in Gujarat in India and the nearby territory in Pakistan is highly saline because of poor rainfall and excessive ET. The naturally growing species on these lands include *Arthrocnemum indicum, Prosopis iulifora, Salvadora persica, Tamarix articulata, T. troupii,* and many other halophytes. A silvopastoral system with suitable salt-tolerant forages such as *Atriplex, Kochia indica, Cressa cretica, Aeluropus lagopoides, Cynodon dactylon, Dactyloctenium sindicum, Bothriochloa pertusa, Dichanthium annulatum, Leptochloa fusca,* and *Sporobolus helvolus* can be created. High-value trees like *Salvadora persica, Pongamia pinnata, Capparis decidua, Carissa carandas,* and *Terminalia catappa*, as well as firewood tree species like *Prosopis juliflora, Acacia nilotica, and Casuarina glauca*, can be grown in furrows, and the forage species mentioned above can be grown in interspaces. In Gujarat, India, a silvopastoral system based on *Salvadora persica* was constructed with fodder grasses *Leptochloa fusca, Eragrostis* sp., and *Dichanthium annulatum* on clay loam saline vertisol (clay 40%, silt 31%, sand 29%; pH 7.2 to 8.9; EC 25 to 70 dS m^{-1}). The subsurface water was 0.5–2 m below the surface and had an EC of 55 to 60 dS m^{-1}. A planting density of 4 m 4 m was shown to be optimal for *Salvadora persica* based on growth pattern in terms of height, canopy spread, and seed yield. At varied salinity levels, the seed output of Salvadora persica ranged from 1.84 to 2.65 Mg ha^{-1} during the fourth year, with oil contents varying from 576 to 868 kg ha^{-1} (Rao et al., 2003). These grasses (*L. fusca, D. annulatum*, and *Eragrostis* sp.) provided 3.17, 1.85, and 1.09 Mg ha^{-1} forage, respectively, when planted on 45 cm high ridges, but 3.75, 1.76, and 0.54 Mg ha^{-1} when planted in furrows, demonstrating their potential for these highly degraded soils.

13.10.5 AFFORESTATION OF COASTAL SALINE AREAS

From the perspective of management, the salinity-related problematic soils of coastal areas were classified as: (i) land impregnated with high salinity and flooded with seawater; (ii) acid sulfate soils; (iii) land impregnated with high salinity and water logging but not flooded with seawater; (iv) land with low salinity and a shallow water table (at 0.5–4.0 m depth) with good quality water but saline; and (v) land with low salinity and shallow water table (at 0.5–4.0 suitable agroforestry systems have the potential to increase income and employment for small farmers while also satisfying local fodder and fuel demands, maintaining biodiversity, improving the coastal environment, protecting soil from erosion, and creating habitat for wildlife. Several agroforestry approaches for coastal and island locations have been explained by Dagar et al. (2014); Nair & Sreedharan (1986); Dagar (1991, 1994, 1995b, 1996, 2000, 2009, 2012); Dagar & Tomar (1998); Pandey et al. (2007); and Kumar & Kunhamu (2011).

A wide range of plant species have been discovered in mangrove habitats, but just 54 species, represented by 20 genera and 16 families, constitute the 'real mangroves,' which exist only in mangrove ecosystems and rarely elsewhere (Hogarth, 1999). However, the International Union for the Conservation of Nature and Natural Resources (IUCN) identified 60 exclusive mangrove species (13 in the western hemisphere and 51 in the eastern hemisphere) and 23 non-exclusive mangrove species spread throughout

16.9 million ha (Saenger et al., 1983). With an exclusive inventory of 62 mangrove species, two species (*Rhizophora x annamalayana* and *Heritiera kanikensis*) have been documented only from India (Banerjee et al., 1989; EISC, 2002). The first 'World Mangrove Atlas' offered an accurate evaluation of the extant mangrove area of 18.23 million hectares (Spalding et al., 1997). With increasing urbanization, particularly in developing countries, mangrove habitats can be prudently explored for meeting demands for food, fuel wood, thatching material, and increasing the productivity of coastal water for marine resources such as fish and shrimp production while causing minimal disruption to the mangrove forests (Dagar, 1982, 2008; Dagar et al., 1991, 1993, 2001a, b, 2009). Mangroves and other useful halophytes can be planted in coastal areas that have become too salty and are often flooded by the sea.

13.10.6 EXPERIENCES ON AFFORESTATION/AGROFORESTRY OF ALKALI/SODIC SOILS

Sodic soils account for approximately 52% of all global salt-affected soils. These soils have a high pH (>8.2), are sodic (ESP > 15), and contain soluble carbonates and sodium bicarbonates such that Na/Cl + SO 4 > 1 (Gupta & Abrol, 1990). Unlike soil reclamation for arable crops, which seeks to improve only the plow layer in the first instance, deep-rooted trees necessitate reclamation of the soil to lower depths. Planting strategies should also ensure optimal use of rainwater and leaching of reaction products following additions. Soil profile and structural improvements for greater water retention and rapid root penetration should be vertical rather than horizontal to reduce direct sodium toxicity threats. With this in mind, various attempts have been made, particularly in India, to create appropriate strategies for growing trees on such soils. Several forest and fruit trees, grasses, medicinal, and fragrant plants, and non-traditional and arable species were tested to see how well they could handle salty soil. The best species for agroforestry were then chosen.

13.10.7 PLANTING TECHNIQUES

A suitable planting approach for sodic lands must assure good rainwater usage, enhance quick root penetration in the vertical rather than horizontal orientation, and limit direct sodium toxicity risks. Previously, pits and trenches of various shapes and sizes were used to raise trees in alkali soils (Khan & Yadav, 1962; Pande, 1967; Yadav, 1972; Yadav et al., 1975), which had some inherent drawbacks such as higher amendment needs, laborious, and time-consuming operation, and leaving the $CaCO_3$ layer at about 0.5–0.7 m depth. The obstacles were overcome by the development of a tree planting technique known as "auger hole planting" (Spalding et al., 1997; Gill, 1985; Singh & Gill, 1992; Singh et al., 1993). The auger is mounted on a tractor and used to make holes 20–25 cm in diameter and 1.2–1.8 cm in depth to pierce the hard calcite layer. This practice recognizes that, because trees have deep root systems, managing the root zone by modifying the soil environment to greater soil depths using a limited quantity of amendments (3–5 kg gypsum, 3 kg FYM, + 20–30 g zinc sulfate, and some insecticide per auger hole) plays an important role in sapling establishment, plantation cost, and practical adaptability. Water stagnation is a regular occurrence during the rainy season. Many forest and fruit tree species can be grown on highly alkaline soil, but some, such as pomegranate (*Punica granatum*) and bael (*Aegle marmelos*), are unable to endure water stagnation. For such situations, the raised and sunken bed agroforestry approach was tested (Dagar et al., 2001a). After filling the auger holes with soil mixture in a leveled field, as described above, the auger holes are marked with sticks. Parallel bunds of 1–2 m height and 1–2 m width is then built, leaving 4–5 m between them to take soil from interspaces. The seedlings are raised in designated areas on the bund, and tiny rings are formed around the seedlings for first irrigation. During the rainy season, the spaces between the rows are planted with water-loving crops like kallar grass (*Leptochloa fusca*) or rice (salt-tolerant types), and during the winter, they are planted with Egyptian clover (*Trifolium alexandrinum*) or salt-tolerant wheat.

13.10.8 AFFORESTATION SPECIES

Several research projects have been undertaken in the Indian subcontinent over the last three decades to establish the beneficial advantages of contemporary planting practices and to identify the most suited plants for alkali soils. At pH 10 compared to pH 7 in tree species such as *Acacia nilotica, Terminalia arjuna*, and *Pongamia pinnata*, but at pH 9.5 *Eucalyptus tereticornis, Casuarina equisetifolia,* and *Acacia nilotica* could grow well in soil with ESP 30.6, whereas Pong Relative tolerance was reported (Singh et al., 1988) in the following order based on the performance of tree seedlings planted in soils of varied pH (7–12): *Acacia nilotica* > *Prosopis juliflora*

> *Haplophragma adenophyllum* > *Albizia lebbeck* > *Syzygium cuminii.* The results of a series of tests conducted on high soil pH (10) (Gill, 1985; Gill & Abrol, 1993; Das & Itnal, 1994; Singh, 1995) revealed that a large number of forest and fruit trees may be effectively cultivated on sodic soils to provide adequate biomass. Three tree species, *Prosopis juliflora, Acacia nilotica*, and *Tamarix articulata*, were found economically suitable out of 30 tree species planted on highly alkaline soil (pH of profile 10.1–10.6) at the Saraswathi in semiarid Haryana, northwest India, with biomass production of 51, 70, and 93 Mg ha^{-1}, respectively, in seven years (Dagar et al., 2001b; Singh & Dagar, 2005). *Tamarix articulata* improved the soil the most by causing the greatest decrease in exchangeable sodium percentage (ESP) and pH values, followed by *P. juliflora* and *A. nilotica*. Organic carbon increased by 0.23, 0.26, and 0.10% in the surface 15 cm layer, respectively. *Prosopis* species such as *P. juliflora, P. alba, P. articulata, P. levigata, P. pallida*, and *P. nigra* produced a lot of biomasses at the same location. All of these can be used successfully as energy crops. Singh et al. (2008) found *Prosopis juliflora* to have the highest biomass of 56.5 Mg ha^{-1}, followed by *Acacia nilotica* (50.8 Mg ha^{-1}), *Casuarina equisetifolia* (42.1 Mg ha^{-1}), *Terminalia arjuna* (41.6 Mg ha^{-1}), *Pithecellobium dulce* (32.3 Mg ha^{-1}), *Eucalyptus* (30.8 Mg ha^{-1}), and Prosopis alba. Maximal biomass production per tree was achieved in *E. tereticornis, A. nilotica, P. juliflora*, and *C. equisetifolia* after 14 years of plantation with 231, 217, 208, and 197 kg bole weight per plant, respectively, whereas *P. alba, P. dulce, T. arjuna, P. pinnata, A. indica*, and *C. siamea* provided relatively lower Bol (Sharma et al., 2010). These tree species greatly enhanced soil in terms of pH and ESP, as well as organic carbon. Organic carbon rose from 0.8 g kg^{-1} to > 3.5 g kg^{-1} under *P. juliflora, P. alba, Pongamia pinnata, A. nilotica, A. indica*, and *C. equisetifolia.* Similarly, Singh et al. (1991, 1995) found that 20–36-year-old plantings improved alkali soil through litter and root decomposition, decreasing soil pH and boosting organic carbon and other accessible nutrients. The amount of the ensuing soil enrichment varied depending on the tree species. Tree species such as *Prosopis juliflora, Azadirachta indica, Salvadora persica*, and *Acacia nilotica* were shown to be particularly effective when planted in auger holes on sodic vertisols. *Cassia siamea* and *Leucaena leucocephala* behaved admirably as well (Sharma et al., 1992). *Acacia auriculiformis, Dalbergia sissoo, Casuarina equisetifolia, Dendrocalamus strictus*, and *Hardwickia binata* did not survive for an extended period of time. Tree species also made clayey vertisols much better by lowering pH and ESP and increasing carbon in the soil.

13.10.9 ARABLE CROP-BASED AGROFORESTRY SYSTEMS

Singh et al. (1995) evaluated this agroforestry approach on moderately alkaline soil (pH 9.2) under irrigated conditions in the Indian state of Haryana with the plantation of three commercial trees, namely poplar (*Populus deltoides*), eucalyptus (*Eucalyptus tereticornis*), and Kikar (*Acacia nilotica*), in association with rice-wheat, rice-Egyptiana, and rice-Egyptiana. During the first three years, intercrops of Egyptian clover, rice, wheat, and mustard were successfully planted alongside these trees. These crops may eventually be replaced by shade-loving crops such as turmeric. These intercrops promote the growth of Populus and Eucalyptus while inhibiting the growth of low-water-demanding plants such as Acacia. Soil improvement, as assessed by lower pH and increased organic carbon, accessible nitrogen, phosphorus, and potassium concentrations, occurred in the following order: Sole crops > *Acacia*-based system > *Populus*-based system > *Eucalyptus*-based system *Poplar* with rice-wheat had the highest benefit-cost ratio (3.30), followed by poplar with rice-Egyptian clover (2.95), rice-wheat (2.79), and Acacia with rice-Egyptian clover (1.76).

13.11 CONCLUSION AND FUTURE PERSPECTIVES

It is obvious that salinity issues are strongly related with land degradation, and as a result of their direct and indirect effects, more than half a billion hectares of land are not being exploited properly for agricultural production. As a result, there is a need to investigate alternate, sustainable, and cost-effective methods of remediating such degraded lands. Despite the abundance of empirical and theoretical data on the adaptability and potential of some halophytes to remediate salty and sodic soils, significant gaps remain in the restoration process of such halophytes, which essentially improves the structure and functional aspects of land. Our understanding of the physiological/biochemical mechanisms underlying halophytes under varying salinities is particularly limited. As a result, physiological and molecular research to uncover the underlying mechanisms of these processes are critical scientific goals. If properly developed, phytoremediation could become a cost-effective and environmentally sound solution for the cleanup of salt-impacted lands. Certain restrictions must be overcome before this plant-based cleanup technique may be widely used. Phytoremediation can be time-consuming because it takes several growing seasons to reduce the level of pollutants in soil. It is also restricted to soil depths in the rooting zone. It is required to locate plants capable of removing the greatest amount of salt by creating more biomass with some economic relevance for phytoremediation, and the selected plant species must survive high salt concentrations. The next step in employing halophytes to repair soil salinity is to create a plant with a diversified salt collecting capacity that is also cost-effective. For this kind of research, it would be important to find new genes that produce a lot of biomass and then make transgenic plants that are better at cleaning up pollution.

KEYWORDS

- **agroforestry**
- **halophytes**
- **Indian arid region**
- **phytoremediation**
- **restoration**
- **saline and sodic soils**
- **soil salinity**

REFERENCES

Abbas, J. A., (2005). Seasonal variations of ash content of the halophyte *Zygophyllum qatarense* Hadidi from saline and non-saline habitats in Bahrain. *Pak. J. Bot., 37*, 853–858.

Ashraf, M. Y., Ashraf, M., & Sarwar, G., (2005). Physiological approaches to improving plant salt tolerance. In: Dris, R., (ed.), *Crops: Growth, Quality and Biotechnology* (pp. 1206–1227).

Ashraf, M. Y., Ashraf, M., Mahmood, K., Akhter, J., Hussain, F., & Arshad, M., (2010). Phytoremediation of saline soils for sustainable agricultural productivity. In: Ashraf, M., Ozturk, M., & Ahmad, M. S. A., (eds.), *Plant Adaptation and Phytoremediation* (pp. 335–355). Springer, Berlin, Germany.

Aslam, Z., Mujtab, M., Alshtar, J., Waheed, R., Malik, K. A., & Naqvi, M., (1993). Biological methods for economically utilizing salt-affected soils in Pakistan. In: Davidson, N., & Galloway, R., (eds.), *Productive use of Saline Land* (pp. 29–31). ACIAR proceedings no. 12, Canberra, Australia.

Ball, M. C., (1988). Salinity tolerance in the mangroves *Aegiceras corniculatum* and *Avicennia marina.* 1. Water-use in relation to growth, carbon partitioning, and salt balance. *Aust. J. Plant Physiol., 15*(3), 447–464.

Banerjee, L. K., Sastry, A. R. K., & Nayar, M. P., (1989). *Mangroves in India: Identify Cation Manual* (p. 113). Botanical Survey of India, Calcutta.

Barnett-Lennand, E. G., (2002). Restoration of saline lands through revegetation. *Agr. Water Manag., 53*, 213–216.

Böer, B., & Al Hajiri, S., (2002). The coastal and sabkha flora of Qatar: An introduction. In: Barth, H. J., & Böer, B., (eds.), *Sabkha Ecosystems: The Arabian Peninsula and Adjacent Countries* (Vol. 1, pp. 63–70). Kluwer, Dordrecht.

Böer, B., (2002). The coastal and sabkha flora of the United Arab Emirates. In: Barth, H. J., & Böer, B., (eds.), *Sabkha Ecosystems: The Arabian Peninsula and Adjacent Countries* (Vol. 1, pp. 303–309). Kluwer, Dordrecht.

Bradshaw, A. D., (2002). The reconstruction of ecosystems. *J. Applied Ecol., 20*, 1–17.

Carrow, R. N., & Duncan, R. R., (2011). Salinity in soils. In: Leinauer, B., & Cockerham, S., (eds.), *Turf Grass Water Conservation* (2nd edn.). Riverside: ANR Communications Service, University of California.

Chaturvedi, A. N., (1984). Firewood crops in areas of brackish water. *Ind. For., 110*(4), 364–366.

Chaudhri, I. I., Shah, B. H., Naqvi, N., & Mallick, I. A., (1964). Investigations on the role of *Suaeda fructicosa* Forsk. in the reclamation of saline and alkaline soils in west Pakistanian plains. *Plant and Soil, 21*, 1–7.

Cortina, J., Maestre, F. M., Vallejo, R., Baeza, M. J., Valdecantos, A., & Perez-Devesa, M., (2006). Ecosystem structure, function, and restoration success: Are they related? *J. Nat. Conser., 14*, 152–160.

CSSRI, (2014). *Annual Reports of Central Soil Salinity Research Institute, Karnal, India – 2002–2003 to 2012–2013.*

Dagar, J. C., & Singh, G., (2007). *Biodiversity of Saline and Waterlogged Environments: Documentation, Utilization and Management* (p. 76). National Biodiversity Authority, Chennai, India.

Dagar, J. C., & Singh, G., (2007). *Biodiversity of Saline and Waterlogged Environments: Documentation, Utilization, and Management.* National Biodiversity Authority, Chennai.

Dagar, J. C., & Tomar, O. S., (1998). Suitable agroforestry systems for coastal saline regions: Case study. *J. Range Manage. Agrofor., 19*, 165–172.

Dagar, J. C., & Tomar, O. S., (2002). Utilization of salt affected soils and poor quality waters for sustainable bio-saline agriculture in arid and semiarid regions of India. In: *Proc. 12th ISCO Conference* (pp. 340–347). Beijing, China.

Dagar, J. C., (1982). Some ecological aspects of mangrove vegetation of Andaman and Nicobar Island. *Sylvatrop. Philipp. For. Res. J., 7*, 177–216.

Dagar, J. C., (1991). Agroforestry practices for coastal Bay Islands. *J. Indian Soc. Coast Agric. Res., 9*(1, 2), 433–439.

Dagar, J. C., (1994). Strategies for sustainable use of island resources of India. *J. Indian Soc. Coast Agric. Res., 12*(1, 2), 187–191.

Dagar, J. C., (1995a). Agroforestry systems for Andaman and Nicobar Islands in India. *Int. Tree Crops J., 8*(2, 3), 107–128.

Dagar, J. C., (1995b). Ecology of halophytic vegetation in India: A review. *Int. J. Ecol. Env. Sci., 21*, 273–296.

Dagar, J. C., (1996). Rehabilitation of coastal saline lands by planting with suitable species. *J. Indian Soc. Coast Agric. Res., 14*, 185–191.

Dagar, J. C., (2000). Agroforestry systems for coastal and island regions. *Indian J. Agrofor., 2*, 59–74.

Dagar, J. C., (2003). Biodiversity of Indian saline habitats and management & utilization of high salinity tolerant plants with industrial application for rehabilitation of saline areas. In: Alsharhan, A. S., Wood, W. W., Goudie, A. S., Fowler, A., & Abdellatif, E. M., (eds.),

Desertification in the Third Millennium (pp. 151–172). Swets & Zeitlinger Publishers, Lisse.

Dagar, J. C., (2005a). Salinity resources in India: An overview. *Bull. Nat. Inst. Ecol., 15*, 69–80.

Dagar, J. C., (2005b). Ecology, management, and utilization of halophytes. *Bull. Nat. Inst. Ecol., 15*, 81–97.

Dagar, J. C., (2008). Indian mangroves: Status, management, and their plausible benefits for livelihood security. *J. India Soc. Coast. Agric. Res., 26*(2), 121–128.

Dagar, J. C., (2009). Opportunities for alternate land uses in salty and water scarcity areas. *Int. J. Ecol Env. Sci., 35*(1), 53–66.

Dagar, J. C., (2012). Utilization of degraded lands/habitats and poor-quality water for livelihood security and mitigating climate change. *Indian J. Agrofor., 14*(1), 1–16.

Dagar, J. C., (2014). Greening salty and waterlogged lands through agroforestry systems. In: Dagar, J. C., Singh, A. K., & Arunachalam, A., (eds.), *Agroforestry Systems in India: Livelihood Security & Environmental Services* (Vol. 10, pp. 333–344). Advances in agroforestry, Springer Publishers, New Delhi.

Dagar, J. C., (2015). Agroforestry for restoration, conservation, and resilience of waste/degraded lands: Challenges and opportunities. In: Dhyani, S. K., Newaj, R., Alam, B., & Dev, I., (eds.), *Agroforestry: Present Status and Way Forward* (pp. 323–362). Biotech Books, New Delhi.

Dagar, J. C., Gururaja-Rao, G., Shukla, Y. K., & Sharma, H. B., (2009). Performance of three flower yielding plants in different sodic soils. *Indian J. Hortic., 66*(3), 404–409.

Dagar, J. C., Mongia, A. D., & Bandyopadhyay, A. K., (1991). *Mangroves of Andaman & Nicobar Islands* (p. 166). Oxford & IBH Publishing Co. Ltd, New Delhi.

Dagar, J. C., Pandey, C. B., & Chaturvedi, C. S., (2014). Agroforestry: A way forward for sustaining fragile coastal and island agro-ecosystems. In: Dagar, J. C., Singh, A. K., & Arunachalam, A., (eds.), *Agroforestry Systems in India: Livelihood Security & Environmental Services* (Vol. 10, pp. 185–232). Advances in agroforestry, Springer, New Delhi.

Dagar, J. C., Sharma, H. B., & Shukla, Y. K., (2001a). Raised and sunken bed technique for agroforestry on alkali soils of northwest India. *Land Degrad. Dev., 12*, 107–118.

Dagar, J. C., Singh, G., & Singh, N. T., (2001b). Evaluation of forest and fruit trees used for rehabilitation of semi-arid alkali/ sodic soils in India. *Arid Land Res. Manage., 15*, 115–133.

Dagar, J. C., Singh, N. T., & Mongia, A. D., (1993). Characteristics of mangrove soils and vegetation of Bay Islands in India. In: Lieth, H., & Al Masoom, A., (eds.), *Towards the Rational Use of High Salinity Tolerant Plants* (Vol. 1, pp. 59–80). Kluwer Academic Publishers, Dordrecht.

Dagar, J. C., Yadav, R. K., Dar, S. R., & Ahamad-Sharif, (2015). Liquorice (*Glycyrrhiza glabra*): A potential salt tolerant highly remunerative medicinal crop which also helps in remediation of alkali soils. *Curr. Sci., 108*(9), 1683–1688.

Das, S. K., & Itnal, C. J., (1994). Capability based land use systems: Role in diversifying dryland agriculture. In: Soil management for sustainable agriculture in dryland area. *Bull. Indian Soc. Soil Sci., 16*, 92–100.

De Lacerda, L. P., Lange, L. C., Giovanni, M., & França, C., (2015). Salinity reduction and biomass accumulation in hydroponic growth of purslane (*Portulaca oleracea*). *Int. J. Phytoremed., 17*, 235–241.

Devi, S., Nandwal, A. S., Angrish, R., Arya, S. S., Kumar, N., & Sharma, S. K., (2015). Phytoremediation potential of some halophytic species for soil salinity, *Int. J. Phytoremed.* doi: 10.1080/15226514.2015.1131229.

Devilliers, A. J., Vanrooyen, M. W., Theron, G. K., & Claassens, A. S., (1995). Removal of sodium and chloride from a saline soil by *Mesembryanthemum barklyi. J. Arid Environ., 29*(3), 325–330.

EISC, (2002). *Mangroves of India: State-of-Art Report* (p. 140). ENVIS publication series, 2/2002. Environmental Information System Center (EISC), Annamalai University, Parangipettai.

FAO-UNESCO, (1974). *Soil Map of the World* (Vol. 1010). 1:5,000,000, UNESCO, Paris, France.

Farzi, A., Borghei, S. M., & Vossoughi, M., (2017). The use of halophytic plants for salt phytoremediation in constructed wetlands. *Int. J. Phytoremed., 19*(7), 643–650.

Fitzpatrick, R. W., (2002). Land degradation processes. *ACIAR Monograph Series, 84*, 119–129.

Flowers, T. J., & Colmer, T. D., (2008). Salinity tolerance in halophytes. *The New Phytologist, 179*, 945–963.

Flowers, T. J., & Flowers, S., (2005). Why does salinity pose such a difficult problem for plants breeders? *Agri. Water Manag., 78*, 15–24.

Flowers, T. J., Galal, H. K., & Bromham, L., (2010). Evolution of halophytes: Multiple origins of salt tolerance in land plants. *Functional Plant Biol., 37*, 604–661.

Flowers, T. J., Troke, P. F., & Yeo, A. R., (1977). Mechanism of salt tolerance in halophytes. *Ann. Rev. Plant Physiol. Plant Mol. Biol., 28*, 89–121.

Gandolfi, S., Joly, C. A., & Rodrigues, R. R., (2007). Permeability-Impermeability: Canopy trees as biodiversity filters. *Scientia Agrícola, 64*(4), 433–438.

Geesing, D., Felker, P., & Bingham, R. L., (2000). Influence of mesquite (*Prosopis glandulosa*) on soil nitrogen and carbon development: Implications for global carbon sequestration. *J. Arid Environ., 46*, 157–180.

Ghosal, T. K., Ghosh, P., & Rezaul, H. A. H. M., (2004). *Cultivating Wetlands in Bangladesh* (pp. 1–6). LEISA Magazine.

Gill, H. S., & Abrol, I. P., (1993). Afforestation and amelioration of salt-affected soils in India. In: Davidson, N., & Galloway, R., (eds.), *Productive Use of Saline Land* (pp. 23–27). ACIAR proceedings no. 42. Canberra.

Gill, H. S., (1985). *Studies on the Evaluation of Selected Tree Species for Their Tolerance to Sodicity and Mechanical Impedance in a Highly Sodic Soil with Particular Reference to Root Growth Behavior*. Ph. D. Dissertation. Kurukshetra University, Kurukshetra, India.

Gleen, E. P., Brown, J. J., & Blumwald, E., (1999). Salt tolerance and crop potential of halophytes. *Critical Reviews in Plant Sciences, 18*(2), 227–255.

Glenn, E. P., Andaya, T., Chaturvedib, R., Martinez-Garciaa, R., Pearlsteina, S., Soliza, D., & Felgera, R. S., (2013). Three halophytes for saline-water agriculture: An oilseed, a forage, and a grain crop. *Environ Exp. Bot., 92*, 110–121.

Glenn, E., Brown, J., & Blumwald, E., (1999). Salt tolerance and crop potential of halophytes. *Critical Rev. Plant Sci., 18*, 227–255.

Greenway, H., & Munns, R., (1980). Mechanisms of salt tolerance halophytes. *Annual Review of Plant Physiology, 31*, 149–190.

Grewal, S. S., & Abrol, I. P., (1989). Amelioration of sodic soils by rain water conservation and Karnal grass grown in interspace of trees. *J. Indian Soc. of Soil Sci., 37*, 371–376.

Gul, B., Weber, D. J., & Han, M. A., (2000). Effect of salinity and planting density on physiological responses of *Allenrolfea occidentalis*. *Western North American Naturalist, 60*(2), 188–197.

Gupta, G. N., Prasad, K. G., Mohan, S., Subramaniam, V., & Manivachakam, P., (1988). Effect of alkalinity on survival and growth of tree seedlings. *J. Indian Soc. Soil Sci., 36*, 537–542.

Gupta, R. K., & Abrol, I. P., (1990). Salt affected soils: Their reclamation and management for crop production. In: Lal, R., & Stewart, B. A., (eds.), *Advances in Soil Science, Soil Degradation* (Vol. III, pp. 223–288). Springer, New York.

Gupta, S. K., & Goyal, M. R., (2017). *Soil Salinity Management in Agriculture: Technological Advances and Applications* (p. 454), Apple Academic Press, Canada.

Hajkowicz, S., & Young, M., (2005). Costing yield loss from acidity, sodicity and dry land salinity to Australian agriculture. *Land Deg. Dev., 16*, 417–433.

Hasanuzzaman, M., Nahar, K., Akam, M. M., Bhowmik, P. C., Hossain, M. A., Rahman, M. M., Prasad, M. N. V., et al., (2014). Potential use of halophytes to remediate saline soils. *Bio. Med Res. Int.* http://dx.doi.org/10.1155/2014/589341.

Heuperman, A. F., Kapoor, A. S., & Denecke, H. W., (2002). *Biodrainage: Principles, Experiences, and Applications* (p. 78). International Program for Technology & Research in Irrigation and Drainage. IPTRID Secretariat, FAO, Rome.

Hogarth, P. J., (1999). *The Biology of Mangroves*. Oxford University Press, Oxford.

Hussain, S., Ahmed, E., Malik, A., Ferheen, S., Jabbar, A., Ashraf, M., Lodhi, M. A., & Choudhary, M. I., (2008). Recursterols A and B, chymotrypsin inhibiting sterols from *Haloxylon recurvum*. *Polish Journal of Chem., 80*, 409–415.

INCID, (2003). *Biodrainage Status in India and Other Countries* (p. 40). Indian National Committee on Irrigation and Drainage (INCID), New Delhi.

Isbell, R. F., (1998). *The Australian Soil Classification* (p. 143). CSIRO, Collingwood.

Jain, B. L., Muthana, K. D., & Goyal, R. S., (1985). Performance of tree species in salt-affected soils in arid regions. *J. Indian Soc. Soil Sci., 33*, 221–224.

Jeet-Ram, D. J. C., Khajanchi, L., Singh, G., Toky, O. P., Tanwar, V. S., Dar, S. R., & Chauhan, M. K., (2011). Bio-drainage to combat water logging, increase farm productivity and sequester carbon in canal command areas of northwest India. *Curr Sci., 100*(11), 1673–1680.

Jianfeng, Z., Shangjun, X., Li, J., Makeschin, F., & Yumin, S., (2004). Agroforestry and its application in amelioration of saline soils in eastern China coastal region. *Forestry Studies in China, 6*(2), 27–33.

Kapoor, A. S., & Denecke, H. W., (2001). *Bio-drainage and Bio-disposal: The Rajasthan Experience*. In: GRID, IPTRID's network magazine no. 17, IPTRID, FAO, Rome.

Kapoor, A. S., (2014). *Managing Groundwater for Irrigated Agriculture – the Relevance of Bio-drainage Model* (p. 341). Rawat Publications, Jaipur.

Khamzina, A., Lamers, J. P. A., Martius, C., Worbes, M., & Vlek, P. L. G., (2006). Potential of nine multipurpose tree species to reduce saline groundwater tables in the lower Amu Darya River region of Uzbekistan. *Agroforest Syst., 68*, 151–165.

Khan, M. A. W., & Yadav, J. S. P., (1962). Characteristics and afforestation problems of saline alkali soils. *Ind. For., 88*(4), 259–271.

Kumar, A., & Abrol, I. P., (1986). *Grasses on Alkali Soils*. Bulletin no. 11, CSSRI, Karnal, India.

Kumar, B. M., & Kunhamu, T. K., (2011). *Quarter Century of Agroforestry Research in Kerala: A Compendium of Research Publications* (p. 404). Kerala Agriculture University, Kerala.

Kushiev, H., Noble, A. D., Abdullaev, I., & Toshbekov, U., (2005). Remediation of abandoned saline soils using *Glycyrrhiza glabra*: A study from the hungry steppes of Central Asia. *International J. Agri. Sustainability, 3*(2), 102–113.

Lewis, M. A., & Devereux, R., (2009). Non-nutrient anthropogenic chemicals in sea grass ecosystems: Fate and effects. *Environ. Toxicology Chem., 28*, 644–661.

Malik, K. A., & Zafar, Y., (1984). Use of 15 N isotope dilution techniques for quantification of associative BNF in Kallar grass. In: Veegar, C., & Newton, W. E., (eds.), *Advances in Nitrogen Fixation Research*. Nijhoff Junk, The Hague.

Malik, K. A., Aslam, Z., & Naqvi, M., (1986). *Kallar Grass-a Plant for Saline Lands*. Nuclear Institute for Agriculture and Biology, Faisalabad.

MandaI, A. K., Sharma, R. C., Singh, G., & Dagar, G., (2010). *Computerized Database on Salt Affected Soil in India*. Technical Bulletin No. CSSRI/Karnal/2/2010.

Mandal, A. K., (2016). Mapping and characterization of salt-affected and waterlogged soils in the Gangetic plain of central Haryana (India) for reclamation and management. *Cogent Geoscience, 2,* 1213689. http://dx.doi.org/10.1080/23312041.2016.1213689.

Manjunatha, M. V., Hebbara, M., Patil, S. G., Kuligod, V. B., & Minhas, P. S., (2005). Effect of trees alone or with grasses on halting canal seepage and shallow water table control in saline vertisols. *J. Indian Soc. Soil Sci., 53*(2), 254–257.

Manousaki, E., & Kalogerakis, N., (2009). Phytoextraction of Pb and Cd by the Mediterranean saltbush (*Atriplex halimus* L.): Metal uptake in relation to salinity. *Environmental Sci. Poll. Res., 16*, 844–854. doi: 10.1007/s11356-009-0224-3.

Marcar, N. E., & Khanna, P. K., (1997). Reforestation of salt affected and acid soils. In: Nambier, E. K. S., & Brown, A. G., (eds.), *Management of Soil, Nutrients and Water in Tropical Plantation Forests* (pp. 481–524). ACIAR monograph 43. ACIAR, Canberra.

Massoud, F. I., (1981). *Salt-Affected Soils at a Global Scale and Concepts of Control*. FAO Land and Water Development Division, Technical Paper, Rome.

Masters, D., Benes, S., & Norman, H., (2007). Bio-saline agriculture for forage and livestock production. *Agr. Ecosys. Environ., 119*, 234–248.

Mathur, B. K., Singh, J. P., Beniwal, R. K., & Singh, N. P., (2007). Utilization of salty shrub-lani (*Salsola baryosma*) of arid region as drought feed for goats. In: *Abstracts of International Tropical Animal Nutrition Conference* (Vol. II, p. 33). National Dairy Research Institute, Karnal, India.

Munns, R., Schachtman, D. P., & Condon, A. G., (1995). The significance of a two-phase growth response to salinity in wheat and barley. *Australian Journal of Plant Physiology, 22*(4), 561–569.

Muscolo, A., Panuccio, M. R., & Piernik, A., (2014). Ecology, distribution, and ecophysiology of *Salicornia Europaea* L. In: Khan, M. A., Böer, B., Öztürk, M., Al Abdessalaam, T. Z., Clüsener-Godt, M., & Gul, B., (eds.), *Sabkha Ecosystems: Cash Crop Halophyte and Biodiversity Conservation* (Vol. IV). Springer, Dordrecht.

Nair, M. A., & Sreedharan, C., (1986). Agroforestry farming systems in the homesteads of Kerala, southern India. *Agrofor Syst., 4*, 339–363.

Nedjimi, B., & Daoud, Y., (2009). Cadmium accumulation in *Atriplex halimus* subsp. *Schwein furthii* and its influence on growth, proline, root hydraulic conductivity and nutrient uptake. *Flora, 204*, 316–324.

Öztürk, M., Altay, V., Gucel, S., & Guvensen, A., (2014). Halophytes in the East Mediterranean-their medicinal and other economical values. In: Khan, M. A., Böer, B., Öztürk, M., Al Abdessalaam, T. Z., Clüsener-Godt, M., & Gul, B., (eds.), *Sabkha Ecosystems: Cash Crop Halophyte and Biodiversity Conservation* (Vol. IV, pp. 247–272). Springer, Dordrecht.

Pande, G. C., (1967). Afforestation of user lands. In: *Proceedings of 11th All India Silviculture Conference*. FRI, Dehra Dun.

Pandey, C. B., Rai, B. R., Singh, L., & Singh, A., (2007). Homegardens of Andaman and Nicobar, India. *Agric. Syst., 92*, 1–22.

Panta, S., Lane, P., Doyle, R. B., & Hardie, M., (2016). Halophytes as a possible alternative to desalination plants: Prospects of recycling saline wastewater during coal seam gas operations. *Halophytes for Food Security in Dry Land*, 317–329.

Pitman, M. G., & Läuchli, A., (2002). Global impact of salinity and agricultural ecosystems. *Salinity: Environment-Plants-Molecules.* Springer Netherlands.

Qadir, M., & Oster, J. D., (2004). Crop and irrigation management strategies for saline-sodic soils and waters aimed at environmentally sustainable agriculture. *Science of the Total Environ., 323*, 1–19.

Qadir, M., Qureshi, R. H., & Ahmad, N., (1996a). Reclamation of a saline-sodic soil by gypsum and *Leptochloa fusca. Geoderma, 74*(3), 207–217.

Qadir, M., Qureshi, R. H., Ahmad, N., & Ilyas, M., (1996b). Salt-tolerant forage cultivation on a saline-sodic field for biomass production and soil reclamation. *Land Degradation & Development, 7*(1), 11–18.

Qadir, M., Tubeileh, A., Akhtar, J., Larbi, A., Minhas, P. S., & Khan, M. A., (2008). Productivity enhancement of salt-affected environment through crop diversification. *Land Degradation and Develop., 19*, 429–453.

Rabhi, M., Hafsi, C., Lakhdar, A., Hajji, S., Barhoumi, Z., Hamrouni, M. H., Abdelly, C., & Smaoui, A., (2009). Evaluation of the capacity of three halophytes to desalinize their rhizosphere as grown on saline soils under non-leaching conditions. *African J. Ecol., 47*(4), 463–468.

Rabhi, M., Talbi, O., Atia, A., Chedly, A., & Smaoui, A., (2008). Selection of halophyte that could be used in the bio reclamation of salt affected soils in arid and semi-arid regions. *Bio-saline Agriculture and High Salinity Tolerance*, 242–246.

Rao, D. L. N., & Gill, H. S., (1990). *Nitrogen Fixation and Biomass Productivity by Annual and Perennial Legumes* (pp. 112–114). *Annual Report on CSSRI*, Karnal.

Rao, G. G., Nayak, A. K., & Chinchmalatpure, A. R., (2003). *Salvadora Persica: A Life Support Species for Salt Affected Black Soils*. Technical bulletin 1/2003. CSSRI, Karnal, India.

Rathore, V. S., Singh, J. P., Bhardwaj, S., Nathawat, N. S., Kumar, M., & Roy, M. M., (2015). Potential of native shrubs *Haloxylon salicornicum* and *Calligonum polygonoides* for restoration of degraded lands in arid western Rajasthan, India. *Environ. Manag., 55*, 205–216.

Ravindran, K. C., Venkatesan, K., Balakrishnan, V., Chellappan, K. P., & Balasubramanian, T., (2007). Restoration of saline land by halophytes for Indian soils. *Soil Biology and Biochemistry, 39*(10), 2661–2664.

Reddy, M. P., Shah, M. T., & Patola, J. S., (2008). *Salvadora persica*, a potential species for industrial oil production in semiarid saline and alkaline soils. *Industrial Crops and Products, 28*, 273–278.

Richards, L. A., (1954). *Diagnosis and Improvement of Saline and Alkali Soils* (pp. 79–81). USDA Handbook no 60. US Department of Agriculture, Washington, DC.

Roy, C. S., Kumar, A., Brahmanand, P. S., Ghosh, S., Mohanty, R. K., Jena, S. K., Sahoo, N., & Panda, G. C., (2011). *Application of Bio-Drainage for Reclamation of Waterlogged Situations in Deltaic Orissa.* Research bulletin 53. Directorate of Water Management (ICAR), Bhubaneswar.

Ruan, C. J., Li, H., Guo, Y. Q., Qin, P., Gallagher, J. L., Seliskar, D. M., Lutts, S., & Mahy, G., (2008). *Kosteletzkya virginica,* an agroeco engineering halophytic species for alternative agricultural production in China's east coast: Ecological adaptation and benefits, seed yield, oil content, fatty acid, and biodiesel properties. *Ecological Engineering, 32*, 320–328.

Rudolf, W. H. V., & Lafferty, D. K., (2011). Stage structure alters how complexity affects stability of ecological networks. *Ecological Letters, 14*(1), 75–79.

Saenger, P., Hegerl, E. J., & Davie, J. D. S., (1983). *Global Status of Mangrove Ecosystems* (Vol. 3, p. 88). Working groups on mangrove ecosystems, IUCN commission on ecology, UNEP/ WWF, Environmentalist.

Sahab, S., Suhani, I., Srivastava, V., Chauhan, P. S., Singh, R. P., & Prasad, V., (2021). Potential risk assessment of soil salinity to agro ecosystem sustainability: Current status, and management strategies. *Science of the Total Environment, 764*, 144–164.

Sandhu, G. R., Aslam, Z., Salim, M., Satter, A., Qureshi, R. H., Ahmed, N., & Wyn, J., (1981). The effect of salinity on the yield and composition of *Diplachne fusca* (Kallar grass). *Plant Cell Environ., 4*, 177–181.

Seaman, J., (2005). *Mechanisms of Salt Tolerance in Halophytes: Can Crop Plants Resistance to Salinity be Improved?* Ph. D. Dissertation, University of Sheffield.

Sen, D. N., Rajpurohit, K. S., & Wissing, F. W., (1982). Survey and adaptive biology of halophytes in western Rajasthan, India. In: Sen, D. N., Rajpurohit, K. S. (eds.). Contributions to the ecology of halophytes. Tasks for vegetation science, vol 2. Dr W. Junk Publishers, The Hague, The Netherlands Springer, Dordrecht. https://doi.org/10.1007/978-94-009-8037-2_5.

Sengbusch, P., (2003). *Halophytes Botanik Online.* University of Hamburg.

Shabala, S., & Mackay, A., (2011). Ion transport in halophytes. *Adv. Bot. Res., 57*, 151–199.

Shahid, S. A., (2013). Developments in soil salinity assessment, modeling, mapping, and monitoring from regional to sub-microscopic scales. In: Shahid, S. A., Abdelfattah, M. A., & Taha, F. K., (eds.), *Developments in Soil Salinity Assessment and Reclamation: Innovative Thinking and Use of Marginal Soil and Water Resources in Irrigated Agriculture* (pp. 3–44). doi 10.1007/978-94-007-5684-7_1. Springer Dordrecht.

Sharma, O. P., Verma, S. K., Kool, Y. M., & Desnonki, S. C., (1992). *Progress Report of the Project Evaluation of Tree Plantations for the Control of Salinity and Water Table.* Jawaharlal Nehru Krishi Viswavidyalaya, Indore.

Sharma, S. K., Dagar, J. C., & Singh, G., (2010). *BIOSAFOR: Bio-saline (Agro) Forestry: Remediation of Saline Wastelands Through Production of Renewable Energy, Biomaterial, and Fodder.* Technical bulletin no. 4/2010. Central Soil Salinity Research Institute, Karnal.

Sheikh, M. I., (1987). *Energy Plantations for Marginal and Problematic Lands, Pakistan.* GCP/RAS/III/NET Fd. Document 5, Bangkok, FAO.

Shekhawat, V. P. S., Kumar, A., & Neumann, K. H., (2006). Effect of sodium chloride salinity on growth and ion accumulation in some halophytes. *Communications in Soil Science and Plant Analysis, 37*(13), 1933–1946.

Singh, B., (1998). Contribution of fine roots in reclamation of semiarid sodic soils. *Arid Soil Res. Rehabilitation, 12*, 207–222.

Singh, G., & Dagar, J. C., (2005). *Greening Sodic Soils: Bichhian Model* (p. 51). Technical bulletin no. 2/2005. Central Soil Salinity Research Institute, Karnal.

Singh, G., & Gill, H. S., (1992). Ameliorative effect of tree species on characteristic of sodic soils at Karnal. *Indian J. Agri. Sci., 62*, 142–146.

Singh, G., (1995). An agroforestry practice for the development of salt lands using *Prosopis juliflora* and *Leptochloa fusca. Agrofor. Syst., 29*, 61–75.

Singh, G., (1995a). An agroforestry practice for the development of salt lands using *Prosopis juliflora* and *Leptochloa fusca. Agroforestry Systems, 27*, 61–75.

Singh, G., (1995b). An agroforestry practice for the development of salt lands using *Prosopis juliflora* and *Laptocloa fusca. Agrofor. Syst., 29*, 61–75.

Singh, G., Abrol, I. P., & Cheema, S. S., (1988). Agroforestry on alkali soil-effect of planting methods and amendments on initial growth, biomass accumulation and chemical composition of mesquite (*Prosopis juliflora* (SW) DC) with inter-space planted with and without Karnal grass (*Diplachne fusca* (Linn) P. Beauv.). *Agrofor. Syst., 7*, 135–160.

Singh, G., Abrol, I. P., & Cheema, S. S., (1989). Effect of irrigation on *Prosopis Juliflora* and soil properties of an alkali soil. *Int. Tree Crops J., 6*, 81–99.

Singh, G., Abrol, I. P., & Cheema, S. S., (1989). Effects of gypsum application on mesquite (*Prosopis juliflora*) and soil properties is an abandoned sodic soil. *For. Ecol. Manage., 29*, 1–14.

Singh, G., Gill, H. S., Abrol, I. P., & Cheema, S. S., (1991). Forage yield, mineral composition, nutrient cycling and ameliorating effects of Karnal grass (*Leptochloa fusca*) grown with *Prosopis juliflora* in a highly alkali soil. *Field Crops Res., 26*, 45–55.

Singh, G., Singh, N. T., & Dagar, J. C., (1993). *Agroforestry in Salt-Affected Soils*. Technical Bulletin No 17. CSSRI, Karnal, India.

Singh, G., Singh, N. T., & Tomar, O. S., (1993). *Agroforestry in Salt-Affected Soils* (p. 65). Technical bulletin no. 17, CSSRI, Karnal, India.

Singh, G., Singh, N. T., Dagar, J. C., Singh, H., & Sharma, V. P., (1995). An evaluation of agriculture, forestry, and agroforestry practices in moderately alkali soil in north-western India. *Agrofor Syst., 37*, 279–295.

Singh, J. P., Mathur, B. K., & Rathore, V. S., (2009). Fodder potential of Lana (*Haloxylon salicornicum*) in hot arid region. *Range Manag. Agrofor., 30*, 34–37.

Singh, Y. P., (2017). Multifunctional agroforestry systems for bio-amelioration of salt affected soils. In: Sanjay, A., Atul, K. S., & Singh, Y. P., (eds.), *Bioremediation of Salt Affected Soils: An Indian Perspective*. doi: 10.1007/978-3-319-48257-6_9.

Singh, Y. P., Mishra, V. K., Sharma, D. K., Singh, G., Arora, S., Dixit, H., & Cerda, A., (2016). Harnessing productivity potential and rehabilitation of degraded sodic lands through Jatropha based intercropping systems. *Agriculture Ecosystem and Environment, 233*, 121–129.

Singh, Y. P., Sharma, D. K., Singh, G., Nayak, A. K., Mishra, V. K., & Singh, R., (2008). *Alternate Land Use Management for Sodic Soils*. CSSRI Technical Bulletin.

Singh, Y. P., Singh, G., & Sharma, D. K., (2011). Ameliorative effect of multipurpose tree species grown on sodic soils of Indo-Gangetic alluvial plains of India. *Arid Land Res. Manag., 25*, 55–74.

Sodha, S., Mathur, M., & Sunndaramoorthy, S., (2018). Temporal assessment of plant community attributes of arid saline habitats. *Int. J. Ecol. Environ. Sci., 44*(3), 251–259.

Spalding, M., Blasco, F., & Field, C., (1997). *World Mangrove Atlas* (p. 180). International Society for Mangrove Ecosystems, Okinawa.

Sprenger, M., Oelmann, Y., Weihermuller, L., Wolf, S., Wilcke, W., & Potvin, C., (2013). Tree species and diversity effects on soil water seepage in a tropical plantation. *For. Ecol Manage, 309*, 76–86.

Stirzaker, R. J., Cook, F. J., & Knight, J. H., (1999). Where to plant trees on cropping land for control of dryland salinity: Some approximate solutions. *J. Agric. Water Manage., 39*, 115–133.

Thorburn, P., (1996). Can shallow water tables be controlled by the revegetation of saline lands? *Aust. J. Soil Water Conserv, 9*(3), 45–50.

Tomar, O. S., & Gupta, R. K., (1984). *Performance of Forest Tree Species in Saline Soils.* Annual reports – 1984–1994. CSSRI, Karnal.

Tomar, O. S., & Gupta, R. K., (1986). Performance of some forest tree species in saline soils under shallow and saline water table conditions. *Plant and Soil, 87*, 329–335.

Tomar, O. S., (1997). Technologies of afforestation of salt affected soils. *Int. Tree Crops J., 9*(2), 131–158.

Turner, N. C., & Ward, P. R., (2002). The role of agroforestry and perennial pasture in mitigating water logging and secondary salinity, summary. *Agric. Water Manage., 53*, 271–275.

Ventura, Y., Eshel, A., Pasternak, D., & Sagi, M., (2015). The development of halophyte-based agriculture: Past and present. *Ann. Bot., 115*, 529–540.

Walter, H., (1961). Salinity problems in the acid zones. The adaptations of plants to saline soils. *Arid Zone Research, 14*, 65–68.

Wang, W., Vinocur, B., & Altman, A., (2003). Plant responses to drought, salinity, and extreme temperatures: Towards genetic engineering for stress tolerance. *Planta, 218*, 1–14.

Yadav, J. S. P., & Singh, K., (1986). Response of *Casuarina equisetifolia* to soil salinity and sodicity. *Journal of Indian Society of Coastal Agriculture Research, 4*, 1–8.

Yadav, J. S. P., (1972). Improvement of saline and alkali soils through biological methods. *Ind. For., 101*, 385–395.

Yadav, J. S. P., (1980). Salt affected soils and their afforestation. *Ind. For., 106*, 259–272.

Yadav, J. S. P., Bhumbla, D. R., & Sharma, O. P., (1975). Performance of certain forest species on a saline sodic soil. In: *Proceedings of International Symposium on New Development in the Field of Salt Affected Soils of Sub Commission of Salt Affected Soils* (pp. 683–690). International Society of Soil Science.

Ye, Y., Tam, N. F. Y., Lu, C. Y., & Wong, Y. S., (2005). Effects of salinity on germination, seedling growth and physiology of three salt-secreting mangrove species. *Aquatic Botany, 83*(3), 93–205.

Yunusa, I. A. M., & Newton, P. J., (2003). Plants for amelioration of subsoil constraints and hydrological control: The prime-plant concept. *Plant and Soil, 257*, 261–281.

CHAPTER 14

Geochemistry and Health Status of Soils in Agroforestry Dominated HRML Regions of Idukki, Southern Western Ghats, India

A. KRISHNAKUMAR,[1] S. K. ADITYA,[1] K. ANOOPKRISHNAN,[1] and K. G. PRIJILAL[2]

[1]*ESSO–National Center for Earth Science Studies (NCESS), Ministry of Earth Sciences, Government of India, Akkulam, Thiruvananthapuram, Kerala, India*

[2]*Ministry of Environment, Forests, and Climate Change, Government of India, Regional Office, Chennai, Tamil Nadu, India*

ABSTRACT

An agroforestry system supports, promotes, and maintains numerous ecosystem functions and agroecological balance by their characteristic multifunctionality feature. They help in improving the quality, fertility status and health of soils through closed nutrient cycling while simultaneously reducing the ecological footprint of agriculture. If the heavy/trace metal content in soils of such system exceeds the natural background levels, it can cause ecological quality deterioration affecting the whole environment. Therefore, in the present investigation geochemical quality of soils in the high range mountain landscape (HRML) regions of Idukki district falling in Periyar River Basin (PRB) in Southern Western Ghats, India has been assessed to determine the concentration of nine major (Si, Al, Ca, Fe, K, Mg, Na, P, and Ti) and seven trace (V, Cr, Ni, Cu, Zn, Sr, Ba) metal ions. A total

Sustainable Management and Conservation of Environmental Resources in India.
Sumit Chakravarty, Vineeta, Jahangeer A. Bhat, Munesh Kumar, Gopal Shukla (Eds.)

of 14 samples were collected from the agroforestry land use (LU) systems and assessed for soil health on the basis of quality and pollution indices viz. Enrichment factor (Ef), contamination factor (Cf), pollution load index (PLI), and degree of contamination (C_{deg}). Spatial variations of metal ion concentrations were mapped using IDW interpolation technique in ArcGIS 10.3 software. From the results it is revealed that the soils of study area are predominantly aluminous type with slight enrichment of silica and the health of the soils in the study area is gradually deteriorating with metals ions like V, Cr, Ni, Cu, and Zn when compared with world shale average values of Turekian & Wedepohl (1961). It is observed that both anthropogenic and geogenic factors were associated with Cr and Ni content, while V, Cu, and Zn concentration is linked to anthropogenic activities. At the same time, Sr and Ba revealed its geogenic origin in the study area. The abundance of concentration of metal ions were in the order of Ba > V > Cr > Ni > Zn > Sr > Cu. The results of Ef and Cf ranged from 0.11–2.7 (low to minor enrichment) and 0.6–2.4 (low to moderate contamination), respectively. PLI values indicated progressive deterioration of the study area (0.97–1.67) and C_{deg} exhibited considerable degree of contamination (15.41–23.86). Based on the results, some recommendations were also made in order to reduce further deterioration of the soil health conditions in the study area.

14.1 INTRODUCTION

Soils are extremely complex and non-renewable natural resource that is fundamental to human existence. The health of the soil that a nation owns determines well-being quality of its people and hence, the proper development of any nation is feasible solely through the sustainable management of its natural resources like soil, water, and protection of its environment. The formation and development of soil depends on climate, soil organisms, landscape, vegetation (organic stand litter), time and lithology (Sucharovà et al., 2012). Change in spatial distribution and vegetation cover, water quality and quantity, water movement, land use (LU)/land cover change all have relationship with the chemistry of soils. Semwal et al. (2009); and Araujo et al. (2012) reported the possible reasons of the soil degradation as unplanned and unscientific agricultural practices, deforestation, over grazing, over-construction, and alignment of roads. Application of commercial fertilizers, sewage sludge and pesticides, which usually contain a wide variety of heavy metals as impurities may initially get accumulated in soil and later on gets readily transferred to soil solution, depending mainly on soil properties. Healthy soil is one of the supreme resources for healthy ecosystems and has been recognized in the context of agro-ecosystems. For the past, several decades health of soil is becoming vulnerable globally, and there has been a renewed attention in guarding and improving this most essential resource for future generations. Agroforestry indeed has a remarkable role in improving soil quality and sustainable land management practice based on the past 40 years of data reported from all over the world. According to Lundgreen & Raintree (1982) "agroforestry is a collective name for land-use systems (LUS) in which woody perennials are deliberately grown on the same piece of land as agricultural crops and/or animals, either in some form of spatial arrangement or in sequence." Agroforestry systems have the potential to facilitate sustainable management, soil quality, soil resilience and soil conservation thus maintaining the protection of other ecosystem services (ES).

Western Ghats with its remarkably high level of biodiversity and endemism is recognized as the world's 8th hotspots of biological diversity and plays several key ecosystem functions. According to NFC report (2016), Western Ghats provide livelihood for over 45 million people directly and around 245 million people receive most of their water needs from rivers of Western Ghats in peninsular India. One of the globally significant biodiversity regions in Western Ghats representing ecological attributes, socio-economic profile and development trajectory is the high range mountain landscapes (hereafter referred as HRML). It is a horseshoe shaped region with high ridges, steep rugged terrain and highly dissected valleys. The key attributes of HRML includes – (i) harbor the largest global population of Nilgiri Tahr (a highly endangered species); (ii) catchment of three major river systems of peninsular India (Periyar, Chalakudy, and Cauvery); (iii) strong eco-cultural affinities; and (iv) support important economic sectors like cardamom, tea, and tourism. In addition to the above, HRML habitats also support key ecosystem functions such as ensuring water security of the plains, irrigational, and hydro-electric projects, supply of raw materials to industries (e.g., Hindustan News Print Ltd. and Kerala State Bamboo Corporation), furnishing food and livelihoods demands for local communities (e.g., NTFPs, artisanal crafts, etc.) and also sites for tourism. Shaded commercial crop systems like cacao (*Theobroma cacao* L.), coffee (*Coffea spp.*), tea (*Camellia sinensis* (L.), tropical homegardens, silvopastoral systems with fodder grasses in association with commercial trees, spices, trees on farm boundaries, and woodlots are the major agroforestry systems observed in HRML (Kumar, 2011). Some of the plant families of HRML include

Myrtaceae, Simplococeae, Lauraceae, Styraceae, Ternstroemiaceae, along with endemic trees such as *Cryptocarya anamalayana*, *Neolitsea fischeri*, *Symplocos anamalayana*, *Eugenia rottleriana*, *Syzygium chandrasekharanii*, *Amoora beddomei*, *Veronia fysonii*, *Sonerilla puleyensis*, and kurinji (*Strobilanthes kunthianus*).

However, plants have the capacity to accumulate toxic metals, when grown on polluted soils representing a threat to higher trophic level organisms consuming those (Smical et al., 2008). Hence, any kind of anthropogenic inputs (agrochemicals) are harmful to all the plant species existing in the region. Reichenberger et al. (2007); and DeLonge et al. (2016) discussed the improper management of agrochemicals resulting in an increase of GHG emissions, soil deterioration, and pollution of water bodies (surface and ground), with direct negative consequences on the health of ecosystems and humans through bioaccumulation. Since HRML is a sensitive zone of Western Ghats and the data is very meager, a preliminary study was conducted in order to assess the geochemical quality of soils in the agroforestry systems in the HRML of Periyar River Basin (PRB) of Southern Western Ghats, India. Major oxides and trace metals were examined with special reference to pedological attributes in terms of quality and chemistry. A River Basin mode of approach is applied in this work since it is holistic and provides a framework for analyzing the effects of human interactions with the environment.

14.2 STUDY AREA

The study area covers 3,226 km^2 and lies between 9°16′45″ and 10°12′35″ and longitude of 76°16′20″–76°56′30″. The rainfall in the basin is dual monsoon dependent; South-West Monsoon (SWM) contributes to 60% and North-East Monsoon (NEM) controls 25% of the total rainfall. The annual average rainfall of the basin is 3,000 mm. The mean monthly maximum temperature of HRML ranges from 25°C to 32°C and minimum temperature from 14°C to 19C. The land use pattern of the area can be broadly divided into agriculture land, forest land and waste land and the major rock types include crystalline rocks of Precambrian age and sedimentaries of tertiary and quaternary periods. According to the United Nations Development Program (UNDP), the HRML in the Western Ghats mountains of peninsular India is a globally notable biodiversity region which is rich in flora and fauna and support important economic sectors including cardamom, tea, and coffee plantations which is indigenous to the medium elevation wet evergreen and semi-evergreen forests of the Western Ghats of peninsular India (UNDP, 2013–2018). The natural vegetation of the study area includes high elevation montane forests and grasslands (GL), humid high elevation forests, humid mid elevation forests, humid low elevation forests and dry forests.

The study area constitutes a traditional agroforestry system called Cardamom Hill Reserves (CHR) which is characteristic to the high-altitude regions of Kerala. It involves growing small cardamom (*Elettaria cardamomum*), a sciophytic commercial crop, under the shade of dominant trees selectively retained by the growers in the evergreen and semi-evergreen natural forest (NF). Black pepper (*Piper nigrum* L.) is a major foreign exchange earning commercial crop cultivated in the study area. *Erythrina indica* forms the dominant support trees in pepper plantations. Although in the traditional system of cardamom production, little or no chemical fertilizers were applied but with the advance of modern technology farmers have started using chemical fertilizers to supplement the inherent site fertility (Kumar et al., 1995). Plantation crops like tea, coffee, cardamom (Plate 14.1(a) and (b)), *Eucalyptus grandis* are the major crops cultivated in the upper catchment area (60,000 ha in 1980 to 42,000 ha in 2010). Highlands of the Periyar river basin (PRB) is one of the largest production centers of cardamom in India with high yielding varieties (HYVs) like Njallani, Kalarikkal, and Panikulangara while, in the lowland regions like Kuttampuzha, Adimali, Mankulam panchayats reed (*Ochlandra travancorica*) collection and allied crafts are the chief income generating activities. Cardamom, being the highest pesticide consuming crop, the environmental risk of soil and water contamination is high (Jacob et al., 2014; Gayathri et al., 2021). In recent times, there has been a perceptible shift towards more varieties of cardamom that require intense farming practices and heavy application of synthetic agrochemicals.

Reckless use of pesticides in the cardamom plantations is reported to be one of the highest in the world and on average in Kerala, farmers use 27 kg/ha of pesticides annually (NIAS, 2010; Krishnakumar et al., 2021). On the basis of the report from UNDP, the HRML comprises largest tea producing area in Southern India (14,000 ha). The higher areas of the landscapes lying to east have vegetable farming as dominant LU. Survey by Murugan (2011) showed that the intensity of pesticide use increased from 27.5 to 35 kg/ha for the 2010 and 2011 in India, and Kerala is a major part in the high usage of agrochemicals. This can possibly trigger the heavy metal concentration in such soils. The plantations and agricultural activities decisively influence the ecological stability of HRML. High application of pesticide has indications

PLATE 14.1 (a) Tea with black pepper and support tree; and (b) tea with silver oak (*Grevillea robusta*).

with increased reports of diseases like dermatitis, asthma, cancer, and reproductive disorders (Misra, 2011).

14.3 SAMPLING AND ANALYSIS

Soil samples were collected systematically from 14 geo-referenced sensitive locations (Figure 14.1) in the highland regions of PRB using a stainless-steel corer.

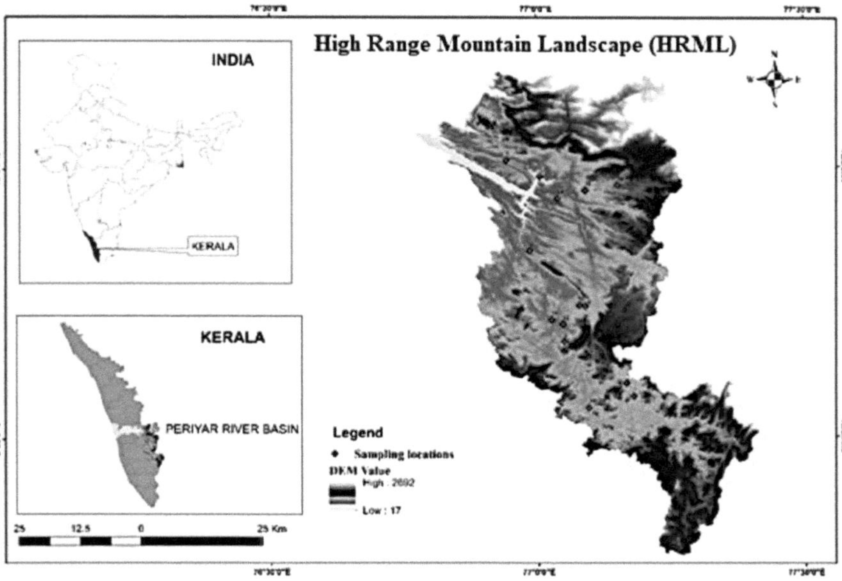

FIGURE 14.1 Location map of HRML region in Periyar river basin.

Top soil, i.e., 5–15 cm depth (A horizon of soil) is used in this study to determine the anthropogenic influence of pollutants. Cone and quartering are used to create homogeneity and reduce the sample size. Soil samples were collected using a stainless-steel corer and transferred into neatly labeled clean zip-lock polythene covers. The soil samples were grinded in an agate mortar and passed through a U.S. Standard Test Sieve Series No. 230. The metal ion concentrations were determined by XRF (Bruker Model S4 pioneer sequential wavelength dispersive X-ray spectrometer) method and the soil quality was assessed employing different contaminant indices (Table 14.1).

14.3.1 ASSESSMENT OF METAL POLLUTION IN SOILS

14.3.1.1 DEGREE OF CONTAMINATION (C_{DEG})

The assessment of soil contamination was carried out using degree of contamination. For this measurement, the level of contamination of soil by metal was calculated and expressed as contamination factor (Cf):

$$Cf = (Cm/Cb)$$

TABLE 14.1 Various Methods Employed for the Determination of Various Geochemical Indices Along with the Range and Category

Indices Studied	Calculation Method	Range	Category	Remarks
Enrichment factor (EF_c) [1]	$EF_C = \frac{\left(\frac{C_X}{C_{Al}}\right)_{sample}}{\left(\frac{C_X}{C_{Al}}\right)_{continental\ crust}}$ where; $(C_x/C_{Al})_{sample}$ is the ratio of the concentration of test element to that of Al; and $(C_x/C_{Al})_{continental\ crust}$ is the same ratio with respect to a reference soil.	$EF_C < 2$ $EF_C = 2–5$ $EF_C = 5–20$ $EF_C = 20–40$ $EF_C > 40$	Deficiency to minimal enrichment. Moderate enrichment. Significant enrichment. Very high enrichment. Extremely high enrichment.	Individual element is taken into consideration for the assessment of enrichment factor. Hence indicating the concentration of a particular element in the location.
Contamination Factor (C_f) [2]	$C_f = \frac{(C_m)_{sample}}{(C_m)_{background}}$ where; $(C_m)_{sample}$ is the metal concentration in the polluted soil; and $(C_m)_{background}$ is the natural background value of that metal. In this study, average shale value suggested by Turekian & Wedpohl (1961) was used as the reference value.	$C_f < 1$ $1 \leq C_f + 3$ $3 \leq C_f \leq 6$ $6 \leq C_f$	Low contamination. Moderate contamination. Considerable contamination. Very high contamination.	C_f denotes the extent of contamination of a metal in relation to the background concentration of that metal.
Degree of contamination (C_{deg}) [2]	$Cdeg = \sum C_{fn}$ The sum of contamination factors of all the elements examined represents the contamination degree (C_{deg}) of the environment.	$C_{deg} < 8$ $8 \leq C_{deg} < 16$ $16 \leq C_{de} < 32$ $32 \leq C_{deg}$	Low degree of contamination. Moderate degree of contamination. Considerable degree of contamination. Very high degree of contamination.	Indicates the strength of contamination by means of all the heavy metals at a particular site.
Pollution load index (PLI) [3]	$PLI = \sqrt[n]{C_{f_1} \times C_{f_2} \times \ldots\ldots\ldots \times C_{f_n}}$ where; C_{fn} is the contamination factor; and n is the number of metals.	PLI = 0 PLI = 1 PLI > 1	Unpolluted Baseline levels of pollution present. Progressive deterioration of site.	Denotes the severity of pollution of a particular site in terms of all the heavy metals present.

Source: Modified after: [1] Buat-Menard & Chesselet (1979); [2] Håkanson (1980); and [3] Tomlinson et al. (1980).

where; 'Cm' is the metal concentration in the polluted soil; and Cb is the natural background value of that metal. In this study, average shale value suggested by Turekian & Wedepohl (1961) was used as the reference value for comparison. Håkanson (1980) defines four categories for Cf based on its ranges.

- $Cf < 1$ low contamination factor;
- $1 \leq Cf \leq 3$ moderate contamination factor;
- $3 \leq Cf \leq 6$ considerable contamination factor;
- $6 \leq Cf$ very high contamination factor.

The sum of Cfs of all the elements represents the contamination degree (C_{deg}) of the environment and the four classes are:

- $C_{deg} < 8$ low degree of contamination;
- $8 \leq C_{deg}$ 6 moderate degree of contamination;
- $16 \leq C_{deg} < 32$ considerable degrees of contamination;
- $32 \leq C_{deg}$ very high degree of contamination.

14.3.1.2 POLLUTION LOAD INDEX (PLI)

To find out the pollution level in the study area the pollution load index (PLI) was calculated using the method proposed by Tomlinson et al. (1980):

$$PLI = (Cf_1 * Cf_2 * \ldots)^{1/n}$$

where; Cf_1 is the contamination factor at site 1; Cf_2 is the contamination factor at site 2; Cf_n is the contamination factor at site 'n;' and n is the number of metals.

Tomlinson et al. (1980) classified PLI as follows:

- PLI = 0 (perfection);
- PLI up to 1 (baseline level of pollution present);
- PLI > 1 (progressive deterioration of site).

14.3.1.3 ENRICHMENT FACTOR (EF)

The extent of soil contamination was assessed using the enrichment factor (EF) suggested by Chen et al. (2001). The crustal EFs was used in the study to assess the relative contribution of natural and anthropogenic heavy metal inputs to the soils (Adamo et al., 2005; Valdes et al., 2005).

To identify anomalous metal concentrations, the geochemical normalization of the metal data to a conservative element, such as Al, Fe, or Si, was employed. Several authors successfully used Fe to normalize metal contaminants (Cevik et al., 2009; Bhuiyan et al., 2010). In the present study, Fe has been used as a conservative tracer to differentiate natural from anthropogenic components. EFc of elements can be calculated using the equation:

$$EF = \frac{(M/Fe)_{sample}}{(M/Fe)_{background}}$$

where; (M/Fe) sample is the ratio of metal and Fe concentrations of the sample; and (M/Fe) background is the ratio of metal and Fe concentrations of a background. The EF values are interpreted as the levels of metal pollution suggested by Chen et al. (2001). Chen et al. (2001) suggested that EF <1 indicates no enrichment, EF = 1–3 indicates minor enrichment, EF = 3–5 indicates moderate enrichment, EF = 5–10 indicates moderately severe enrichment, EF = 10–25 indicates severe enrichment, EF = 25–50 indicates very severe enrichment, and EF >50 indicates extremely severe enrichment.

14.4 RESULTS AND DISCUSSION

The ranges and average concentration of nine major metal oxides which fall in the first transition series in the periodic table are presented in Table 14.2. Results of the major oxides in the present study were compared with the International Standards for Major Oxides in Soil (Bohn et al., 2001) and Composition of parent rocks (Charnockites) from Kerala (Soman, 2002) to assess the enrichment or depletion of these major metal oxides.

14.4.1 MAJOR OXIDES

Major oxides are essential components of rock forming minerals. Major element concentration in soil is usually expressed in terms of oxide weight percentages. Results of the major oxides (SiO_2, TiO_2, Al_2O_3, Fe_2O_3, CaO, MgO, Na_2O, K_2O, and P_2O_5) from the present study are given in Table 14.2.

TABLE 14.2 Major Oxides in the Collected Soil Samples of the Study Area Along with the Reference Value Reported

Sample	Location	Long	Lat	SiO_2	TiO_2	Al_2O_3	Fe_2O_3	CaO	MgO	Na_2O	K_2O	P_2O_5
S1	Thekkady	77.18	9.58	40.89	1.23	27.37	13.19	0.26	0.53	0.08	0.91	0.15
S2	Grampithodu	77.09	9.55	36.13	1.02	29.62	10.8	0.25	0.71	0.1	0.8	0.23
S3	Kumali	77.16	9.64	36.29	0.92	33.42	12.43	0.3	0.92	0.14	1.1	0.3
S4	Marykulam	77.05	9.67	31.32	1.03	29.72	10.52	0.35	2.07	0.17	1.89	0.4
S5	Ayyapankovil	77.04	9.70	35.15	0.86	35.52	11.8	0.16	0.29	0.06	0.69	0.16
S6	Nariampara	77.08	9.74	38.23	0.75	35.2	8.45	0.17	0.66	0.1	1.3	0.26
S7	Sengulam	77.03	10.01	27.72	0.98	32.88	9.38	0.71	2.06	0.25	1.72	0.64
S8	Idavanakandathil	77.00	9.98	33.47	1.28	30.41	13.59	0.48	1.09	0.1	1.33	0.25
S9	Idukki	77.00	9.85	38.48	0.96	27.92	10.31	0.67	1.49	0.26	2.6	0.33
S10	Parathodu	77.03	9.94	29.47	1.39	34.29	11.17	0.15	0.74	0.06	0.79	0.4
S11	Kanchiyar	77.07	9.74	32.99	0.89	33.24	9.51	0.64	0.44	0.04	0.24	0.25
S12	Cheripuram	77.09	9.95	24.32	1.68	35.79	14.5	0.48	0.64	0.06	0.54	0.34
S13	Kattapana	77.12	9.75	30.75	1.45	33.79	12.88	0.45	1.07	0.17	1.42	0.37
S14	Rajakumari	77.15	9.96	31.52	1.06	31.82	9.72	1.62	0.77	0.09	1.01	0.68
Average				33.33	1.10	32.21	11.30	0.47	0.96	0.12	1.16	0.34
Felsic charnockite	–	–	–	70.29	0.88	>13.2	5.77	1.44	0.99	0.99	1.2	0.18
[4, 5]				68.8	0.79	14.7	5.77	2.52	1.0	3.15	5.04	0.27

Source: [4] Bohn et al. (2001); and [5] Soman (2002).

From the results it is observed that the SiO_2 concentration ranged from 24.32–40.89% indicating high quartz content in the study area. All samples were found to be within the reference limit of 70.29% established by Bohn et al. (2001). Slightly elevated concentration of TiO_2, MgO, and P_2O_5 were seen in the study area. The concentration of TiO_2 ranged between 0.75 and 1.68%, MgO ranged between 0.29 and 2.7% and P_2O_5 between 0.15 and 0.68%. Compared to the standard value, the soil samples exhibited higher concentration of Al_2O_3 and Fe_2O_3. The concentration of Al_2O_3 and Fe_2O_3 in the study area ranged between 27.37 and 35.79% and 8.45–14.50%, respectively. It is seen that soluble or exchangeable Al levels increased as a consequence of increasing acidity and from the results it is observed that the soils of PRB are acidic to neutral in nature (Krishnakumar et al., 2019). The other major elements studied were found to be depleted in concentration. Based on the distribution of major oxides in the study area it is observed the soils are predominantly aluminous type with slight enrichment of silica (Krishnakumar et al., 2017). These soils are typical of semi-arid and tropical mode of weathering in granites and charnockites and display the inherent characteristic of alumina and silica of the parent rock. Main contribution may be of chemical weathering due to the presence of tropical humid climate.

14.4.2 METAL IONS

The mean concentration of metal ions along with their comparison to Igneous Basaltic rocks (Turekian & Wedephol, 1961) are represented in Table 14.3.

The concentration of vanadium (V) in the soil samples ranged from 146–268 mg/kg with an average of 195.85 mg/kg. Vanadium inputs to soils related to anthropogenic activities can be derived from phosphate fertilizers (Molina, 2009) and in the present study, phosphorous and potassium is applied, respectively in the form of rock phosphate and muriate of potash (Murugan, 2007). Hence the Vanadium content in the present study may be of the phosphate containing fertilizers. The nickel (Ni) and chromium (Cr) concentration in the study area ranged from 54–191 mg/kg with an average of 116.93 mg/kg and 84–270 mg/kg with an average of 166.78 mg/kg, respectively. It was observed that Ni and Cr concentration in HRML was high and the major input of Cr content to the soil might be from the use of phosphatic fertilizers in the study area. The status of Ni in soil is a function of parent rock while in surface soils, it is a reflection of soil forming processes and pollution due to the over use of chemical fertilizers (Kabata-Pendias & Pendias, 1992).

TABLE 14.3 Heavy Metal Concentration of the Study Area (in mg/kg)

Sample	V	Cr	Ni	Cu	Zn	Sr	Ba
S1	207	152	108	131	118	57	474
S2	189	167	126	82	70	103	232
S3	166	270	191	38	76	110	576
S4	188	181	146	88	118	128	891
S5	160	114	54	29	44	85	254
S6	146	126	127	39	33	160	835
S7	183	164	120	79	84	155	900
S8	213	84	64	60	68	17	307
S9	173	172	115	75	93	96	638
S10	236	224	139	74	80	56	ND
S11	171	111	90	150	113	64	168
S12	268	254	133	40	30	59	ND
S13	231	149	94	79	131	88	114
S14	193	157	115	107	262	106	391
Average	194	166	115	76.5	94.2	91	481
Igneous basaltic rocks	250	170	130	87	105	465	330
Continental crust value	60	35	20	10.4	71	350	550

Franklin (2005) reported that potassic fertilizers contained 2.7–16 mg/kg Ni as impurity while, commercial phosphatic fertilizers contain 19–24 mg/kg. Hence, the source of Ni might be mainly because of the use of chemical fertilizers. The solubility and mobility of Ni increased with decrease in pH; hence, soil pH is the major factor regulating Ni solubility, mobility, and sorption (Anderson & Christensen, 1988; Ge et al., 2000; Sauve et al., 2000; Tye, 2004). Soil in the present study area was acidic with pH range of 3.69–5.88.

Cu and Zn are considered as anthropogenic metals and in the present study the concentration of Cu ranged from 29–150 mg/kg while, Zn concentration varied from 30–262 mg/kg. The major source for both of these elements were liquid manure, agricultural use of sewage sludge, composted materials, fertilizers, and pesticides. Cu in the form of CuO and $Cu(OH)_2$ was applied extensively in the study area to control various diseases of cardamom during the rainy seasons.

The concentration of Strontium (Sr) in the study area was normal and was quantified in the range of 17–160 mg/kg which was within the permissible levels. Generally, strontium was reported associated with calcium carbonate fraction (Elbaz-Poulichet, 1997). Since calcium carbonate was comparatively lesser in the study area, the respective values for strontium were also less, similarly observed else wherein earlier studies (Elbaz-Poulichet, 1997; Ge et al., 2000). An increased concentration of Barium (Ba) was also observed (168–900 mg/kg) in soils of the study area due to presence of old crystalline rocks in HRML which generally corresponds to high Ba (Sucharovà et al., 2012). Figure 14.2 represents the spatial distribution of V, Cr, Ni, and Cu in the study area and Table 14.4 shows the environmental and health consequences due to increased concentration of metal ions, which in turn results in their bioaccumulation.

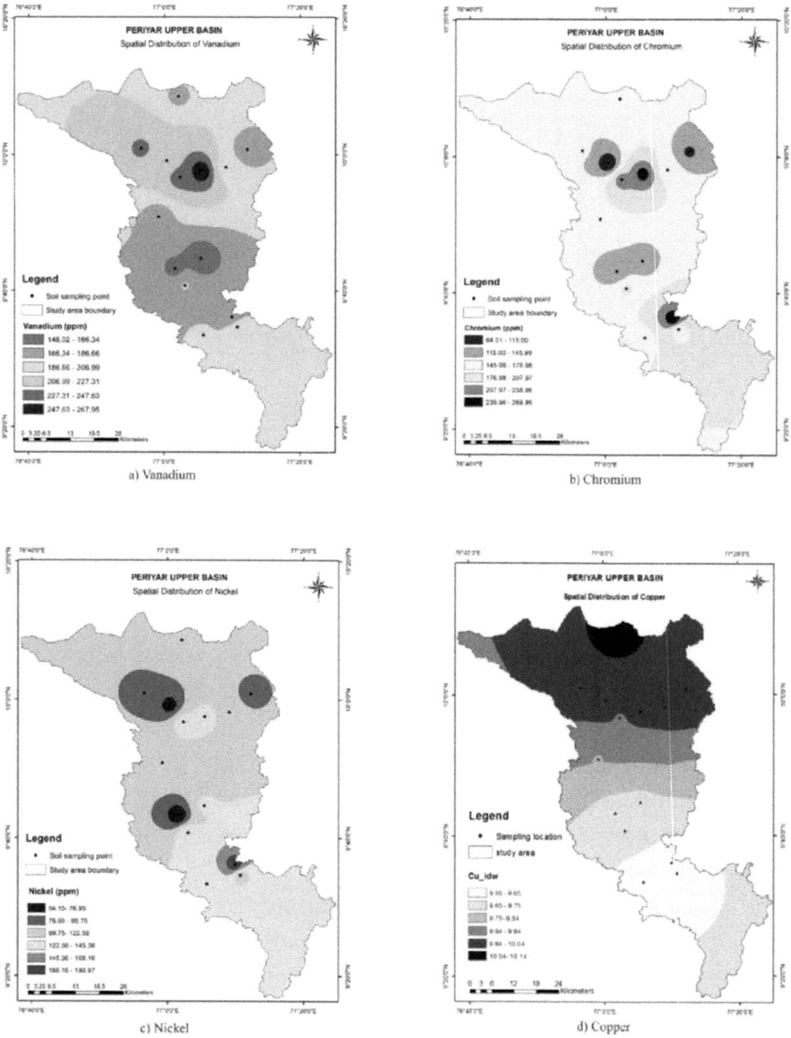

a) Vanadium b) Chromium c) Nickel d) Copper

FIGURE 14.2 Spatial distributions of major ions in the study area.

TABLE 14.4 Environmental and Health Effects of Each Metal Ions

Element	Toxicity	Bioaccumulation Risk
Barium (Melo et al., 2011)	Short term exposure causes vomiting, abdominal cramps, diarrhea, difficulties in breathing, increased or decreased blood pressure, numbness around the face, and muscle weakness. Large amounts of barium intake can cause, high blood pressure, changes in heart rhythm or paralysis and possibly death.	Accumulation in plants is rare, but accumulates in certain plants, seaweed, and fish.
Chromium (Jaishankar et al., 2014)	Breathing problems – irritation to the lining of the nose; nose ulcers, asthma, cough, shortness of breath, or wheezing. Skin contact causes skin ulcers. Long term exposure can cause damage to liver, kidney circulatory and nerve tissues.	High potential for uptake by aquatic life, especially bottom-feeders.
Nickel (Engwa et al., 2019)	Cancer, dermatitis, skin irritations. Respiratory diseases, asthma, congenital malformations, cancers.	–
Vanadium (Chen et al., 2021)	Stomach discomfort, diarrhea, and nausea. It can also cause a greenish tongue, loss of energy.	Problems with the nervous system, and kidney damage.
Copper (Engwa et al., 2019)	Irritation of the nose, mouth, and eyes. Headaches, stomachaches, dizziness, vomiting, and diarrhea.	When found excess in soils, interrupts the microbial activity and accumulates in plants and animals.

14.4.3 CONTAMINANT GEOCHEMISTRY

The Cf values of metals such as Sr and Ba estimated are low (< 1) which indicate the study area as not polluted by these metals (Table 14.5) whereas, V, Cr, Ni, Cu, and Zn falls in the moderate contamination category. It was also observed that the quantified amount of the entire analyzed elements was in low to moderate contamination category. Mean values of V, Cu, Zn, Sr, and Ba were observed in the deficiency to minimal category whereas; Cr and Ni showed a marginal increase and were observed in the moderate enrichment category. The assessment of overall contamination and pollution of the soil was based on the PLI and C_{deg}.

TABLE 14.5 Enrichment Factor and Contamination Factor of the Study Area

Elements	Enrichment Factor			Contamination Factor		
	Minimum	Maximum	Mean	Minimum	Maximum	Mean
V	1.05	1.91	1.48	1.23	2.06	1.645
Cr	1.19	3.51	2.35	0.93	3	1.965
Ni	1.15	4.34	2.745	1.08	3.8	2.44
Cu	0.49	2.91	1.7	0.8	3.3	2.05
Zn	0.17	1.76	0.965	0.32	2.76	1.54
Sr	0.02	0.2	0.11	0.3	0.9	0.6
Ba	0.09	0.82	0.455	0.2	1.5	0.85

C_{deg} values of the soil samples ranged from 15.41 to 23.86 (Table 14.6) with an average value of 20.59 which indicates considerable degree of contamination (Håkanson, 1980).

TABLE 14.6 Levels of Contamination and PLI with Respect to the Various Locations in the Study Area

Sample	DoC	Category	PLI	PLI Category
S1	21.3	Considerable degree of contamination	1.38	Progressive deterioration of site
S2	21.6	Considerable degree of contamination	1.37	Progressive deterioration of site
S3	22.6	Considerable degree of contamination	1.4	Progressive deterioration of site
S4	23.86	Considerable degree of contamination	1.67	Progressive deterioration of site
S5	18.71	Considerable degree of contamination	1.09	Progressive deterioration of site
S6	19.91	Considerable degree of contamination	1.31	Progressive deterioration of site
S7	20.91	Considerable degree of contamination	1.48	Progressive deterioration of site
S8	15.41	Moderate degree of contamination	0.97	Baseline level of pollution
S9	18.26	Considerable degree of contamination	1.03	Progressive deterioration of site
S10	20.74	Considerable degree of contamination	1.39	Progressive deterioration of site
S11	22.02	Considerable degree of contamination	1.25	Progressive deterioration of site
S12	19.96	Considerable degree of contamination	1.24	Progressive deterioration of site
S13	20.64	Considerable degree of contamination	1.26	Progressive deterioration of site
S14	22.4	Considerable degree of contamination	1.52	Progressive deterioration of site

Abbreviations: DoC: Degree of contamination; and PLI: pollution load index.

Similarly, PLI values estimated was in the range of 0.97–1.67 (Table 14.6) with an average value of 1.31 which indicates progressive deterioration of the sites in the study area (Tomlinson et al., 1980). High degree of contamination and PLI values indicates high anthropogenic activities in the study area with higher inputs of chemical fertilizers, pesticides, weedicides and improper irrigation methods.

14.5 CONCLUSION AND RECOMMENDATIONS

In the present study, environmental geochemical analyzes of soils were carried out in the HRML regions of Idukki district falling in PRB of Southern Western Ghats, India. Major oxide and trace metal ion concentrations were determined using XRF technique. Geochemical analysis of the average concentration of major oxides revealed that TiO_2, Al_2O_3, Fe_2O_3, MgO, P_2O_5 were above the International Standards for major oxides in Soils. Slightly greater values of Cr and Ni were observed due to the geogenic and anthropogenic activities (application of agrochemicals) in the study area. Elements like V, Cu, and Zn exhibited anthropogenic origin while a similar chemical composition of genesis and development was shown by Sr and Ba. The mean values of Cf and EF were in the low to moderate contamination category and deficiency to moderate enrichment category, respectively for all elements. Average values of C_{deg} and PLI for metal ions revealed considerable degree of contamination and progressive deterioration of site. Thus, the results of the present analysis highlight the requirement for spatially intensified and broadened monitoring of the soil resources in the agriculture dominated areas of the HRML. As per the existing scenario, the planters of the study area intermittently use various agrochemicals especially harmful synthetic agrochemical inputs which are easily available from Kerala and adjoining regions of Tamil Nadu for better crop yield without any kind of scientifically prescribed norms and standards.

Present day agroforestry practices in HRML region resulted in an increase of the agricultural production and thereby the standard of living of local people (tribals) and farmers. But sustainable plans have to be derived for preserving the health of the soils in HRML. Some of the suggestions and recommendations from the study to improve the soil quality of HRML are:

- Excessive use of agrochemicals can alter and degrade the innate chemical and physical properties of soil leading to its poor quality. Hence, proper measures have to be implemented to check the surplus use of these chemicals in agriculture.
- There is a lack of clear database regarding agrochemical usage and no such data is available with any local body or department till date. It is now, the need of the hour to create such a database to protect soil deterioration or else, the negative externalities may lead to barren/ wasteland in the long run affecting the livelihood of farmers.

- There should be provision for periodic monitoring of soil quality (after each harvest season) by a competent govt agency and the result should be available to the farmers in order to take remedial measures for enhancing the soil fertility status. For this, the advice of in line R&D institutes may be sought through the concerned local bodies.
- In fact, the both the health and economic status of residents of HRML are very much linked directly or indirectly linked with the usage of agrochemicals. Synthetic agrochemicals including pesticides, weedicides, and fungicides retain in soil for a longtime and enter into the food chain through plants. It is noted that the rate of cancer patients is increasing in some of the Grama panchayats falling in the HRML region which may be due to processes like bioaccumulation. Therefore, a thorough study is required to correlate the health status of LSGDs in HRML with the inputs of agrochemicals. No such studies have been done so far as per the available literature.
- It is observed that money invested by the Govt in the health sector is much more than the income generated by the farmers through their agricultural practices. Gradually, it becomes very difficult to meet even the health expenses from their income. A database related to the relative comparison of income and expenditure of farmers have to be compiled by the concerned local bodies and can be made available and based on this, new farming practices/techniques and pertinent biofertilizers and insect repellents can be chosen which can provide them with more income. The local bodies may constitute a team for this task and incorporate Scientists from the relative areas for proper advice and thereby enhance good soil health conditions.
- Enlighten the people of HRML region through creating better awareness, imparting knowledge, skills, and inculcate the benefits of better LU practices, and improved soil quality. This may result in better productivity and can improve soil health, as well.
- Soil is a thin layer between man and starvation and the soil is formed as a result of various earth processes over the hundreds of years. Being an important element in *Panchabhootas*, as closely linked with our day-to-day life, it is the responsibility of each individual to protect and safeguard this resource. Often, we forget the fact that it is for the future generation to come. Hence, new rules and regulations have to be framed by the government in order to preserve the health and quality of the precious natural resource.

ACKNOWLEDGMENTS

This study has been conducted as part of NCESS core program (W.P.3B.4) 'Assessment of Global Environmental Changes in Sahyadri.' The authors are very much thankful to the Director, NCESS for providing research facilities, encouragement, and support. The financial support of MoES, Government of India for travel expenses to conduct field survey is greatly acknowledged.

KEYWORDS

- **degree of contamination**
- **metal ions**
- **metal pollution**
- **Periyar river basin**
- **pollution load index**
- **soil geochemistry**
- **western ghats**

REFERENCES

Adamo, P., Arienzo, M., Imperato, M., Naimo, D., Nardi, G., & Stanzione, D., (2005). Distribution and partition of heavy metals in surface and sub-surface sediments of Naples city port. *Chemosphere, 61*(6), 800–809.

Anderson, P. R., & Christensen, T. H., (1988). Distribution coefficients of Cd, Co, Ni, and Zn in soils. *J. Soil Sci., 39*(1), 15–22.

Araujo, A. S. F., Leite, L. F. C., Iwata, B. F., Lira Jr, M. A., Xavier, G. R., & Figueiredo, M. V. B., (2012). Microbiological process in agroforestry systems: A review. *Agro. Sustain. Develop., 32*, 215–226.

Bhuiyan, M. A. H., Parvez, L., Islam, M. A., Dampare, S. B., & Suzuki, S., (2010). Heavy metal pollution of coalmine-affected agricultural soils in the northern part of Bangladesh. *J. Hazard. Mater., 173*, 384–392.

Bohn, H., McNeal, B., & O'Connor, G., (2001). *Soil Chemistry* (3rd edn.). John Wiley and Sons. New York.

Buat-Menard, P., & Chesselet, R., (1979). Variable influence of the atmospheric flux on the trace metal chemistry of oceanic suspended matter. *Earth Planetary Sci. Lett., 42*(3), 399–411.

Cevik, U. Ğ. U. R., Celik, N., Celik, A., Damla, N., & Coskuncelebi, K., (2009). Radioactivity and heavy metal levels in hazelnut growing in the Eastern Black Sea Region of Turkey. *Food. Chem. Toxicol., 47*(9), 2351–2355.

Chen, H. M., Zheng, C. R., Tu, C., & Zhou, D. M., (2001). Studies on loading capacity of agricultural soils for heavy metals and its applications in China. *Appl. Geochem., 16*, 1397–1403.

DeLonge, M. S., Miles, A., & Carlisle, L., (2016). Investing in the transition to sustainable agriculture. *Environ. Sci. Pollut., 55*, 266–273.

Elbaz-Poulichet, F., (1997). Biogeochemistry of trace metals (Mn, Sr, Rb, Ba, Cu, Zn, Pb and Cd) in a river -wetland -lake system (Balaton Region, Hungary). *Aquat. Geochem., 2*, 379–402.

Engwa, G. A., Ferdinand, P. U., Nwalo, F. N., & Unachukwu, M. N., (2019). Mechanism and health effects of heavy metal toxicity in humans. *Poisoning in the Modern World-New Tricks for an Old Dog,* 10.

Franklin, R. E., (2005). Trace element content of selected fertilizers and micronutrient source materials. *Comm. Soil Sci. Plant Analy., 36*, 1591–1609.

Gayathri, S., Vinu, V. D., Shiny, R. R., Krishnakumar, A., Vishnu, M. T. M., & Anoop, K. K., (2021). Spatiotemporal evaluation of hydrochemical facies and pesticide residues in the cardamom plantations of Southern Western Ghats, India. *Environmental Nanotechnology, Monitoring & Management, 16*, 100599.

Ge, Y., Murray, P., & Hendershot, W. H., (2000). Trace metal speciation and bioavailability in urban soils. *Environ. Pollut., 107*(1), 137–144.

Håkanson, L., (1980). An ecological risk index for aquatic pollution control: A sedimentological approach. *Water Res., 14*, 975–1001.

Jacob, S., Resmi, G., & Mathew, P. K., (2014). Environmental pollution due to pesticide application in Cardamom Hills of Idukki, District, Kerala, India. *Intern. J. Basic. Appl. Sci. Res., 1*, 27–34.

Jaishankar, M., Tseten, T., Anbalagan, N., Mathew, B. B., & Beeregowda, K. N., (2014). Toxicity, mechanism, and health effects of some heavy metals. *Interdiscip. Toxicol., 7*(2), 60.

Kabata-Pendias, A., & Pendias, H., (1992). *Trace Elements in Soils and Plants*. New York. doi: 10.1201/b10158-25.

Krishnakumar, A., Aditya, S. K., Pavitra, P. V., & Anoop, K. K., (2017). Spatial distribution of heavy metals in the soils of Periyar River Basin. In: *Proc. National Seminar on Geospatial Information Systems: Emerging Trends and Utilities* (pp. 11–21).

Krishnakumar, A., Anoop, K. K., & Aditya, S. K., (2019). Impact of climate change on land resources: Assessment of soil geochemical variation in Periyar Basin, Southern Western Ghats, India on August 2018 Flooding. In: *Proc. National Symposium on Innovations in Geospatial Technology for Sustainable Development with Special Emphasis on NER* (pp. 180–182).

Krishnakumar, A., Revathy, D., Aditya, S. K., & Anoop, K., (2021). Enrichment of potential toxic elements and environmental health implications: A study of the tropical agricultural soils in southern Western Ghats, India. *Environ Qual Manage., 1*, 10.

Kumar, B. M., (2011). Quarter century of agroforestry research in Kerala: An overview. *J. Trop. Agri., 49*(1, 2), 1–18.

Kumar, B. M., Sajikumar, V., & Mathew, T., (1995). Floristic attributes of small cardamom (*Elettaria cardamomum* (L.) Maton) areas in the Western Ghats of peninsular India. *Agroforest. Syst., 31*, 275–289.

Lundgreen, B. O., & Raintree, J. B., (1982). Sustained agroforestry. In: Nestel, B., (ed.), *Agri. Res Develop: Potentials and Challenges in Asia* (pp. 37–49). ISNAR.

Misra, S. S., (2011). *Kerala Gets Cautious, Down to Earth.* Available at: https://www.downtoearth.org.in/coverage/kerala-gets-cautious-33029 (accessed on 28 December 2023).

Molina, M., (2009). Trace element composition of selected fertilizers used in Chile: Phosphorus fertilizers as a source of long -term soil contamination. *Soil Sediment Cont., 18*(4), 497–511.

Murugan, M., (2007). Yield of small cardamom (*Elettaria cardamomum* M) variety PV1 as influenced by levels of nutrients and neem cake under rain fed condition in Southern Western Ghats, India. *Caspian. J. Environ. Sci., 5*(1), 19–25.

Murugan, M., (2011). Environmental impacts of intensive cardamom (small) cultivation in Indian Cardamom Hills: The need for sustainable and efficient practices. *Recent Res. Sci. Techn., 3*(2), 9–15.

National Institute of Advanced Studies, (2010). Cross reference from Misra, S. S., (2011) Kerala gets cautious. Down to Earth.

NFC, (2006). *National Forest Commission Report*. Ministry of Environment and Forest.

Reichenberger, S., Bach, M., Skitschak, A., & Frede, H. G., (2007). Mitigation strategies to reduce pesticide inputs into ground- and surface water and their effectiveness: A review. *Sci. Tot. Environ., 384*(1–3), 1–35.

Sauve, S., Hendershot, W., & Allen, H. E., (2000). Solid solution partitioning of metals in contaminated soils: Dependence on pH, total metal burden and organic matter. *Environ. Sci. Technol., 34*(7), 1125–1131.

Semwal, D. P., Uniyal, P. L., Bahuguna, Y. M., & Bhatt, A. B., (2009). Soil nutrient storage under different forest types in a part of Central Himalayas, India. *Annals For., 17*(1), 43–52.

Smical, A. I., Hotea, V., Oros, V., Juhasz, J., & Pop, E., (2008). Studies on transfer and bioaccumulation of heavy metals from soil into lettuce. *Environ. Eng. Manag. J., 7*(5), 609–615.

Soman, K., (2002). *Geology of Kerala* (p. 335). Geological Society of India, Bangalore.

Sucharovà, J., Suchara, I., Hola, M., Marikova, S., Reimann, C., Boyd, R., & Englmaier, P., (2012). Top-bottom soil ratios and enrichment factors: What do they really show. *Appl. Geochem., 27*(1), 138–145.

Tomlinson, D. L., Wilson, J. G., Harris, C. R., & Jeffrey, D. W., (1980). Problems in the assessment of heavy-metal levels in estuaries and the formation of a pollution index. *Helgoländer Meeresuntersuchungen, 33*, 566–575.

Turekian, K. K., & Wedepohl, K. H., (1961). Distribution of the elements in some major units of the earth's crust. *Geo. Soc. America Bullet., 72*(2), 175–192.

Tye, A. M., (2004). Speciation and solubility of Cu, Ni and Pb in contaminated soils. *Europ. J. Soil Sci., 55*(3), 579–590.

UNDP, (2013–2018). Report India High Range Landscape Project, Implemented by Department of Forest and Wildlife, Government of Kerala, India, Page 1–81.

Valdés, J., Vargas, G., Sifeddine, A., Ortlieb, L., & Guinez, M., (2005). Distribution and enrichment evaluation of heavy metals in Mejillones Bay (23 S), Northern Chile: Geochemical and statistical approach. *Mar. Pollut. Bull., 50*(12), 1558–1568.

CHAPTER 15

Sustainable Soil Health Management Through Agroforestry in the Soils of Arunachal Pradesh, India

JITENDRA KUMAR,[1] H. KALITA,[2] NIRMAL,[3] NISHANT SINHA,[1] DHIRAJ KUMAR,[1] AVINASH PANDEY,[4] R. S. CHAUDHARY,[1] and A. K. PATRA[1]

[1]*ICAR–Indian Institute of Soil Science, Nabibagh, Bhopal, Madhya Pradesh, India*

[2]*ICAR–RC for NEH Region Arunachal Pradesh Center, Basar, Arunachal Pradesh, India*

[3]*ICAR–Indian Institute of Farming System Research, Modipuram, Meerut, Uttar Pradesh, India*

[4]*ICAR–Indian Institute of Agricultural Biotechnology, Ranchi, Jharkhand, India*

ABSTRACT

Jhum cultivation is a conventional agricultural exercise in the north-eastern state of India. Jhum cultivation is the shifting cultivation in which land is cleared by deforestation and subsequent burning of demonstrated vegetation for crop cultivation then abandoned after 2–3 cultivation cycles and select to another site for further subsistence of livelihood. In Arunachal Pradesh, once it was considered a sustainable system due to sufficient time, it got to regenerate. However, due to increased population pressure, and unplanned and faulty agricultural practices, the jhum cycle is reduced as a result soil is degrading day by day. Along with these, the steep slope and high rainfall

Sustainable Management and Conservation of Environmental Resources in India.
Sumit Chakravarty, Vineeta, Jahangeer A. Bhat, Munesh Kumar, Gopal Shukla (Eds.)

during a short period make the soil more vulnerable to erosion through runoff. This made researchers rethink these problems and their solutions. The agroforestry-based systems have shown considerable perspective as a sustainable soil and land use (LU) management exercise for the improvement of soil health. Integration of trees along with crops has improved soil parameters that define soil health through different processes like the addition of organic matter via litterfall, underground root decomposition, twig, flowers, and fruits biomass to the soil. Decomposition, nutrient release and return to the soil (nutrient cycling), biological nitrogen fixation (BNF), nutrient pumping and reduction of soil erosion improve soil properties under agroforestry systems. The policymakers, broader agricultural community and other stakeholders must take agroforestry into their account as a feasible approach to recovery and sustaining soil health.

15.1 MINTRODUCTION

In the northeastern part of the Himalayan region, Arunachal Pradesh is one of the biggest hilly states of India, the state's high annual rainfall, large forest vegetation, variability in the National Bureau of Soil Survey and Land Use Planning (NBSS and LUP) divided the states into four distinct zones. The zones are characterized as 1. Ice-capped mountains (5,500 m AMSL); 2. Inner Himalayan ranges (3,500 m AMSL); 3. Outer Himalayas Siwalik hills (700 m AMSL); and 4. The eastern Brahputra plains (Maji et al., 2001). The traditional communities of the state depend on forest resources and perform *jhum* cultivation (slash and burn cultivation) for generations (Singh et al., 2009). These communities have diverse food habits, cultures, and languages (Mishra et al., 2009). Biodiversity and culture have been interwoven in a unique manner that has formed a bio-cultural diversity. Every festival and social occasion of these tribes linked with agroforestry, forest resources and *jhum* cultivation. Initially, the jhum cultivation system was low input-driven and sustainable due to low population pressure as well as sufficient area of undisturbed forest. However, due to increased population pressure, the period of jhum cycle got reduced, making the soil health unsustainable.

The soil is an essential natural deposit on earth; therefore, its function and its complexities are needed to quantify in order to sustainably manage the overall soil health and their quality. Soil health assessment is gaining importance over the last several years, particularly since the inception of the 1990s, for the for-ecosystem services (ES) (Sinha et al., 2014, 2019). Soil health management is gaining momentum because it plays a vital driving role in the mitigation of climate change (CC) and food availability. However, we are still finding how we can precisely estimate soil health for application in agriculture, economic, and other social sectors. Researchers debating on the appropriate methodology for precisely estimating soil health due to various complexities involved in different measurements and indicators selected that qualify a soil to be "healthy." When the awareness of soil health is improved, it is important to understand how soil health is estimated, What soil health entails, and how to control it for optimum and tenable delivery of the services that the ecosystem provides through the soil. Soil health is made of two-word – soil and health. To define soil health properly, we have to think about both words one by one that form "soil health." As per the SSSA (The Soil Science Society of America) "soil" is defined as "the unconsolidated mineral or organic material on the immediate surface of the earth that serves as a natural medium for the growth of land plants" and health is defined by Merriam-Website, as "the condition of being sound in body, mind, or spirit." Now, integrate the definition of both Soil and health now the "soil health" can be defined as "the state of the soil being in sound physical, chemical, and biological condition, having the capacity to tenable the growth and development of land plants."

15.2 MEASUREMENT OF SOIL HEALTH

There is no single and direct method to judge the degree of soil health just like the health of human beings, there are no single instruments that can be used to implement to evaluate whether an individual is healthy or not. However, we can make some observations and measure different parameters to evaluate individual health; it is also feasible to estimate specific soil properties to estimate soil health. Such estimation may be quantitative or subjective, or a blend of both. Soils include their physical, chemical, and biological aspects (Table 15.1).

To formulate the precise methods for estimating soil health, we have to work on all the dimensions and select specific quantifiable parameters that reflect our intended use goal of the soil application. The best set of measurements that will allow us to judge soil health status is called the minimum data set (Rezaei et al., 2006). The measurements that make up the minimum data set is called soil quality. Dataset selected for Arunachal Pradesh is given in Table 15.1. The soils of Arunachal Pradesh are entirely

TABLE 15.1 Soil Health Indicators for the Hilly Soil of Arunachal Pradesh

Soil Physical Indicators	Soil Chemical Indicators	Soil Biological Indicators
Soil compaction.	N, P, and K content.	Soil organic matter.
Water holding capacity.	Soil micronutrients (Zn and Fe).	Nitrogen mineralization.
Soil depth.	Exchangeable Ca and Al.	Dehydrogenase activities.
Bulk density.	Soil pH.	Urease activities.
Percentage of slope.	Base saturation.	Soil microbial carbon.
Soil structure.	Cation exchange capacity (CEC).	Total microbial population.
Dry aggregate stability.		
Wet aggregate stability.		
Infiltration.		Earthworm abundance.
Hydraulic conductivity.		
Gravel content.		Microbial diversity.

different owing to the topography and the weather situation prevailing. Agroforestry systems have the potential to improve soil health and crop productivity. "The agroforestry practices have been promoted for decades across the world for their perceived benefits of improving soil health and providing other ES" (Jose, 2009). Nevertheless, the advantages and ES looking for from agroforestry would not eventuate unless these exercises enhance the capability of soils to be healthy and fertile over the long term. Healthy soil is potentially one of the most precious resources for the health of natural and agro-ecosystems to sustain food production and the provision furnishing of ES.

15.3 TRADITIONAL AGROFORESTRY SYSTEMS IN ARUNACHAL PRADESH

Owing to agroclimatic and other prevailing conditions, agroforestry has a long tradition in the northeastern hill (NEH) of India, where plantation trees are incorporated with the agronomical crop along with the livestock production system. Many cereal crops, rhizomatous crops, Pineapple, plantation crops like coffee, tea, and vegetables are grown with trees like pine, pear, plum, apple, coconut, jackfruit, areca-nut, orange, guava, banana, large cardamom, *Alnus nepalensis*, *Schima wallichii, Erythrina spp., Ficus spp., Bauhinia spp., Terminalia* spp., etc., in different agroclimatic zones (Table 15.2).

TABLE 15.2 Agroforestry Systems in Arunachal Pradesh

Zone	Altitude (m)	Existing System	Agroforestry Practices
Alpine	3,500–5,500	Non-cultivable	Pine tree, kiwi, walnut, almond, apricot.
Temperate and sub-alpine	1,500–3,500	Agrisilviculture	Pine with field and vegetable crops.
		Agri-horticulture	Field crops like finger millet, barley, buckwheat, and paddy with apple tree.
Lower Himalayas	1,500–3,500	Silvipasture	Pine with broom grass, vegetables, or cabbage, cauliflower, bean, and broom grass.
		Horti-pasture	Plum with vegetable crops like chili, potato, bean, cabbage, radish, and broom grass.
		Agri-horti-silviculture	*Ficus* spp. with food crops; *Ficus* with maize, black gram, soybean, and pine.
		Homegardens	Mixtures of crops, vegetables, fruit trees, fodder crops and trees.
Sub-tropical hill and plain zone	700–1,500	Agri-silviculture	Alnus/Alder with large cardamom *Albizia/Ficus/Grewia/Artocarpus* with maize and taro.
		Agri-horticulture	Khasi mandarin with pineapple/cow pea/pineapple/beans/radish/ginger/turmeric/Cole crops.
Himalayan Siwalik hills	400–700	Silvi-pasture	*Alder* with broom or guinea grass/Toko palm.
		Horti-pasture	*Khasi Mandarin* with broom or guinea grass.
		Agri-horti-silviculture	Crops ginger/buckwheat/millet/pulses/sesbania/gliricidia/alnus vegetables; ginger/rice bean; maize/sweet potato guava, banana, and drumstick homestead farming.
		Homegarden	Mix of crops, vegetables, fruit trees, fodder crops, palms, and trees
Mild tropical hill or Eastern Assam plains	0–400	Agri-silviculture	Coconut, areca nut, jackfruit, and banana and Toko palm (*Livistona jenkinsiana Griff*).
		Agri-horticulture	Areca nut with betel vine; mandarin with crops intercrops *viz.* beans, chili, ginger, and plain zone and turmeric.
		Silvi-pasture	Albizia, areca nut with pineapple, broom/guinea grass.
		Horti-pasture	Areca nut with black pepper and/or pineapple.
		Agri-horti-silviculture	Banana, arecanut around fishpond in homestead.
		Homegarden	*Erythrina* with coffee and black pepper *Terminalia myriocarpa* with coffee and black pepper.

15.3.1 AGROFORESTRY IN SHIFTING CULTIVATED AREAS

"Around 2 million population of tribal people are practicing *Jhum* cultivation on 11 million hectares of forest land in India, the north-eastern state of India accounts for 7.76% of the total *jhuming* area of the country" (Singh et al., 2001). Jhum triggers the loss of soil nutrients, and soil organic matter (SOM) thereby affecting soil health (Kumar et al., 2017a). The Jhum cultivation involves transforming secondary forests generally and, less often, primary forests into clearings (Plate 15.1) subsequently start cultivating the food crops like rice along with maize, vegetable crop and tuber crop.

PLATE 15.1 Jhum cultivated land of Lepa Rada district of Arunachal Pradesh.
Source: Kumar et al. (2020).

This form of agricultural practice is considered untenable unsustainable and accountable for the loss of the upper layer of soil, deforestation, and is a major contributor to global warming (Houghton, 2012). The implications of jhum practices on soil vary depending on the different stages of the practice and also on the type of soil (Ribeiro et al., 2013). In the burning stage, the physical characteristics of the soil are adversely impacted as a result loss of SOM whereas the chemical properties are positively impacted due to ash deposition (Giardina et al., 2000; Filho et al., 2013) further the ash deposition increased the pH of the soil (Kumar et al., 2017b) and hence this is acts as a natural fertilizer. During the cropping year, the fertility of soil declines as the root of the crop extracts the nutrients that are then exported in the harvest and restricted vegetative cover leads to some erosion losses. These agricultural practices provide food security to far remotely located farming families in the region and are also interlaced, with their tradition (Ramakrishnan, 1992). The practice varies from short fallow in some parts of Tripura (Gupta, 2000) to longer fallow in the Siang region of Arunachal Pradesh (Teegalapalli & Datta, 2016). Agroforestry, therefore, can be an alternative to the hilly areas where the unscientific *jhuming* method is practiced. In such modern cultivation techniques, the top area of the hill along the vegetation is allowed to grow without disturbance. The anti-erosive cultural practices like hillside contour ditches, terracing, straw mulch treatment, grass-strip barriers and alley cropping can be adopted (Kumar et al., 2020). *Sesbania* and bamboos can also be planted to arrest soil and water loss to sustain soil health. The foothill area can be utilized to cultivate Citrus along with pineapple (Plate 15.2) or crops such as pineapple (Plate 15.3), ginger, turmeric, and vegetables (Bhabesh, 2011).

PLATE 15.2 Khasi mandarins and pine apple-based agroforestry system in jhum land.

Source: Kumar et al. (2020).

15.3.2 BAMBOO-BASED AGROFORESTRY AND ITS EFFECT ON SOIL HEALTH

Arunachal Pradesh is one of the extensive bamboo growing states in the country. Every tribe of Arunachal Pradesh has its customs and traditions, and each uses bamboo to fulfill its everyday needs. Arunachal is known to use cane and bamboo to build houses and suspension bridges. Bamboos can grow

PLATE 15.3 Pineapple cultivation in the foothill.

Source: Kumar et al. (2020).

in various types of soils and weather conditions. It can grow in marginal to semi-arid conditions. It also plays important role in soil rehabilitation (Nath et al., 2009). The abundant litterfall and high biomass accumulation help in enhancing and maintaining the physical, biological, and chemical characteristics of the soil, as a result, soil health improves and have multifaceted uses in all the spheres of life (Figure 15.1).

Bamboos are rich in silica; therefore, their litters slowly decompose, returning substantial amounts of N, K, Mg, Ca, and P (Shanmughavel et al., 2000). The bamboo biomass growing across the degraded lands of the hill slopes (e.g., abandoned shifting agricultural land, locally called 'jhum' cultivation) holds together the loose sandy–loamy soil, and helps in slope stability, reducing run-off and soil erosion (Toky & Ramakrishnan, 1982). The high volume of fine roots up to the deeper layer of soil helps in recovering most of the nutrients leached up to the deeper soil layer (Christanty et al., 1996). A high density of leaves and their architecture is like an umbrella canopy that intercepts the rainfall, decreases runoff and helps in reducing the soil erosion on the steep slope of the hilly soil. It further assists in saving the soil moisture for a longer duration in jhum area due to its shading effect that facilitates the growth and activities of soil microbial activities and maintains soil organic carbon (SOC) that enhances soil health. Improvements are also reflected in greater aggregate stability, increased porosity, decreased bulk density,

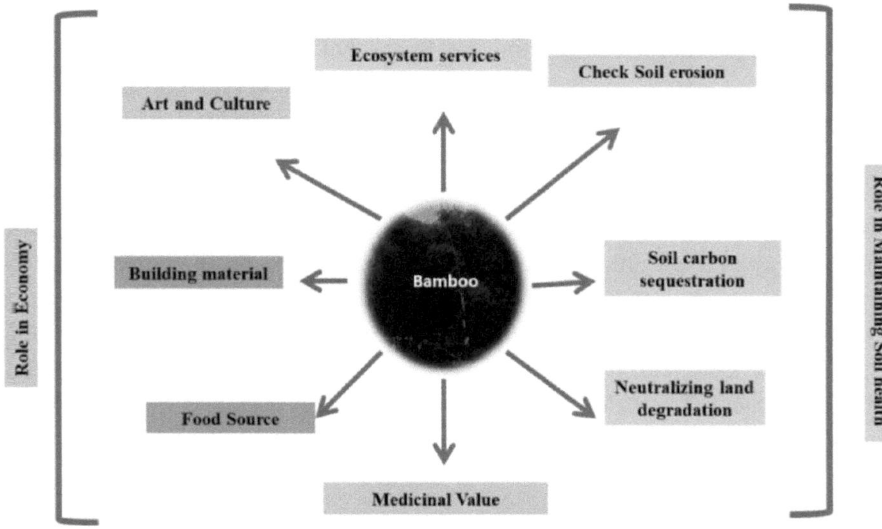

FIGURE 15.1 The benefit of bamboo-based agroforestry system on the economy and soil health.

enhanced rainwater infiltration and decreased surface resistance to penetration. The tree canopy shade also alters soil conditions to enhance microbial activity and the mineralization processes (Arunachalam & Arunachalam, 2002). Singh & Singh (1999) reported that *Dendrocalamus strictus* plantation has an efficient restoration potential and positive rehabilitation effect on mine soils and in a dry tropical region in India. It has also been reported that the bamboo-based agroforestry system boosts the ecological properties of the intensively degraded basaltic soil of Jabalpur. Also, a bamboo-based agroforestry system r the biodiversity under its habitat (Bipin et al., 2000). Singh et al. (1992) studied the impact of 25 to 30 years old *Bambusa nutans* clump cultivation in agrisilviculture system on the chemical properties of soil and reported that the availability of phosphorus enhanced whereas exchangeable potassium, magnesium, and calcium declined with increased distance from the bamboos rows, however, the soil pH and SOM remain intact.

15.3.3 TOKO PALM-BASED AGROFORESTRY SYSTEM AND ITS EFFECT ON SOIL HEALTH

Toko palm tree or Assam Fan Palm (family Arecaceae) seems to play a great role in the social, traditional, ceremonial, and economic activities of human life. In northeastern states like Arunachal Pradesh, Assam, and Nagaland Toko tree (*Livistona jenkinsiana*) is known to have its special space in the traditional milieu but this has been declared a threatened species (Singh et al., 2010). It is distributed in the tropical and subtropical belts and grown on lower plains and hill slopes. The ethnic communities of Arunachal Pradesh have been traditionally using this species. This palm has a multipurpose application and provides different utility items like leaves, fiber, and fruits. Toko garden is considered immovable property for the Adi ethnic community of Siang district in Arunachal Pradesh. In the course of family property sharing, it is taken into account and is inherited as ancestral property. The Toko palm is also being cultivated as a mixed crop in the jhum cultivated area. This is beneficial in checking the soil erosion due to its leaf architecture like an umbrella canopy that intercepts the rainfall and decreases runoff and helps in reducing soil erosion on the steep slope of the hilly soil. It further assists in saving the soil moisture for a longer duration in jhum area due to its shading effect that facilitates the growth and activities of soil microbial activities, maintaining SOC that enhances soil health.

15.4 MEFFECT OF AGROFORESTRY SYSTEM ON SOIL HEALTH

The agroforestry systems in several ways help in enhancing the soil fertility and its quality. Young (1991) also reported the influence of agroforestry systems, on the soil he reiterated in his report that agroforestry system maintains SOM, reduce soil erosion, promotes desirable soil microbial activity, maintains soil physical properties, reduces soil toxic material, enhances the rate of nitrogen fixation, improves soil nutrient inputs, accelerates nutrient cycling, enhance soil water availability to crops, through root systems of agroforestry (Figures 15.2 and 15.3).

Nair (1984) found that agri-horticultural, agroforestry, and agropastoral systems have a great prospect to control soil loss by reducing runoff volume and improving SOM, maintaining soil physical condition and increasing the rate of nitrogen fixation and promoting nutrient cycling by mineralization. Many other researchers have also highlighted the role of agroforestry, ley farming (Jaradat, 1990), alley cropping (Kang et al., 1985), and agroforestry systems and agri-horticultural (MacDicken, 1990; Das et al., 1993) in increasing income and improving soil health. Nutrient cycling (Figure 15.3) is one of the most predominant benefits the soil system gets out of the several positive impacts from agroforestry systems. The plant nutrients present in the soil are in a state of the continuous and dynamic soil-plant system. Plants take up nutrients through an extensive root system from the soil solution and utilize them for their metabolic processes. In turn, these nutrients come back to the soil through the decomposition of plant parts after litterfall. The concentration of different nutrients in different forestry systems is presented in Table 15.3.

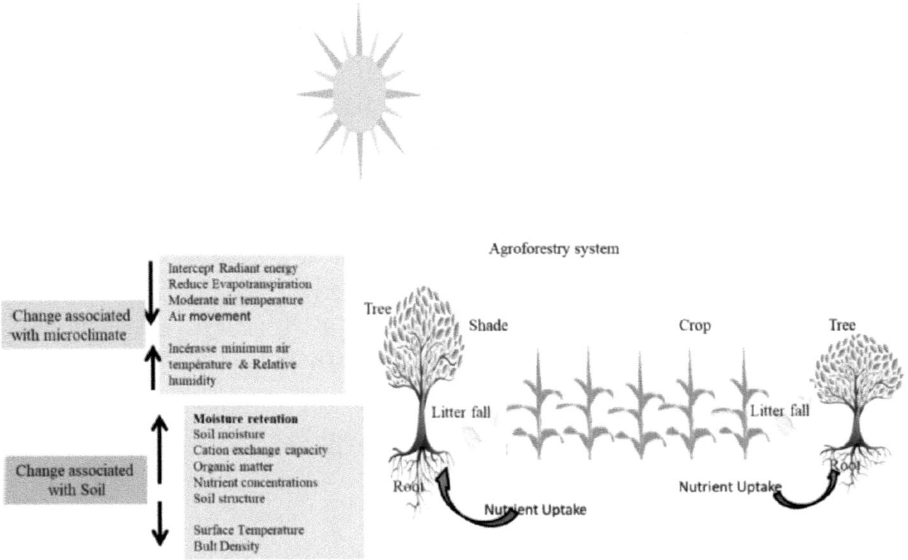

FIGURE 15.2 Influence of agroforestry systems on soil, alley crop, and microclimate.

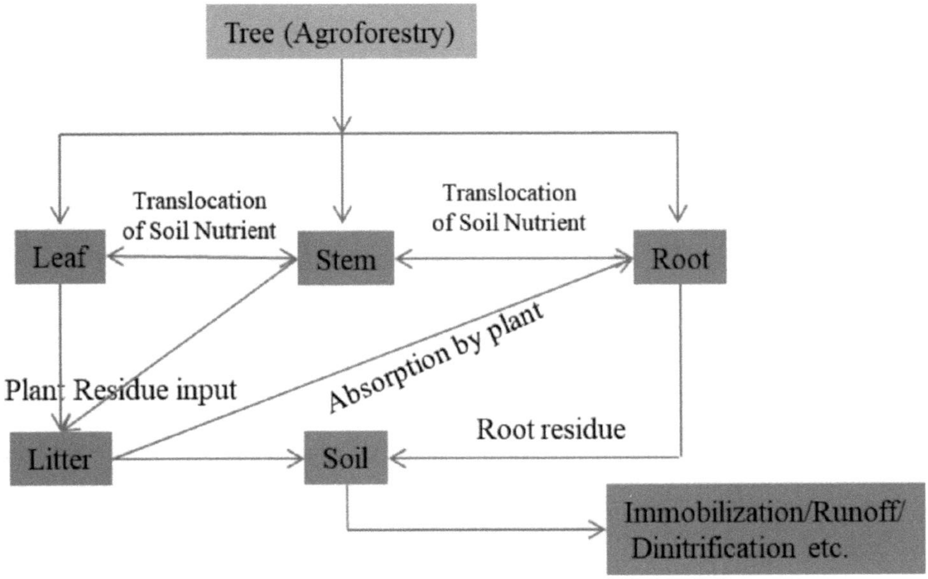

FIGURE 15.3 Soil nutrient dynamics under agroforestry system.

Soil microorganisms decompose the organic matter and regulate the cycling of the nutrients the activity levels of these microbes are regulated by the food web relationship within the soil community (Susilo et al., 2004). As a result, the nutrient is recycled and is once more available to plants (Nair et al., 1999).

TABLE 15.3 Concentration (mg/kg) of Different Nutrient in Different Forestry System

Tree	N	K	P	Ca	Mg
Pinus spp. (U.P. India)	47.78	16.86	28.81	70.97	–
Butea spp. (U.P. India)	40.69	9.0	15.16	52.78	–
Piecea smithiana (H.P. India)	24.45	6.03	7.93	34.93	–
Cedrus deodara (H.P. India)	120.0	101.0	10.0	60.0	66.0
Abies pindrow (H.P. India)	141.0	121.0	13.0	72.0	79.9
Eucalyptus tereticornis	14.2–34.5	–	8.6–20.8	98.2–334.5	–

Source: Modified from Seth et al. (1963).

The integrated efforts of plant roots and soil microorganisms are known as "soil ecosystem engineers" that consistently alter the soil by forming channel pores biological aggregated this leads to the modification in the physical properties of the soil that create favorable microhabitats for other soil organisms and maintaining the soil structure (Six et al., 2002). These microorganisms act as biological control agents for pests and diseases mostly through parasitism, competition, and predation (Susilo et al., 2004). These aggregated ecosystem functions participate in more than one soil-based delivery process. Agroforestry system can also play a very important role in maintaining and enhancing the SOM. Young (1991) found that agroforestry systems have a great prospect to control both erosion due to water and wind, which eventually save the loss of SOC and nutrients which is washed away along with the eroded particle. SOM has played many ways in enhancing soil fertility. These include the positive impacts on soil physical properties water-holding capacity, the decreased rate of nutrient release, especially important in low input agricultural systems, an increase in the cation exchange capacity (CEC), significance where fertilizers are applied, and they provide a favorable environment for soil flora and faunal activity. The physical properties of soil are affected not due to the presence of organic matter but also by the vigorous roots system and exudates within the soil. It is a fact that the physical and environment of soils, are dependent on nutrients present in the soil and it can considerably influence fertility (Lal & Greenland, 1979). Such kinds of systems maintain more convenient soil physical conditions than an unsustainable agricultural system like monoculture through carbon sequestration and the effects on the soil. Tree roots can improve both physical properties and nutrient intake from the B/C horizons. Through penetrating compact soil layers that fracture soil layers into small aggregates lines and rocky laterite and these phenomena improve both physical condition and nutrient intake from the B/C horizons. Further scientific pieces of evidence on the enhancement of the physical condition of the soil by agroforestry systems are still required whereas the effects of tree residue and litter fall on soil physical properties are substantial it gets through mulching and zero tillage studies.

The fixation of nitrogen by the tree components indicates a clear gain to the nutrient economy in agroforestry systems, with high economic value. The effectiveness of the agroforestry system is proven and research into the enhancement in the rates of nitrogen fixation, through suitable species selection and inoculation is continued. Along with the enhancement of nitrogen fixation rate the agroforestry systems, can result in more efficient cycling of nutrients, as a result, the crop yield can be sustained over some time or lead to a steady-state in low-input systems or making more efficient use of fertilizers application in high-input systems. Under low-input agricultural systems like jhum and subsistence farming, crop yields normally decline without inorganic fertilizers, leading either to an abandonment of the land, as is happening in the various forms of jhum cultivation, or unsatisfactory low yields and conditions of low level with a stable equilibrium. The advantages of low-input systems would be significant if, under specific circumstances, nutrient cycling was so effective that natural inputs would make up for harvest removal. In agroforestry systems, there is a substantial amount of circulation of basic cations. The circulation of bases from tree litter is brought into the soil to assist in the amelioration of soil acidity and checking further acidification processes (Table 15.4) and reclaiming saline or alkaline soils.

TABLE 15.4 Effect of Different Agroforestry System on Soil Health Parameters in the Soil of Arunachal Pradesh

Land Use	pH	OC (%)	N (Kg ha^{-1})	P_2O_5 (Kgha^{-1})	K_2O (Kg ha^{-1})
Jhum cultivation	4.6	1.15	266	15.98	263
Maize	5.1	1.73	227	17.09	216
Fallow land	4.8	0.93	275	8.51	221
Guava orchard	5.6	2.52	344	25.34	204
Citrus orchard	5.4	2.28	314	20.36	188
LSD (0.05)	0.47	0.72	36.21	10.48	33.78

Source: Kumar et al. (2017b).

Trees crops have been successfully incorporated to reduce saline and alkaline soils as an associated intercropping Central Soil Salinity Research Institute (CSSRI), Karnal, Haryana successfully demonstrated the potential of agrihorticulture system-based intercrop of *Emblica officinallis* (Anola) *Aegle marmelos* (Bael), and *Carissa congesta* (karonda) with Cluster bean during Rabi and barley during rabi season. They perform very well during the initial four years, suggesting the affinity of fruit species with intercrop components and the likely absence of allelopathic relationships under saline conditions. Agroforestry systems have shown multiple positive effects on soil flora and fauna with subsequent improvements in soil productivity and soil health. Furthermore, it is also probable that trees, integrated with crops will increase the total water supply by lowering evaporation by changing the micro-climate around the canopy (Figure 15.2). The significance of roots in preserving the productivity of soil in agroforestry systems is equally vital as that of shoot (above ground) biomass. Roots contribute a vital role in nearly most of the soil's physicochemical and biological processes (Figure 15.2), particularly in SOM addition, soil physical environment, fixation of nitrogen and retrieval of nutrients and their recycling. At the same time, fighting for nutrients between tree and crop roots is a potentially poor attribute of agroforestry for crop production and soil fertility, However, no such report has yet been reported (Young, 1991). Understanding the behavior of soils over time under particular agricultural practices is essential for effective soil management that maintains soil quality and soil health. Due to this reason, there is growing interest in measuring the quality of our soil resources, which has been activated by increasing recognition that soil is a crucial component of the earth's biosphere, performing not only in the creation of food and fiber but also in the restoration of indigenous, regional, and worldwide ecological quality (Doran & Parkin, 1994). At the same time, feeding off the ever-increasing human population is a great challenge for the farming community in developing countries because of shrinking soil resources and soil health deterioration. Nearly the whole world's feed is produced in the soil-directly or indirectly yet it a calculated that 40% of it is deteriorated. Loss of soil is one of the serious threats in hilly terrain; hence, erosion and resistance cultural practices may be suggested for the tenable agricultural production system.

15.5 CONCLUSION

special attention and a holistic approach are required for the sustainable management of soil health in the hilly fragile land of Arunachal Pradesh to intact the traditional tribal agriculture system at the same time they can avail the healthy ecosystem, meet out good and services and tenable agricultural production to supply food for the increasing population. In Arunachal Pradesh Horticultural crop-based agroforestry such as citrus, pineapple, and bamboo are found suitable for sustainable soil health because this will not only reduce the new area to bring into jhum cultivation and also generate a source of income for the livelihood. This system has also shown good evidence of improving soil quality by addition of organic matter, improving soil nutrients, enhanced soil physical properties and reducing soil erosion owing to vigorous roots holding the soil particle which is highly susceptible to washing away with runoff due to the steep slope. The viable option is bamboo-based agroforestry which produces abundant litterfall and high biomass accumulation helping in restoring and enhancing the soil's physicochemical and biological properties. Therefore, integrating different agroforestry systems sustains production and fulfills the required timber, fodder, fuelwood, and medicines. The policymakers, broader agricultural community and other stakeholders must initiate the settled agriculture of horticulture crops, bamboo, Toko palm and other suitable traditional tree-based agroforestry systems as a viable mechanism to restore and enhance the agricultural productivity and sustain the soil health of Arunachal Pradesh.

KEYWORDS

- **agricultural productivity**
- **agroforestry**
- **bamboo**
- **jhum or shifting cultivation**
- **soil health**
- **soil nutrients**
- **sustainable management**

REFERENCES

Arunachalam, A., & Arunachalam, K., (2002). Evaluation of bamboos in eco-restoration of jhum fallows in Arunachal Pradesh: Ground vegetation, soil, and microbial biomass. *Forest Ecology and Management, 159*(3), 231–239.

Bhabesh, G., (2011). Rejuvenation of degraded land through bamboos: A biological approach. In: Singh, S., & Das, R., (eds.), *Productivity Enhancement and Value Addition of Bamboo* (pp. 68–74). Excel India Publishers, New Delhi.

Bipin, B., Rashmi, A., Singh, A. K., & Banerjee, S. K., (2000). Vegetation development in a degraded area under bamboo based agro-forestry system. *Indian Forester, 126*(7), 710–720.

Christanty, L., Mailly, D., & Kimmins, J. P., (1996). Without bamboo, the land dies: Biomass, litterfall, and soil organic matter dynamics of a Javanese bamboo Talun-Kebun system. *Forest Ecology and Management, 87*(1–3), 75–88.

Das, S. K., Sharma, S., Sharma, K. L., Saharan, N., Nimbole, N. N., & Reddy, Y. V. R., (1993). Land use options on a semi-arid Alfisol. *American Journal of Alternative Agriculture,* 34–39.

Doran, J. W., & Parkin, T. B., (1994). Defining and assessing soil quality. *Defining Soil Quality for a Sustainable Environment, 35*, 1–21.

Filho, A. A. R., Adams, C., & Murrieta, R. S. S., (2013). The impacts of shifting cultivation on tropical forest soil: A review. *Boletim do Museu Paraense Emılio Goeldi. Ciencias Humanas, 8*, 693–727.

Giardina, C. P., Sanford, R. L., Døckersmith, I. C., & Jaramillo, V. J., (2000). The effects of slash burning on ecosystem nutrients during the land preparation phase of shifting cultivation. *Plant and Soil, 220*(1), 247–260.

Gupta, A. K., (2000). Shifting cultivation and conservation of biological diversity in Tripura, Northeast India. *Human Ecology, 28*(4), 605–629.

Houghton, R. A., (2012). Carbon emissions and the drivers of deforestation and forest degradation in the tropics. *Current Opinion in Environmental Sustainability, 4*(6), 597–603.

Jaradat, A. A., (1990). Farming systems around the Mediterranean exploitation versus sustainability. *Sustainable Agriculture: Issues, Perspectives and Prospects in Semi-arid Tropics, 1*, 313–336.

Jose, S., (2009). Agroforestry for ecosystem services and environmental benefits: An overview. *Agrofor. Syst., 76*(1), 1–10.

Kang, B. T., Grimme, H., & Lawson, T. L., (1985). Alleu cropping sequentially cropped maize and cowpea with Leucaena on a sandy soil in Southern Nigeria. *Plant and Soil, 85*(2), 267–277.

Kumar, J., Chandra, A., Alone, R. A., Bhagawati, R., Rekhung, W., & Borah, M. C., (2017b). Effect of different land use systems on selected soil parameters in mid hill of West Siang district, Arunachal Pradesh. In: Arora, S., & Bhan, S., (eds.), *A Soil Conservation Society of India Manual* (pp. 1–13). New Delhi.

Kumar, J., Kalita, H., Sinha, N., Mohanty, M., Somasundaram, J., Rani, A. R., & Chaudhary, R. S., (2020). Hill agriculture prospect and problem in the soil of Arunachal Pradesh. *Harit Dhara, 3*(1), 26–29.

Kumar, J., Shukla, K., Sen, A., Chandra, A., Bhagawati, R., & Rekhung, W., (2017a). Effect of different processes of jhuming on soil properties in mid hills of Arunachal Pradesh. *Environment & Ecology, 35*(2A), 999–1003.

Lal, R., & Greenland, D. J., (1979). *Soil Physical Properties and Crop Production in the Tropics* (No. 631.41 S67). John Wiley & Sons, New York.

MacDicken, K. G., (1990). Multipurpose nitrogen fixing trees and their role in agroforestry systems. *Indian Society of Agronomy, 11*, 141–157.

Maji, A. K., Nayak, D. C., Krishna, N. D., Srinivas, C. V., Kamble, K., Reddy, G. P. O., & Velayutham, M., (2001). Soil information system of Arunachal Pradesh in a GIS environment for land use planning. *International Journal of Applied Earth Observation and Geoinformation, 3*(1), 69–77.

Mishra, S., Singh, R. K., & Singh, A., (2009). A dynamics of Adi women's traditional foods and livelihoods in varying socio-ecological systems of Arunachal. *The New Cultures of Food: Marketing Opportunities from Ethnic, Religious and Cultural Diversity* (p. 203).

Nair, P. K. R., (1984). *Soil Productivity Aspects of Agroforestry: Science and Practices of Agroforestry* (p. 85). ICRAF, Nairobi, Kenya.

Nair, P. R., Buresh, R. J., Mugendi, D. N., & Latt, C. R., (1999). Nutrient cycling in tropical agroforestry systems: Myths and science. *Agro Forestry in Sustainable Agricultural Systems.* CRC Press, Boca Raton, FL, USA.

Nath, S., Das, R., Chandra, R., & Sinha, A., (2009). *Bamboo Based Agroforestry for Marginal Lands with Special Reference to Productivity, Market Trend and Economy* (pp. 80–96). Agroforestry in Jharkhand, Envis. Jharkhand News.

Ramakrishnan, P. S., (1992). *Shifting Agriculture and Sustainable Development- an Interdisciplinary Study from Northeastern India; Man and Biosphere Series* (Vol. 10).

Rezaei, S. A., Gilkes, R. J., & Andrews, S. S., (2006). A minimum data set for assessing soil quality in rangelands. *Geoderma, 136*(1, 2), 229–234.

Ribeiro, F. A. A., Adams, C., & Murrieta, R. S. S., (2013). The impacts of shifting cultivation on tropical forest soil: A review. *Boletim do Museu Paraense Emílio Goeldi. Ciências Humanas, 8*(3), 693–727.

Seth, S. K., Kaul, O. N., & Gupta, A. C., (1963). Some observations on nutrition cycle and return of nutrients in plantations at new forest. *Indian Forester, 89*(2), 90–102.

Shanmughavel, P., Peddappaiah, R. S., & Muthukumar, T., (2000). Litter production and nutrient return in *Bambusa bambos* plantation. *Journal of Sustainable Forestry, 11*(3), 71–82.

Singh, A. N., & Singh, J. S., (1999). Biomass, net primary production, and impact of bamboo plantation on soil redevelopment in a dry tropical region. *Forest Ecology and Management, 119*(1–3), 195–207.

Singh, J., Bora, I. P., Barua, K. N., & Baruah, A., (2001). *Project Report on Shifting Cultivation.* RFRI, Jorhat, Assam.

Singh, K. A., Patiram, L., Singh, N., & Roy, R. N., (1992). Effect of bamboo (*Bambusa nutans* Wall. Ex Munro) shade on the yield of some agricultural crops at mid hills of eastern Himalaya. *Indian Journal of Forestry, 15*(4), 339–341.

Singh, R. K., Srivastava, R. C., & Mukherjee, T. K., (2009). Meeting report: Community-based sustainable natural resources management and development in Northeast India. *Current Science, 96*(1), 19–21.

Singh, R. K., Srivastava, R. C., & Mukherjee, T. K., (2010). Toko-Patta (*Livistona jenkinsiana Griff*): Adi community and conservation of culturally important endangered tree species in eastern Himalaya. *Indian Journal of Traditional Knowledge, 9*(2), 231–241.

Sinha, N. K., Chopra, U. K., & Singh, A. K., (2014). Cropping system effects on soil quality for three agro-ecosystems in India. *Experimental Agriculture, 50*(3), 321.

Sinha, N. K., Chopra, U. K., Singh, A. K., Kumar, N., Mohanty, M., & Somasundaram, J., (2019). Soil physical quality as quantified by S index under different management system in rice-wheat system. *Journal of Soil and Water Conservation, 18*(1), 27–34.

Six, J., Feller, C., Denef, K., Ogle, S., De Moraes Sa, J. C., & Albrecht, A., (2002). *Soil Organic Matter, Biota, and Aggregation in Temperate and Tropical Soils-Effects of No-Tillage.* Soils of Arunachal Pradesh. NBSSLUP.

Susilo, F. X., Neutel, A. M., Van, N. M., Hairiah, K., Brown, G., & Swift, M. J., (2004). Soil biodiversity and food webs. In: Van, N. M., Cadisch, G., & Ong, C. K., (eds.), *Below-Ground Interactions in Tropical Agro-ecosystems: Concept and Models with Multiple Plant Components* (pp. 285–308). CAB International Publishing.

Teegalapalli, K., & Datta, A., (2016). Field to a forest: Patterns of forest recovery following shifting cultivation in the Eastern Himalaya. *Forest Ecology and Management, 364*, 173–182.

Toky, O. P., & Ramakrishnan, P. S., (1982). Role of bamboo (*Dendrocalamus hamiltonii* Nees and Arn.) in the conservation of potassium during slash and burn agriculture (Jhum) in north-eastern India. *Journal of Tree Sciences, 1*(2), 17–26.

Young, A., (1991). Soil fertility. In: Avery, M. E., Cannel, M. G. R., & Ong, C. K., (eds.), *Biophysical Research for Asian Agroforestry* (pp. 187–208). Winrock International USA and South Asia Books, USA.

CHAPTER 16

Large Cardamom-Based Agroforestry System for Livelihood Security in Sikkim Himalayas

SANTOSH K. CHETTRI, ASEESH PANDEY, KAILASH S. GAIRA, and RAJESH JOSHI

G. B. Pant National Institute of Himalayan Environment (NIHE), Sikkim Regional Center, Pangthang, Gangtok, Sikkim, India

ABSTRACT

The tiny Himalayan state Sikkim is the leading producer of large cardamom in India, contributing nearly 84% of country's production which is cultivated in agroforestry system. Realizing the potentiality of the crop in improving the rural livelihood and economy, this is our gentle attempt to highlight the current developments in large cardamom cultivation system and the possible impacts on economy. Apart from this, large cardamom based agroforestry system, its uniqueness have also been presented in this chapter. Additionally, diversity of large cardamom endemic to the region, present status, production trends of declining phase, problems, and issues highlighting changing socio-cultural scenario and *chirkey/foorkey* diseases, and future opportunities under large cardamom farming have been discussed. This information base would be useful in understanding the potentiality of large cardamom in improving the economy of rural farmers for long term livelihood sustainability in the region.

16.1 INTRODUCTION

Large cardamom (*Amomum subulatum* Roxb. under family Zingiberaceae) is a cash crop of Indian Himalayan states Sikkim, Arunachal Pradesh, West

Sustainable Management and Conservation of Environmental Resources in India.
Sumit Chakravarty, Vineeta, Jahangeer A. Bhat, Munesh Kumar, Gopal Shukla (Eds.)

Bengal (Darjeeling hills) and Uttarakhand and neighboring countries of Nepal and Bhutan (Negi et al., 2018). The cultivation of large cardamom is not much labor intensive and after plantation, it requires two times weeding in a year before it's harvesting (Singh et al., 2000). However, in current changing scenario with spread of knowledge, large cardamom plantation has reached to the near farms from the naturally managed agro-ecosystems due to intensive use of organic manure and improved irrigation for better yield. Cardamom cultivation is an integral part of mountain farming systems. Even a small quantity of large cardamom capsules can generate high incomes. In Sikkim Himalaya cardamom-based agroforestry system comprise of nitrogen fixing Himalayan alder (*Alnus nepalensis*) 70% and mixed-tree agroforestry system 30% (Negi et al., 2018). The alder plant is commonly grown in association with cardamom for managing shade and nitrogen (100–155 kg/ha/yr.) and the alder litter is rich in nutrients and decomposes at faster rate (Sharma et al., 2009; Negi et al., 2018). The association occurs as cardamom-siris (*Albizia procera*) in lower elevations (600–1,200 m) (Sharma et al., 2007, 2009). Since the cardamom based-agroforestry systems are considered ecologically more viable and sustainable as compared to the rain-fed agriculture systems. As in case of carbon sequestration, the cardamom based-agroforestry system can stock up 3.5 times more carbon compared to the rainfed agriculture system (Negi et al., 2018). In a similar fashion, farmers may get 5.7 times higher benefits by managing cardamom based-agroforestry systems compared to the rainfed systems (Negi et al., 2018). The agroforestry system is considered a potential management system that requires fewer inputs having with output to input ration of more than 13% as compared to rain-fed agriculture (Negi et al., 2018). This system also significantly contribute in sustainable development of the mountain region benefitting the disadvantaged dwelling farming communities where the flow of multiple ES namely, conservation of water and soil, controlling the erosion, rejuvenation of soil, nutrient recycling, conservation of biological diversity, maintaining watershed cycles and esthetic values will be ensured (Sharma et al., 2007). However, over last few decades, the declining trend in its production and fluctuation in price is affecting the economy of cardamom growers. Hence, to understand the current development of change in Large Cardamom based agroforestry system and associated socio-cultural changes in the Himalayan region is highly essential. Considering the knowledge gap, this chapter mainly deals with the description on large cardamom diversity, current production trend, the most recent problems and issues including disease of *chirkey* and *foorkey*, possible opportunities for deriving suggestive recommendations. It is expected that the information base provided in this chapter, would be helpful to farming communities in better management of agriculture practices and supporting the policy planning process for improved research and development based interventions leading to livelihood security of large cardamom farmers in the Himalaya.

16.2 AGROFORESTRY SYSTEMS: SIKKIM PERSPECTIVE

Status quo of agroforestry practices in the Sikkim Himalaya is encouraging. It is conserved through *in situ* measures helping in promotion and preservation of traditional ecological knowledge and gene. Apart from the crop base (e.g., cultivation of large cardamom), the agroforestry system is maintained mainly for acquiring fuel and fodder species. Most of the large cardamom plantations are affected from hail strom, diseases, and pests, due to the lack of adequate shade trees, and non-adoption of scientific cultivation practices. Sikkim is well known for large cardamom based agroforestry system. Many crops are practiced in the state since centuries among which large cardamom is the prominent spice crop (third highly priced spice of the world after saffron and vanilla). It generally grows in a unique agroforestry system. In past, either slash and burn methods were used to manage the system or by lopping branches of trees and shrubs. However, after ban on agriculture waste burning by Government of Sikkim in the year 2015, this practice has been stopped. Herbs are cleared from the farmland twice a year and are used for mulching and bio-composting. Irrigation in the region is carried out in two ways: (i) through water channels; and (ii) using water sprinkler. Large cardamom is cultivated in fields along with other shade tree species viz., *Measa chisia*, *Leucosceptrum canum*, *Engelhardtia spicata*, *Macaranga* spp. and many other species. *Artemisia vulgaris* is important to provide moisture and manure to large cardamom plantations. *Thysanolaena* spp. has emerged as important broom grass cultivated at hill slopes. Additionally, fodder trees such as *Ficus spp.,* fodder grasses are also grown in agricultural fields depicting a unique traditional farming knowledge base inherited from the ancestors.

16.3 TYPES OF AGROFORESTRY SYSTEM

In Sikkim, the traditional farming system encompasses a diversity of agroforestry practices such as farm-based system, forest-based system, *Albizia* based system, mandarin based system and forest based cardamom system (Sharma et al., 2016a). Agroforestry based farming system is managed by growing multipurpose tree species (fodder, fuel, and timber) for soil stabilization (Plates 16.1–16.3). Whereas, forest-based agroforestry system is managed using bamboo groves. Multipurpose trees (MPTs) (including timber species) are grown on such systems. Besides Albizia-mixed system mandarin based system is another promising agroforestry system. Under the system, Sikkim mandarin, a high-value cash crop is intercropped with maize, pulses, etc., on non-irrigated land (locally'*sukha bari'*).

PLATE 16.1 Farm-based agroforestry.

PLATE 16.2 Forest-based agroforestry.

PLATE 16.3 Large cardamom-based agroforestry.

The agroforestry systems here in Sikkim and Darjeeling Himalayas stands for growing crops together with trees, large cardamom with shade trees namely *Artemisia vulgaris*, *Leucosceptrum canum*, etc. (Sharma et al., 2016a). Large cardamom, Sikkim mandarin, ginger, broom-grass are commonly grown in above stated diversified agroforestry systems (Sharma et al., 2016a). Forest based cardamom system is divided into two viz., categories: alder-based and forest-based systems (Plate 16.3).

16.4 LARGE CARDAMOM CULTIVATION

Large cardamom is an important perennial cash crop sympatric to sub-tropical and temperate agro-climatic zones of the Eastern Himalayan Region (EHR) including Nepal (Eastern part) and Bhutan (Sharma et al., 2016b). It is a cash incentive crop of the world and third most expensive crop that has high value in international markets. This crop is widely grown in agroforestry sites, forests, and in agricultural fields in association with shade trees. Traditional cultivation of large cardamom is a good example of controlling highland niche. Its cultivation is indigenously evolved agroforestry practices, an exceptional example of the ecological and economic viability of traditional farming system (Subba, 1984; Singh et al., 1989). Production option of large cardamom is perfectly matched with the mountain specificity that completes the conditions of sustainable farming (Singh et al., 2005).

India is the highest producer of large cardamom where Sikkim contributes most proportion (about 80% of India's total production). It is also cultivated to some extent in some of South Asian countries such as Thailand, Indonesia, and Laos besides India, Nepal, and Bhutan (Naik et al., 2004), central part of China (Gautam et al., 2016), and Myanmar where limited species of *Amomum* are grown (Pandey et al., 2020). Recently, Indonesia has also emerged as the producer of large cardamom (ACCESO, 2011). In Sikkim, it is growing in varying elevations ranging between 500 and 2,300 m asl. It grows wild in forest ecosystem and also domesticated in the sub-Himalayan region (Rao et al., 1993a). It is grown in all the four districts of Sikkim. Places such as Hee-Gyathang, Hee-Martam, Hee-Patal, and Uttarey possess abundant abundance of large cardamom plantations in West Sikkim. Similarly, Dzongu and Mangan is the best hub of large cardamom in North Sikkim. In addition to this, enormous large cardamom plantations are found in Sumik, Bara Pathing, Regu, and Pemlakha in East Sikkim and Ravangla, Phamtam, Damthang, and Bhanjyang in South Sikkim. The major farm based variety of large cardamom growing in agricultural fields are Ramsey and Seremna whereas, its forest based variety include Ramsey, Varlangey, Dzongu Golsey and green cardamom. Around 15 trees belonging to 12 genera and 11 families were identified as the commonly used shade trees in large cardamom plantation. The botanical name, local name, common name, and family and flowering season of each species is given in Table 16.1. Documentation of ethnobotanical information of 15 tree species is also completed. Besides all, large cardamom is also largely is grown in open habitats.

TABLE 16.1 List of Common Shade Plants of Large Cardamom in Sikkim

Sl. No.	Common Name	Scientific Name	Family
1.	Argeli	*Edgeworthia gardneri* (Wall.) Meisn.	Thymelaeaceae
2.	Asare	*Viburnum erubescens* Wall	Adoxaceae
3.	Aule Phaledo	*Erythrina stricta* Roxb.	Leguminosae
4.	Bilaune	*Maesa chisia* Buch-Ham.ex D.Don	Primulaceae
5.	Chamlane	*Symplocos abietorum* Standl. and Steyerm.	Symplocaceae
6.	Chilaune	*Schima wallichii* Choisy	Theaceae
7.	Hario siris	*Albizia lebbeck* (L.) Benth.	Leguminosae
8.	Jhingani	*Eurya japonica* Thumb.	Pentaphylacaceae
9.	Kalo siris	*Albizia odratissima* (L.f.) Benth.	Leguminosae
10.	Malato	*Macaranga denticulata* (Blume) Muuell-Arg.	Euphorbiaceae
11.	Panisaj	*Terminalia myriocarpa* Van Heurck and Müll. Arg.	Combretaceae
12.	Pati	*Artemisia roxburghiana* Wall. ex Besser	Compositae
13.	Phaledo	*Erythrina arborescens* Roxb.	Leguminosae
14.	Pipli	*Exbucklandia populnea* Roland W. Brown	Hamamelidaceae
15.	Seto siris	*Albizia procera* (Roxb.) Benth	Leguminosae
16.	Uttis	*Alnus nepalensis* D. Don	Betulaceae

16.4.1 TRADITIONAL PRACTICES

Besides agriculture based agroforestry areas, large cardamom was once cultivated in forest areas on royalty basis (on paying taxes). Initially shrubs and herbs are cleared before seedling plantation, during the months of May and June. Labor input is required for caring, once the crop is planted in forest areas. Recently it is also being planted in agricultural field after replacing agriculture crops. Clearing of weeds from the large cardamom field, irrigating through canal (locally *kulo*), removal of litter during flowering time, manual harvesting of fruits using locally designed harvesting knives and drying through local drying system (locally *Bhatti*, i.e., furnace) are some of the traditional farming practices involved.

16.4.2 MODERN PRACTICES AND INITIATIVES OF LARGE CARDAMOM CULTIVATION

In Sikkim Himalaya, modern practices of large cardamom cultivation are not established very prominently and followed. However, Government under the MGNREGA scheme is supporting farmers in large cardamom seedling distribution and plantation, weeding, manure preparing and soil testing. Additionally, Spices Board of Government of India regularly provides trainings and other incentives to cardamom growers for skill development and management of large cardamom across the growing areas. Spices Board has developed a modified large cardamom dryer (*bhatti*) containing Flue pipe. Drying through using this system may help reduce problems of energy and quality loss in traditional system of drying (Spices Board of India, 2020). Very recently a network of large cardamom farmers' has been established in Sikkim under KLCDI India program of Sikkim Regional Center of G.B. Pant Institute of Himalayan Environment (GBPNIHE). The aim of the network is to share problems and opportunities in the field including accessing better marketing channel and higher rates of large cardamom.

16.5 MORPHOLOGICAL CHARACTERIZATION AND DIVERSITY OF LARGE CARDAMOM IN SIKKIM

So far, seven varieties of large cardamom in Sikkim have been characterized on the basis of leaf arrangement and pseudostem color. Leaf arrangement showed straight and dropping type (Table 16.2). The common cultivars of

Sikkim are Ramsey, Golsey, Chibesey, Dambersey, Sawney, and Kantidar. The prominent varieties are Golsey, Dzongu Golsey, Ramsey, Ramla, Sawney, and Seremna (Vijayan et al., 2018). Based on the climatic conditions of Sikkim and Darjeeling Himalayas, Indian Cardamom Research Institute (ICRI), Regional Station under Spices Board of India, Sikkim Center has developed two high yielding varieties (HYVs) cardamom (ICRI Sikkim 1 and ICRI Sikkim 2) during 2004. These two varieties namely – ICRI Sikkim 1 and ICRI Sikkim 2 are selection from the cultivar Sawney (Vijayan et al., 2018). Later Negi et al. (2018) also reported six commercially popular cultivars of large cardamom of Sikkim viz., Ramsey, Ramla, Sawney, Varlangey, Seremna, and Golsey (Table 16.3).

TABLE 16.2 Morphological Characters of Large Cardamom Cultivars

Sl. No.	Cultivar	Color of Pseudostem	Leaf Character	Leaf Dimensions (cm)	Capsule Size
1.	Chibbey	Green	Straight	7.94–9.36 × 54.6–71.84	Medium
2.	Dzongu Golsey	Green	Drooping	8.8–10.04 × 55.8–61.26	Medium
3.	Hangey	Full green	Straight	7.28–8.9 × 51.6–78.6	Large
4.	Ramsey	Greenish maroon	Drooping	6.5–9.56 × 48.4–73.8	Medium
5.	Sawney	Greenish maroon	Straight	6.5–8.76 × 54.5–68.8	Medium
6.	Seremna	Green	Straight	7.84–8.44 × 51.44–56.66	Medium
7.	Varlangey	Greenish maroon	Straight	7.6–9.74 × 54.1–84.8	Large

TABLE 16.3 MDistribution of Large Cardamom Varieties Across Sikkim

Variety	Elevation (m)	Location
Ramsey	1,209–1,800	Rumbuk Dickchu, East Sikkim
	1,210–1,212	Rakshey Dikchu, East Sikkim
	1,605–1,700	Upper Khesey, East Sikkim
	1,397–1,532	Panney Phensang, North Sikkim
Varlangey	1,492–1,700	Dickhu, East Sikkim
Sawney	1,561–1,564	Kabi Lungchok, North Sikkim
Seremna	1,600–1,800	Dikchu, East Sikkim
Green cardamom (Hangey)	1,600–1,800	Pasting (Lingtam), North Regu (Dokchin)
Dzongu Golsey	1,500–1,800	Dongu (Lower to Upper)
Chibsey	1,550–1,600	Mid elevation areas

Chempakam & Sindhu (2008) reported *Ramla*, *Chivey Ramsey*, *Garday*, *Seto Ramsey*, *Ramnang*, *Seto Golsey*, Slant *Golsey*, Red *Sawney*, Green

Sawney and *Mingley* cultivars of large cardamom. Rao et al. (1990) reported *Varlangey* from higher altitudes. *Bebo, Boklok Tali, Jaker*, and *Belak* are cultivars from this region (Pathak, 2008). Generally, Golsey is grown at lower altitudes whereas, Sawney at varying altitudes (Biswas et al., 1986). The cultivar *Dzongu Golsey* grows well in forests whereas, Varlangey is the dominant cultivar grown in the region. Cultivar Seremna is also growing well in mid-altitudes. The following are 18 cultivars of large cardamom *A. spiceum, A. xanthophlebium, A. macranthum, A. macrodeus, A. hastilabium, A. cylindraceum, A. vivale, A. testaceum, A. squarrosum, A. biforum, A. macroglossa, A. citrinum, A. uticolosum, A. ochreum, A. uphalotes, A. leppaceum, A. aculeatum, A uliginosum* (Anon., 1950). Certain varieties of large cardamom have been found growing in agricultural fields of low altitude (Table 16.2). Distribution of these large cardamom varieties is given in Table 16.3 (Vijayan et al., 2018; Pandey et al., 2020).

16.6 MPRODUCTION TREND OF LARGE CARDAMOM IN SIKKIM

Study on the area and production of large cardamom showed a steady decline in India. The cultivation area of large cardamom in Sikkim increased by 2.3 times in 20 years (from 2004–2005 to 2011–2012) despite regular gap filling and replantation drive (Sharma et al., 2009). During 2004–2005, most of the leased plantations under reserve forest and protected areas (PA) (approx. 1,316 ha) were withdrawn by the Department of Forest Wildlife and Environment, Government of Sikkim. The productivity of cardamom in Sikkim had increased during 2007–2008 with the area under large cardamom cultivation 26,734 ha and the production 4,305 tons (1.61 quintals/ha). Thereafter, the trend in area of cultivation and production showed a steady decline and reduced to 23,679 ha with the production of 3,310 tons (1.40 quintals/ha) in 2010–2011, respectively. In 2011–2012, production of 3,234 tons (1.39 quintals/ha) was recorded from the cultivated area of 23,155 ha (Spices Board of India, 2011). The production of large cardamom in Sikkim, however, improved to 4,780 MT in 2020 (Spices Board of India, 2020).

Total production of large cardamom across the World was about 12,278 MT during the year 2012–2013 (Pothula & Singh, 2013). While, Nepal maintains a significant growth trend in large cardamom production and became the leading producer (Avasthe et al., 2011; Pothula & Singh, 2013; Kafle, 2013). The eastern development region of Nepal accounts for 97% of the total national production. Whereas, the four districts (Taplejung, Illam, Sankhuwasabha, and Panchthar) of Nepal account for 81% of the national production (MoAC,

2010). In India the plantation area of large cardamom is approximately 34,252 ha (Sharma et al., 2009), wherein current production rate is 8,600 MT during 2019–2020 (Spices Board of India, 2020). Over the years the plantations of large cardamom in agriculture land has increased but due to decline in market price and disease outbreaks, the agroforestry system of the region is seriously affected. As adaptive measures and to cope with the changes, the large cardamom which used to be planted in nearby natural forest (NF) stands, is now being gradually shifted to agricultural fields and the use of manure to enhance yield is gaining the momentum (Pandey et al., 2020).

16.7 PERSPECTIVES OF GEOGRAPHICAL INDICATORS (GI) TAG

Geographical indicator helps in recognizing and protecting the items belonging to a particular place along with preserving to traditional knowledge associated with them (Kishore, 2018). Tapping the untapped potential of the registered GI products must be initiated immediately (Sharma & Rajan, 2018). However, North East India (NEI) despite its rich horticultural diversity fails to take benefit of GI tagging (Datta et al., 2020). Large cardamom comes under the list of GI registered products of India. Higher market price, remunerative price for producers, consumers, market penetration, traditional knowledge, rural development and legal protection are the benefits of GI (Datta et al., 2020). In a similar manner, among the large cardamom cultivars grown in Sikkim Himalaya; *Varlangey*, *Seremna*, and *Sawney* cultivars have been registered recently under geographical indicators (GI) tagging of the region by the State Biodiversity Board of Sikkim. This initiative will be helpful in locating these cultivars at global level.

16.8 ISSUES ASSOCIATED WITH LARGE CARDAMOM AGROFORESTRY SYSTEM

16.8.1 ELUDING AGROFORESTRY SYSTEMS

In some districts of Sikkim Himalaya, e.g., South and West faces prolonged dry months of about seven months (Mandal et al., 2004, 2012) leading to the moisture stress to the crop causing spread of diseases. Availability of appropriate shade trees is a real challenge. Shade of *Artimisia vulgaries* (locally *Titeypati*), a shrub is considered as good shade plant for large cardamom plantation in managing greenery and moisture. To retain moisture in the LC fields, trenches across the slope of the large cardamom fields would be very helpful. The water thus retained provides moisture to the crop during the dry months.

16.8.2 DECLINE IN MARKET PRICE

Large cardamom cultivation is in practice since time immortal. It is generating a good income for rural households. But yearly fluctuation in price is a matter of concern. The local retailer informs the farmers about the daily market rate of large cardamom. But at presently daily rate can be understood from accessing cardamom business website (or in the website of the Spices Board of India). The trend shows that price may go much lower to the standard price thereby creating negative impacts among the cardamom growers. Although the growth and production of large cardamom revived during the year 2011–2012 which helped in fetching higher selling prices all times (i.e., INR ₹ 1,800 per kg) but unfortunately the prices further decreased in the recent years. Due to lack of financial resources and future scope, farmers are failing to focus on productivity of the crop. Also on account of improper management in marketing of large cardamom, local farmers are getting relatively lower prices. For price realization, the multiple intermediaries viz., North Eastern Agricultural Marketing Corporation (NERAMAC), SIMFED, Sikkim, etc., are working, however, the farmers are yet to receive good prices for the produce. In such a situation, farmers' co-operative organization could be very helpful towards price realization and price fixation.

16.8.3 PESTS AND DISEASES

Unpredictable rainfall patterns, drying up of springs and streams, species migration to higher elevations, shift in crop sowing and harvesting time, emergence of invasive species, and increase in diseases/pests outbreak in crops are some climate change (CC) impacts (Sharma et al., 2009). Due to the changes in climate, large cardamom in the region has been impacted by leaf eating caterpillar (e.g., *Clelea* spp.), borer (*Glyphepterix* spp.), leaf eating beetles and weevils (*Chrysomela chlorine*), aphids (*Pentalonia nigronervosa* Cock.), storage pests (*Lepidoptera*), root grubs (*Polyphagus* white grubs), minor pests (*Kolla mimica*), nematodes (*Meloidogyne* spp.), rodents and mammals (monkey) are common (Rao et al., 1993b).

Amongst them, leaf caterpillar (*Artona chorista* Jordon) and stem borer (*Glyphipterix* spp.) are the main pests infecting the leaves and stems, respectively. Infestation by white grubs (*Anomala spp.*) and aphids are some other limiting factors. Aphids are considered rather more infectious as compared to the others cause *Chirke* and *Foorkey*. The spread of fungal disease *Colletotrichum* blight begins by approaching pre-monsoon showers (April-May) and progresses rapidly during the rainy season (June-August) that hampered plantation loss in 10 years (Sharma & Dhakal, 2011). Likewise *Chirke* and *Foorkey* have drastically reduced the production (60%) and plantation area (almost 50%) of large cardamom in Sikkim Himalaya during the starting of eighties (Sharma & Dhakal, 2011). *Chirkey* is identified as spot lesions that appear on the leaves whereas *Foorkey* claims multiple stunted growths of the tillers/shoots. The viral infection in large cardamom was first studied by Raychaudhuri & Chatterjee (1958). The *Foorkey* disease virus is recognized as Nanovirus under family Nanoviridae (Mandal et al., 2004). A new virus (LCCV) of large cardamom has been characterized by Mandal et al. (2012). Due to rapid spread of *Chirke* and *Foorkey* in the Sikkim Himalaya, the crop has been immensely damaged causing a huge gap in farmers' income since early eighties (Sharma et al., 2016a, b).

16.8.4 MANAGEMENT CONSTRAINT

The large cardamom cultivation is in continuous decline due to problems related to crop management, include: (i) lack of awareness, improper infrastructure, lack of incentives mechanism for replantation; (ii) impediments in gap filling of older plantation; (iii) inadequacies in dissemination of research and development activities and scientific backups in terms of improved varieties (disease resistant and HYVs); and (iv) lack of pollination management and post harvesting innovations. Additionally, the progressive shift from traditional commodities to other farm produce and adoption of new income generation activities also caused serious setbacks to large cardamom farming (Spices Board of India). Another major constraint for its production is the lack of extension, infrastructure, and research support which include lack of active support from the State Horticulture and Agriculture Departments in order to make available quality planting materials and providing knowledge on better post harvest practices. Further, inadequate infrastructure for quality testing, packing, and storage and unavailability of organic certification facilities.

16.8.5 PRODUCTION CONSTRAINT

Change in land use (LU) patterns, agricultural practices and landscape transitions, traditionally managed agro-ecosystem across the region is severely affected. The system has been impacted with the ongoing transformations which lead them to homogeneity in the system. The changes and practices towards the expansion of agricultural area and productivity have also resulted into many unforeseen implications on ecosystem functioning as well as in livelihood of the people. In the process, the systems become feeble with diminished ecosystem resilience, making them more prone to diseases and pests. The major factors impacting the large cardamom farming include: (i) unaltered old cardamom plantations of 1980s (15,000 ha, 50%) resulting to low productivity; (ii) diversification of agricultural system; (iii) decline in pollinator abundance (diversity and density), altered flowering period, change in climatic patterns; and (iv) other natural factors namely soil erosion, draft, hailstorms, snowfall in higher altitudes, etc. (Negi et al., 2018; Gaira et al., 2016).

16.8.6 CHANGING LIVELIHOOD OPTIONS

The ability of cardamom based agroforestry system to sustain and to meet the growing demands could be addressed with the optimization of land for availability of regular incomes. In the changing scenario of modern economic structures, farmers are shifting their reliance on traditional commodities and opting for improved options of income generation. It includes: (i) technology-driven changing farming practices and production of seasonal commodities, other spice crops namely turmeric, ginger plantation, etc.; and (ii) non-agricultural occupations such as merchandise shops, hotels/home-stays/restaurants, salaried jobs mainly in private sectors and village tourism.

16.8.7 POLLINATOR CONSTRAINTS

Large cardamom possesses zygomorphic bisexual flowers with heterostylic (pin type) nature and essentially a cross-pollinated (allogamous) crop. The most common insect visitors to large cardamom are bumble bees (*Bombus braviceps* Smith and *B. haemorrhoidalis*), honey bee (*Apis cerena*), fruit fly (*Bactocera spp.*) and skipper (*Udaspes folus*). However, as far as the pollination is concerned only bumble bee is considered as effective pollinator of large cardamom while others are just pollen robbers (Gaira et al., 2016). The act of pollen robbery by visiting insects, especially, *A. cerana* brought out competition for the resources and often deprived the plant with pollen availability causing pollination syndrome (Negi et al., 2018). In support, the impact of pollinator abundance in large cardamom was studied by Gaira et al.

(2016) and portrays that the bumble bee visits resulted to more yield of crop (i.e., 17–41 gm/plant) and the increasing abundance of all bees including honey bee (21–41 gm/plant) was significant.

The highly specialized mode of pollination in large cardamom (Bumble bees mediated pollination) makes the crop highly vulnerable as most of the plants remained un-pollinated due to low pollinator density. The diversity of pollinator guilds hardly impacts the pollination efficiency in large cardamom rather hampers the process by competing for the same resources (Palatty & Shivanna, 2007). Pandey et al. (2019) demonstrated a model to ensure year-round availability of foraging resources to maintain the pollinator density in agriculture farms. The traditional approach of promoting foraging resources including planting and maintenance of several flowering plants in large cardamom fields essential for promotion of pollinators (Gaira et al., 2015).

Realizing the importance of effective pollination in large cardamom for increasing the yield, there is a need to promote pollination services but the knowledge base on pollination aspects is quite lower among the farmers (Gaira et al., 2015; Pandey et al., 2019), only 30–40% of rural population is aware about the pollinators and pollination process. The diversity of pollinators known to farmers is bumble bee, honey bee, stingless bee and *Trigona* spp. In addition, traditional bee keeping practices in log hives depicts that significant decline in pollination services and resource dearth for pollinators. Reduction in pollinator's population is either due to apathy of farmers towards bee keeping or due to intensive agriculture (industrial agriculture) techniques (Negi et al., 2018).

16.9 CONCLUSION

Sikkim Himalayan agroforestry system is highly significant in terms of preserving agro-biodiversity including the large cardamom. Agroforestry based large cardamom cultivation is a socio-culturally managed agriculture system of the Himalaya. Increasing trend in large cardamom production can be achieved through sustainable management mechanisms in cultivation, post-harvest technology and capacity building and institutional support. Quality of the product is important in order to fetch higher prices in global market. There is necessity of forming greater Farmers' Federations so that control on price could be maintained by the Federation itself. Associated agencies should support farmers for the promotion of large cardamom. Diseases must be tackled skillfully which are causing decline in cardamom productions. Farmers must be trained and equipped for proper cultivation of large cardamom. Pollination management may provide fruitful gain in the year-round availability of the pollinations by sustaining them during the dearth period which may result in increased production. If there is a price hike then productions may increase through building trusts on farm productions. It is commonly used in food stuffs, beverages, and medicines. However, over past several years its production is in declining trend that has largely affected livelihood of the farmers cultivating this niche crop. The improved post-harvest management approaches and development of value chain will ensure sustainability of the crop. The postharvest includes scientifically refined traditional methods followed during several post-harvest steps including marketing. The value chain system involves growers, collectors, traders, and exporters. There is need to provide the growers with the necessary training on scientific methods of cultivation, identification, and control of insects-pests and disease incidence at an early stage. In addition to all, farmers need to be trained in post-harvest handlings such as in sorting, grading, and curing for checking post-harvest losses. Along with the proper cultivation methods, it is very essential to raise awareness of the farmers about the up to date market information, technology, and technical knowhow through various means in broadcasting in local radios and television channels. It is imperative to develop sustainable and best suitable storage, improved drying facilities, better transportation facilities and road network; communication means and promotes cooperative methods of collection and marketing among the growers in order to reduce marketing cost of the producers. To increase the production and productivity of the crop a functional network of stakeholders has to be established.

KEYWORDS

- **agroforestry**
- **functional network**
- **large cardamom**
- **livelihood**
- **Sikkim Himalayas**
- **stakeholders**
- **value addition**

REFERENCES

ACCESO, (2011). *The World Market for Cardamom*. Market Survey #02. Product for the USAID.

Anonymous, (1950). *Amomum. Garden Bulletin, 13*, 192–214.

Avasthe, R. K., Singh, K. K., & Tomar, J. M. S., (2011). Large cardamom (*Amomum subulatum* Roxb.) based agroforestry systems production, resource conservation and livelihood security in the Sikkim Himalaya. *Indian J. Soil Conserve., 39*(2), 155–160.

Biswas, A. K., Gupta, R. K., & Bhutia, D. T., (1986). Characteristics of different plant parts of large cardamom. *Cardamom, 19*(2), 7–11.

Chempakam, B., & Sindhu, S., (2008). Large cardamom. In: Parthasarathy, V. A., Vhempakam, B., & Zachariah, T. J., (eds.), *Chemistry of Spices* (pp. 59–69). CAB International. doi.10.1079/9781845934057.0059.

Datta, H. S., Sharma, G., & Bora, S. S., (2020). Geographical indicators in horticulture: North East India perspective. *Int. J. Curr. Microbiol. App. Sci., 9*(1), 1207–1221.

Gaira, K. S., Rawal, R. S., & Singh, K. K., (2016). Variations in pollinator density and impacts on large cardamom (*Amomum subulatum* Roxb.) crop yield in Sikkim Himalaya, India. *Journal of Asia-Pacific Biodiversity, 9*, 17–21.

Gaira, K. S., Singh, K. K., Rai, L. K., Rawal, R. S., & Dhyani, P. P., (2015). *Large Cardamom Farming in Sikkim-Pollination Needs*. G.B. Pant Institute of Himalayan Environment & Development (GBPIHED), Kosi-Katarmal, Almora, India.

Gautam, N., Bhattarai, R. R., Khanal, S. B. K., & Oli, P., (2016). Technology, chemistry, and bioactive properties of large cardamom (*Amomum subulatum* Roxb.) An overview. *International J. Appl. Sci. Biotech., 4*(2), 139–149. doi: 10.3126/ijasbt.v4i2.15104.

Kafle, B., (2013). *Cardamom Farming in Nepal*. http://kaflebishal.blogspot.com/2013/03/cardamom-farming-in-nepal.html (accessed on 28 December 2023).

Kishore, K., (2018). Geographical indicators in horticulture: An Indian perspective. *J. Int. Prop. Rights, 23*, 159–166.

Mandal, B., Mandal, S., Pun, K. B., & Verma, A., (2004). First report of association of nanovirus with *Foorkey* disease of large cardamom in India. *Plant Dis., 88,* 428. doi: 10.1094/PDIS.2004.88.4.428A.

Mandal, B., Vijayanandraj, S., Shilpi, S., Pun, K. B., Singh, V., Pant, R. P., Jain, R. K., et al., (2012). Diseases distribution and characterization of new *Maclura* virus associated with *chirke* disease of large cardamom. *Ann. Appl. Biol., 160*, 225–236.

MoAC, (2010). *Summer Crop Nepal 2009/2010: Crop and Food Security Update* (p. 20). Ministry of Agriculture and Cooperatives, Government of Nepal and World Food Program, Nepal.

Naik, J. P., Jangan, M. R. L., Mohan, K. T. M., & Sampathu, S. R., (2004). Chemical composition of the volatile oil from the pericarp (husk) of large cardamom (*Amomum subulatum* Roxb.). *Flavor Fragr. J., 19*(5), 441–444. doi: 10.1002/ffj.1336.

Negi, B. K., Joshi, R. K., & Pandey, A., (2018). Status of large cardamom (*Amomum Subulatum* Roxb.) farming systems in the changing scenario of modern economics of Sikkim, Himalaya. *Global J. of Biosci and Biotech., 7*(1), 189–199.

Palatty, A. S., & Shivanna, K. R., (2007). Pollination biology of large cardamom (*Amomum subulatum*). *Curr. Sci., 93*, 548–552.

Pandey, A., Gaira, K. S., Chettri, G., Chettri, S., Sharma, P., Lepcha, J., Joshi, R., & Chettri, N., (2020). *Proceedings of the Workshop cum Brain Storming on Post Harvest and Value Chain Management of Large Cardamom in the Khangchendzonga Landscape*. NIHE.

Pandey, A., Joshi, R. K., & Negi, B. K., (2019). Winter season bloomer hairy *Bergenia ciliata* (Haw.) sternb. (*Saxifragales: Saxifragaceae*), an important winter forage for diverse insect groups. *Journal of Threatened Taxa, 11*(7), 13937–13940.

Pathak, A., (2008). Cultivation of large cardamom in Sikkim. *Ishani, Bimonthly J. Mahatma Gandhi Ishani, Foundation, 2*(6), 1–9.

Pothula, A. K., & Singh, A. I., (2013). Postharvest processing of large cardamom in the Eastern Himalaya. *M. Res. Dev., 33*(4), 453–462. doi: 10.1659/MRD/-Journal-D-12-00069.1.

Rao, Y. S., Gupta, U., Kumar, A., & Naidu, R., (1990). Phenotypic variability in large cardamom. In: *Proc. Nat. Symp. New Trends in Crop Improvement of Perennial Species*. Kottayam.

Rao, Y. S., Gupta, U., Kumar, A., & Naidu, R., (1993b). A note on large cardamom (*Amomum subulatum* Roxb.) Germplasm collection. *Journal of Spices and Aromatic Crops*, *2*(1, 2), 77–80.

Rao, Y. S., Kumar, A., Chateerjee, S., Naidu, R., & George, C. K., (1993a). Large cardamom (*A. subulatum* Roxb.): A review. *Journal of Spices and Aromatic Crops*, *2*(1, 2), 1–15.

Raychaudhuri, S. P., & Chatterjee, S. N., (1958). A preliminary note on the occurrence of a new virus disease of large cardamom (Roxb.), in Darjeeling Dist. In: *Proc. Mycological Research Worker's Conference, ICAR*.

Sharma, G., & Dhakal, T., (2011). Opportunities and challenges of the globally important traditional agriculture heritage systems of the Sikkim Himalaya. In: Arrawatia, M. L., & Tambe, S., (eds.), *Biodiversity of Sikkim: Exploring and Conserving a Global Hotspot* (pp. 381–403). Gangtok, Sikkim: Information and Public Relations Department, Government of Sikkim.

Sharma, G., Hunsderfer, B., & Singh, K. K., (2916b). Comparative analysis on the socio-ecological and economic potentials of traditional agroforestry systems in the Sikkim Himalaya. *Trop. Ecol., 57*(4), 751–764.

Sharma, G., Partap, U., Sharma, E., Rasul, G., & Awasthe, R. K., (2016a). *Agrobiodiversity in the Sikkim Himalaya: Sociocultural Significance, Status, Practices, and Challenges* (p. 64). ICIMOD working paper 2016/5 Kathmandu: ICIMOD.

Sharma, G., Sharma, R., & Sharma, E., (2009). Traditional knowledge system in large cardamom farming: Biophysical and management diversity in Indian mountainous regions. *Indian J. of Tradi. Know., 8*(1), 17–22.

Sharma, M., & Rajan, S., (2018). Improving GIs significance in fruit crops for doubling farmer's income. *J. Pharm. Phytochem., 7*(2), 1903–1905.

Sharma, R., Jianchu, X., & Sharma, G., (2007). Traditional agroforestry in the Eastern Himalayan region: Land management system supporting ecosystem services. *Tropical Ecology, 48*(2), 189–200.

Singh, K. A., Rai, R. N., Patira, & Bhutia, D. T., (1989). Large cardamom (*Amomum subulatum* roxb.) plantation: An age-old agroforestry system in Eastern Himalaya. *Agroforestry Systems, 9*, 241–257.

Singh, K. K., Krishna, A. P., & Chettri, S., (2005). Large cardamom agroforestry based sustenance in the fringe areas of *Khangchendzonga* biosphere reserve, Sikkim Himalaya. In: Bhatt, B. P., & Bujarbaruah, K. M., (eds.), *Agroforestry in North East India:*

Opportunities and Challenges, (pp. 71–78). ICAR Research Complex for NEH Region, Umiam, Meghalaya.

Singh, K. K., Krishna, A. P., & Sharma, E., (2000). Effects of altitudes and shade-tree types on large cardamom, nitrogen, and spectral properties in the Sikkim Himalaya. *Int. J. Ecol. Environ. Sci., 26*, 139–147.

Spices Board of India, (2011/2020). *Spices Board Statistics*. Spices Board, Kochi, India.

Subba, J. R., (1984). *Agriculture in the Hills of Sikkim* (p. 299). Sikkim Science Society.

Vijayan, A. K., Bhat, S. K., Gudade, B. A., Bora, S. S., Mathew, Mary, K., & Remashree, A. B., (2018). Crop cultivars and varieties with emphasis on nursery propagation of large cardamom (*Amomum subulatum* Roxb.) in India. *International Journal of Agriculture Sciences, 10*(20), 73647368. https://www.bioinfopublication.org/jouarchive.php?opt=&jouid=BPJ0000217 (accessed on 28 December 2023).

CHAPTER 17

Traditional Knowledge, Ethnic Foods, and Their Conservation of Indigenous Lepcha Community of Darjeeling and Sikkim Himalaya

GEETAMANI CHHETRI, ASEESH PANDEY, KAILASH S. GAIRA, and RAJESH JOSHI

G. B. Pant National Institute of Himalayan Environment, Sikkim Regional Center, Pangthnag, Gangtok, Sikkim, India

ABSTRACT

The Lepchas are one of the most primeval indigenous tribal communities dwelling in the foothills of the Mt. Kangchenjunga since time immemorial. They have own distinctive culture, religion, and language and hence have their unique ethnic identity in the region. Originally, they were nomads and lived in herding, hunting, fishing, as well as slash and burn agriculture system. Thus, they have got substantial knowledge about wild resources around them and utilize these resources efficiently in their day today life for the sustenance. Lepchas have unique traditional crafts and fabrics; their traditional head gear "*Sumok thyaktuk*" made up of local canes and bamboos is one of excellent and oldest craft forms of the tribe to be worth mentioning. This community specific knowledge is restricted within the Lepchas. Likewise, they have got good knowledge about local medicinal plants to treat common diseases/ailments of the community and sustainable utilization of the natural resources is evident in their culture. Therefore, conservation and promotion of such nature based indigenous traditions can be achieved by encouraging younger generations to take up traditional practices as means of their livelihood. This may be further strengthened by organizing capacity

Sustainable Management and Conservation of Environmental Resources in India.
Sumit Chakravarty, Vineeta, Jahangeer A. Bhat, Munesh Kumar, Gopal Shukla (Eds.)

building programs for younger generations by value additions with technological intervention leading towards sustainable livelihood of the community with the involvement of government and non-government organizations.

17.1 INTRODUCTION

The Lepchas are indigenous tribe living in the Himalayas on the southern and eastern slopes of Mount Kangchenjunga having their own unique indigenous religion, language, culture, and custom. They are considered as aboriginal inhabitants of this large tract of mountainous land stretches from Ilam area of Nepal to hilly areas of West Bengal (Darjeeling and Kalimpong districts), Sikkim, Chumbi valley of Tibet, Har Chu valley and Ammo Chu valley of Bhutan. Thus, they extend from mid-Bhutan on east to central Nepal on the west. However, the origin of Lepcha tribe is a topic of controversy. Historians and anthropologists are still debating whether the tribe had originated from Nagas or were associated with the Jimdars and Mech in their eastward migration from somewhere in central Nepal whereas, some scholars have also found similarities between Lepchas and tribes of Arunachal Pradesh. Some others believe that they are related to the Khasis of Meghalaya (Sharma, 2013). According to some historians they are said to be of Tibeto-Burmese and Chinese origin and some modern scholars relate them with Japanese. Some even believe that the place of Lepcha origin is on the holy mountain Kailash (*Tise*) way up in Tibet. However, Lepchas refer to their homeland as "*ne mayel lyang*" (hidden paradise) or "*ne mayel maluk lyang*" (land of eternal purity). The Lepchas call themselves "*mutanchi rongkup rumkup*" meaning children of Mother of creation and Almighty God. The name "Lepcha" has been derived from Nepali word *Lapche* meaning "vile speakers." They call themselves *Rong* or the 'people of the valley.' Tibetans call them *Mon* and Bhutanese *Meri*. Their tradition however, states that their original home was in the neighborhood of the great mountain, *Kang-chen-mdzod-lnga* (Tibetan name for Mt. Kangchenjunga) means 'Great Glacier of the Five Store-Rooms. The Lepcha name for the mountain is '*King-tzum-song-bu'* means highest over our head (Subba, 2008). There is evidence to depict that they have lived in this region since early times, as most of the mountains, rivers, and other prominent places in the region have Lepcha names and have narrations in Lepcha folklore. In the Census 2001, the population of Lepchas in Sikkim was 40,000; West Bengal was 33,000 and Ilam was 3000 (Lepcha, 2013). As per census 2011, total Lepcha population in Sikkim and northern part of West Bengal (Darjeeling and Kalimpong districts) is 76,871. The decadal population growth rate indicated was low (i.e., 5.3%) as compared to the national decadal population growth rate of 17.64% (Census of India, 2001, 2011).

The Lepcha divide themselves into four groups according to the region they inhabit. The Lepcha from Sikkim are called *Renjóngmú*. The Lepcha from Kalimpong, Kurseong, Mirik, and Darjeeling in West Bengal are known as *Támsángmú*. There is some debate over whether the Lepcha from Kurseong, Darjeeling, and Mirik should belong to the *Renjóngmú* or the *Támsángmú*, as some people use the name *Támsángmú* strictly for Lepcha living in and around Kalimpong. Lepcha living in Ilam district of eastern Nepal are known as *Ilammu* and in Bhutan are referred to as *Promu* (Plaisier, 2005). In India, Lepchas are concentrated in Sikkim and northern part of West Bengal, i.e., Darjeeling and Kalimpong districts. In Sikkim, higher concentration of Lepcha population is found in Dzongu, Mangan, and Kabi-Tingda in North district of Sikkim. The Dzongu area has been treated as Lepcha reserve and lies in the lower ranges of Kanchenjunga where outsiders are not allowed to settle. In East district Lepchas are found in Ragdong-Tintek, Rumtek, Ranka, and Assam Lingjey; in South district they are concentrated in Wok and Rateypani area; whereas in West district Lepchas have settled in Rinchengpong, Daramdin, Dentam, and Tashiding areas (Subba, 2008). In Kalimpong district, Lepchas are found to settle in villages of Bong busty, Gitdabling, Sindeybung, Nassey busty, Lingsey, Lingsekha, and Tandrabong. In Darjeeling district, they are found in villages of Maney Dara-Singritam, Lower, and Middle Lingding, Chegra, Lanku, Rolak, and Sittong areas.

They are mongoloid in appearance with oblique eyes, short in stature and fair in complexion. They are amiable, cheerful, hospitable, shy, good humored, sociable, and polite and peace-loving people. They were basically hunters and agriculturalist and herdsmen. In past they used to take large herds of cattle in search of good pastures hence they used to live nomadic lives. Their traditional method of cultivation was slash and burn or shifting cultivation. Later, during mid-19th-century they started practicing settled agriculture (Das, 1978). They lived in hunting, fishing, trade, and later agriculture; paddy, oranges, cardamom are their preferred crops. Being the forest dwellers since antiquity, Lepchas have developed an intrinsic relationship with nature, and they live in perfect harmony with it (Singh & Chakraborty, 2014). Lepchas are well acquainted about wild resources and its uses therefore they are mostly dependent on wild edibles and are rich in traditional knowledge of wild plant resources as food, medicines, and other livelihood uses (Chhetri & Rai, 2018). Realizing the importance of the Lepcha community in the Himalayan ecosystem, there is an urgent need to improve the understanding about the indigenous Lepcha community and their traditional livelihood and bio-cultural interactions. In this context, the present chapter is an attempt to highlight the traditional practices of the Lepcha community in the Kangchenjunga landscape (India).

17.2 MATERIALS AND METHODS

Lepcha dominated villages in far flung areas of Darjeeling and Kalimpong districts of West Bengal and Dzongu region of Sikkim were chosen in order to document traditional practices of the community. Traditional life style of the community still prevails in pure form in these remote villages. A total of seven Lepcha villages covering 30–40% of population from each village were surveyed. The information on Lepcha traditional knowledge system was gathered through household survey with the help of structured questionnaire, semi-directive interviews and group discussion with elderly persons having rich indigenous knowledge (IK) as well as other knowledgeable persons including Lepcha medicine men '*Maondaoks*' and Priest-cum-healers '*Boongthings.*' The questionnaire covered various traditional aspects of the community including bioresource utilization and its conservation, rituals, folk traditions, ethnic foods and other ethno-biological information. Mostly elderly respondents were chosen for interview intending that they have acquired adequate traditional knowledge which has been passed on through generations. Accordingly, in our survey, 65% of total respondents were aged above 50 years. Effort was also made to gather information on uses of local plant species in preparing traditional dishes and preparation details especially from elderly women of the community as they do acquire such knowledge through years of trial-and-error methods, experiences, and some obtain from their elders as well. The gathered information was verified by consulting available literature from the region (Tamsang, 2004; Pradhan & Badola, 2008; Sharma, 2013; Chhetri & Rai, 2018; Pandey et al., 2019, 2020; Sinha et al., 2019) and also crosschecked among different surveyed villages to examine whether their practices diverge. Correct botanical names of the plant species used by the tribe were updated using online web portals like http://www.tropicos.org and http://www.plantsoftheworldonline.org.

17.3 RESULTS AND DISCUSSION

17.3.1 CRAFTS AND FABRICS

Bamboos are used commonly and exquisitely by the Lepchas for various purposes ranging from house and bridge construction, articles of routine requirement to artifacts, water distribution network, utensils, musical instruments, hunting devices and food since their existence in the foothills of the Himalayas. Such bamboo technology can be adopted from this community (Pradhan & Badola, 2008). The spectacular variety and diverse range of the bamboo crafts demonstrate their expertise as well as the traditional knowledge. Lepcha traditional hat or head gear "*Sumok thyaktuk*" is recognized as one of their best and oldest craft forms. This knowledge is essentially community specific and restricted within the Lepcha community. In ancient times they used to wear own woven clothes from yarn out of nettle plants *Girardinia diversifolia* or *Kuzoo* (Pandey et al., 2020). Now-a-days cotton, woolen, and silk yarn is used by the community to weave clothes using vegetable dyes and modern synthetic colors as well. Lepcha women play a predominant role in handloom weaving from their young ages. Numerous items such as shoulder bags, table mats, ladies' purse and cushion cover were also made beside traditional dresses from this woven fabric. The use of nettle bark for cloth and bags weaving by the primitive Lepchas is an excellent example of their skill (Pandey et al., 2020). In order to strengthen this skill of crafts and fabrics, trainings on value addition to their bamboo crafts and nettle fiber products were organized in Dzongu area of Sikkim, intending to provide them good livelihood opportunity (Pandey et al., 2019, 2020).

17.3.2 MEDICINAL PLANTS

Lepchas have enormous understanding and knowledge on the use of plants. Their herbal traditional medicinal treatment to various ailments is quite prominent and so far, there is no evidence of epidemic in Lepcha history (Roy, 2011). Lepcha medicine man '*Maon-dok*' and priest-cum-healer '*Boongthing*' possess huge knowledge on the use of plants for curing various ailments, though senior members of the family can also treat common ailments like cold, cough, indigestion, dysentery, diarrhea, skin diseases, jaundice, cuts, and wounds using the surrounding locally available herbs. Health management system of Lepchas is purely nature based. The Lepchas have got adequate knowledge on medicinal uses of local plants which are found to grow around their dwellings, farmlands, and wastelands. During such studies it was found that, out of 75 species of medicinal plants reported to be used by the tribe from Kalimpong district of West Bengal (Table 17.1), 72% were found to be wild (Chhetri & Rai, 2018).

TABLE 17.1 List of Medicinal Plants Used by Lepcha Indigenous Community

Sl. No.	Botanical Name with Family	Local Names	PF	Parts Used	Uses	C/W
1.	*Abelmoschus esculentus* (L.) Moench; Malvaceae	*Tarya bee* (L) *Bhindi* (N)	H	Seeds, leaves	Seeds used as purgative and leaves juice used in conjunctivitis.	C
2.	*Acorus calamus* L.; Acoraceae	*Ruk laop* (L) *Bojo* (N)	H	Rhizome	Grinded paste used to cure skin disease and taken orally to relive bowel complaints.	W
3.	*Albizia* sp.; Mimosaceae	*Tuk chyer Koong* (L) *Siris* (N)	T	Leaves, bark, seeds	Bark and seeds used against piles and diarrhea. Powdered bark of roots used to strengthen gums; juice of young leaves applied on eyes to cure night blindness.	W
4.	*Allium cepa* L.; Liliaceae	*Munggu* (L) *Piyaz* (N)	H	Bulb	Stimulant and expectorant, juice of bulbs applied over bee sting to relieve pain.	C
5.	*Allium sativum* L.; Liliaceae	*Oo Tsong/Payk* (L) *Lasun* (N)	H	Bulb	Taken in gastritis, indigestion, and high-altitude sickness. Juice used in skin diseases and as ear drops in ear ache, mixed with water and sprinkled around houses to drive out snakes, scorpions, and other insects.	C
6.	*Aloe vera* (L.) Burm.f; Liliaceae	*More* (L) *Ghiu Kumari* (N)	H	Leaves	Jelly pulp from leaves applied over burnt skin and taken orally during stomach problem and gastritis.	C
7.	*Alstonia scholaris* (L.) R.Br.; Apocynaceae	*Puru Koong* (L) *Chatiwan* (N)	T	Bark and root	Bark paste used in skin diseases and rheumatism. Juice extracted from root mixed with milk taken orally to cure leprosy.	W
8.	*Amomum subulatum* Roxb.; Zingiberaceae	*Lenzee* (L) *Alaichi* (N)	H	Seeds and roots	Decoction of seeds used in teeth and gum infection. Powdered roots are used to cure urinary diseases in cattle.	C
9.	*Ananas comosus* (L.) Merr.; Bromeliaceae	*Kaong Chyey* (L) *Bhui katahar* (N)	H	Leaves and fruits	Leaf juice used for expelling intestinal worms of children. Fruit juice to cure swollen and bleeding gums, fruits taken for improving digestion.	C

TABLE 17.1 *(Continued)*

Sl. No.	Botanical Name with Family	Local Names	PF	Parts Used	Uses	C/W
10.	*Ageratum conyzoides* (L.) L.; Asteraceae	*Nambyu* (L) *Elamey* (N)	H	Leaves	Juice applied on cuts and wounds to stop bleeding.	W
11.	*Artemisia vulgaris* L.; Asteraceae	*Tuk nyil* (L) *Titey pati* (N)	S	Leaves	Crushed and rubbed on diseased or infected areas of skin. Juice used in nose bleeding. Decoction of young shoot taken to increase appetite and promote digestion. Fumigation of leaves used for repelling insects.	W
12.	*Bauhinia variegata* L.; Fabaceae	*Kachyik Koong* (L) *Taaki* (N)	T	Roots, leaves, barks, flowers, and flower buds	Paste made of bark used in ulcers, inflammation of joints and skin disease. Dried flower buds taken orally in diarrhea, dysentery, and piles. Tender leaves and young shoots eaten as vegetables for maintaining good health.	W
13.	*Bombax ceiba* L.; Bombacaceae	*Tunglu koong/ Singlu* (L) *Simal* (N)	T	Gum and roots	Gum extracts used in dysentery and young root in gonorrhea.	W
14.	*Buddleja asiatica* Lour.; Scrophulariaceae	*Pondam Koong* (L) *Bhimsen pati* (N)	T	Leaves	Leaf paste applied on skin diseases.	W
15.	*Callicarpa arborea* Roxb.; Lamiaceae	*Sa-ngaa Kung* (L) *Guyelo* (N)	T	Bark	Bark crushed and soak in water and liquid taken to cure pneumonia.	W
16.	*Carica papaya* L.; Caricaceae	*Naaroo Paot* (L) *Mewa* (N)	H	Fruits	Taken ripe or unripe matured fruits for better digestion.	C
17.	*Centella asiatica* (L.) Urban; Apiaceae	*Kahley Nyaok* (L) *Gol patta* (N)	H	Leaves	Chewed raw during sore throat and lungs infection.	W
18.	*Cheilocostus speciosus* (J.Koenig) C.D. Specht; Costaceae	*Ruyang* (L) *Betlauri* (N)	H	Rhizome and stems	Decoction prepared from rhizome taken against burning sensation during urination and juice extracted from stem taken to cure jaundice.	W

TABLE 17.1 *(Continued)*

Sl. No.	Botanical Name with Family	Local Names	PF	Parts Used	Uses	C/W
19.	*Cinnamomum glaucescens* (Nees) Hand. Mazz.; Lauraceae	*Ruhun Koong* (L) *Malagiri* (N)	T	Bark and leaves	Bark used to cure gonorrhea and leaves paste used in rheumatism.	W
20.	*Citrus medica* L.; Rutaceae	*Kachyo Kung* (L) *Bimbira* (N)	T	Fruits	Taken orally for better digestion. Dried rind powder taken during indigestion and dysentery.	W
21.	*Clematis buchananiana* DC.; Ranunculaceae	*Chyeelaok rik* (L) *Pinasey Lahara* (N)	Cl	Roots, leaves	Root juice taken for curing headache and leaves burnt and inhale to cure sinus.	W
22.	*Colocasia esculenta* (L.) Schott; Araceae	*Sungte/Pazaok luktuk* (L) *Pindalu* (N)	H	Tuber/corm	Crushed and cooked with butter and taken for curing tuberculosis.	W
23.	*Coriandrum sativum* L.; Apiaceae	*Oo Su* (L) *Dhaniya* (N)	H	Leaves	Fresh leaves taken orally in gastritis, helps in digestion.	C
24.	*Cucumis sativus* L.; Cucurbitaceae	*Saret* (L) *Kakra* (N)	Cl	Fruits and seeds	Taken raw against urinary trouble.	C
25.	*Cucurbita pepo* L.; Cucurbitaceae	*Tung zaong* (L) *Pharsi* (N)	Cl	Seeds and leaves	Seeds taken orally to kill intestinal worms; leaves juice applied externally over burnt skin.	C
26.	*Curcuma caesia* Roxb.; Zingiberaceae	*Gey sying* (L) *Kalo hardi* (N)	H	Rhizome	Chewed fresh or dried rhizome orally against stomach troubles and to get relieve from flatulence.	W
27.	*Curcuma longa* L.; Zingiberaceae	*Mung gaa* (L) *Hardi* (N)	H	Rhizome	Paste made of rhizome applied over sprains and wounds. Decoction taken against cough and cold. Fresh paste used as antiseptic on skin infection.	C
28.	*Cynodon dactylon* (L.) Pers.; Poaceae	*Paong mook* (L) *Dubo* (N)	H	Whole plant	Fresh roots taken orally in piles. Juice of aerial parts applied over cuts and wounds.	W

TABLE 17.1 *(Continued)*

Sl. No.	Botanical Name with Family	Local Names	PF	Parts Used	Uses	C/W
29.	*Datura metel* L.; Solanaceae	*Rinchen Nyongboou* (L) *Dhatura* (N)	S	Roots, leaves, and seeds	Leaves paste used in rheumatic swelling, lumbago, sciatica, and inflammation. Seeds taken orally as anti-rabies during dog bites and its smoke inhaled against asthmatic troubles.	W
30.	*Dichroa febrifuga* Lour.; Hydrangeaceae	*Geybu Khanaok* (L) *Basak* (N)	S	Root	Root decoction used in fever and malaria.	W
31.	*Diploknema butyracea* (Roxb.) Lam.; Sapotaceae	*Yel Paot* (L) *Chiwri* (N)	T	Seed	Oil extracted from seeds used as ointment for rheumatism. Used to get rid of seasonal damage/wrinkles during winter.	W
32.	*Drymaria cordata* (L.) Willd. ex Schult.; Caryophyllaceae	*Eayok Pundim* (L) *Abijalo* (N)	H	Aerial part	Smoke by burning aerial parts inhaled during headache and chronic sinus problem. Juice extract taken orally for curing diphtheria and pneumonia.	W
33.	*Entada gigas* (L.) Fawc. and Rendle; Mimosaceae	*Kooluk Paot* (L) *Pangra* (N)	Cl	Bark and seeds	Bark juice used for skin diseases. Paste made of seeds used for washing hairs to remove dandruff, also applied on mumps and swelling.	W
34.	*Eryngium foetidum* L.; Apiaceae	*Seyd Oosu* (L) *Bhotey Dhaniya* (N)	H	Roots and leaves	Roots cut into pieces and soaked overnight in water and taken that water at morning in urine trouble and kidney problems. Leaves taken during stomach disorder.	W
35.	*Ficus semicordata* Buch.-Ham. ex Sm.; Moraceae	*Beesu Paot/Toksot Paot* (L) *Khasrey Khaniu* (N)	S	Fruits	Taken for better digestion. Dried rind used in dysentery.	W
36.	*Fraxinus paxiana* Lingelsh.; Oleaceae	*Payjew* (L) *Lakuri* (N)	T	Bark	Paste is applied over sprain, bone fracture and dislocation. Decoction taken orally during body pain.	W

TABLE 17.1 *(Continued)*

Sl. No.	Botanical Name with Family	Local Names	PF	Parts Used	Uses	C/W
37.	*Gynocardia odorata* R. Br.; Flacourteaceae	*Tuk Koong* (L) *Gantey* (N)	T	Seeds	Oil extracted from seeds used against leprosy, eczema, and other skin diseases and chronic rheumatism.	W
38.	*Helianthus annus* L.; Asteraceae	*Satsuk Rip* (L) *Gham Phul* (N)	H	Seeds	Taken orally in urine trouble, cough cold, bronchitis, and lung diseases.	C
39.	*Heracleum wallichii* DC.; Apiaceae	*Samben* (L) *Chimphing* (N)	H	Seeds	Fresh or dried seeds taken orally in stomach disorder and gastric troubles.	W
40.	*Hordeum vulgare* L.; Poaceae	*Kachyer* (L) *Jau/uwa* (L)	H	Seeds	Powdered grains used against indigestion. Given to sick as it is easy to digest.	C
41.	*Ipomoea batatas* (L.) Lam.; Convolvulaceae	*Moongur Book* (L) *Sakar-Khanda* (N)	H	Tuber	Boiled and taken orally in constipation.	C
42.	*Juglans regia* L.; Juglandaceae	*Kaol Koong* (L) *Okhar* (N)	T	Bark and fruits	Bark decoction use in expelling intestinal worms. Fruits used in rheumatism.	W
43.	*Justicia adhatoda* L.; Acanthaceae	*Boosyika* (L) *Asuro/Basak* (N)	S	Root, bark, leaves, flowers.	Root, bark, leaves, and flowers decoction used to remove phlegm. Flowers and leaves taken as vegetables in hypertension. Leaves juice taken in diabetes and used as antiseptic.	W
44.	*Kaempferia rotunda* L.; Zingiberaceae	*Sodong Maon* (L) *Bhui champa* (N)	H	Rhizome	Paste applied on bone fracture and rheumatism.	C
45.	*Lantana camara* L.; Verbenaceae	*Kadao rip* (L) *Kanrhe banmara* (N)	S	Petals and roots	Paste made of petals applied to cuts to stop bleeding. Root decoction taken in tuberculosis.	W
46.	*Lycopersicon esculentum* Mill.; Solanaceae	*Byooru Paot* (L) *Tamatar/ Rambhera* (N)	H	Leaves	Leaf juice applied on animals wound to extract maggots.	C

TABLE 17.1 *(Continued)*

Sl. No.	Botanical Name with Family	Local Names	PF	Parts Used	Uses	C/W
47.	*Macrotyloma uniflorum* (Lam.) Verdc.; Fabaceae	*Fyetlaasyee* (L) *Gahat* (N)	H	Seeds	Decoction taken orally to cure measles.	C
48.	*Magnolia champaca* (L.) Baill. ex Pierre; Magnoliaceae	*Gok rip* (L) *Chaap* (N)	T	Bark, flower, and fruit	Bark used in fever and cough. Flowers and fruits in nausea, fever, and urine trouble.	W
49.	*Mentha spicata* L.; Lamiaceae	*Oosu Daong* (L) *Padina* (N)	H	Leaves and young shoots.	Taken fresh against headache and cholera. Stimulant and digestive.	C
50.	*Mimosa pudica* L.; Mimosaceae	*Aaook Mook/ Tuksaam mook* (L) *Buhari jhar/ Lajjawati* (N)	H	Leaves and roots	Crushed into paste and applied on piles.	W
51.	*Momordica charantia* L.; Cucurbitaceae	*Khaaktik* (L) *Titey karela* (N)	Cl	Leaves and fruits	Leaves juice rubbed in burning of feet soles. Fruits taken in gastritis and better digestion.	C
52.	*Morus australis* Poiret; Moraceae	*Mikrap Koong* (L) *Kimbu* (N)	S	Bark and leaves.	Bark used as purgative. Leaves decoction used for gargling to cure hoarse voice and inflammation of vocal cords.	W
53.	*Oroxylum indicum* (L.) Kurz; Bignoniaceae	*Phago Rip* (L) *Totola* (N)	T	Bark and seeds	Paste made of bark applied on cuts and wounds. Bark ashes applied to cure chronic wounds. Seed paste taken in sores of throat and lungs and in typhoid. Good for liver.	W
54.	*Pandanus furcatus* Roxb.; Pandanaceae	*Bor Koong* (L) *Tarika* (N)	T	Bark	Juice prepared from bark taken against gonorrhea.	W
55.	*Persea americana* Hill.; Lauraceae	*Kafaam/Phaamsyi* Koong (L) *Phamphal* (N)	T	Fruit and bark	Fruit juice along with water taken to cure cough. Paste made of bark and leaves to stop bleeding.	W

TABLE 17.1 *(Continued)*

Sl. No.	Botanical Name with Family	Local Names	PF	Parts Used	Uses	C/W
56.	*Phyllanthus emblica* L.; Phyllanthaceae	*Saa aom Paot* (L) *Amala* (L)	T	Bark and fruits	Taken orally in diarrhea and dysentery. Fruit juice taken to cure urine trouble.	W
57.	*Piper longum* L.; Piperaceae	*Kuntim Paot* (L) *Pipla* (N)	Cl	Roots and fruits	Powdered fruits prescribed with sugar or molasses for inducing sound sleep. Fruits taken in asthma, cough, rheumatism, gonorrhea, and piles.	W
58.	*Prunus cerasoides* Buch.-Ham. ex D. Don; Rosaceae	*Kaong kee Koong* (L) *Paiyun* (N)	T	Bark	Paste made of bark applied over bone fracture.	W
59.	*Rhododendron arboreum* Sm.; Ericaceae	*Eetok Koong* (L) *Lali gurans* (N)	T	Flowers	Fresh or dried flower petals chewed to cure dysentery and diarrhea.	W
60.	*Rubia cordifolia* L.; Rubiaceae	*Vyumrik/Cho-ek* (L) *Majito* (N)	H	Root and fruits	Paste prepared from root and fruits applied over skin diseases.	W
61.	*Rubus ellipticus* Sm.; Rosaceae	*Kasyum Koong* (L) *Aiselu* (N)	S	Roots and young shoots	Decoction given to children in colic pain and for killing intestinal worms.	W
62.	*Schima wallichii* Choisy; Theaceae	*Sungbrang Koong* (L) *Chilauney* (N)	T	Bark	Paste used for curing gonorrhea.	W
63.	*Shorea robusta* Gaern.; Dipterocarpaceae	*Tuk tal Koong* (L) *Saal/Sakhuwa* (N)	T	Resin	Resin fumes used for repelling mosquitoes and other insects and fumes much used for fumigating sick person's room.	W
64.	*Stephania glabra* (Roxb,) Miers.; Menispermaceae	*Mungkyo Rik* (L) *Tamarkey* (N)	Cl	Leaves, tubers	Leaves paste used for treating boils and tubers taken against gastritis, diabetes, ulcers, and its paste applied over burnt skin.	W

TABLE 17.1 *(Continued)*

Sl. No.	Botanical Name with Family	Local Names	PF	Parts Used	Uses	C/W
65.	*Swertia chirayita* H. Karst.; Gentianaceae	*Rung Ken* (L) *Chirauto* (N)	H	Whole plant	Decoction taken against fever, acidity dyspepsia and skin diseases, also taken as liver tonic.	W
66.	*Terminalia bellerica* (Gaertn.) Roxb.; Combretaceae	*Kutnaom Paot* (L) *Barra* (N)	T	Fruits	Fresh or dried fruits taken during diarrhea, sore throat, fever, and to cure piles, good in constipation.	W
67.	*Terminalia chebula* Retz.; Combretaceae	*Salim Paot* (L) *Harra* (N)	T	Fruits	Raw fruits chewed in cough, sore throat, tonsillitis, and diabetes, good in constipation.	W
68.	*Tetradium fraxinifolium* (Hook.f.) T.G. Hartley; Rutaceae	*Kanu Paot* (L) *Khanakpa* (N)	T	Fruits	Fresh or dried fruits taken during indigestion, gastritis food poisoning and dysentery.	W
69.	*Thysanolaena latifolia* Roxb. ex Hornem.) Honda; Poaceae	*Pasyor* (L) *Amliso* (N)	H	Roots	Root paste applied around boils to rupture easily.	C
70.	*Toona ciliata* Roem.; Meliaceae	*Samaal Koong* (L) *Tuni* (N)	T	Bark	Powdered bark is astringent, used for curing ulcers.	W
71.	*Urtica parviflora* Roxb.; Urticaceae	*Kuzoo* (L) *Sisnu* (N)	H	Roots, leaves, flowers	Root paste applied on fracture and joint dislocation. Root juice taken against gonorrhea. Young leaves and flowers used as vegetables for controlling high BP.	W
72.	*Viscum articulatum* Burm. f.; Santalaceae	*Sun tee pro/ Singhthutmu* (L) *Harchur* (N)	S	Whole plant	Paste is used as poultice for treating muscular pain, bone fracture, etc. Decoction taken orally in body and muscle pain.	W
73.	*Zanthoxylum acanthopodium* DC.; Rutaceae	*Naong ryu* Paot (L) *Bokey timbur* (N)	S	Fruits and bark	Fruits and bark used as tonic in fever and cholera. Oil extracted from dry fruits used in gout, rheumatism, and toothache.	W

TABLE 17.1 *(Continued)*

Sl. No.	Botanical Name with Family	Local Names	PF	Parts Used	Uses	C/W
74.	*Zingiber officinale* Roscoe; Zingiberaceae	*Heng* (L) *Aduwa* (N)	H	Rhizome	Rhizome juice mixed with turmeric powder and honey used to cure cold, cough, and asthma; rhizome juice mixed with lemon juice taken to cure dyspepsia, taken with butter to cure rheumatism. Fresh or dried rhizome peeled off and chewed as an appetizer and to cure throat troubles.	C
75.	*Zingiber zerumbet* (L.) Roscoe ex Sm.; Zingiberaceae	*Salek* (L) *Phachyang* (N)	H	Rhizome	Fresh or dried rhizome chewed in nausea, gastritis, and stomach ache.	W

Abbreviations: L: Lepcha; N: Nepali; C: cultivated; W: wild; PF: plant form (H: herb; S: shrub; Cl: climber; and T: tree.

Source: Adapted from Chhetri & Rai (2018).

17.3.3 WILD EDIBLES

Lepchas are living with nature since time immemorial; consequently they have got significant knowledge of wild edible plants and suitable parts to consume either in raw form or after cooking, boiling, and fermenting. These edible plant parts include roots, tubers, rhizomes leaves, tender shoots, flowers, berries, etc., they also know the indigenous technique of freeing poison from some poisonous roots and make them edible (Roy, 2011). Moreover, they are also good in identifying the names and behavior of the wild animals, birds, insects, fishes, and frogs for eating purpose.

17.3.4 ETHNOECOLOGICAL KNOWLEDGE

Since antiquity Lepcha community perform their agricultural activities based on some indications from nature such as blooming of cherries and appearance of certain birds. They start sowing the seeds on arrival of certain birds believing the birds were sent to them every year to remind them when to sow the seeds. On arrival of Migratory bird '*Demoiselle Cranes' called "Kurnoak fuo*" in Lepcha, they started sowing seeds of cucumber, pumpkin, and other Kharif crops. The appearance of migratory bird indicates the onset of spring after winter which is the right time to start growing different vegetable crops. Likewise, during harvesting also they perform their activities on such indications from nature. When wild Himalayan cherry or *Prunus cerasoides* (*Kaong Kee Koong* in Lepcha), starts blooming they realize that the harvesting time of their cereal crops have come. *Prunus cerasoides* blooms during October-November in the region and by that time cereal crops shown during monsoon get ready for harvesting. In this way Lepcha people by observing a natural phenomenon like of cherry blossoms get the indication for harvesting of their cereals and pulses and started preparing for celebrating their New Year festival "Namsoong" which falls during December-January. Such IK on natural indications for performing agricultural activities used by Lepcha tribe since their existence in the foothills of the Himalayas is noteworthy to be documented intriguing their inner scientific value based on close attachment to environment, a lesson to be learned by modern society for preservation/conservation of nature.

17.3.5 ETHNIC FOOD

Lepcha people have huge knowledge of wild eatables like berries, roots, mushrooms, and veggies and hence shortage of food to the Lepcha tribe is scarce. In past, Lepchas were hunters and used to hunt down animals and birds for meat.

They used to trap fishes and toads using unique bamboo net, hence fishes were also a major part of their diet. Overall, rice, meat, and vegetables form the staple food items of the Lepcha people. Those living in the remote ends still eat roots, frogs, toads, and other plant and animal resources. They drink tea and milk as well as traditional beverage '*Chi*' prepared from finger millet, which is intoxicating, and they consider it as holy brew. With time and religious transformation, the tribe is gradually changing their food habit as well. At present times they eat fishes, meat of domestic animals like ox, buffalos, pigs, goats, and domestic fowls. Though they prepare and eat their traditional dishes occasionally, mostly during traditional rituals and festivals. Originally, being the tribe of forest, they are fond of tubers and roots. Even today, they often eat *Moongurbuk* (sweet potato) and *Kasalok* (yams) after boiling as main course (Plate 17.1).

PLATE 17.1 Some common tubers eaten by Lepchas: (a) wild yam (*Pojokbuk*); (b) sweet potato (*Moongurbuk*); (c) cassava (*Tunglubuk*); (d) sweet potato; yam (*Moongurbuk, Leebuk*); (e) taro (*Singtee*); and (f) purple yam (*Kafferbuk*).

Major traditional dishes of Lepcha tribe prepared with cereals are *Sarongbee tuluk* (rice and nettle cooked as porridge), *Cho-nyok* (rice, butter, and vegetables cooked together as porridge), *Po Nguzom* (preparation of rice, fish, and vegetables grilled in bamboo), *Khu zom* (baked wheat, millet, and maize on hot stone) and *Khu-ree* (prepared from finger millet or buckwheat flour with leafy vegetables). *Khu-ree* is now served only during special occasions like *Namsoong*, i.e., Lepcha New Year, or hosting special guests. The dish has now become special culinary attraction for the tourists in Lepcha home-stays and restaurants indicating commercial importance with socio-economic significance of Lepcha tradition.

Plate 17.2 illustrates the *Khu-ree* preparation by the Lepchas. Finger millet (*Mong*) or buck wheat (*Kushru*) is sun dried for a few days and then smashed to remove the husk and ground to make powder (*Mongtee*). The flour then mixed with a little bit of wheat flour for consistency, in a bamboo container (*Po puthut*) and mixed with water and stirred to make a slightly thick, consistent batter. A flat pan is heated (earlier in the absence of utensils flat stones were used) in fire and after the pan is heated properly a thin layer of batter is spread over to bake g a round shaped bread. Now-a-days butter is used to add flavor and texture to the bread. Both sides of the bread are properly cooked in a low flame and kept aside. In the meantime, locally available fresh wild green vegetables especially the stalk less *Elatostema* spp. (*Kanchel Bee, Kamchol Bee*) is cooked along with the crushed *shiso* seeds, i.e., *Perilla frutescens* (*Nuhum*) for about 15 minutes then properly cooked vegetables wrapped in already cooked finger millet or buckwheat bread and served with chutney (Sinha et al., 2019).

PLATE 17.2 Khu-Ree preparations.

Source: © Sinha et al. (2019).

17.3.6 TRADITIONAL BEVERAGES

'Chi' the traditional fermented beverage is prepared from finger millet. It is not just an alcoholic beverage to the Lepchas but also a culinary tradition to them. Therefore, *Chi* is also called *Chi Zaom* in Lepcha. The word *Zaom* has been derived from the word *Azaom* meaning food (Tamsang, 1997). This mild brew is so popular that it has imbibed itself into every social and cultural life of the Lepchas that without it their social and religious rituals are incomplete. From cradle to grave this traditional beverage has significant role in Lepcha culture as they consider it a holy brew. They believe that their creator God *'Itbu Deboo Rum'* and other Gods/Goddesses as well as evil spirits accept the offering of *Chi*, hence, before drinking it a Lepcha always offers *Chi* to his God, Guardian deity and spirits with prayers, this offering ritual is called *'Chi Faat.'* The priest (*Boongthing*) and priestess (*Mun*) have to be served with *Chi* when they come to their house to perform any social and religious rituals. Moreover, it is difficult to find workers for their farm works on wages if *Chi* is not served to them. It is said to be a medicine as it helps in confronting the Lepchas in their day to day struggles to make ends meet and hardships of living with lots of physical exercises in difficult terrain in the forests (Molommu, 2017). It is refreshing as well as nourishing (Plate 17.3).

PLATE 17.3 Lepcha traditional drinks *'Chi.'*

17.3.7 CONSERVATION STRATEGIES

Agroforestry system is common in Lepchas residing localities across Sikkim and parts of West Bengal, which helps conserving multipurpose plant species including medicinal plants, wild edibles, along with a large number of underutilized potential crops, semi-domesticated crops with their wild relatives. Most of the Lepchas being cardamom growers, thus cardamom-based agroforestry system along the sloppy area is common among the tribe. Ecologically adaptive multipurpose tree species like *Albizia spp.*, *Alnus nepalensis* along with large cardamom is common in Dzongu (North Sikkim) and Darjeeling and Kalimpong areas of West Bengal. Sikkim Mandarin based agroforestry along with multipurpose trees (MPTs) is popular in Dzongu area (Pandey et al., 2020). This system supports in conserving a large number of economically important crops along with numerous native food crops. Since, Lepchas were food gatherers from the wild in past, but after starting of settled agriculture since mid-19th-century many wild edible species have been domesticated and multiplied for self-consumption as well as for commercial purposes which further helped in preserving these plant species.

Tuber crops being the important source of food to the Lepcha tribe, they either harvest it from wild or from their farmlands where they cultivate for their easy harvest. If harvesting is not done annually for tuber crops, they fail to regenerate in the next season, thus annual harvesting of these tuber crops like *Dioscorea hamiltonii* helps in regenerating them in next season, such indigenous tradition of the ethnic communities contribute in safeguarding these plant species from being extinct (Bhujel et al., 2018, 2020). The tribe is not only conserving the tuber crops but also conserving other crops like finger millet (*Mong*) and buckwheat (*Kushru*) without which their traditional recipes are incomplete. Hence, they play a significant role in preserving native plant resources as a part of their traditional lifestyle.

In order to safeguard Lepchas age old traditional arts and crafts G.B. Pant National Institute of Himalayan Environment (NIHE), Sikkim Regional Center, Gangtok under International Center for Integrated Mountain Development (ICIMOD) funded Kangchenjunga Landscape Conservation and Development Initiative (KLCDI)–India program, took an initiative from a small pocket of Lepcha reserve of Dzongu in North Sikkim to reinforce and empower this community specific IK into a viable livelihood opportunity. Therefore, capacity building programs for the Lepcha youths were organized in Dzongu to strengthen bamboo crafts and nettle fiber based traditional knowledge of the tribe (Pandey et al., 2019, 2020). The training was imparted emphasizing on value addition to their traditional crafts and handlooms to make these crafts more attractive and presentable so that the eco-friendly products will grasp the market in local, national, as well as international level providing the tribe a better livelihood opportunity in wider level. Consequently, the valuable traditional knowledge of the tribe will be preserved and get promoted. This step towards reviving and preserving IK linking with livelihood of the Lepcha tribe was vitally important which was gradually fading away with the wave of modernization.

17.4 CONCLUSION AND WAY FORWARD

In modern era of unsustainable practices/activities such indigenous traditions need to be promoted and preserved, since their socio-cultural heritage has great significance not only in biodiversity conservation but also for a healthy sustainable society. However, increasing anthropogenic pressure in the process of modernization and development is causing tremendous impact to the nature at present, which may lead to extinction of many plant/animal species in near future. Moreover, lack of interest among young generation on age-old traditions/practices and continuous impact of modern and unsustainable developmental activities, changed lifestyles and livelihoods the traditional knowledge system of this tribe on sustainable utilization and management of bioresources is in the verge of extinction. Therefore, it is an urgent need of the hour to conserve both traditional knowledge system and natural resources for which generating awareness and encouraging the young generation is crucial. Hence KLCDI program's initiative to revitalize and restore Lepchas indigenous arts and crafts through trainings and capacity building programs will certainly contribute to preserving their traditional skill, which is one of the most precious indigenous technologies of the Lepcha tribe. Furthermore, promotion of their traditional cuisines through trade fairs and festivals, as well as popularizing them in tourist homestays will definitely facilitate in reviving Lepchas diminishing culture. Efforts therefore should be made on this direction as well, so that the indigenous Lepcha community could make their traditional lifestyle as means of their sustainable livelihood. Thus, their eco-friendly cultural heritage will be protected and preserved for long.

ACKNOWLEDGMENTS

Authors wish to thank Director, GBPNIHE for providing necessary facility, encouragement, and guidance. The financial support received from ICIMOD; Nepal through KLCDI-India program is acknowledged. Reymit Lepcha, Sikkim University, Gangtok, and community people of surveyed villages across Sikkim and parts of West Bengal states are gratefully acknowledged for their support.

KEYWORDS

- **bioresource conservation**
- **ethnoecological knowledge**
- **Kangchenjunga**
- **Lepcha community**
- **medicinal plants**
- **traditional knowledge**

REFERENCES

Bhujel, D., Chhetri, G., & Rai, Y. K., (2018). Wild edible plants used by ethnic communities in Kalimpong district of West Bengal, India. *NeBIO, Intern. J. Environ. Biodiv., 9*(4), 314–326.

Bhujel, D., Chhetri, G., & Rai, Y. K., (2020). Notes on ethno-biological aspect of Makar Sankranti celebrated by Nepali community of Darjeeling and Sikkim Himalayas. *NeBIO, Intern. J. Environ. Biodiv., 11*(1), 20–26.

Chhetri, G., & Rai, Y. K., (2018). Ethno-medicinal practices of the Lepcha Tribe in Kalimpong District of West Bengal, India. *NeBIO, Intern. J. Environ. Biodiv., 9*(1), 158–167.

Das, A. K., (1978). *The Lepchas of West Bengal*. Calcutta, India. Indian Edition.

Lepcha, C. K., (2013). *Religion, Culture, and Identity: A Comparative Study of the Lepchas of Dzongu, Kalimpong and Ilam*. Ph. D. Dissertation. Dept. of Anthropology, NEHU, Shillong.

Molommu, S. P., (2017). Chi- the traditional drink of the Lepchas: A case study. *J. Human. Soc. Sci., 22*(11), 11–17.

Pandey, A., Lepcha, N., Sinha, S., Gaira, K. S., Lepcha, U. P., Singh, M., Joshi, R., Rawal, R. S., & Chettri, N., (2019). *Bamboo Craft: Strength and Opportunity in Dzongu, A Pilot Site of KLCDI-India.* GBPNIHE.

Pandey, A., Lepcha, U. P., Gaira, K. S., Joshi, R., & Chettri, N., (2020). *Nettle Plant (Girardinia Diversifolia) Fiber Extraction and Yarn Making*. NIHE.

Plaisier, H., (2005). A brief introduction to Lepcha orthography and literature. *Bull. Tibetol., 41*(1).

Pradhan, B. K., & Badola, H. K., (2008). Ethnomedicinal plant use by Lepcha tribe of Dzongu valley, bordering Khangchendzonga biosphere reserve, in North Sikkim, India. *J. Ethnobiol. Ethnomed., 4,* 22. doi: 10.1186/1746-4269-4-22.

Roy, D. C., (2011). Lepcha-nature relation. *Aachuley, A Bilingual Journal Illustrating the Lepcha Way of Life.*

Sharma, A., (2013). *The Lepchas of Dzongu Region in Sikkim, a Narrative of Cultural Heritage and Folklore*. INTACH and Aryan New Delhi.

Singh, P. K., & Chakraborty, S., (2014). Munism- a traditional face of Lepcha religion. A study on preservation of cultural heritage. *Asian Mirror- Intern. J. Res., 1*(1).

Sinha, S., Lepcha, N., Pandey, A., Gaira, K. S., Lepcha, U. P., Singh, M., Joshi, R., Rawal, R. S., & Chettri, N., (2019). *Khu-Ree: A Traditional Lepcha Cuisine, Its Ingredients and Recipe.* GBPNIHESD.

Subba, J. R., (2008). *History, Culture and Customs of Sikkim.* Gyan Publishing House, New Delhi.

Tamsang, K. P., (1997). *Lepcha Folklore.* Mani Printing Press Kalimpong.

Tamsang, K. P., (2004). *Glossary of Lepcha Medicinal Plants.* Kalimpong.

CHAPTER 18

Remote Sensing and Machine Learning Applications for Smart Precision Agriculture: A Review

SIVAKUMAR RAJAGOPAL, BISWAJITA MOHANTY, A. MADHUMATHI, and B. PRABADEVI

Vellore Institute of Technology, Vellore, Tamil Nadu, India

ABSTRACT

The soil fertility is declining slowly with an increasing population which is not recognizable with naked eye like disasters such as floods and droughts. Therefore, it does not get the as much as attention it requires. Soil pollution emerging as an important threat to soil function for which specific accelerated actions are needed, to maintain nutrient balance, offset the effects of soil pollution, salinization, and soil erosion, and sustains overall soil quality. The physical, chemical, and mineralogical composition of soil is greatly influenced by the processes like weathering, erosion, transportation, and deposition. Thus, an understanding of the processes that controls the behavior of the source material is essential for qualitative as well as quantitative characterization of soil. The sustainable management of soil is directly or indirectly crucial for mineral resource management, nutrient balance, crop production, environmental sustainability, and on human health. Thus, soil heath status study not only will provide an early indication of the problems associated with soil nutrient surplus or a deficit but also it helps in crop yields management. This chapter provides a depth knowledge about the relation between the soil health and nutrient flow by identifying the root cause from the system, i.e., at the field level, farm level, and regional level. This process is done with the help of traditional knowledge and high-resolution

Sustainable Management and Conservation of Environmental Resources in India.
Sumit Chakravarty, Vineeta, Jahangeer A. Bhat, Munesh Kumar, Gopal Shukla (Eds.)

data like multispectral data, hyperspectral data, and geographic information system (GIS) approach for spatial variability-based methods proposed in this study. This chapter's main focus is to manage the spatial variability of soils and nutrients availability by using the techniques like machine learning (ML) and the internet of things (IoT). Digital information extraction method can be broadly used in agricultural field to help farmers to increase productivity, pest management and land sustainability practices. Agriculture can be integrated into urban and peri-urban areas which can contribute to the self-sufficient neighborhood while, reconnecting people with nature and consumption with production. This chapter delivers a comprehensive knowledge of the current and potential applications of modern technologies, its challenges, and possible solutions implementations for soil fertility.

18.1 INTRODUCTION

Agriculture is the backbone of a country and its economy. In a country like India, agriculture accounts for almost 30% of GDP and employs 70% of the population. The concern now is: How long will it continue to supply the population's expanding needs and demands? If we simply rely on traditional farming, it will be difficult to answer this question. This prompts us to consider new ways to address the upcoming needs and challenges by shifting toward new technologies to increase our agricultural productivity. As a result of population growth and industrialization, agricultural production has stagnated since the green revolution and the horizontal expansion of arable land has been constrained. India had 0.33 ha of land available per person in 1952, but that number has since dropped to 0.15 ha. Therefore, creating eco-friendly technologies is crucial for preserving crop productivity. It has long been understood that different fields have different crops and soils. Technical approaches have been developed during the past 10 years to use contemporary electronics to adapt to field variability. Such methods are known as spatially variable crop production, geographic positioning system (GPS). Traditional practices in agriculture are tedious, time consuming, laborious, costly. With the application of modern technologies like machine learning (ML) and the internet of things (IoT) can assist the farmers in various stages of crop production, thereby making agriculture more sustainable. Traditional method of farming follows a crop rotation pattern, intercrop pattern which makes the soil more fertile and nourished, thereby yielding better production in the next yield of the same crop. The land must be made reusable by planting the appropriate crop that nourishes the soil. Sustainable land management practices not only help in yielding good productivity but also it will help to maintain the soil biota or in other words soil ecosystem. Therefore, sustainable land management practices or Precision farming depends on the factors such as: soil physicochemical behavior, soil biological properties, climatic condition, and greenhouse gas (GHG) emission as described in Figure 18.1.

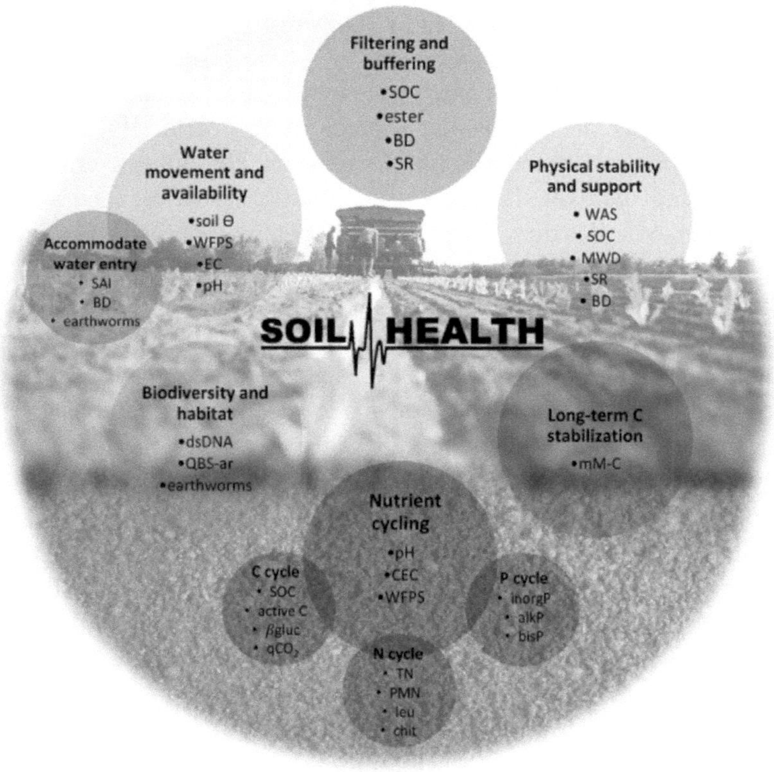

FIGURE 18.1 Diagrammatic representation of soil health.

Source: Modified from Andrea et al. (2017).

The major concern in sustainable precision farming is soil fertility. Soil fertility is defined as the capacity of the soil to absorb the supplied energy from varied sources (macro- and micro-nutrients), store it and transmit it adequately for plant growth to provide consistently excellent quality yields. Sustainability in crop production primarily depends on maintaining appropriate levels of soil fertility and soil health (i.e., physical, chemical, and biological properties of soil). This requires a regular look up in the soil health status to get a proper yield which is a difficult task in traditional method of crop production and analysis. Therefore, there is the need of modern technologies which not only helps in sustainable land management but also will increase the productivity.

18.2 REMOTE SENSING (RS) AND GEOGRAPHIC INFORMATION SYSTEM (GIS) IN SOIL FERTILITY IDENTIFICATION

Soil, the unconsolidated organic or mineral matter is one of the essential natural resources present on the surface of the earth. Its characteristics vary depending on the factors like its genesis, chronology, environmental condition and also on the biotic and abiotic activities. The sustainable management of soil is directly or indirectly crucial for crop production, environmental sustainability, mineral resource management, and also on human health. Thus, there is a requirement to manage the land resources so that the soil nutrient does not become deficit or toxic. Therefore, it is essential to understand nature of the soil and its spatio-temporal changes. However, the conventional method of soil analysis is expensive, time-consuming, and tedious. There is a need for technologies for rapid and accurate measurement of soil properties.

The increasing population growth and increasing demand for food with limited resources necessitate developing such technology, which improves the agricultural products profitable and environmentally sustainable (Zhang & Kovacs, 2012). That leads to the formation of advance technologies termed as precision agriculture (PA), precision farming, site-specific crop management or even site-specific farming. Generally, PA is a combination of agricultural management practices with technologies like remote sensing (RS), global positioning system (GPS), and geographical information systems (GIS). RS has emerged as an alternative for the estimation of soil variability analysis in a large spatial domain compared to the traditional method (Galvao et al., 1997). It provides the landscape information synoptically, repetitively, and objectively with reliably more accuracy (Pandey et al., 2011).

Studies have comprehended that RS based analysis are the basis for a new pattern of rapid and non-destructive methods of quantifying the soil properties (Coleman et al., 1991; Ben-Dor & Banin, 1995). In contrast, the spectroscopic approach for estimations of soil properties is an emerging tool for updating the soil status that can be used in biogeochemical models and monitoring the soil degradation (Dunne & Willmott, 1996; Hartemink, 2008; Stevens et al., 2013) with precise and cost-effective manner. Thus, RS and GIS is evolving as a promising tool for mapping and monitoring of natural resources including the soil properties characterization (Kasturirangan et al., 1996; Saxena et al., 2000; Srivastava & Saxena, 2004). Condit (1970) and Stoner & Baumgardner (1981) were the first two scientists who developed a relationship between the spectra and soil properties and stored them as a spectral library which was used as a standard for estimating the soil properties in a proximal mode. For the last few decades, diffuse reflectance spectroscopy (DRS) is being used as an analytical tool for soil surveys as an alternative to conventional laboratory measurements (Janik et al., 1998). Spectroscopic approach predicts the soil properties either by direct absorption of the light associated with functional groups such as organic carbon, total nitrogen, and clay composition (Van der Marel & Beutelspacher, 1976) or by inter-relationship between the soil physicochemical properties and the mineral composition of the soil. This approach helps in exposing the potential as well as its limitation in the soil management practices in rapid and non-invasive manner.

Over the recent decades, high-resolution data has been used for the mineralogical studies (Hunt & Ashley, 1979; Clark et al., 1990; Boardman & Kruse, 1994; Crowley & Zimbelman, 1997). The spectral region over visible-near infrared–shortwave infrared (Vis-NIR-SWIR) (400–2,500 nm) characterized as a significant spectral region for the quantitative analysis of several soil properties (Ben-Dor & Banin, 1995; Ben-Dor et al., 1999; Ben-Dor, 2002; Shepherd & Walsh, 2002; Nanni & Dematte, 2006). It is believed that hyperspectral data is useful for soil attributes mapping due to its contiguous spectral band covering from VIS to IR spectral region of the electromagnetic spectrum (EMS) with a spectral resolution of 5–10 nm. Thus, it can classify the object in a scene more accurately based on their spectral properties (Goetz et al., 1985; Vane & Goetz, 1988; Plaza et al., 2009).

RS data was used for salt-affected areas mapping (Hunt & Salisbury, 1976; Goetz & Herring, 1989; Hick & Russell, 1990; Mougenot et al., 1993; Verma et al., 1994; Dehaan & Taylor, 2003; Madani, 2005). RS data was also used for soil erosion map of Southern France by developing a mixture modeling using end member spectra (Hill et al., 1995). Moreover, minerals like kaolinite, muscovite, alunite, calcite, and dolomite were identified by the spectra derived from AVRIS data (Green et al., 1998) while, prepared a mineral map of clay minerals like kaolin, smectite, and illite for Australian soil was prepared using NIR spectroscopy at different soil depth (0–20 cm, 60–80 cm) by applying model tree algorithm (Viscarra-Rossel, 2011).

18.3 HYPERSPECTRAL REMOTE SENSING (HRS) SPECTROSCOPY

The term "hyper" is derived from the Greek word "over," "above," or "exaggerated amount" and "spectra" means color, thus "hyperspectral" means "many colors." Hyperspectral remote sensing (HRS) is the science of studying a portion of the electromagnetic (EMR) spectrum by breaking it into a number of pieces for computational analysis. Generally, spectroscopy is the study of reflected, absorbed, or emitted light energy the material it interacts with as a function of wavelength (Clark, 1995). In particular, the scattering of light reflected at all angles from a surface is measured which provides spectral information of the material's physical and chemical composition. The remotely sensed images are characterized by their spectral and spatial resolutions which signify the geometric relationship with one another and variations within a pixel as a function of wavelength, respectively. The absorption features are created by the transmission of surface scattered lights through the particles. This interaction between the crystal lattice and photons were studied under two headings, i.e., electronic transition (ET) and vibrational transition which causes diagnostic absorption features in the reflected and emitted spectra through the optical region between 400 and 1,400 nm. These spectral absorption peaks at NIR (NIR, 700–2,500 nm) are observed due to the vibrational stretching and bending behavior associated with N–H, –OH, and C–H functional group and features in the visible (400–700 nm) region is due to the ETs of atoms (Ben-Dor et al., 1999). The fundamental vibrations, because of these energy interactions occur along the mid-infrared (MIR) region. In contrast, overtones and combinations observed over the visible (VIS) and near-infrared (NIR) regions (Janik et al., 1998). This approach is appropriate for the quantitative analysis of a spectrum rather than a qualitative analysis.

The imaging spectrometer (IS) sensors covers the entire region starting from the reflective VIS (0.4–0.7 μm), IR (0.75–2.5 μm) to the emissive thermal (TIR; 3–14 μm) spectral regions to generates a mixed spectra which is the result of photon-matter interactions across these spectral regions (Figure 18.2).

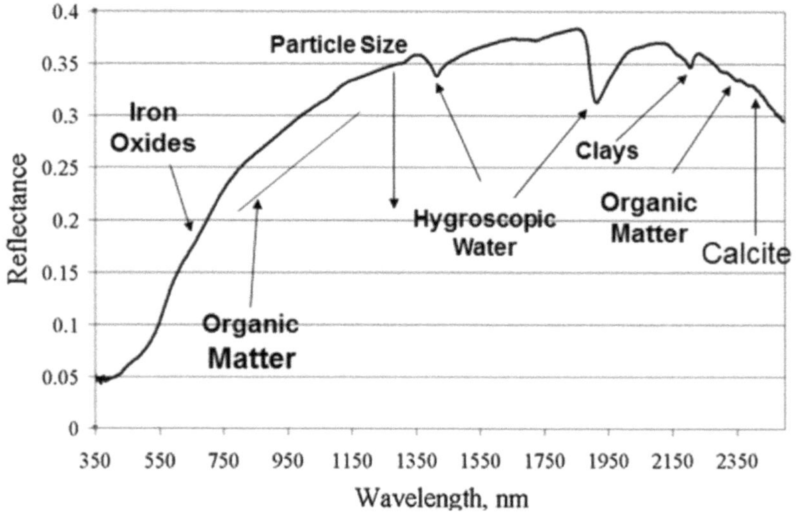

FIGURE 18.2 Soil spectrum representing the major chromophores (Ben-Dor et al., 2008).
Source: Modified from Ben-Dor et al., 2008

18.4 APPLICATION OF HRS SPECTROSCOPY ON SOIL SCIENCE

Spectroscopic approach over VIS, NIR, and MIR (2,500–25,000 nm) regions is measured as a promising new analytical tool for qualitative and quantitative analysis of soil properties. This approach may be adapted to estimate multiple soils attribute both in proximal and remote modes. The method is in development as an improvement to multispectral RS with the availability of hundreds of narrow, contiguous bands produced by hyperspectral imaging cameras onboard RS platforms and is called HRS method. HRS technology involves both optical and thermal sensors to create a repository of reflectance spectra under controlled laboratory conditions (Cohen et al., 2007).

HRS spectroradiometers has been developed to measure the soil properties across the electromagnetic spectrum (EMS) at a resolution of 5–10 nm. The identification of material (i.e., soil, rocks, minerals) using HRS data based on the concept that the nature, structure, and chemistry of crystal lattice of individual mineral in each material exhibits a characteristic absorption or reflection in the visible and infrared (IR) portion EMS (Goetz et al., 1985; Clark et al., 1990; Nidamanuri & Ramiya, 2014). The organized database of measured spectral reflectance of known materials is termed as the Spectral library. The generation of the spectral library not only helps in identifying the unknown content, but it also quantifies the material by comparing it with the existing spectral libraries.

In a large spatial domain several soil properties have been estimated using DRS in the Vis-NIR (350–2,500 nm) region (Ben-Dor & Banin, 1995; Chang et al., 2001; Islam et al., 2003; Viscarra-Rossel et al., 2006; Waiser et al., 2007; Sarathjith et al., 2014a, b; Vohland et al., 2014). The spectral signature is an inherent soil property that varies across different soils (Ben-Dor et al., 2009). The spectral features observed in the visible region (i.e., 400–700 nm) is due to the ET associated with the iron bearing minerals such as hematite and goethite (Sherman & Waite, 1985; Viscarra-Rossel & Behrens, 2010). The absorption features in the NIR region (700–2,500 nm) is considered as the influence of the overtones and combinations of covalent bonds between O-H, C-H, and N-H atoms (Malley & Martin, 2003; Workman & Shenk, 2004). Researchers like Reeves III et al. (1999) and Viscarra-Rossel & McBratney (1998) reported that the vibrational absorption feature in the NIR region is due to the presence of OH groups and OM in minerals. The absorption that peaks at around 1,400 nm, 1,900 nm, and 2,200 nm are the typical characteristic feature of soil spectra. The characteristic absorption peaks at 1,400 nm and 1,900 nm are known as water absorption band which is the result of first overtone of O-H stretch and the combination of H-O-H bend with O-H stretches (Clark, 1999) whereas, the absorption feature at 2,200 nm and 2,300 nm are associated with the clay mineral. This spectral absorption feature is due to the combination of metal-OH bend and O-H stretch (Chabrillat et al., 2002; Viscarra-Rossel et al., 2006). Absorption features between 870 and 1,000 nm is due to iron oxides; between 2,200 and 2,500 nm is due to carbonates in the NIR region (Clark et al., 1990; Chang et al., 2001). Along with NIR absorption spectra, carbonates can be detected by an overtone between 2,300 and 2,350 nm and three combination bands near 2,120–2,160 nm, 1,970–2,000 nm, and 1,850–1,870 nm (Clark, 1999). Clay minerals have also been assessed using Vis-NIR spectra (Hunt et al., 1973; Dufrechou et al., 2015). Spectroscopic approach has also been implemented in the estimation of SOC (Stevens et al., 2010; Bellon-Maurel & McBratney, 2011; Deng et al., 2013), texture (Knadel et al., 2013; Curcio et al., 2013), soil particle size distribution (Rawlins et al., 2011; Sarathjit et al., 2014a, b), EC (Shrestha, 2006), bulk density (Al-Asadi & Mouazen, 2014), hydraulic conductivity (Santra et al., 2009; Babaeian et al., 2015), soil moisture (Whiting et al., 2004; Lakhankar et al., 2009; Wu et al., 2010; Zhu et al., 2010; Younis & Iqbal, 2015), soil quality (Cohen et al., 2005; Cecillon et al., 2009; Paz-Kagan et al., 2014). The composition of mineral-organic mixes can also be predicted using UV-VIS-NIR spectra (Viscarra-Rossel et al., 2006). The MIR region of the spectra (2,500–25,000 nm) was also used for the quantitative assessment of clay minerals (Saikia & Parthasarathy, 2010; Djomgoue & Njopwouo, 2013; Peltre et al., 2014). Many worker used airborne and satellite hyperspectral data quantitative analysis of soil properties (Selige et al., 2006; Gomez et al., 2008; Stevens et al., 2010; Anne et al., 2014; Franceschini et al., 2015).

18.5 PHYSICOCHEMICAL AND MINERALOGICAL PROPERTIES OF SOIL USING REFLECTANCE SPECTROSCOPY

The scattering of light reflected from a surface provides spectral information of the material's physical and chemical composition. Spectroscopy covering the entire VIS, NIR, and MIR (350–2,500–25,000 nm) regions of the EMS are being used for the characterization of several soil attributes. VIS range widely used for the characterization of field soil color (Viscarra-Rossel et al., 2006) and oxide and hydroxides of iron (Ben-Dor et al., 1999). Iron oxide was predicted with a characteristic absorption feature at 550–650 nm, 750–950 nm, 1,406 nm, and 1,449 nm (Ben-Dor & Banin, 1995; Summers et al., 2011). The presence of clay minerals at 1,900–2,200 nm is studied by Gomez et al. (2008). The absorption features near 1,900–2,300 nm shows the presence of carbonates (Lagacherie et al., 2008). The spectral region most sensitive to soil organic matter (SOM) includes 960 nm, 1,100 nm, 1,400 nm, 1,720 nm, 1,744 nm, 1,870 nm, 1,900 nm, 2,052 nm, 2,180 nm, and 2,309 nm (Dalal & Henry, 1986; Sudduth & Hummel, 1991; Palacios-Orueta & Ustin, 1998; Shepherd & Walsh, 2002; Daniel et al., 2004). The estimation of SOC using VIS spectroscopy was found more accurate than the traditional laboratory dichromate digestions method (McCauley et al., 1993). SOC is determined as it decreases the reflectance of the 900 nm absorption band, which is influenced by iron oxide (Minasny & Hartemink, 2011; Cambule et al., 2012). Prediction of soil properties like pH, clay, SOC, inorganic carbon, total nitrogen, and carbonates over the MIR is more accurate than the NIR (Janik & Skiemstad, 1995; McCarty & Reeves, 2006). Quantitative analysis of the mineral abundance of complex soil mixtures can be best predicted by the reflectance spectra near 2,100–2,400 nm of VIS-NIR (Mulder et al., 2013).

18.6 SPECTRAL CHARACTERIZATION INFLUENCED BY BASIC SOIL PROPERTIES AND MINERALOGY

Fundamental energy interactions possible with incidence of electromagnetic energy (EME) on any material are reflection, absorption, and transmittance.

These energy interactions vary depending on the material type and condition and the wavelength. As soil is a complex material with extreme variables in its physical and chemical properties, spatial variation in soil properties plays a significant role in precision agricultural farming, land degradation, and soil erosion study. The spectroscopic approach is a rapid, non-destructive, and cost-effective analytical approach for studies of soil properties (Sheppard et al., 1985; Ciurczak, 2001; Geladi, 2003; Ben-Dor et al., 2009). The parameters that affect the shape and nature of the spectra are known as "chromophores." It may be physical or chemical chromophores (Ben-Dor et al., 1999; Ben-Dor & Thenkabail, 2011). Physical chromophores are those which affect the shape as well as the position of waveband in spectra. In contrast, chemical chromophores are those that not only affect the shape but also affect the absorption peak of the spectra, depending on the chemical composition of the substance. In soil, the chemical chromophores that affect the spectra are mineral (clay minerals, iron oxides), water (moisture content), and organic matter. The physical chromophores affecting spectra are particle size and instrumental errors.

18.6.1 CLAY MINERALS

Clay minerals are the hydrated alumino-silicates with a very fine particle size (< 2 mm) having a layered structure formed by the bonding between the tetrahedral and octahedral sheet sharing optical oxygen. The presence of OH functional group in clay minerals may be due to the mineral structure or due to the water molecules absorbed by the minerals. Thus, clay minerals have a significant influence on spectral reflectance. Spectral responses of clay minerals are a result of combined effect of structural water, hydroxyl molecules, silicate ions and the octahedral and interlayer cations (Farmer & Russell, 1964; Hunt, 1979; Clark et al., 1990; Bishop et al., 1994).

Generally, three significant portions of the spectra are greatly influenced by the clay minerals, i.e., 1,400 nm, 1,900 nm, and 2,200 nm. The absorption peak around 1,400 nm is due to the kaolinite clay minerals and 1,900 nm signifies illite and smectite type of clay minerals whereas, presence of sharp absorption peak at both 1,400 nm and 1,900 nm indicates montmorillonite clay minerals. Spectral peak in the above-mentioned absorption bands are due to overtones of O-H stretching in 1,400 nm, and a combination of the H-O-H bending and O-H stretching in 1,900 nm whereas, the spectral peaks at 2,200–2,300 nm are due to combination of metals (i.e., Al, Fe, and Mg) OH bend and O-H stretch (Clark, 1999). Thus, the presence of an absorption peak around 1,400 nm and 2,200 nm indicates the presence of kaolinite clay (Hunt & Salisbury, 1970). The peculiar behavior of different clay minerals on spectra is described in subsections in detail.

18.6.1.1 KAOLINITE

Spectral behavior showing sharp doublet at around 2,705 nm and 2,762 nm, combination, and overtones about near 10,660 nm and 10,952 nm, 12,594 nm and 14,326 nm is a diagnostic feature of kaolinite clay minerals. The sharp doublet at around 2,705 nm and 2,762 nm on the spectra is due to the presence internal surface –OH group (Balan et al., 2001), spectral peak at 10,660 nm and 10,952 nm is due to the deformed OH band while, 12,594 nm and 14,326 nm are due to Si-O bond (Vaculikova & Plevova, 2005; Viscarra-Rossel & Behrens, 2010; Bayer et al., 2012).

18.6.1.2 ILLITE AND MUSCOVITE

The broad OH stretching band near 2,762 nm combined with 12,062 nm and 13,333 nm doublets is indicative of illite or a muscovite clay mineral. This particular behavior of the spectra is due to the Al-Mg-OH deformation and Al-O-Si vibration (Farmer, 1974). Both the kaolinite and illite have a strong absorption near 2,762 nm, the presence of 12,062 nm and 13,333 nm doublets confirms the presence illite.

18.6.1.3 MONTMORILLONITE

The spectral peak near 2,759 nm looks similar to kaolinite clay, but it has a broader peak than kaolinite which helps in separating it from kaolinite clay minerals. This doublet peaks are broad OH-stretching at 2,759 nm, peaks at 10,928 nm and 11,862 nm are due to the influence of Al-Al-OH and Al-Mg-OH, respectively. The diagnostic features near 10,928 nm and 11,862 nm confirms the presence of montmorillonite (Russell et al., 1970) whereas the absorption peak at 16,129 nm is a characteristic feature of Al-rich montmorillonite.

18.6.1.4 SMECTITE

It is a 2:1 type of clay mineral. In smectite, the presence of –OH functional group is due to the molecular water present in metal (i.e., Mg, Ca, and Na) structure. Smectite shows strong overtones near 1,410 nm and a combination near 1,920 nm for water while, a strong mix of –OH occurs near 2,250–2,500 nm region (Gates, 2005). A strong doublet near 22,421 nm and 18,867 nm indicates the presence of smectite with (due to Si-O stretching) and OH bending (Salisbury et al., 1991; Bishop et al., 2008). Some of the vibrational features observed are 10,900 nm (for Al_2OH), 11,363 nm (for $AlFe^{+3}OH$), 11,800 nm (for AlMgOH), 12,200 nm (for $Fe^{+2}OH$), 12,600 nm (for $MgFe^{+3}OH$) and 14,000–15,000 nm (for Mg_3OH).

18.6.1.5 NONTRONITE

Nontronite reflectance spectra display absorption centered in the extended visible region, i.e., near 500 nm, 650 nm, and 900 nm due to the electronic vibration. A sharp overtone appears near 1,430 nm due to Fe_2OH. AlFeOH, and Fe_2OH combination bands observed at 2,240 nm and 2,290 nm. The intense absorption peak appears in the 2,800 nm region, confirming the nontronite is due to OH stretching vibrations. Similarly, the Fe-rich nontronite shows a Christiansen feature near 8,400 nm and a doublet near 20,000 nm and 23,500 nm (Bishop et al., 2008).

18.6.2 IRON OXIDES

Iron oxides play an essential role in the soil profile during pedogenesis by producing intense color to the soil. The yellowish-brown color of land is due to goethite while, hematite provides red color to the ground (Cornell & Schwertmann, 1996). The iron (ferrous and ferric oxides) content has a significant influence in the shape of the soil reflectance spectra (Ben-Dor, 2002). Generally, the spectral activity iron oxides occur in the visible spectrum 400–1,000 nm range. According to the presence of particular absorption peaks, the iron oxides are identified.

18.6.2.1 HEMATITE

Diagnostic absorption features near 550 nm, 630 nm, and 860 nm indicate the iron oxide present in the soil is hematite. According to Schodlok & Ramanaidou (2011), the presence of hematite in a mixture display increases reflectance between 6,000 and 7,500 nm, and beyond 7,500 nm the reflectivity of spectra decreases.

18.6.2.2 GOETHITE

Absorption peak near 480 nm, 650 nm, and 920 nm confirm the presence of goethite. Goethite exhibits two prominent peaks near 11,200–11,400 nm and 12,400–12,600 nm due to the functional OH bending vibration (Farmer, 1974; Kustova et al., 1992; Ruan et al., 2002; Glotch et al., 2004; Glotch & Rossman, 2009).

18.6.2.3 LIMONITE

Magnetite spectra show a characteristic feature in the SWIR region, i.e., 1,100 nm.

18.6.3 MOISTURE CONTENT

Wigneron et al. (1998) described that the soil moisture content (SMC) is a vital factor in plant growth, soil biogeochemistry, soil weathering and erosion, and water holding capacity (WHC). As a result, regular and accurate SMC analysis is necessary for a better comprehension of the soil. Angstrom (1925) collected for the first time the spectral reflectance of bare soils at different moisture contents in the 400–2,500 nm region under laboratory condition. He explained an inverse relationship between spectral reflectance with SMC, i.e., there is a decrease of reflectance with an increase in SMC. Some research experiments on bare soils confirmed the approaches of spectral reflectance in SMC estimation (Lobell & Asner, 2002; Weidong et al., 2002; Liu et al., 2003; Whiting et al., 2004; Khanna et al., 2007; Haubrock et al., 2008).

According to some researchers (Bowers & Hanks, 1965; Skidmore et al., 1975; Clark et al., 1990, 1999; Whiting et al., 2004) there is a dominant influence of SMC in VNIR (400–1,100 nm) and SWIR (1,100–2,500 nm) regions of the electromagnetic spectrum (EMS). Generally, the –OH bond present in the water is strongly absorbed in the NIR region. Some of the fundamental absorption of water in the liquid phase is observed at 3,106 nm, 6,079 nm, and 2,903 nm due to the symmetric O-H stretching, H-O-H bending, and asymmetric O-H stretching, respectively. Similarly, absorption peak near 2,738 nm, 6,270 nm, and 2,663 nm was observed for water in the vapor phase. The three significant absorption peaks influenced by the moisture content of the soil are 1,400 nm, 1,900 nm, and 2,200 nm.

The absorbance occurs near 1,400 nm is due to the overtones of O-H stretching a peak near 1,900 nm is a result of a combination of the H-O-H bending and O-H stretching whereas, the combination of metal-OH bend and O-H stretch results in a sharp absorption peak near 2,200–2,300 nm (Clark et al., 1990).

18.6.4 SOIL ORGANIC CARBON (SOC)

Soil organic carbon (SOC) is an important attribute for the soil quality assessment. SOC has a potential influence in the physical, chemical, and biological properties of soils. Management practices like tillage, residue management and fertilization not only affect SOC quality but also alter the vertical distribution of SOC quantity in the soil profile (Angers et al., 1997). Thus, a rapid, non-destructive, and inexpensive method is required for a better understanding of this property of soil. In this context, method of characterization will help in understanding the quality of the land. The characteristic bands for SOC in the VNIR spectral region are 800 nm, 1,000 nm, 1,400 nm, and 1,900–2,400 (Reeves III et al., 2002; Vasques et al., 2008; Ge et al., 2011). SOC is also confirmed in the visible region with an absorption peak at 600–700 nm (Yang et al., 2012).

According to Peng et al. (2014), the wavelength region which is greatly influenced by SOC mostly found in the association with other soil properties spectral areas. Generally spectral region of water, i.e., 1,915 nm, 1,455 nm, 1,380 nm, 1,135 nm, and 940 nm show characteristic deviations due to the influence of SOC (Viscarra-Rossel & Behren, 2010). Organic components near 2,381 nm, 2,147 nm, 2,084 nm, 1,967 nm, 1,931 nm, 1,517 nm, 1,454, nm and 860 nm (Viscarra-Rossel & Behren, 2010) shows a significant rise in spectral reflectance. Similar influence can be observed for clay minerals (2,444 nm, 2,201 nm, and 2,174 nm), regions related to the iron oxides goethite (920 nm), hematite (884 nm). Significant area for SOC absorption is 500–700 nm and 850–900 nm (Galvao & Vitorello, 1998), 1,702–2,052 nm (Dalal & Henry, 1986), and 2,300–2,400 nm (Ben-Dor et al., 1997). Generally, an increase in SOC of soil causes a decrease in the reflectance over the entire spectrum. SOC less than 1.7% decreases the influence of iron oxide on the spectra around 900 nm (Galvao & Vitorello, 1998).

18.6.5 PARTICLE SIZE

There exists an inverse relationship between particle size and reflectance spectra, i.e., with an increase in particle size, the spectral reflectance decreases. Particle size also influences the smoothness of the spectra as there is rapid increase in spectral reflectance for particle size less than 0.4 mm diameter (Bowers & Hanks, 1965). Mohanty et al. (2016) reported important spectral ranges and wave lengths in VIS-NIR for soil mineralogical properties citing many earlier studies (Tables 18.1 and 18.2).

TABLE 18.1 List of Important Spectral Ranges in VIS-NIR Region for Soil Mineralogical Properties (Mohanty et al., 2016)

Vis-NIR Region*	Soil Constituent	References
4086–4301	SiO_2, Al_2O_3, Fe_2O_3, CaO, Na_2O, K_2O	Madejova et al. (2009); Viscarra-Rossel et al. (2006)
4508–4617	SiO_2, Al_2O_3, Fe_2O_3, MgO, CaO, Na_2O, K_2O	Viscarra-Rossel et al. (2006); Madejova et al. (2009)
5175–5258	SiO_2, Al_2O_3, Fe_2O_3, CaO, K_2O	Hunt & Salisbury (1970)
7087–7235	Al_2O_3, Fe_2O_3, MgO	Hunt & Salisbury (1970); Madejova et al. (2009)
20535–21142	Al_2O_3, MgO	Hunt & Salisbury (1970)

*Wave number, cm^{-1}.

TABLE 18.2 List of Important Wavelengths for Soil Mineralogical Properties in the MIR Region (Mohanty et al., 2016)

MIR Region*	Soil Constituent	References
399	SiO_2	Clark et al. (1990)
536–548	Cao, Na_2O, K_2O SiO_2	Madejova (2003)
582–612	SiO_2, Na_2O, Al_2O_3	Madejova (2003)
600–700	Fe_2O_3, K_2O, Na_2O	Nayak & Singh (2007)
763–808	SiO_2, Al_2O_3, CaO, Na_2O, K_2O	Clark et al. (1990); Djomgoue & Njopwouo (2013)
808–845	SiO_2, Al_2O_3, Fe_2O_3, MgO, Na_2O, K_2O	Clark et al. (1990); Djomgoue & Njopwouo (2013)
863–871	Fe_2O_3	Clark et al. (1990); Djomgoue & Njopwouo (2013)
1079–1101	Fe_2O_3	Saikia & Parthasarathy (2010)
3581–3648	SiO_2, Al_2O_3, Fe_2O_3, MgO	Madejova (2003); Vaculikova & Plevova (2005)
3669–3712	SiO_2, Al_2O_3, MgO, CaO, Na_2O, K_2O	Madejova (2003); Vaculikova & Plevova (2005)

*Wave number in cm^{-1}.

18.7 RS APPLICATIONS FOR URBAN AND PERI-URBAN AGRICULTURE (UPA)

The global transformation of natural and agricultural land into urban land is very rapid (Deelstra & Girardet, 2000; Pickett et al., 2001). Effects of conversion from natural and agricultural vegetation to urban infrastructure include soil sealing by impervious surfaces, hydrological cycle changes (through decreased evapotranspiration (ET) and soil absorption, increased storm water runoff and depleted rivers and streams), and urban heat island effects (higher air temperatures than neighboring rural areas) (Pickett et al., 2001). Around 50% of the population now lives in urban areas (United Nations, 2014). The rate of conversion to urban land far exceeds in comparison to urban populations (Lincoln Institute of Land Policy, 2015). Environmental effects of urban areas extend well beyond urban boundaries, as basic needs of urban populations require the importation of food, energy, and clean water and exportation of waste products by air, water, and land (Deelstra & Girardet, 2000; Pickett et al., 2001; Aitkenhead-Peterson et al., 2010). According to UN projections, the global urban population will increase by more than a billion people between 2010 and 2025, a trend that is unlikely to occur among the rural population (United Nations, 2008). These increase populations worldwide not only lead to food deficiency but also it will require a proper management from urban to rural areas and vice versa. To meet the demand, it requires significant changes in agriculture food supply chain.

Accurate awareness of the spatio-temporal pattern of urbanization is essential for policy making, natural resource management, and decision-making to reduce the negative impacts of rapid urbanization (Hirschman & Battiata, 2016). The basic idea to reversing the effects of urbanization on the environment, ecology, and food production has been universally regarded as the (re-)introduction of productive landscapes in city design and development planning. Urban and peri-urban agriculture (UPA), rural-urban links and landscape development, urban food systems, and city area food systems are all in line with the reintroduction of landscapes into city design.

18.7.1 URBAN AND PERI-URBAN AGRICULTURE (UPA)

Urban agriculture, urban gardening or urban farming can be defined as an agricultural enterprise situated within or on the outskirts of a town, city or metropolis that grows or produces, processes, and distributes a variety of food and non-food items (re-)uses, mainly human and material resources, products, and services found in and around the particular urban area and, in effect, provides human and material resources, products, and services (Mougeot, 1999). The practice has become an evident economic activity in urban (within the built-up area, open spaces within the urban areas) and peri-urban regions of cities globally. It engages about 800 million people globally and utilizes available urban/peri-urban empty spaces (Smit et al., 1996). Peri-urban agriculture makes a major contribution to environmental services, acting as a drain for storm water runoff, a recharge zone for the groundwater table, and contributing to esthetic beauty and food health (Parece & Campbell, 2017). Urban agriculture is a competitive urban green space that has several different forms; for example, homegardens which typically constitute food for consumption next to their homes (Drescher et al., 2006); urban farms – usually commercial operations including greenhouses, rooftop gardens, and community supported agriculture (Smit et al., 2001); orchards – consisting of trees providing a source of food (fruit or nut-bearing trees); fuelwood, or building materials (Smit et al., 2001); and community gardens, informal gardens, school gardens, or allotment gardens – all characterized by land divided into plots and the garden site is usually not owned by the people working on it (Lawson, 2005). Vertical farming is also one of the types of agriculture which are gaining momentum globally throughout several urban cities due to its impacts on the agricultural sector. Agriculture offers a green infrastructure that serves urban communities by providing food, cultural, and social services and mitigating adverse environmental impacts (World Bank, 2013; McClintock & Simpson, 2014; Huang & Drescher, 2015). Urban agriculture will minimize the carbon footprint of a city by reducing the amount of transportation that happens to carry goods to the customer. The unsustainable ecological growth in large urban agglomerations can be addressed through urban/peri-urban farming practices. Such practices are seen as a crucial component of the urban environment. Over the past 10 years, urban food systems and sustainable urban agriculture noticed to have a quick shift from being a "fringe interest" to attracting the attention of politicians and planners in many cities, in both developed and developing nations. Urban planning frequently undervalues and underestimates the advantages of urban farming, which include greater access to fresh products, particularly vegetables, and contributions to food security and sustainable livelihoods.

18.7.2 SIGNIFICANCE OF RS FOR URBAN AND PERI-URBAN AGRICULTURE (UPA)

The capacity of urban agriculture to supply food to urban poor people on a continuous basis, especially in developing countries, will rely on better planning based on accurate geospatial data to allow sustainable management of the activity. Reliable data on the scale of urban/peri-urban areas used for cultivation, the spatial distribution of these areas, the types of crops and proximity to marketplaces are deficit in most of the developing countries. Recently, increased attention has been given to urban land use (LU)/land cover transition, as ecosystems in urban areas are significantly influenced by the activities of humankind. Adopting ground-based survey methods for capturing of data in order to detect and estimate changes is relatively expensive and time-consuming (Appeaning Addo, 2009). Because of this, the actual area used for urban farming in most of the developing countries has not been sufficiently assessed. Little is also known for the spatial distribution of urban agriculture in cities (Bryld, 2003). RS, however, offers spatially coherent data sets covering vast areas with both high spatial precision and high temporal frequency. Thereby this technique allows for the rapid determination of the current size of urban agricultural land at a relatively low cost. The method of RS can also produce reliable historical time series data. Repeated observations in the future would allow, over time, a thorough quantification of the changes in the size of the farmland and the types of crops produced. RS also offers a comprehensive understanding of the spatial as well as temporal complexities of urban development and LU transition processes. RS techniques using satellites have therefore been widely practiced detecting and monitor land cover transition, including urban farming, on a range of scales with useful results (Stefanov et al., 2001). Geographical Information Systems (GIS) and GPS have recently been used in conjunction with RS to quantify land cover change more accurately than just using RS data alone (Weng, 2002). It has already shown its potential in mapping urban areas and serving as a data source for research on urban expansion and changes in LU and land cover (Herold et al., 2003). In the meantime, improvements in urban farming in developing countries can be quantified by linking remotely sensed data with available historical information from archival area photography and other sources in the GIS context.

Various ecological applications, meteorological, and climatic data, LU, and LU data from a wide range of spatial scales are required which cannot be collected using field-based methods. RS data and techniques address the need to recognize and detail the biophysical environment of plant species ecosystems and to detect natural and human-caused changes in scales ranging from individual landscapes to the world. Urban farmland can be easily established and monitored using RS techniques at comparatively low costs.

This part of the chapter depicts an overview of the advances in the use of RS technology to develop an integrated technique for identifying and monitoring urban farmlands and how the information obtained can be utilized to create sustainable urban farming. RS data has been used for decades to identify and map urban areas, beginning with aerial photography and shifting to satellite RS. The various approaches to identify suitable locations within urban limits include using very high-resolution satellite images along with moderate and coarse resolution imagery and information from social media and the web (Xia et al., 2019). A mixed-method approach is necessary to locate the presence of urban farming and to monitor its change over time. Although the vegetation and water indices based on RS are useful, the ancillary data would be sufficient to support mapping the locations of urban agricultural farms. RS is an excellent way to monitor vegetation conditions in areas where actively managed urban farms are known to exist (Ozdogan et al., 2010). RS combined with GIS and Web sources can help to plan the easiest route to transport and market agricultural products.

18.8 TECHNIQUES FOR ACQUIRING SPATIAL INFORMATION

Land use (LU) and land-use change, the critical driving force behind the global environmental transition, are central to the discussion on sustainable development. Rapid changes in LU and LU, particularly in developing nations, are often characterized by rapid urban sprawl, land degradation, or agricultural land transformation. Human-induced changes in land cover, for example, affect the global carbon (GC) cycle and lead to the rise in atmospheric carbon monoxide. It is therefore essential to analyze changes in LU/cover so that their effect on the terrestrial ecosystem can be distinguished, and sustainable LU planning can be developed (Muttitanon & Tripathi, 2005). Identifying the peri-urban locations is also significant, as rural areas at one point become peri-urban. It is necessary to record the extent of the conservation of agricultural activities in the peri-urban area.

Urban growth monitoring is a method of analyzing variations in the status of an object or phenomenon by remotely observing it at various times. Growth monitoring has become a significant application of the GIS and RS data. Growth monitoring is the process of assessing and defining changes in land-covering and land-use properties based on co-registered, multi-temporal RS data. The basic principle of using RS data for the identification of difference is that the method will distinguish variations between two or more dates that are uncharacteristic of the regular variety (Shalaby & Tateishi, 2007). In particular, LU practices and urban growth in RS require an examination of two recorded, aerial, or multispectral satellite bands from the same geographical area collected at two different times. Such analysis aims at identifying variabilities that occurred in the same geographic region between the two times considered (Radke et al., 2005). RS data is beneficial due to its synoptic vision, repeated coverage, and real-time data collection (Mukherjee, 1987). Such strategies would help to identify the amount of potential agricultural area lost due to urbanization and would also help to determine the existing open spaces in the city that have the requirements for urban agriculture to be developed.

Applications of coarse-resolution imagery in combination with the observations of anthropogenic lights, Schneider et al. (2009) identified urban areas with a supervised decision tree (DT) classification algorithm. Part of this study was to determine what urban meant. In their analysis, all impervious surfaces, including highways, houses, runways, parking lots, and other human-constructed components which constituted more than 50% of the landscape unit were considered urban. With that in mind, an overview of the city can be manifested inside Google Earth Pro using VHR data to classify areas of dense to relatively dense residential structures and road networks, taking care to avoid the transition to rural areas on the edges of cities. When urban spaces are established, the second challenge is to separate parks, undeveloped land, and backyards from gardens. Thebo et al. (2014) present a global analysis of UPA including its irrigated area. Another primary way to enhance our awareness of ongoing UPA activities is through the use of data mining techniques. The scrapping of social networking sites such as Facebook, Twitter, and Instagram, as well as intersecting economic, community, and agricultural data with geospatial information will lead to a significant improvement in our knowledge of UPA location and development. There is multi-source for identifying and mapping of UPA. The mixed-methods approach is possible to overcome the issues of classifying and monitoring urban farming in cities. Very high resolution (VHR) and moderate resolution imagery may be used along with non-imagery information discovered through data science methods on where farms are located, which could be one of the methods used to classify UPA remotely.

18.9 APPLICATIONS OF SATELLITE IMAGERY IN THE PLANNING OF UPA

GIS is a decision-making support network that can promote urban planning. The use of GIS modeling has become prevalent in the field of urban planning science. Some civic planning work uses GIS as a method to understand the effect of urban sprawl on the natural environment. For example, GIS shows spatial patterns of urban sprawl by measuring the distances of new urban growth areas from city centers and roads (Xia et al., 2019). GIS simulates potential land development that will enable the planning of new agricultural growth areas and the recovery of open and green areas in the city.

18.9.1 MEDIUM RESOLUTION DATA

Medium-resolution satellite data can be used to conduct space searches in urban environments. The extensive area coverage allows large geographical areas to be covered by a single scene, making it a cost-effective evaluation tool. Satellite data will enable decision-makers to determine the pace of change in vegetative regions relative to the rate of new developed and infrastructural growth, as well as to envision patterns of urban distribution in relation to new construction projects. It also enables to perform an overview of models in LU and to recognize areas of encroachment and environmental sensitivity.

18.9.2 HIGH-RESOLUTION DATA

The high-res imagery enables for spatial updates, classification of areas of UPA based on species of plants, the ecology of the area, soil conditions, and status of water resources. Planning for action scenarios is made possible by the improved spatial accuracy, which also creates an accurate database of spatial environments.

18.9.3 PANCHROMATIC DATA

Panchromatic imagery provides the basis on which powerful tools can be developed, particularly, environmental planning and monitoring, asset management, urban, and peri-urban agricultural diversity, mapping, and analysis of ecological conditions, disaster management, and access to GIS-based property information on the internet.

18.10 REMOTE SENSING (RS) APPLICATIONS

18.10.1 VEGETATION

The normalized difference vegetation index (NDVI) can be considered as an environmental metric. NDVI is commonly used in forestry and agriculture to monitor changes in plant phenology, to assess plant stress and health, and to distinguish vegetation from impervious surfaces in urban areas. NDVI has a favorable relationship to net primary growth. The NDVI satellite product can be used to assess urban community garden locations. The research can be carried out with time series over years to evaluate the rising seasons. The NDVI profiles show the robust essence of urban agriculture (Parece & Campbell, 2017). The satellite imagery of many earth observation satellites currently orbiting the earth can be used for research. Satellite imagery gives researchers the ability to make period distinctions between only a few days and decades (USGS, 2016).

Sensors aboard the satellites measure electromagnetic (EMR) radiation from the sun that is reflected off objects/features on the earth's surface. Sensors aboard different satellites differ in their sensitivity to regions of the electromagnetic spectrum; they can be cross-calibrated, enabling comparison of vegetation indices over time (Teillet et al., 2001). Healthy vegetation indices offer insights into the vigor of the ecosystem, defined by vegetation indices – ratios determined from pixel values in different satellite imagery bands. The pixel represents the footprint of a specific area on the surface of the earth and the size of the area varies by satellite system. The value of each pixel provides a digital number based on the observed brightness of the landscape within that pixel, and the set of values across all bands reflects the spectral signature(s) of that particular region. However, an individual pixel could be a homogeneous land cover (e.g., a particular species in the forest, the top of a house, or a green cover) or a heterogeneous land cover (e.g., a mixture of sidewalk, grass, and a portion of a building roof). NDVI is commonly used to monitor plant phenology, to identify plant stress, and to assess vegetation health in forestry and agriculture (Huete et al., 1994) and to differentiate vegetation from impervious surfaces in urban areas (Weng, 2012).

18.10.2 METEOROLOGY

Urban agriculture is heavily influenced by climatic and metrological phenomena. Metrological data can be obtained through a separate point station observation spatial network. Conventional agrometrological techniques have a significant restriction on the use of their data for real-time agricultural monitoring. Satellite metrology has provided precise and regular measurements of many basic agrometrological parameters (e.g., surface albedo, solar radiation, surface temperature, precipitation, and evapotranspiration). Agrometeorological inputs for the correlation-weighted regression model included significant precipitation at biweekly intervals, minimum, and maximum temperatures. Geostationary satellites RS of weather and environment is considered to be the single most important advance for urban vegetation monitoring.

18.10.3 WATER RESOURCES MANAGEMENT

Remote sensing (RS) is one of the most powerful methods for evaluating and tracking water quality. HRS is emerging as a more in-depth means of examining spatial, spectral, and temporal changes in order to provide more precise estimates of the information needed for water management applications. The advent of remote microwave sensing has made it possible to determine the availability of soil moisture using RS data. One of the most important natural resources is groundwater, which encourages ecological diversity. The applications of GIS and RS techniques in groundwater hydrology have gained cursor treatment. Good knowledge of the geographical area and related spatial information such as water supplies, watersheds, land usage, agricultural usage, rainfall, temperature, humidity, soil quality and composition, geology, atmospheric conditions, human activities, and environmental data are critical for water and agriculture management. GIS and RS technologies also identify challenges, importance, and sustainable management of groundwater and freshwater (Gebeyehu, 2019). The application of the GIS and RS techniques has made it possible to analyze vegetation production, water quality and floodplain disruptions over time and provided due consideration is given to the development of source materials and databases for urban agriculture.

18.10.4 SPECTRAL ALGORITHMS FOR QUALITATIVE AND QUANTITATIVE ANALYSIS OF SOIL PROPERTIES

The main idea behind the reflectance spectroscopy is to develop a relationship between different soil properties and its reflectance spectra. Therefore, several multivariate calibrations and validation algorithms, statistical algorithms have been developed to extract different soil properties using DRS (Ben-Dor & Banin, 1995; Shepherd & Walsh, 2002; Brown et al., 2006). With the advent of time a number of multivariate calibration techniques such as multiple linear regression (MLR) (Dalal & Henry, 1986; Ben-Dor & Banin, 1995), stepwise multiple linear regression (SMLR) (Shibusawa, 2001), multivariate adaptive regression spline (MARS) (Shepherd & Walsh, 2002), principal components regression (PCR) (Chang et al., 2001) and partial-least-squares regression (PLSR) (McCarty & Reeves, 2006; Viscarra-Rossel et al., 2006; Mouazen et al., 2010) are being used for the development of HRS algorithms to extract different soil attributes.

Initially, MLR was being used to develop modeling in the soil spectroscopic studies. It showed a problem of multi-collinearity associated with the reflectance spectra. To overcome this issue, SMLR comes into the soil attribute studies. But a very few researchers worked with this SMLR technique. In 1991, Friedman developed MARS, an accurate and fast algorithm for continuous and binary variables. It automatically detects the non-linear relationship in an extensive database and creates an explicit model. However, this model could not handle large datasets. Thus, PCR comes into existence, which not only solves the multi-collinearity problem associated with the spectral data but also helps in the reduction in the data size. PCR developed a relationship among the variables by an orthogonal transformation of the attributes. Similar to PCR, PLSR has been used extensively in the prediction of soil characteristics using spectral data. In PLSR, both the response and predictor variables are projected to orthogonal transformation to get maximum covariance between them, providing an accurate estimation of attributes with the help of HRS.

18.11 MACHINE LEARNING (ML) IN SOIL FERTILITY ANALYSIS

Machine learning (ML) creates an intelligent model that can learn by itself through the experience without human intervention. ML uses supervised and unsupervised algorithms to train the model with training data, validate the data using the validation data set, and test the model using the test dataset. ML algorithms can be used for soil health monitoring to provide real-time up to minute information on soil health. They are learning done through the data acquired from IoT devices, assist in maintaining soil health, and provide recommendation on appropriate fertilizers for nourishing the soil.

A ML algorithm used soil fertility prediction on the intrinsic properties of a particular soil. Based on the results, the soil type graded, and a suitable crop for that particular land was recommended (Keerthan et al., 2019). They have considered the electrical conductivity (EC), power of hydrogen, potassium, magnesium, zinc, sulfur, phosphorus, boron, and organic carbon as feature variables for grading the soil. They have used a regression algorithm, namely multivariate linear regression, for predicting the soil fertility (on the scale of 5). The gradient descent algorithm used to attain a global minimum. Support vector machine (SVM), DT, Gaussian Naïve Bayes (NB), and random forest (RF) classification algorithms evaluated, and RF proves to be the best fit for the model. Kesavaraj & Sukumaran (2013) classified algorithms and concluded that the performance of the algorithms depends on the dataset chosen. The workers surveyed various algorithms in DT, SVM, Neural Networks, Bayesian Network, and Rule-based Learning and couldn't suggest a particular algorithm best suited for classifying the entire data set. The necessary data for evaluation obtained from soil analysis carried out in the laboratory. Figure 18.3 shows the stages of soil fertility analysis.

The outcome of this analysis will provide the suitable parameters required for prediction. Appropriate analytical methods will be adopted to acquire the estimated values of micronutrients available, soil texture, cation exchange capacity (CEC), soil structure, water-holding capacity, soil moisture, pH, EC, lime requirement, gypsum requirement, total nitrogen, mineralizable nitrogen, organic carbon, inorganic nitrogen, available potassium, available phosphorus, available sulfur, calcium, and magnesium. Depending on the prediction, suitable parameters can be chosen from the dataset selected. Soil grids are designed to provide an accurate forecast of soil properties across the globe. The predictions made through soil grids can be considered as the standard values for the soil type in any location. Hengl et al. (2017) developed a version of Soil Grids using an ensemble of three ML algorithms, namely multi-nominal logistic regression, gradient boosting, and RF. They have compared the results of ensemble methods with that of the previous version of Soil Grids and suggested improvements for variation of results with Soil Grids prior version. Automation in agriculture through artificial intelligence, its subsets ML, and deep learning, the IoT for agriculture, RS, and wireless sensor systems will increase crop production by nourishing the soil periodically (Jha et al., 2019). These technologies are an improvement over one another but still provide effective results in the automation when integrated to work together.

Wireless sensor systems and communication will help in better connection and interaction of IoT devices in the application domain, the real-time data sensed through IoT devices will be given as input to the ML algorithms for make wise predictions and classification. Deep learning outperforms ML algorithms in terms of massive data or image datasets. Automation through

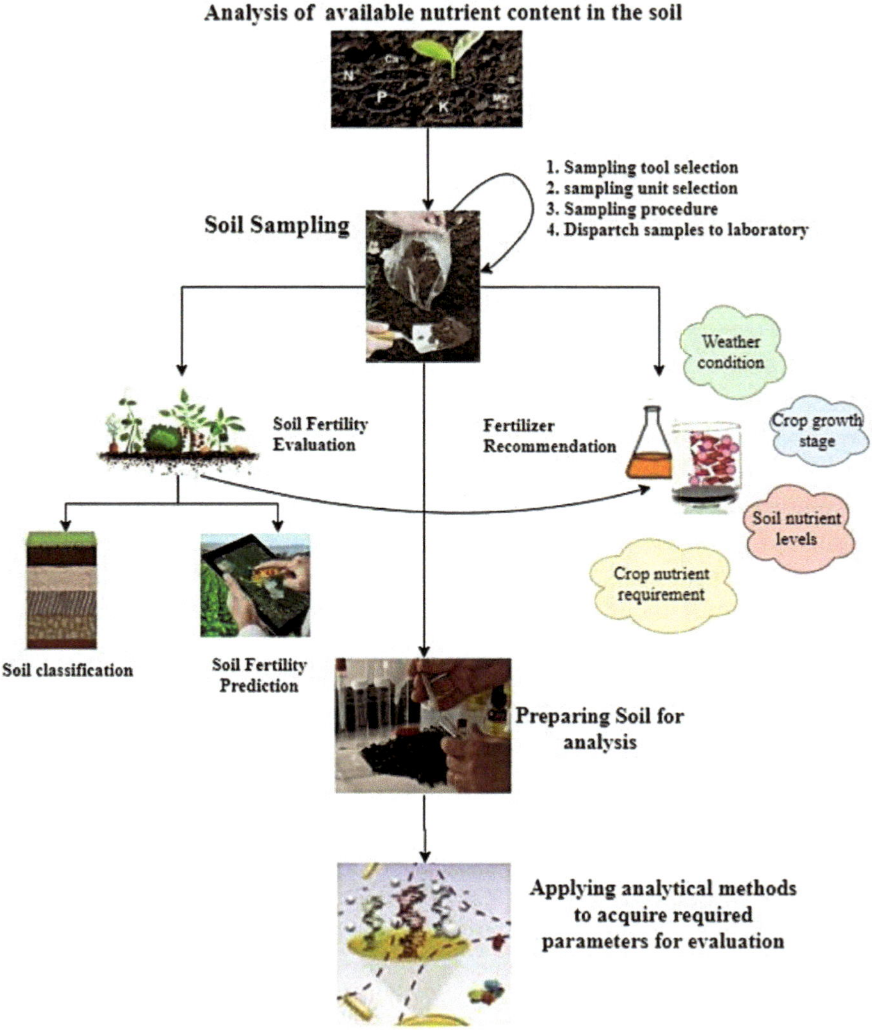

FIGURE 18.3 Stages in soil fertility analysis.

these technologies made in all stages of crop management from seed selection to post-harvest seed processing. A predictive model for soil fertility identification to suggest appropriate crops that can be planted in the land selected developed (Vijayabaskar et al., 2017). This model also recommends suitable fertilizer for retaining the soil fertility and nourishment of the plant growth.

Soil management is the most prominent role in determining a region's suitability for sustainable agriculture. Especially soil's temperature provides the best indication of a region's impact on the climatic changes. For last few decades chemometric analysis such as principal component analysis (PCA), discriminant analysis (DA), principal component regression (PCR), MLR are being used as quantitative indicator of the composition of soil (Haberhauer et al., 1999, 2000). However, ML approach is emerging as a new tool for prediction on soil properties (Liakos et al., 2018). The ML algorithms used for the measurement of different soil parameters are: prediction of soil drying considering the precipitation and ET for assisting the remote agriculture (Coopersmith et al., 2014); self-adaptive evolutionary-extreme learning machine model was developed to estimate the soil temperature from different depths from different regions utilizing the daily weather report (Nahvi et al., 2016); soil moisture estimation based on sensor data using Artificial Neural Network models (Johann et al., 2016). Prakash et al. (2018) estimated the soil moisture on three different soil data set using support vector regression, MLR and recurrent neural network. Dimitriadis & Goumopoulos (2008) implemented ML approach to irrigate and supply of fertilizers to crop based on the soil condition.

18.12 IOT IN SOIL FERTILITY ANALYSIS

IoT plays a prominent role almost in all the fields of science and technology. As IoT makes it possible for any kind of object to communicate, it has become mandatory for using it in all aspects, demanding information, and communication technology. Seamlessly it also has a greater role in modern agriculture practices. With the help of the IoT devices, agriculturalists can get up-to-minute information about soil health, crop growth, intruders, and rainfall. Nevertheless, AI technologies acquire the data from these devices for making a valuable prediction on weather, fertilizers, and water supply, thus making the task simpler. Jaishetty & Patil (2016) have developed a 4G compatible IoT based system for monitoring the agricultural field and irrigation control. They have used various sensors to monitor the soil moisture, weather condition, and rainfall prediction. The results are very optimal with high packet delivery rates and effectively predict abnormal behavior in plant growth. A real-time soil monitoring and pest warning system for mart agriculture developed using IoT enabled wireless sensor networks. The system acquires the data on soil properties, plant growth from the sensor nodes and provides SMS alerts on pest warning, crop disease based on images captured through camera sensors (Patil & Kale, 2016; Sreekantha & Kavya, 2017). IoT based crop monitoring system using SMC, temperature, and humidity was developed to monitor crop growth and provides periodical SMS updates to the concerned person (Balaji, G. Naveen et al., 2018).

Kondalli & Sohu (2016) have developed an IoT based crop monitoring system using Losant IoT – a cloud platform. This system will provide alerts to the farmer whenever the moisture content in the soil is low. It is an automate irrigation process based on temperature and humidity. Soil is the main factor that will determine the yield. An IoT based crop yield monitoring system with an automated irrigation facility implemented (Rajalakshmi & Mahalakshmi, 2016). This system can provide precise information on crop health obtained through various sensors in the field. Also, depending on the SMC and weather condition, this system reports the areas under a scarcity of water followed with automated water supply. Ananthi et al. (2017) have proposed an IoT based soil monitoring system with assisted irrigation facility and pests warning. Soil fertility can be predicted based on the values retrieved from the moisture sensor, humidity sensor, pH sensor, and temperature sensor. This system can suggest appropriate vegetation that can be grown on the available soil health status. Also, can plan the irrigation system using automation technology based on soil temperature. With the help of this IOT based system not only we can measure the plant growth but also, we can manage the pest attack on plants with the images of the crop captured at different stages of plant. So, an IoT based soil monitoring system was proposed to track excess water stagnation in the agriculture farm and controlling the motor switch through an application from the mobile (Ayyasamy et al., 2020). Also, the prediction of micronutrients in the soil made to ensure soil fertility. DHT sensors are used for predicting weather conditions. Zhang et al. (2017) have used IoT based decision support system for monitoring the soil moisture, temperature, and humidity detection in the citrus orchard. This system had helped the farmers to improve the citrus production by recommending proper fertilizer input accessing the appropriate level of chemical fertilizer and proper water supply. Remote monitoring of soil nutrient level to estimate the moisture content of the soil by considering the inverse relation between soil moisture and its resistance using data acquired through IoT sensors, namely, DS18B20 temperature sensor (Na et al., 2016). This system was deployed on the STM32 Nucleo platform. Thus, IoT integrated with wireless sensor networking technologies like ZigBee and Bluetooth deployed on various open-source platforms assist in developing automated models for smart agriculture.

18.13 CONCLUSION

It has been concluded from this study that the RS and GIS based approaches are the most promising tools for PA. The applications of ML and the IoT approaches in agriculture are the emerging techniques not only to maintain soil health but also increase in the crop yield. Soil nutrient management, crop health monitoring, pest control, land management and water contamination analysis becomes easy with the help of these spatial/statistical technologies. This chapter also highlights urban agriculture, which indicates the social awareness towards food demand. This chapter discussed the importance of different modern technologies in the field of agriculture like measuring soil health status using spatial technologies and provides an outline of the current and potential future applications and possible solutions of implementations.

KEYWORDS

- **decision-making**
- **GIS**
- **internet of things**
- **meteorology**
- **precision agriculture**
- **remote sensing applications**
- **smart agriculture**
- **vegetation**

REFERENCES

Aitkenhead-Peterson, J. A., Steele, M. K., & Volder, A., (2010). Services in natural and human dominated ecosystems. *Urban Ecosystem and Ecology, 55*, 373–390.

Al-Asadi, R. A., & Mouazen, A. M., (2014). Combining frequency domain reflectometry and visible and near infrared spectroscopy for assessment of soil bulk density. *Soil and Tillage Research, 135*, 60–70.

Ananthi, N., Divya, J., Divya, M., & Janani, V., (2017). IoT based smart soil monitoring system for agricultural production. In: *Proc. IEEE Technological Innovations in ICT for Agriculture and Rural Development (TIAR)*.

Andrea, F., Bini, C., & Amaducci, S., (2017). Soil and ecosystem services: Current knowledge and evidences from Italian case studies. *Applied Soil Ecology, 123*, 693–698.

Angers, D. A., Bolinder, M. A., Carter, M. R., Gregorich, E. G., Drury, C. F., Liang, B. C., Voroney, R. P., et al., (1997). Impact of tillage practices on organic carbon and nitrogen storage in cool, humid soils of eastern Canada. *Soil and Tillage Research, 41*(3, 4), 191–201.

Angstrom, A., (1925). The albedo of various surfaces of ground. *Geografiska Annaler, 7*(4), 323–342.

Anne, N. J., Abd-Elrahman, A. H., Lewis, D. B., & Hewitt, N. A., (2014). Modeling soil parameters using hyperspectral image reflectance in subtropical coastal wetlands. *International Journal of Applied Earth Observation and Geoinformation, 33*, 47–56.

Appeaning, A. K., (2009). *Detection, Measurement and Prediction of Shoreline Change in Accra, Ghana*. Lambert Academic Publishing, Saarbrücken, Germany.

Ayyasamy, S., Eswaran, S., Manikandan, B., Solomon, S. M., & Kumar, S. N., (2020). IoT based agri soil maintenance through micro-nutrients and protection of crops from excess water. In: *Proc. Fourth International Conference on Computing Methodologies and Communication (ICCMC)*.

Babaeian, E., Homaee, M., Montzka, C., Vereecken, H., & Norouzi, A. A., (2015). Towards retrieving soil hydraulic properties by hyperspectral remote sensing. *Vadose Zone Journal, 14*(3).

Balaji, G. N., Nandhini, V., Mithra, S., Priya, N., & Naveena, R., (2018). IoT based smart crop monitoring in farm land. *Imperial Journal of Interdisciplinary Research (IJIR), 4*, 88–92.

Balan, E., Saitta, A. M., Mauri, F., & Calas, G., (2001). First-principles modeling of the infrared spectrum of kaolinite. *American Mineralogist, 86*(11, 12), 1321–1330.

Bayer, A., Bachmann, M., Müller, A., & Kaufmann, H., (2012). A comparison of feature-based MLR and PLS regression techniques for the prediction of three soil constituents in a degraded South African ecosystem. *Applied and Environmental Soil Science*. doi: 10.1155/2012/971252.

Bellon-Maurel, V., & McBratney, A., (2011). Near-infrared (NIR) and mid-infrared (MIR) spectroscopic techniques for assessing the amount of carbon stock in soils- Critical review and research perspectives. *Soil Biology and Biochemistry, 43*(7), 1398–1410.

Ben-Dor, E., & Banin, A., (1995). Near-infrared analysis as a rapid method to simultaneously evaluate several soil properties. *Soil Science Society of America Journal, 59*(2), 364–372.

Ben-Dor, E., & Thenkabail, P. S., (2011). Characterization of soil properties using reflectance spectroscopy. *Hyperspectral Remote Sensing of Vegetation*, 513–558.

Ben-Dor, E., (2002). Quantitative remote sensing of soil properties. *Advances in Agronomy, 75*, 173–244.

Ben-Dor, E., Chabrillat, S., Demattê, J. A. M., Taylor, G. R., Hill, J., Whiting, M. L., & Sommer, S., (2009). Using imaging spectroscopy to study soil properties. *Remote Sensing of Environment, 113*, S38–S55.

Ben-Dor, E., Inbar, Y., & Chen, Y., (1997). The reflectance spectra of organic matter in the visible near-infrared and short wave infrared region (400–2500 nm) during a controlled decomposition process. *Remote Sensing of Environment, 61*(1), 1–15.

Ben-Dor, E., Irons, J. R., & Epema, G. F., (1999). Soil REFLETTANTE. Manual of remote sensing. *Remote Sensing for the Earth Sciences, 1*, 111.

Ben-Dor, E., Taylor, G. R., Hill, J., Demattê, J. A. M., Whiting, M. L., Chabrillat, S., & Sommer, S., (2008). Imaging spectrometry for soil applications. *Advanced Agronomy, 97*, 321–392.

Bishop, J. L., Lane, M. D., Dyar, M. D., & Brown, A. J., (2008). Reflectance and emission spectroscopy study of four groups of phyllosilicates: Smectites, kaolinite-serpentines, chlorites, and micas. *Clay Minerals, 43*(1), 35–54.

Bishop, J. L., Pieters, C. M., & Edwards, J. O., (1994). Infrared spectroscopic analyses on the nature of water in montmorillonite. In: *Clays and Clay Minerals.*

Boardman, J. W., & Kruse, F. A. (1994). Automated Spectral Analysis: A Geological Example Using AVIRIS Data, North Grapevine Mountains, Nevada. In: ERIM, Ed., *Proc. 10th Thematic Conference on Geological Remote Sensing,* San Antonio, 9–12 May 1994, 407–418.

Bowers, S. A., & Hanks, R. J., (1965). Reflection of radiant energy from soils. *Soil Science, 100*(2), 130–138.

Brown, D. J., Shepherd, K. D., Walsh, M. G., Mays, M. D., & Reinsch, T. G., (2006). Global soil characterization with VNIR diffuse reflectance spectroscopy. *Geoderma, 132*(3, 4), 273–290.

Bryld, E., (2003). Potentials, problems, and policy implications for urban agriculture in developing countries. *J. Agr. Human Values, 20*, 79–86.

Cambule, A. H., Rossiter, D. G., Stoorvogel, J. J., & Smaling, E. M. A., (2012). Building a near infrared spectral library for soil organic carbon estimation in the Limpopo National Park, Mozambique. *Geoderma, 183*, 41–48.

Cecillon, L., Barthes, B. G., Gomez, C., Ertlen, D., Genot, V., Hedde, M., Stevens, A., & Brun, J. J., (2009). Assessment and monitoring of soil quality using near-infrared reflectance spectroscopy (NIRS). *European J. of Soil Sci., 60*(5), 770–784.

Chabrillat, S., Goetz, A. F., Krosley, L., & Olsen, H. W., (2002). Use of hyperspectral images in the identification and mapping of expansive clay soils and the role of spatial resolution. *Remote Sensing of Environment, 82*(2, 3), 431–445.

Chang, C. W., Laird, D. A., Mausbach, M. J., & Hurburgh, C. R., (2001). Near-infrared reflectance spectroscopy–principal components regression analyses of soil properties. *Soil Sci. Soc. America J., 65*(2), 480–490.

Ciurczak, E. W., (2001). Principles of near-infrared spectroscopy. *Practical Spectroscopy Series, 27*, 7–18.

Clark, R. N., (1995). Reflectance spectra. *Rock Physics and Phase Relations: A Handbook of Physical Constants* (Vol. 3, pp. 178–188). AGU Reference Shelf.

Clark, R. N., (1999). Spectroscopy of rocks and minerals, and principles of spectroscopy. *Manual of Remote Sensing, 3*(3–58), 2–12.

Clark, R. N., King, T. V., Klejwa, M., Swayze, G. A., & Vergo, N., (1990). High spectral resolution reflectance spectroscopy of minerals. *Journal of Geophysical Research: Solid Earth, 95*(B8), 12653–12680.

Cohen, M. J., Mylavarapu, R. S., Bogrekci, I., Lee, W. S., & Clark, M. W., (2007). Reflectance spectroscopy for routine agronomic soil analyses. *Soil Science, 172*(6), 469–485.

Cohen, M. J., Prenger, J. P., & DeBusk, W. F., (2005). Visible-near infrared reflectance spectroscopy for rapid, nondestructive assessment of wetland soil quality. *Journal of Environmental Quality, 34*(4), 1422–1434.

Coleman, T. L., Agbu, P. A., Montgomery, O. L., Gao, T., & Prasad, S., (1991). Spectral band selection for quantifying selected properties in highly weathered soils. *Soil Science, 151*(5), 355–361.

Condit, H. R., (1970). The spectral reflectance of American soils. *Photogrammetric Engineering.*

Coopersmith, E. J., Minsker, B. S., Wenzel, C. E., & Gilmore, B. J., (2014). Machine learning assessments of soil drying for agricultural planning. *Computers and Electronics in Agriculture, 104*, 93–104.

Cornell, R. M., & Schwertmann, U., (1996). *The Iron Oxides*. VCH-Verlagsgesellschaft.

Crowley, J. K., & Zimbelman, D. R., (1997). Mapping hydrothermally altered rocks on Mount Rainier, Washington, with airborne visible/infrared imaging spectrometer (AVIRIS) data. *Geology, 25*(6), 559–562.

Curcio, D., Ciraolo, G., D'Asaro, F., & Minacapilli, M., (2013). Prediction of soil texture distributions using VNIR-SWIR reflectance spectroscopy. *Procedia Environmental Sciences, 19*, 494–503.

Dalal, R. C., & Henry, R. J., (1986). Simultaneous determination of moisture, organic carbon, and total nitrogen by near infrared reflectance spectrophotometry 1. *Soil Sci. Soc. America J., 50*(1), 120–123.

Daniel, K. W., Tripathi, N. K., Honda, K., & Apisit, E., (2004). Analysis of VNIR (400-1100 nm) spectral signatures for estimation of soil organic matter in tropical soils of Thailand. *International Journal of Remote Sensing, 25*(3), 643–652.

Deelstra, T., & Girardet, H., (2000). *Growing Cities, Growing Food: Urban Agriculture on the Policy Agenda* (p. 531). A Reader on Urban Agriculture.

Dehaan, R., & Taylor, G. R., (2003). Image-derived spectral end members as indicators of salinization. *International Journal of Remote Sensing, 24*(4), 775–794.

Deng, L., Wang, K. B., Chen, M. L., Shangguan, Z. P., & Sweeney, S., (2013). Soil organic carbon storage capacity positively related to forest succession on the Loess Plateau, China. *Catena, 110*, 1–7.

Dimitriadis, S., & Goumopoulos, C., (2008). Applying machine learning to extract new knowledge in precision agriculture applications. In: *Proc. Pan-Hellenic Conference on Informatics*. IEEE.

Djomgoue, P., & Njopwouo, D., (2013). FT-IR spectroscopy applied for surface clays characterization. *Journal of Surface Engineered Material Advance Technology, 3*(4), 275.

Drescher, A., Holmer, R., & Iaquinta, D., (2006). Urban homegardens and allotment gardens for sustainable livelihoods: Management strategies and institutional environments. *Tropical Homegardens* (pp. 317–338). Springer, the Netherlands.

Dufrechou, G., Grandjean, G., & Bourguignon, A., (2015). Geometrical analysis of laboratory soil spectra in the short-wave infrared domain: Clay composition and estimation of the swelling potential. *Geoderma, 243*, 92–107.

Dunne, K. A., & Willmott, C. J., (1996). Global distribution of plant-extractable water capacity of soil. *International Journal of Climatology: A Journal of the Royal Meteorological Society, 16*(8), 841–859.

Farmer, V. C., (1974). *Infra-red Spectra of Minerals*. Mineralogical Society.

Farmer, V. T., & Russell, J. D., (1964). The infra-red spectra of layer silicates. *Spectrochimica Acta, 20*(7), 1149–1173.

Franceschini, M. H. D., Dematte, J. A., Da Silva, T. F., Vicente, L. E., Bartholomeus, H., & De Souza, F. C. R., (2015). Prediction of soil properties using imaging spectroscopy: Considering fractional vegetation cover to improve accuracy. *International Journal of Applied Earth Observation and Geoinformation, 38*, 358–370.

Galvao, L. S., Vitorello, I., & Formaggio, A. R., (1997). Relationships of spectral reflectance and color among surface and subsurface horizons of tropical soil profiles. *Remote Sensing of Environment, 61*(1), 24–33.

Galvao, L. S., Vitorello, I., (1998). Role of organic matter in obliterating the effects of iron on spectral reflectance and color of Brazilian tropical soils. *International Journal of Remote Sensing, 19*(10), 1969–1979.

Gates, W., (2005). Infrared spectroscopy and the chemistry of dioctahedral smectites. *The Application of Vibrational Spectroscopy to Clay Minerals and Layered Double Hydroxides* (pp. 126–168). Clay Minerals Society.

Ge, Y., Morgan, C. L., Grunwald, S., Brown, D. J., & Sarkhot, D. V., (2011). Comparison of soil reflectance spectra and calibration models obtained using multiple spectrometers. *Geoderma, 161*(3, 4), 202–211.

Gebeyehu, M. N., (2019). Remote sensing and GIS application in agriculture and natural resource management. *Int. J. Environ. Sci. Nat. Res., 19*(2).

Geladi, P., (2003). Chemometrics in spectroscopy. Part 1. Classical chemometrics. *Spectrochimica Acta Part B: Atomic Spectroscopy, 58*(5), 767–782.

Glotch, T. D., & Rossman, G. R., (2009). Mid-infrared reflectance spectra and optical constants of six iron oxide/oxyhydroxide phases. *Icarus, 204*(2), 663–671.

Glotch, T. D., Morris, R. V., Christensen, P. R., & Sharp, T. G., (2004). Effect of precursor mineralogy on the thermal infrared emission spectra of hematite: Application to Martian hematite mineralization. *Journal of Geophysical Research: Planets, 109*(E7).

Goetz, A. F., & Herring, M., (1989). The high-resolution imaging spectrometer (HIRIS) for Eos. *IEEE Transactions on Geoscience and Remote Sensing, 27*(2), 136–144.

Goetz, A. F., Vane, G., Solomon, J. E., & Rock, B. N., (1985). Imaging spectrometry for earth remote sensing. *Science, 228*(4704), 1147–1153.

Gomez, C., Rossel, R. A. V., & McBratney, A. B., (2008). Soil organic carbon prediction by hyperspectral remote sensing and field vis-NIR spectroscopy: An Australian case study. *Geoderma, 146*(3, 4), 403–411.

Green, R. O., Eastwood, M. L., Sarture, C. M., Chrien, T. G., Aronsson, M., Chippendale, B. J., Faust, J. A., et al., (1998). Imaging spectroscopy and the airborne visible/infrared imaging spectrometer (AVIRIS). *Remote Sensing of Environment, 65*(3), 227–248.

Haberhauer, G., & Gerzabek, M. H., (1999). Drift and transmission FT-IR spectroscopy of forest soils: An approach to determine decomposition processes of forest litter. *Vibrational Spectroscopy, 19*, 413–417.

Haberhauer, G., Feigl, B., Gerzabek, M. K., & Cerri, C., (2000). FT-IR spectroscopy of organic matter in tropical soils: Changes induced through deforestation. *Applied Spectroscopy, 54*, 221–224.

Hartemink, A. E., (2008). Soils are back on the global agenda. *Soil Use and Management, 24*(4), 327–330.

Haubrock, S. N., Chabrillat, S., Lemmnitz, C., & Kaufmann, H., (2008). Surface soil moisture quantification models from reflectance data under field conditions. *International Journal of Remote Sensing, 29*(1), 3–29.

Hengl, T., De Jesus, J. M., Heuvelink, G. B., Gonzalez, M. R., Kilibarda, M., Blagotić, A., Shangguan, W., et al., (2017). Soil grids 250 m: Global gridded soil information based on machine learning. *PLoS One, 12*(2).

Herold, M., Goldstein, N., & Clarke, K., (2003). The spatiotemporal form of urban growth: Measurement, analysis, and modeling. *Remote Sensing of Environment, 86*, 286–302.

Hick, P. T., & Russell, W. G. R., (1990). Some spectral considerations for remote sensing of soil salinity. *Soil Research, 28*(3), 417–431.

Hill, J., Megier, J., & Mehl, W., (1995). Land degradation, soil erosion and desertification monitoring in Mediterranean ecosystems. *Remote Sensing Reviews, 12*(1, 2), 107–130.
Hirschman, D., & Battiata, J., (2016). Urban storm water management: Evolution of process and technology. Sustainable water management in urban environments. *The Handbook of Environmental Chemistry* (Vol. 47, pp. 83–120). Springer, Switzerland.
Huang, D., & Drescher, M., (2015). Urban crops and livestock: The experiences, challenges, and opportunities of planning for urban agriculture in two Canadian provinces. *Land Use Policy, 42*, 1–14.
Huete, A., Justice, C., & Liu, H., (1994). Development of vegetation and soil indices for MODIS-EOS. *Remote Sensing of Environment, 9*, 224–234.
Hunt, G. R., & Ashley, R. P., (1979). Spectra of altered rocks in the visible and near infrared. *Economic Geology, 74*(7), 1613–1629.
Hunt, G. R., & Salisbury, J. W., (1970). Visible and near infrared spectra of minerals and rocks. I. Silicate minerals. *Modern Geology, 1*, 283–300.
Hunt, G. R., & Salisbury, J. W., (1976). *Mid-Infrared Spectral Behavior of Metamorphic Rocks (No. AFCRL-TR-76-0003).* Air force Cambridge research Labs Hanscom AFB Mass.
Hunt, G. R., (1979). Near-infrared (1.3–2.4) μm spectra of alteration minerals- potential for use in remote sensing. *Geophysics, 44*(12), 1974–1986.
Hunt, G. R., Logan, L. M., & Salisbury, J. W., (1973). Mars: Components of infrared spectra and the composition of the dust cloud. *Icarus, 18*(3), 459–469.
Islam, K., Singh, B., & McBratney, A., (2003). Simultaneous estimation of several soil properties by ultra-violet, visible, and near-infrared reflectance spectroscopy. *Soil Research, 41*(6), 1101–1114.
Jaishetty, S. A., & Patil, R., (2016). IoT sensor network based approach for agricultural field monitoring and control. *International Journal of Research in Engineering and Technology, 6.*
Janik, L. J., & Skjemstad, J. O., (1995). Characterization and analysis of soils using mid-infrared partial least-squares. 2. Correlations with some laboratory data. *Soil Research, 33*(4), 637–650.
Janik, L. J., Merry, R. H., & Skjemstad, J. O., (1998). Can mid infrared diffuse reflectance analysis replace soil extractions? *Animal Production Science, 38*(7), 681–696.
Jha, K., Doshi, A., Patel, P., & Shah, M., (2019). A comprehensive review on automation in agriculture using artificial intelligence. *Artificial Intelligence in Agriculture.*
Johann, A. L., De Araújo, A. G., Delalibera, H. C., & Hirakawa, A. R., (2016). Soil moisture modeling based on stochastic behavior of forces on a no-till chisel opener. *Computers and Electronics in Agriculture, 121*, 420–428.
Kasturirangan, K., Aravamudan, R., Deekshatulu, B. L., Joseph, G., & Chandrasekhar, M. G., (1996). Indian remote sensing satellite IRS-1C- the beginning of a new era. *Curr. Sci., 70*(7), 495–500.
Keerthan, K. T. G., Shubha, C., & Sushma, S. A., (2019). Random forest algorithm for soil fertility prediction and grading using machine learning. *International Journal of Innovative Technology and Exploring Engineering, 9*(1), 1301–1304.
Kesavaraj, G., & Sukumaran, S., (2013). A study on classification techniques in data mining. In: *Fourth International Conference on Computing, Communications and Networking Technologies (ICCCNT).*

Khanna, S., Palacios-Orueta, A., Whiting, M. L., Ustin, S. L., Riano, D., & Litago, J., (2007). Development of angle indexes for soil moisture estimation, dry matter detection and land-cover discrimination. *Remote Sensing of Environment, 109*(2), 154–165.

Knadel, M., Viscarra-Rossel, R. A., Deng, F., Thomsen, A., & Greve, M. H., (2013). Visible-near infrared spectra as a proxy for topsoil texture and glacial boundaries. *Soil Sci. Soc. America J., 77*(2), 568–579.

Kodali, R. K., & Sahu, A., (2016). An IoT based soil moisture monitoring on LOSANT platform. In: *Proc. 2nd International Conference on Contemporary Computing and Informatics (IC3I)*.

Kustova, G. N., Burgina, E. B., Sadykov, V. A., & Poryvaev, S. G., (1992). Vibrational spectroscopic investigation of the goethite thermal decomposition products. *Physics and Chemistry of Minerals, 18*(6), 379–382.

Lagacherie, P., Baret, F., Feret, J. B., Netto, J. M., & Robbez-Masson, J. M., (2008). Estimation of soil clay and calcium carbonate using laboratory, field, and airborne hyperspectral measurements. *Remote Sensing of Environment, 112*(3), 825–835.

Lakhankar, T., Ghedira, H., Temimi, M., Sengupta, M., Khanbilvardi, R., & Blake, R., (2009). Non-parametric methods for soil moisture retrieval from satellite remote sensing data. *Remote Sensing, 1*(1), 3–21.

Lawson, L. J., (2005). *City Bountiful: A Century of Community Gardening in America.* University of California Press, Berkeley, CA, USA.

Liakos, K. G., Busato, P., Moshou, D., Pearson, S., & Bochtis, D., (2018). Machine learning in agriculture: A review. *Sensors, 18*(8), 2674.

Lincoln Institute of Land Policy, (2015). *Atlas of Urban Expansion.* Lincoln Institute of Land Policy, Cambridge, MA, USA. Available at: http://www.lincolninst.edu/subcenters/atlas-urban-expansion/ (accessed on 28 December 2023).

Liu, W., Baret, F., Gu, X., Zhang, B., Tong, Q., & Zheng, L., (2003). Evaluation of methods for soil surface moisture estimation from reflectance data. *International Journal of Remote Sensing, 24*(10), 2069–2083.

Lobell, D. B., & Asner, G. P., (2002). Moisture effects on soil reflectance. *Soil Sci. Soc. America J., 66*(3), 722–727.

Madani, A. A., (2005). Soil salinity detection and monitoring using Landsat data: A case study from Siwa Oasis, Egypt. *GI Science & Remote Sensing, 42*(2), 171–181.

Madejova, J., (2003). FTIR techniques in clay mineral studies. *Vibrational Spectroscopy, 31*(1), 1–10.

Madejova, J., Pentrak, M., Palkova, H., & Komadel, P., (2009). Near-infrared spectroscopy: A powerful tool in studies of acid-treated clay minerals. *Vibrational Spectroscopy, 49*(2), 211–218.

Malley, D. F., & Martin, P. D., (2003). The use of near-infrared spectroscopy for soil analysis. In: Currie, L., & Hanly, J., (eds.), *Tools for Nutrient and Pollutant Management: Application to Agriculture and Environmental Quality* (pp. 371–404). Fertilizer and Lime Research Center, New Zealand.

McCarty, G. W., & Reeves, J. B., (2006). Comparison of near infrared and mid infrared diffuse reflectance spectroscopy for field-scale measurement of soil fertility parameters. *Soil Science, 171*(2), 94–102.

McCauley, J. D., Engel, B. A., Scudder, C. E., Morgan, M. T., & Elliott, P. W., (1993). *Assessing the Spatial Variability of Organic Matter.* In: American Society of Agricultural Engineers Meeting. USA.

McClintock, N., & Simpson, M., (2014). *A Survey of Urban Agriculture Organizations and Businesses in the US and Canada: Preliminary Results*. Urban Studies and Planning Faculty Publications and Presentations. Portland State University.

Minasny, B., & Hartemink, A. E., (2011). Predicting soil properties in the tropics. *Earth-Science Reviews, 106*(1, 2), 52–62.

Mohanty, B., Gupta, A., & Das, B. S., (2016). Estimation of weathering indices using spectral reflectance over visible to mid-infrared region. *Geoderma, 266*, 111–119.

Morellos, A., Pantazi, X. E., Moshou, D., Alexandridis, T., Whetton, R., Tziotzios, G., Wiebensohn, J., et al., (2016). Machine learning based prediction of soil total Nitrogen, organic carbon, and moisture content by using VIS-NIR spectroscopy. *Biosystems Engineering, 152*(1), 104–116.

Mouazen, A. M., Kuang, B., De Baerdemaeker, J., & Ramon, H., (2010). Comparison among principal component, partial least squares, and back propagation neural network analyses for accuracy of measurement of selected soil properties with visible and near infrared spectroscopy. *Geoderma, 158*(1, 2), 23–31.

Mougenot, B., Pouget, M., & Epema, G. F., (1993). Remote sensing of salt affected soils. *Remote Sensing Reviews, 7*(3, 4), 241–259.

Mougeot, L. J. A., (1999). Urban agriculture: Definition, presence, potentials and risks, main policy challenges. In: *Proceedings of International Workshop on Growing Cities Growing Food: Urban Agriculture on the Policy Agenda*. La Habana, Cuba.

Mukherjee, S., (1987). Land use maps for conservation of ecosystems. *Geog. Rev. India, 3*, 23–28.

Mulder, V. L., Plotze, M., De Bruin, S., Schaepman, M. E., Mavris, C., Kokaly, R. F., & Egli, M., (2013). Quantifying mineral abundances of complex mixtures by coupling spectral deconvolution of SWIR spectra (2.1–2.4 μm) and regression tree analysis. *Geoderma, 207*, 279–290.

Muttitanon, W., & Tripathi, N., (2005). Land use/cover changes in the coastal zone of bay don bay, Thailand using landset 5 TM data. *Int. J. Remote Sen., 26*(11), 2311–2323.

Na, A., Isaac, W., Varshney, S., & Khan, E., (2016). An IoT based system for remote monitoring of soil characteristics. In: *Proc. International Conference on Information Technology (InCITe)-the Next Generation IT Summit on the Theme-Internet of Things: Connect Your Worlds*.

Nahvi, B., Habibi, J., Mohammadi, K., Shamshirband, S., & Al Razgan, O. S., (2016). Using self- adaptive evolutionary algorithm to improve the performance of an extreme learning machine for estimating soil temperature. *Computers and Electronics in Agriculture, 124*, 150–160.

Nanni, M. R., & Dematte, J. A. M., (2006). Spectral reflectance methodology in comparison to traditional soil analysis. *Soil Sci. Soc. America J., 70*(2), 393–407.

Nayak, P. S., & Singh, B. K., (2007). Instrumental characterization of clay by XRF, XRD and FTIR. *Bulletin of Materials Science, 30*(3), 235–238.

Nidamanuri, R. R., & Ramiya, A. M., (2014). Spectral identification of materials by reflectance spectral library search. *Geocarto International, 29*(6), 609–624.

Ozdogan, M., Yang, Y., Allez, G., & Cervantes, C., (2010). Remote sensing of irrigated agriculture: Opportunities and challenges. *Remote Sens., 2*(9), 2274–2304.

Palacios-Orueta, A., & Ustin, S. L., (1998). Remote sensing of soil properties in the Santa Monica Mountains I. Spectral analysis. *Remote Sensing of Environment, 65*(2), 170–183.

Pandey, A., Chowdary, V. M., Mal, B. C., & Dabral, P. P., (2011). Remote sensing and GIS for identification of suitable sites for soil and water conservation structures. *Land Degradation & Development, 22*(3), 359–372.

Parece, T. E., & Campbell, J. B., (2017). Assessing urban community gardens' impact on net primary production using NDVI. *Urban Agric. Reg. Food Syst., 2.*

Parece, T. E., & Campbell, J. B., (2017). Geospatial evaluation for urban agriculture land inventory: Roanoke, virginia USA. *Int. J. Appl. Geospat. Res., 8*, 43–63.

Patil, K. A., & Kale, N. R., (2016). A model for smart agriculture using IoT. In: *Proc. International Conference on Global Trends in Signal Processing, Information Computing and Communication (ICGTSPICC).*

Paz-Kagan, T., Shachak, M., Zaady, E., & Karnieli, A., (2014). A spectral soil quality index (SSQI) for characterizing soil function in areas of changed land use. *Geoderma, 230*, 171–184.

Peltre, C., Bruun, S., Du, C., Thomsen, I. K., & Jensen, L. S., (2014). Assessing soil constituents and labile soil organic carbon by mid-infrared photoacoustic spectroscopy. *Soil Biology and Biochemistry, 77*, 41–50.

Peng, X., Shi, T., Song, A., Chen, Y., & Gao, W., (2014). Estimating soil organic carbon using VIS/NIR spectroscopy with SVMR and SPA methods. *Remote Sensing, 6*(4), 2699–2717.

Pickett, S. T. A., Cadenasso, M. L., Grove, J. M., Nilon, C. H., Pouyat, R. V., Zipperer, W. C., & Costanza, R., (2001). Urban ecological systems: Linking terrestrial ecological, physical, and socioeconomic components of metropolitan areas. *Annu. Rev. Ecol. Syst., 32,* 127.

Plaza, A., Benediktsson, J. A., Boardman, J. W., Brazile, J., Bruzzone, L., Camps-Valls, G., Chanussot, J., et al., (2009). Recent advances in techniques for hyperspectral image processing. *Remote Sensing of Environment, 113,* S110–S122.

Prakash, S., Sharma, A., & Sahu, S. S., (2018). Soil moisture prediction using machine learning. In: *Proc. Second International Conference on Inventive Communication and Computational Technologies (ICICCT).*

Radke, J., Andra, S., Al-Kofani, O., & Roysan, B., (2005). Image change detection algorithms: A systematic survey. *IEEE Trans. Image Process., 14*(3), 291–307.

Rajalakshmi, P., & Mahalakshmi, S. D., (2016). IoT based crop-field monitoring and irrigation automation. In: *Proc. 10th International Conference on Intelligent Systems and Control (ISCO).*

Rawlins, B. G., Kemp, S. J., & Milodowski, A. E., (2011). Relationships between particle size distribution and VNIR reflectance spectra are weaker for soils formed from bedrock compared to transported parent materials. *Geoderma, 166*(1), 84–91.

Reeves, III. J. B., McCarty, G. W., & Meisenger, J. J., (1999). Near infrared reflectance spectroscopy for the analysis of agricultural soils. *J. Near Infrared Spectroscopy, 7*, 179–193.

Reeves, III. J. B., Mccarty, G., Reeves, V., Follett, R., & Kimble, J., (2002). Mid- versus near-infrared diffuse reflectance spectroscopy for the quantitative analysis of organic matter in soils and other biological materials. In: *Proceedings of American Chemical Society National Meeting.*

Ruan, H. D., Frost, R. L., Kloprogge, J. T., & Duong, L., (2002). Infrared spectroscopy of goethite dehydroxylation: III. FT-IR microscopy of in situ study of the thermal transformation of goethite to hematite. *Spectrochimica Acta Part A: Molecular and Biomolecular Spectroscopy, 58*(5), 967–981.

Russell, J. D., Farmer, V. C., & Velde, B., (1970). Replacement of OH by OD in layer silicates, and identification of the vibrations of these groups in infra-red spectra. *Mineralogical Magazine, 37*(292), 869–879.

Saikia, B. J., & Parthasarathy, G., (2010). Fourier transform infrared spectroscopic characterization of kaolinite from Assam and Meghalaya, Northeastern India. *Journal of Modern Physics, 1*(4), 206.

Salisbury, J. W., D'Aria, D. M., & Jarosewich, E., (1991). Mid-infra-red (2.5–13.5 μm) reflectance spectra of powdered stony meteorites. *Icarus, 92*(2), 280–297.

Santra, P., Sahoo, R. N., Das, B. S., Samal, R. N., Pattanaik, A. K., & Gupta, V. K., (2009). Estimation of soil hydraulic properties using proximal spectral reflectance in visible, near-infrared, and shortwave-infrared (VIS–NIR–SWIR) region. *Geoderma, 152*(3, 4), 338–349.

Sarathjith, M. C., Das, B. S., Vasava, H. B., Mohanty, B., Sahadevan, A. S., Wani, S. P., & Sahrawat, K. L., (2014a). Diffuse reflectance spectroscopic approach for the characterization of soil aggregate size distribution. *Soil Sci. Soc. America J., 78*(2), 369–376.

Sarathjith, M. C., Das, B. S., Wani, S. P., & Sahrawat, K. L., (2014b). Dependency measures for assessing the co-variation of spectrally active and inactive soil properties in diffuse reflectance spectroscopy. *Soil Sci. Soc. America J., 78*(5), 1522–1530.

Saxena, R. K., Verma, K. S., Chary, G. R., Srivastava, R., & Barthwal, A. K., (2000). IRS- 1C data application in watershed characterization and management. *International Journal of Remote Sensing, 21*(17), 3197–3208.

Schneider, A., Friedl, M. A., & Potere, D., (2009). A new map of global urban extent from MODIS satellite data. *Environmental Research Letters, 4*(4), 044003.

Schodlok, M. C., & Ramanaidou, E. R., (2011). Mapping quartz, carbonates and riebeckite in banded iron formation iron ore drill core using CSIRO's TIR-HyLogging system. *Proc. Iron Ore*, 147–153.

Selige, T., Böhner, J., & Schmidhalter, U., (2006). High resolution topsoil mapping using hyperspectral image and field data in multivariate regression modeling procedures. *Geoderma, 136*(1, 2), 235–244.

Shalaby, R., & Tateishi, (2007). Remote sensing and GIS for mapping and monitoring land-cover and land-use changes in the Northwestern coastal zone of Egypt. *Appl. Geogr., 27*, 28–41.

Shepherd, K. D., & Walsh, M. G., (2002). Development of reflectance spectral libraries for characterization of soil properties. *Soil Sci. Soc. America J., 66*(3), 988–998.

Sheppard, N., Willis, H. A., & Rigg, J. C., (1985). Names, symbols, definitions, and units of quantities in optical spectroscopy (recommendations 1984). *Pure and Applied Chemistry, 57*(1), 105–120.

Sherman, D. M., & Waite, T. D., (1985). Electronic spectra of Fe^{3+} oxides and oxide hydroxides in the near IR to near UV. *American Mineralogist, 70*(11, 12), 1262–1269.

Shibusawa, S., (2001). Real-time soil spectrophotometer for precision farming. *Journal of Spectroscopic Research, 50*(6), 251–260.

Shrestha, R. P., (2006). Relating soil electrical conductivity to remote sensing and other soil properties for assessing soil salinity in northeast Thailand. *Land Degradation and Development, 17*(6), 677–689.

Skidmore, E. L., Dickerson, J. D., & Schimmelpfennig, H., (1975). Evaluating surface-soil water content by measuring reflectance 1. *Soil Sci. Soc. America J., 39*(2), 238–242.

Smit, J., Nasr, J., & Ratta, A., (2001). *Urban Agriculture: Food, Jobs, and Sustainable Cities*. Publication Series for Habitat II. United Nations Publications.

Smit, J., Ratta, A., & Asr, J., (1996). *Urban Agriculture, Food, Jobs, and Sustainable Cities*. UNDP Habitat II Series; United Nations Development Program (UNDP): New York, NY, USA.

Sreekantha, D. K., & Kavya, A. M., (2017). Agricultural crop monitoring using IoT-a study. In: *Proc. 11th International Conference on Intelligent Systems and Control (ISCO)*.

Srivastava, R., & Saxena, R. K., (2004). Technique of large-scale soil mapping in basaltic terrain using satellite remote sensing data. *International Journal of Remote Sensing, 25*(4), 679–688.

Stefanov, W. L., Ramsey, M. S., & Christensen, P. R., (2001). Monitoring urban land cover change: An expert system approach to land cover classification of semiarid to arid urban centers. *Remote Sensing of Environment, 77*, 173–185.

Stevens, A., Nocita, M., Toth, G., Montanarella, L., & Van, W. B., (2013). Prediction of soil organic carbon at the European scale by visible and near infrared reflectance spectroscopy. *PLoS One, 8*(6), 66409.

Stevens, A., Udelhoven, T., Denis, A., Tychon, B., Lioy, R., Hoffmann, L., & Van, W. B., (2010). Measuring soil organic carbon in croplands at regional scale using airborne imaging spectroscopy. *Geoderma, 158*(1, 2), 32–45.

Stoner, E. R., & Baumgardner, M. F., (1981). Characteristic variations in reflectance of surface soils 1. *Soil Science Society of America Journal, 45*(6), 1161–1165.

Sudduth, K. A., & Hummel, J. W., (1991). Evaluation of reflectance methods for soil organic matter sensing. *Transactions of the ASAE, 34*(4), 1900–1909.

Summers, D., Lewis, M., Ostendorf, B., & Chittleborough, D., (2011). Visible near- infrared reflectance spectroscopy as a predictive indicator of soil properties. *Ecological Indicators, 11*(1), 123–131.

Teillet, P. M., Barker, J. L., Markham, B. L., Irish, R. R., & Fedosejevs, G., (2001). *Radiometric Cross-Calibration of the Landsat-7 ETM+ and Landsat TM Sensors Based on Tandem Data Sets*. NASA Publications 13.

Thebo, A. L., Drechsel, P., & Lambin, E. F., (2014). Global assessment of urban and peri-urban agriculture: Irrigated and rainfed croplands. *Environ. Res. Lett.*, 11114002.

United Nations, (2008). *World Urbanization Prospects: The 2007 Revision*. CD-ROM Edition United Nations Department of Economic and Social Affairs, Population Division, New York.

United Nations, (2014). *World Urbanization Prospects: The 2014 Revision*. United Nations Department of Economic and Social Affairs, Population Division, New York.

USGS, (2016). *Landsat Missions*. U.S. Geological Survey. http://landsat.usgs.gov/ (accessed on 28 December 2023).

Vaculikova, L., & Plevova, E., (2005). Identification of clay minerals and micas in sedimentary rocks. *Acta Geodynamica et Geomaterialia, 2*(2), 163.

Van Der, M. H. W., & Beutelspacher, H., (1976). *Atlas of Infrared Spectroscopy of Clay Minerals and Their Admixtures*. Elsevier Publishing Company, New York.

Vane, G., & Goetz, A. F., (1988). Terrestrial imaging spectroscopy. *Remote Sensing of Environment, 24*(1), 1–29.

Vasques, G. M., Grunwald, S. J. O. S., & Sickman, J. O., (2008). Comparison of multivariate methods for inferential modeling of soil carbon using visible/near-infrared spectra. *Geoderma, 146*(1, 2), 14–25.

Verma, K. S., Saxena, R. K., Barthwal, A. K., & Deshmukh, S. N., (1994). Remote sensing technique for mapping salt affected soils. *International Journal of Remote Sensing, 15*(9), 1901–1914.

Vijayabaskar, P. S., Sreemathi, R., & Keertanaa, E., (2017). Crop prediction using predictive analytics. In: *Proc. International Conference on Computation of Power*. Energy Information and Communication (ICCPEIC).

Viscarra-Rossel, R. A., & Behrens, T., (2010). Using data mining to model and interpret soil diffuse reflectance spectra. *Geoderma, 158*(1, 2), 46–54.

Viscarra-Rossel, R. A., & McBratney, A., (1998). Laboratory evaluation of a proximal sensing technique for simultaneous measurement of soil clay and water content. *Geoderma, 85*, 19–39.

Viscarra-Rossel, R. A., (2011). Fine-resolution multi-scale mapping of clay minerals in Australian soils measured with near infrared spectra. *Journal of Geophysical Research: Earth Surface, 116*(F4).

Viscarra-Rossel, R. A., Walvoort, D. J. J., McBratney, A. B., Janik, L. J., & Skjemstad, J. O., (2006). Visible, near infrared, mid infrared or combined diffuse reflectance spectroscopy for simultaneous assessment of various soil properties. *Geoderma, 131*(1), 59–75.

Vohland, M., Ludwig, M., Thiele-Bruhn, S., & Ludwig, B., (2014). Determination of soil properties with visible to near-and mid-infrared spectroscopy: Effects of spectral variable selection. *Geoderma, 223*, 88–96.

Waiser, T. H., Morgan, C. L., Brown, D. J., & Hallmark, C. T., (2007). *In situ* characterization of soil clay content with visible near-infrared diffuse reflectance spectroscopy. *Soil Sci. Soc. America J., 71*(2), 389–396.

Weidong, L., Baret, F., Xingfa, G., Qingxi, T., Lanfen, Z., & Bing, Z., (2002). Relating soil surface moisture to reflectance. *Remote Sensing of Environment, 81*(2, 3), 238–246.

Weng, Q., (2002). Land use change analysis in the Zhujiang delta of China using satellite remote sensing, GIS, and stochastic modeling. *J. Environ. Manag., 64*, 273–284.

Weng, Q., (2012). Remote sensing of impervious surfaces in the urban areas: Requirements, methods, and trends. *Remote Sensing of Environment, 117*, 34–49.

Whiting, M. L., Li, L., & Ustin, S. L., (2004). Predicting water content using Gaussian model on soil spectra. *Remote Sensing of Environment, 89*(4), 535–552.

Wigneron, J. P., Schmugge, T., Chanzy, A., Calvet, J. C., & Kerr, Y., (1998). Use of passive microwave remote sensing to monitor soil moisture. *Agronomie, 18*(1), 27–43.

Workman, Jr. J., & Shenk, J., (2004). Understanding and using the near-infrared spectrum as an analytical method. *Near-infrared Spectroscopy in Agriculture, 44*, 1–10.

World Bank, (2013). *Urban Agriculture: Findings from Four City Case Studies*. The World Bank, Washington, DC.

Wu, D., Yan, B., Cui, Y., Fan, W., & Xu, X., (2010). Retrieve soil moisture from mixed- pixels based on scale transformation using hyperspectral data. In: *Geoscience and Remote Sensing Symposium (IGARSS),* (pp. 3875–3878). IEEE International.

Xia, N., Cheng, L., & Li, M. C., (2019). Mapping urban areas using a combination of remote sensing and geolocation data. *Remote Sens., 11,* 1470.

Yang, X. M., Xie, H. T., Drury, C. F., Reynolds, W. D., Yang, J. Y., & Zhang, X. D., (2012). Determination of organic carbon and nitrogen in particulate organic matter and particle size fractions of Brookston clay loam soil using infrared spectroscopy. *European Journal of Soil Science, 63*(2), 177–188.

Younis, S. M. Z., & Iqbal, J., (2015). Estimation of soil moisture using multispectral and FTIR techniques. *The Egyptian Journal of Remote Sensing and Space Science, 18*(2), 151–161.

Zhang, C., & Kovacs, J. M., (2012). The application of small unmanned aerial systems for precision agriculture: A review. *Precision Agriculture, 13*(6), 693–712.

Zhang, X., Zhang, J., Li, L., Zhang, Y., & Yang, G., (2017). Monitoring citrus soil moisture and nutrients using an IoT based system. *Sensors, 7*(3), 447.

Zhu, Y., Weindorf, D. C., Chakraborty, S., Haggard, B., Johnson, S., & Bakr, N., (2010). Characterizing surface soil water with field portable diffuse reflectance spectroscopy. *Journal of Hydrology, 391*(1, 2), 133–140.

Zuazo, V. H. D., Pleguezuelo, C. R. R., Flanagan, D., Tejero, I. G., & Fernández, J. L. M., (2011). Sustainable land use and agricultural soil. In: *Alternative Farming Systems, Biotechnology, Drought Stress and Ecological Fertilization* (pp. 107–192). Springer, Dordrecht.

Index

B

C

D

E

F

G

H

M

N

Q

R

T

U

V

W